Progress in Mathematics

Volume 239

Number Fields and Function Fields—
Two Parallel Worlds

Gerard van der Geer
Ben Moonen
René Schoof
Editors

Birkhäuser
Boston • Basel • Berlin

Gerard van der Geer and Ben Moonen
Universiteit van Amsterdam
Korteweg-de Vries Instituut
1018 TV Amsterdam
The Netherlands

René Schoof
Università di Roma "Tor Vergata"
Dipartimento di Matematica
I-00133 Roma
Italy

AMS Subject Classifications (2000): 14Gxx, 11Fxx, 11Rxx, 11Mxx, 11F52

ISBN-10 0-8176-4397-4   e-IBSN 0-8176-4447-4   Printed on acid-free paper.
ISBN-13 978-0-8176-4397-3

©2005 Birkhäuser Boston   *Birkhäuser*

Printed in the United States of America.   (JLS/EB)

9 8 7 6 5 4 3 2 1

www.birkhauser.com

Contents

Preface

Rien n'est plus fécond, tous les mathématiciens le savent,
que ces obscures analogies, ces troubles reflets d'une théorie
à une autre, ces furtives caresses, ces brouilleries inexplicables;
rien aussi ne donne plus de plaisir au chercheur.

André Weil (1960)

The analogy between number fields and function fields in one variable was observed at least as early as the second half of the 19th century. It became an important heuristic principle in the hands of Dedekind. It led him to set up a completely algebraic theory of algebraic curves as a counterpart to the transcendental theory of algebraic curves established by Riemann. In this analogy, a prime of a number field is viewed as being similar to a point on an algebraic curve. Indeed, each gives rise to a valuation. This point of view led, for instance, to the introduction of the infinite primes of a number field.

In the famous paper by Dedekind and Weber, published in 1882 in *Crelle's Journal*, the analogy between number fields and function fields is worked out in detail. A few years later, in his once-famous "Grundzüge" paper, Kronecker made an attempt to develop a general theory encompassing both function theory and arithmetic. In a related development, Weil and van der Waerden established in the first half of the 20th century the foundations of algebraic geometry along algebraic rather than analytic lines, but it was not until the 1960s that Grothendieck's theory of schemes provided a satisfactory framework for the unified point of view envisioned by Kronecker.

In the meantime, E. Artin had discovered that the zeta functions of Riemann and Dedekind have analogues for function fields over finite fields and that Riemann's hypothesis concerning the zeroes of his zeta function translates into a conjecture about the absolute values of the eigenvalues of the Frobenius endomorphism associated to curves over finite fields. This revealed a much deeper analogy than the one presented by Dedekind and Weber. Once the Riemann hypothesis for curves over finite fields was proved by Hasse for curves of genus 1 and by Weil for curves of any genus, a better understanding of the analogy became a Holy Grail, holding a promise for a proof of the Riemann hypothesis.

Weil himself popularized this point of view on various occasions. In the 1960s, perhaps inspired by a remark in a 1942 letter of Weil to Artin, Iwasawa developed his theory of $\mathbb{Z}_p$-extensions. Guided by the analogy, he was led to the so-called Main Conjecture in Iwasawa Theory. It was proved in 1982 by Mazur and Wiles. Rather unexpectedly, the analogy also provided to be an inspiration in the other direction when the theory of cyclotomic fields led to the development of an analogous theory for Drinfeld modules. Similarly, Deligne's mixed Hodge theory was largely inspired by the properties of the Frobenius endomorphism on the cohomology of varieties in characteristic p.

A new exciting chapter was written in the 1970s when Arakelov, at the instigation of Shafarevich, showed how to compactify a curve over a number field. This led to the concept of an arithmetic surface, which is the analogue of a compact surface fibered over a curve. Arakelov invoked the differential geometry of Riemann surfaces and constructed a good intersection theory on arithmetic surfaces that mimicked the intersection theory of compact algebraic surfaces. The analogy between the finite places and the infinite places of a number field is extended to a more subtle analogy between curves over p-adic fields on the one hand and Riemann surfaces over the complex numbers on the other. Arakelov's point of view has become a guiding principle. Basic results such as the Riemann–Roch Theorem and Noether's Theorem have been obtained by Faltings. They have been generalized to higher dimensions by Gillet, Soulé, and others.

Perhaps Arakelov theory has not yet fulfilled its initial promises. But its philosophy has had a profound influence on the recent development of number theory. Deepening the analogy between geometry and arithmetic as much as possible seems to be a worthwhile enterprise. The same can be said of other, perhaps less traditional, analogies. Here one may think of the analogy between number fields and knot theory as observed by Mazur, or the existence of the mysterious mathematical object that plays the role of a "field with one element."

In this volume we present different aspects of this parallelism. It is published on the occasion of the 4th Texel Conference, which was devoted to the analogy and held during the last week of April 2004.

We would like to take the opportunity to thank the participants and speakers who made the conference a success. We also would like to thank the institutions that financed the conference: the Korteweg-de Vries Instituut at the Universiteit van Amsterdam, NWO, the Thomas Stieltjes Instituut, and the Koninklijke Nederlandse Akademie van Wetenschappen.

Gerard van der Geer
Ben Moonen
René Schoof
Amsterdam
March, 2005

Participants

The following is the list of participants of the conference *The Analogy between Number Fields and Function Fields, Texel* 2004, which was held on Texel Island, The Netherlands, during the last week of April 2004.

Gebhard Böckle (Essen)
Theo van den Bogaart (Leiden)
Florian Breuer (Stellenbosch)
Björn Buth (Essen)
Robert Carls (Groningen)
Keith Conrad (Storrs)
Christopher Deninger (Münster)
Bas Edixhoven (Leiden)
Carel Faber (Stockholm)
Gerd Faltings (Bonn)
Carlo Gasbarri (Rome)
Gerard van der Geer (Amsterdam)
Fahid Hajir (Amherst)
Günter Harder (Bonn)
Urs Hartl (Freiburg)
Robin de Jong (Amsterdam)
Wilberd van der Kallen (Utrecht)
Mikhail Kapranov (New Haven)

T. Katsura (Tokyo)
Kai Köhler (Düsseldorf)
Ulf Kühn (Berlin)
Klaus Künnemann (Regensburg)
Jeffrey Lagarias (Florham Park)
M. Morishita (Kanazawa)
Y. Miyaoka (Tokyo)
Ben Moonen (Amsterdam)
Niko Naumann (Regensburg)
T. Oda (Tokyo)
Jordan Rizov (Utrecht)
Damian Roessler (Zürich)
René Schoof (Rome)
Jan Stienstra (Utrecht)
Christophe Soulé (Bures-sur-Yvette)
Lara Thomas (Toulouse)
Valerio Talamanca (Rome)
Alexei Zaitsev (Amsterdam)

List of Contributors

Gebhard Böckle
Institut für Experimentelle Mathematik, Universität Duisburg-Essen, Campus Essen, Ellernstraße 29, 45326 Essen, Germany.
boeckle@iem.uni-due.de

Theo van den Bogaart
Mathematical Institute, University of Leiden, P. O. Box 9512, 2300 RA Leiden, The Netherlands.
bogaart@math.leidenuniv.nl

Holger Brenner
Department of Pure Mathematics, University of Sheffield, Hicks Building, Hounsfield Road, Sheffield S3 7RH, U.K.
H.Brenner@sheffield.ac.uk

Florian Breuer
Department of Mathematics, University of Stellenbosch, Stellenbosch 7602, South Africa.
fbreuer@sun.ac.za

Keith Conrad
Department of Mathematics, University of Connecticut, Storrs, CT 06269-3009, USA.
kconrad@math.uconn.edu

Anton Deitmar
Mathematisches Institut, Universität Tübingen, Auf der Morgenstelle 10, 72076 Tübingen, Germany.
deitmar@uni-tuebingen.de

Christopher Deninger
Westfälische Wilhelms-Universität Münster, Mathematisches Institut, Einsteinstraße 62, 48149 Münster, Germany.
deninger@math.uni-muenster.de

Bas Edixhoven
Mathematical Institute, University of Leiden, P. O. Box 9512, 2300 RA Leiden, The Netherlands.
edix@math.leidenuniv.nl

Gerd Faltings
Max-Planck-Institut für Mathematik, Vivatsgasse 7, D-53111 Bonn, Germany.
gerd@mpim-bonn.mpg.de

Urs Hartl
University of Freiburg, Institute of Mathematics, Eckerstraße 1, D-79104 Freiburg, Germany.
urs.hartl@math.uni-freiburg.de

Robin de Jong
Universiteit Leiden, Mathematisch Instituut, Niels Bohrweg 1, 2333 CA Leiden, The Netherlands.
rdejong@math.leidenuniv.nl

Kai Köhler
Mathematisches Institut, Heinrich-Heine-Universität Düsseldorf, Universitätsstraße 1, Gebäude 25.22, D-40225 Düsseldorf, Germany.
koehler@math.uni-duesseldorf.de

Ulf Kühn
Institut für Mathematik, Humboldt Universität zu Berlin, Unter den Linden 6, D-10099 Berlin, Germany.
kuehn@math.hu-berlin.de

Jeffrey C. Lagarias
University of Michigan, Department of Mathematics, Ann Arbor, MI 48109-1043, USA.
lagarias@umich.edu

Vincent Maillot
Institut de Mathématiques de Jussieu, Université Paris 7 Denis Diderot, C.N.R.S., Case Postale 7012, 2 place Jussieu, F-75251 Paris Cedex 05, France.
vmaillot@math.jussieu.fr

Richard Pink
Department of Mathematics, ETH Zentrum, 8092 Zürich, Switzerland.
pink@math.ethz.ch

Damian Roessler
Institut de Mathématiques de Jussieu, Université Paris 7 Denis Diderot, C.N.R.S.,
Case Postale 7012, 2 place Jussieu, F-75251 Paris Cedex 05, France.
dcr@math.jussieu.fr

Annette Werner
Universität Stuttgart, Fachbereich Mathematik, Institut für Algebra und Zahlentheo-
rie, Pfaffenwaldring 57, 70569 Stuttgart, Germany.
Annette.Werner@mathematik.uni-stuttgart.de

Arithmetic over Function Fields:
A Cohomological Approach

Gebhard Böckle

Institut für Experimentelle Mathematik
Universität Duisburg-Essen
Campus Essen
Ellernstraße 29
45326 Essen
Germany
boeckle@iem.uni-due.de

1 Introduction

The present article is a survey of some recent developments on a particular aspect in the arithmetic of function fields. It is not intended to be a survey on all recent developments, of which there are many, nor on all the foundations of the subject, for which there is a number of good references available, such as [Al96], [Ge86], [Go96]. The main emphasis, as expressed by the subtitle, is to advertise some developments that are based on a cohomological theory, introduced by R. Pink and the author [BP04].

This survey is aimed at a reader who has some familiarity with the arithmetic of elliptic curves over number fields, with algebraic geometry and étale sheaves, and has perhaps some (vague) ideas about motives, *and* who wants to learn more about parallel aspects in the arithmetic of functions fields.

Our starting point in Section 2 is a short review of the similarities between elliptic curves on the one hand and Drinfeld modules on the other hand. We emphasize motivically interesting information that is encoded in them, namely their analytic and étale realizations. The subsequent section gives a rapid introduction into some aspects of Anderson's theory of t-motives. It generalizes the theory of Drinfeld modules and thereby provides some additional flexibility.

Section 4 introduces the cohomological viewpoint introduced by R. Pink and the author; cf. [BP04]. It is a natural generalization of Anderson's theory. The construction starts with a theory that looks very much like the theory of coherent sheaves, where as an additional piece of data the sheaves are equipped with an endomorphism. (The endomorphism itself needs the absolute Frobenius endomorphism which is present in the characteristic p situation we consider. A similar type of object in characteristic zero would be a vector bundle with a connection.)

Then comes a crucial point. We localize this first category at a suitable multiplicatively closed subset. The resulting category is called the "category of crystals over function fields." The reader should not be deterred by this two-step construction, but instead realize all the improvements to the theory that result from the localization. The motivation for the construction is that in [BP04] we realized that in the localized category there would exist a functor "extension by zero" which is not present in the original category. The introduction of crystals is perhaps the main novelty in comparison to some earlier concepts generalizing Drinfeld modules.

As a test case, we compare the theory of crystals over $\mathbb{F}_p$ with the theory of constructible étale sheaves of $\mathbb{F}_p$-modules over schemes of characteristic p, and establish an equivalence of categories.

The following section gives two applications of the cohomological theory. The first is a rationality proof for L-functions that can be attached to families of t-motives or for Drinfeld modules. An analytic proof had previously been given by Taguchi and Wan (cf. [TW96]) using Dwork-style methods. A similar development had taken place in the rationality proofs of L-functions of varieties over finite fields. Again, it was first Dwork who gave an analytic and then Grothendieck who gave an algebraic proof. It was precisely this parallel that motivated the construction of the theory of crystals and led to [BP04].

The second application in Section 5 is the proof of the existence of a meromorphic continuation of global L-functions attached to t-motives by Goss to a mod p analogue of the complex plane. Here again, the pioneering work [TW96] of Taguchi and Wan had yielded a first p-adic analytic proof (at least in an important special case).

In the last section, we explain yet another application, namely the construction of something that could be called a motive for Drinfeld modular forms. This "motive" is an arithmetic object whose analytic realization is the space of Drinfeld cusp forms for a fixed weight and level. Its étale realization allows one to attach Galois representations to Drinfeld cusp forms.

There are many interesting open problems in this subject, and throughout the last two sections, we describe a number of them. It is hoped that they will further stimulate the interest in the arithmetic of function fields.

2 The basic example

Let us first recall some arithmetic properties of an elliptic curve E over $\mathbb{Q}$ which are basic for its realization as a motive:

1. Via $\mathbb{Q} \hookrightarrow \mathbb{C}$ the curve E becomes an elliptic curve over $\mathbb{C}$. Its first Betti homology is $\Lambda := H_1(E(\mathbb{C}), \mathbb{Z})$, and one has a pairing

$$\Lambda \times H^0(E(\mathbb{C}), \Omega_{E(\mathbb{C})/\mathbb{C}}) \to \mathbb{C} : (\lambda, \omega) \mapsto \int_\lambda \omega$$

   so that $E(\mathbb{C}) \cong \mathrm{Hom}_\mathbb{C}(H^0(E(\mathbb{C}), \Omega_{E(\mathbb{C})/\mathbb{C}}), \mathbb{C})/\Lambda \cong \mathbb{C}/\Lambda$.
2. Furthermore for any rational prime ℓ of $\mathbb{Z}$ one has the ℓ-adic Tate module $\mathrm{Tate}_\ell(E)$ of E. As a group it is isomorphic to $\mathbb{Z}_\ell^2$, and it provides us with a representation of the absolute Galois group $G_\mathbb{Q} := \mathrm{Gal}(\mathbb{Q}^{\mathrm{sep}}/\mathbb{Q})$ of $\mathbb{Q}$ on $\mathrm{Tate}_\ell(E)$, i.e., a homomorphism $\rho_{E,\ell} : G_\mathbb{Q} \to \mathrm{GL}_2(\mathbb{Z}_\ell)$.

The Betti cohomology of E may be viewed as its *analytic realization*, the Tate module of E as its ℓ-*adic étale realization*. (The motive of E in the sense of Grothendieck also contains parts in degree 0 and 2—but these are not particular to E, and we therefore do not consider them.)

Let $\mathbb{F}_q$ be the finite field of q elements, and let F be a function field of transcendence degree 1 over its constant field $\mathbb{F}_q$. We fix a place, denoted by ∞, of F, and let A denote the ring of functions in F which have a pole at most at ∞. Then A is a Dedekind domain.

The completion F_∞ of F at ∞ is a discretely valued field. By q_∞ the cardinality of its residue field is denoted and by val_∞ its normalized valuation. The latter extends uniquely to a valuation on the algebraic closure F_∞^{alg} of F_∞. The completion $\mathbb{C}_\infty$ of F_∞^{alg} with respect to val_∞ remains algebraically closed. Denoting the extended valuation again by val_∞, the expression $|.|_\infty := q_\infty^{-\mathrm{val}_\infty}$ defines a norm on $\mathbb{C}_\infty$ which is often abbreviated $|.|$. Finally, ι denotes the composite homomorphism $A \hookrightarrow F \hookrightarrow F_\infty \hookrightarrow \mathbb{C}_\infty$.

We fix a finite extension L of F and a Drinfeld A-module ϕ of rank r on L of characteristic $\iota_L : A \hookrightarrow F \hookrightarrow L$. (Below we recall them and some further definitions.) With ϕ one associates

1. its *analytic realization*, which is a discrete A-lattice $\Lambda \subset \mathbb{C}_\infty$ of rank r, such that ϕ base changed to $\mathbb{C}_\infty$ arises from Λ;
2. for every place v of L which is not above ∞, its v-*adic étale realization*, which is a representation of the absolute Galois group G_L of L on the v-adic Tate module $\mathrm{Tate}_v(\phi)$ of ϕ. If A_v is the completion of A at v, then $\mathrm{Tate}_v(\phi)$ is isomorphic to A_v^r. The action of G_L yields an A_v-linear representation on $\mathrm{Tate}_v(\phi)$.

We owe some definitions, at least to the novice in function field arithmetic. For any ring R over $\mathbb{F}_q$ one defines $R\{\tau\}$ as the noncommutative polynomial ring subject to the noncommutation rule $\tau r = r^q \tau$ for any $r \in R$. An alternative description of $R\{\tau\}$ is as follows. Let $R[z]_{\mathbb{F}_q}$ denote the set of $\mathbb{F}_q$-linear polynomials in the (commutative) polynomial ring $R[z]$, i.e., of polynomials of the form $\sum_i \beta_i z^{q^i}$. These are precisely the polynomials p with $p(\alpha x + y) = \alpha p(x) + p(y)$ for all $\alpha \in \mathbb{F}_q$ and all x, y in some R-algebra, i.e., they define $\mathbb{F}_q$-linear maps on R-algebras. Then $R[z]_{\mathbb{F}_q}$ is a ring under addition and composition of polynomials. Moreover, the substitution $\tau^i \mapsto z^{q^i}$ yields an isomorphism of rings $R\{\tau\} \longrightarrow R[z]_{\mathbb{F}_q}$. In the sequel, we will always use a small Greek letter to denote a polynomial in $R\{\tau\}$, and the corresponding capital letter to denote its image in $R[z]_{\mathbb{F}_q}$—for instance, $\phi_a \leftrightarrow \Phi_a$.

A *Drinfeld A-module ϕ on some field K* is a ring homomorphism $\phi : A \to K\{\tau\}$: $a \mapsto \phi_a$ such that its image $\phi(A)$ contains some nonconstant polynomial of $K\{\tau\}$. There is a unique positive integer $r \in \mathbb{N}$, called the *rank of ϕ* such that for any $a \in A$ the highest nonzero coefficient of ϕ_a occurs in degree $r \deg(a)$.

The composition $\iota_K : A \to K$ of ϕ with the projection $K\{\tau\} \to K$ onto the zeroth coefficient is a ring homomorphism which is called the *characteristic of ϕ*. The Drinfeld A-module ϕ is called of *generic characteristic* if ι_K is injective. Otherwise, it is called of *special characteristic*. The case of generic characteristic is the one analogous to that of an elliptic curve over a number field. The other case corresponds to the case of an elliptic curve over a field of positive characteristic.

If $K \hookrightarrow \mathbb{C}_\infty$ is a field homomorphism, the base change of ϕ to $\mathbb{C}_\infty$ is obtained by composing ϕ with the induced monomorphism $K\{\tau\} \hookrightarrow \mathbb{C}_\infty\{\tau\}$. We will only consider homomorphisms $\alpha : K \hookrightarrow \mathbb{C}_\infty$ such that $\alpha\iota_K$ agrees with the homomorphism $\iota : A \to \mathbb{C}_\infty$ defined above.

A *discrete A-lattice* $\Lambda \subset \mathbb{C}_\infty$ is a finitely generated projective A-submodule of $\mathbb{C}_\infty$ (via ι) such that the set $\{\lambda \in \Lambda \mid |\lambda| \leq c\}$ is finite for any $c > 0$. Given such a lattice Λ of rank r, there is a unique power series e_Λ of the form $\sum_{n=0}^\infty a_n z^{q^n}$, $a_n \in \mathbb{C}_\infty$, with infinite radius of convergence, with $a_0 = 1$ and with simple zeros precisely at the points of Λ. The entire function e_Λ is $\mathbb{F}_q$-linear and surjective and hence one has a short exact sequence of $\mathbb{F}_q$-vector spaces

$$0 \longrightarrow \Lambda \longrightarrow \mathbb{C}_\infty \overset{e_\Lambda}{\longrightarrow} \mathbb{C}_\infty \longrightarrow 0.$$

Thus if we let $a \in A$ act by multiplication on the middle and left term, there exists a unique $\mathbb{F}_q$-linear function $\Phi_a(z)$ on the right such that the diagram

$$
\begin{array}{ccc}
\mathbb{C}_\infty & \xrightarrow{\;z \mapsto e_\Lambda(z)\;} & \mathbb{C}_\infty \\
{\scriptstyle z \mapsto az}\Big\downarrow & & \Big\downarrow{\scriptstyle z \mapsto \Phi_a(z)} \\
\mathbb{C}_\infty & \xrightarrow{\;z \mapsto e_\Lambda(z)\;} & \mathbb{C}_\infty
\end{array}
$$

commutes. One verifies that Φ_a lies in $\mathbb{C}_\infty[z]_{\mathbb{F}_q}$, and denotes by ϕ_a the corresponding polynomial in $\mathbb{C}_\infty\{\tau\}$. Then $A \to \mathbb{C}_\infty\{\tau\} : a \mapsto \phi_a$ defines a Drinfeld A-module of rank r. It is known that any Drinfeld A-module on $\mathbb{C}_\infty$ of characteristic ι arises in this way.

Suppose now that ϕ is a Drinfeld A-module on L of rank r and generic characteristic $\iota_L : A \hookrightarrow F \hookrightarrow L$. Fix a place v of A, and denote by $\mathfrak{p}_v$ the corresponding maximal ideal of A. Then one defines

$$\phi[v^n] := \{\lambda \in L^{\mathrm{alg}} \mid \forall a \in \mathfrak{p}_v^n : \Phi_a(\lambda) = 0\}.$$

The derivative of $\Phi_a(z)$ is the constant $\iota_L(a)$. Since ϕ is of generic characteristic, $\iota_L(a)$ is nonzero, and so the polynomial Φ_a is separable. From this one deduces that $\phi[v^n]$ defines a finite separable Galois extension of L. The *module of v-primary torsion points of ϕ* is

$$\phi[v^\infty] := \bigcup_n \phi[v^n] \subset L^{\mathrm{sep}}.$$

It is stable under the action of G_L and, as an A-module, isomorphic to $(F_v/A_v)^r$, where F_v is the fraction field of A_v. Finally by $\mathrm{Tate}_v(\phi) := \mathrm{Hom}_A(F_v/A_v, \phi[v^\infty])$ one denotes the v-adic Tate module of ϕ.

Example 1. Let $F := \mathbb{F}_q(t)$, $A := \mathbb{F}_q[t]$ and L a finite extension of F. The point ∞ therefore corresponds to the valuation v_∞ which to a quotient f/g of polynomials assigns $\deg g - \deg f$. A Drinfeld A-module $\phi: A \to L\{\tau\}$ is uniquely determined by the image of t. If $\iota_L: A \hookrightarrow F \hookrightarrow L$ is the characteristic of ϕ and if ϕ is of rank r, then ϕ_t must be of the form

$$t + a_1\tau + \cdots + a_r\tau^r \in L\{\tau\}$$

with $a_r \in L^\times$, and where we identify $t = \iota_L(t)$. Conversely any such polynomial defines a Drinfeld A-module of characteristic ι_L and rank r.

Suppose v is the place corresponding to the maximal ideal (t). Then $\Phi_t(z) = tz + a_1 z^q + \cdots + a_r z^{q^r}$ and

$$\phi[v^\infty] = \{\lambda \in L^{\mathrm{sep}} \mid \exists n : \underbrace{\Phi_t \circ \cdots \circ \Phi_t}_{n}(\lambda) = 0\}.$$

To generalize the notion of a Drinfeld A-module to arbitrary schemes over $\mathbb{F}_q$, we now give an (equivalent) alternative definition of a Drinfeld A-module on K. Let $\mathbb{G}_{a,K}$ be the additive group (scheme) on K. By $\mathrm{End}_{\mathbb{F}_q}(\mathbb{G}_{a,K})$ we denote the ring of $\mathbb{F}_q$-linear endomorphisms of the group scheme $\mathbb{G}_{a,K}$. This ring is known to be generated over K by the Frobenius endomorphism τ of $\mathbb{G}_{a,K}$, and thereby isomorphic to $K\{\tau\}$. Thus we may define a Drinfeld A-module as a nonconstant homomorphism

$$\phi: A \to \mathrm{End}_{\mathbb{F}_q}(\mathbb{G}_{a,K}).$$

Since the elements of $a \in A$ act as endomorphisms of the group scheme $\mathbb{G}_{a,K}$, they induce an endomorphism $\partial\phi_a$ on the corresponding Lie algebra. This is called the derivative of ϕ and yields a ring homomorphism

$$\partial\phi: A \to \mathrm{End}(\mathrm{Lie}\,\mathbb{G}_{a,K}) \cong \Gamma(\mathrm{Spec}\,K, \mathcal{O}_{\mathrm{Spec}\,K}) = K. \tag{1}$$

This homomorphism is precisely the characteristic of ϕ.

Over an arbitrary $\mathbb{F}_q$-scheme X one defines a Drinfeld A-module as follows: Let $\mathcal{L}$ denote a line bundle on X (i.e., a scheme which is a $\mathbb{G}_a$-bundle on X), and $\mathrm{End}_{\mathbb{F}_q}(\mathcal{L})$ the ring of $\mathbb{F}_q$-linear endomorphisms of the group scheme $\mathcal{L}$ over X. A *Drinfeld A-module of rank r on $(X, \mathcal{L})$* is a homomorphism

$$\phi: A \to \mathrm{End}_{\mathbb{F}_q}(\mathcal{L}),$$

such that for all fields K and all morphisms $\pi: \mathrm{Spec}\,K \to X$ the induced homomorphism $\phi: A \to \mathrm{End}_{\mathbb{F}_q}(\pi^*\mathcal{L})$ is a Drinfeld A-module of rank r on K.

In the same way as (1) was constructed, ϕ induces a homomorphism $\partial\phi: A \to \Gamma(X, \mathcal{O}_X)$. The corresponding morphism of schemes $\mathrm{char}_\phi: X \to \mathrm{Spec}\,A$ is called the characteristic of ϕ. Via the characteristic, X becomes an A-scheme (i.e., a scheme with a morphism to $\mathrm{Spec}\,A$).

A *morphism from a Drinfeld A-module* ϕ on $\mathcal{L}$ to ϕ' on $\mathcal{L}'$ is a morphism $\alpha \in \mathrm{Hom}_{\mathbb{F}_q}(\mathcal{L}, \mathcal{L}')$ which is A-equivariant, i.e., such that for all $a \in A$ one has $\alpha\phi_a = \phi'_a\alpha$.

3 Anderson's motives

In the preceding section we encountered an object defined over function fields, namely a Drinfeld A-module, which has properties seemingly similar to those of an elliptic curve. Therefore, it seemed natural to look for a category of motives which would naturally contain all Drinfeld A-modules (of generic characteristic). In particular, this category should be A-linear, it should allow for constructions from linear algebra such as sums, tensor, exterior and symmetric products, and it should have étale as well as analytic realizations. It was Anderson in his seminal paper [An86] who first realized how to construct such a theory, perhaps motivated by the earlier definition of *shtuka* due to Drinfeld. Some further details can be found in [vdH03, Chapter 4].

We fix a subfield K of $\mathbb{C}_\infty$ containing F, denote by σ the Frobenius on K relative to $\mathbb{F}_q$, and set $\iota_K : A \hookrightarrow F \hookrightarrow K$.

Definition 1. *An* abelian A-motive *on* K *consists of a pair* (M, τ) *such that we have the following:*

1. *M is a finitely generated projective $K \otimes A$-module.*
2. *$\tau : M \to M$ is an injective σ-semilinear endomorphism of M, i.e., for all $m \in M$, $x \in K$ and $a \in A$ one has $\tau((x \otimes a)m) = (x^q \otimes a)\tau(m)$.*
3. *The module $M/K\tau(M)$ is of finite length over $K \otimes A$, and annihilated by a power of the maximal ideal generated by $\{\iota(a) \otimes 1 - 1 \otimes a \mid a \in A\}$.*
4. *$M \otimes_K K^{\mathrm{perf}}$ is finitely generated over $K^{\mathrm{perf}}\{\tau\}$, where K^{perf} denotes the perfect closure of K.*

Using that $K^{\mathrm{perf}}\{\tau\}$ is a left principal ideal domain, it is shown in [An86] that for any abelian A-motive M the module $M \otimes_K K^{\mathrm{perf}}$ is free over $K^{\mathrm{perf}}\{\tau\}$. The rank of $M \otimes K^{\mathrm{perf}}$ over $K^{\mathrm{perf}}\{\tau\}$ is called the *dimension of M*, its rank over $K \otimes A$ is called the *rank of M*.

If K is perfect, condition 3 can be simplified using $M/K\tau(M) = M/\tau(M)$. The maximal ideal in 3 defines a K-rational point of $\mathrm{Spec}(K \otimes A)$.

Definition 2. *A pair (M, τ) which satisfies conditions 1–3 only, is called an A-motive on K.*

This definition differs significantly from [Go96, Definition 5.4.2], while Definition 1 is the same as in [Go96] and [An86]. One has the following obvious result.

Proposition 1. *If (M, τ) and (M', τ') are abelian A-motives on K, then so is $(M \oplus M', \tau \oplus \tau')$.*

If (M, τ) and (M', τ') are A-motives on K, then so is their tensor product, as well as all tensor, exterior, and symmetric powers of (M, τ).

To define the analytic realization of an A-motive, we have to introduce some further notation. The Tate algebra over $\mathbb{C}_\infty$ is defined as

$$\mathbb{C}_\infty\langle t \rangle := \left\{ \sum a_n t^n \mid a_n \in \mathbb{C}_\infty, |a_n| \to 0 \text{ for } n \to \infty \right\}.$$

Any monomorphism $\mathbb{F}_q[t] \hookrightarrow A$ is finite and flat. Fixing one, any module M underlying some A-motive can be regarded as a (free and finitely generated) module over $K[t]$. We set

$$M\langle t\rangle := M \otimes_{K[t]} \mathbb{C}_\infty\langle t\rangle,$$

and define $M\langle t\rangle^\tau$ as the $\mathbb{F}_q[t]$-module of τ-invariant elements of $M\langle t\rangle$. The module of τ-invariants is a projective (left) A-module and satisfies

$$\operatorname{rank}_A M\langle t\rangle^\tau \le \operatorname{rank}_{K \otimes A} M. \tag{2}$$

Definition 3. *If equality holds in* (2), *then* (M, τ) *is called* uniformizable.

Anderson gave examples of abelian A-motives of dimension greater than 1 which are not uniformizable. If (M, τ) is uniformizable, we regard the A-module $M\langle t\rangle^\tau$ as its *analytic realization*. It is also shown in [An86], that for any uniformizable motive M of dimension d and rank r, there is a related short exact sequence

$$0 \longrightarrow \Lambda_M \longrightarrow \mathbb{C}_\infty^d \xrightarrow{\ e_M\ } \mathbb{C}_\infty^d \longrightarrow 0,$$

where Λ_M is a discrete A-lattice in $\mathbb{C}_\infty^d$ of rank r. Let Ω_A denote the module of Kähler differentials of A over $\mathbb{F}_q$. Then by [An86, Section 2], there is the following isomorphism which relates Λ_M with the analytic realization of M:

$$\operatorname{Hom}_A(M\langle t\rangle^\tau, \Omega_A) \cong \Lambda. \tag{3}$$

Following Anderson [An86, Section 2], one has a fully faithful functor from Drinfeld A-modules to uniformizable A-motives of dimension 1: Let ϕ be a Drinfeld A-module on K of generic characteristic ι_K. Via left multiplication by K and right multiplication by ϕ_a for $a \in A$, we regard $M(\phi) := K\{\tau\}$ as a module over $K \otimes A$. Left multiplication by τ defines a σ-semilinear endomorphism $\tau : M(\phi) \to M(\phi)$.

Theorem 1. *The assignment* $\phi \mapsto (M(\phi), \tau)$ *defines a functor which identifies the category of Drinfeld A-modules of rank r with the category of abelian A-motives M of rank r that satisfy $M \cong K\{\tau\}$. Any such A-motive is uniformizable. Moreover, if $K = \mathbb{C}_\infty$ and ϕ arises from a lattice Λ, then Λ and $M\langle t\rangle^\tau$ are related via* (3).

To define *étale realizations*, one proceeds as follows: Let $\mathfrak{p}_v$ be the prime ideal of A corresponding to the place v of A. Then

$$M \otimes_{K \otimes A} (K^{\mathrm{sep}} \otimes A/\mathfrak{p}_v^n)$$

is free and finitely generated over $K^{\mathrm{sep}} \otimes A/\mathfrak{p}_v^n$. By condition 3 for a motive, the induced τ-action is, in fact, bijective. From Lang's theorem [An86, 1.8.2], one easily deduces that

$$M_{v,n} := \left(M \otimes_{K \otimes A} (K^{\mathrm{sep}} \otimes A/\mathfrak{p}_v^n) \right)^\tau$$

is a free $A/\mathfrak{p}_v^n$-module of rank r. Because M is defined over K, the actions of G_K and of τ commute on $M \otimes_{K \otimes A} (K^{\mathrm{sep}} \otimes A/\mathfrak{p}_v^n)$, and so G_K acts $A/\mathfrak{p}_v$-linearly on

$M_{v,n}$. The inverse limit $M_{v,\infty} := \varprojlim M_{v,n}$ is thus a free A_v-module of rank r with a continuous linear action of G_K. This we regard as the *v-adic étale realization of M*. It exists independently of the uniformizability of M. One has the following result due to Anderson.

Proposition 2. *Suppose ϕ is a Drinfeld A-module of rank r and generic characteristic, and $(M, \tau) := (M(\phi), \tau)$. Then there is a canonical isomorphism*

$$\mathrm{Hom}_{A_v}(M_{v,\infty}, \varinjlim \mathfrak{p}_v^{-n} \Omega_A / \Omega_A) \cong \phi[v^\infty].$$

4 A cohomological framework

In the previous section, we have seen that Anderson's category of A-motives has a number of very useful properties. It allows constructions from linear algebra, it has realizations as one would expect from motives, and it contains the category of Drinfeld A-modules. Also pullback along morphisms $Y \to X$ for families of A-motives on a scheme X is easily defined.

Anderson's theory does not, however, provide a cohomological theory of A-motive like objects, which is also desirable and by which we mean the following: On every base scheme X over $\mathbb{F}_q$ we would like to have a category of objects similar to families of A-motives. For any morphism $f : Y \to X$, there should be (derived) functors $R^i f_*$ between these categories, and perhaps also other ones such as f^*, $\otimes$, Hom, $Rf_!$, etc.

In [BP04] such a theory is developed in joint work with R. Pink. Much of the material described in the previous section was inspirational for this. The main motivation for the work in [BP04] was to give a cohomological proof of a rationality conjecture by Goss, that had, by analytical methods, previously been established in work of Taguchi and Wan; cf. [TW96]. This is discussed in greater detail in Subsection 5.1.

An alternative construction of such a cohomological theory was recently also described by M. Emerton and M. Kisin in [EK04]. The main reference for the present section is [BP04].

Conventions. Throughout, X, Y, etc. will be noetherian schemes over $\mathbb{F}_q$. By σ_X or simply σ we denote the absolute Frobenius endomorphism of X relative to $\mathbb{F}_q$. We fix a morphism of schemes $f : Y \to X$.

The symbol B (or B') will always denote an $\mathbb{F}_q$-algebra which arises as a localization of an $\mathbb{F}_q$-algebra of finite type. Typically, B will be A, or $A/\mathfrak{a}$ for some ideal of A, or the fraction field F of A.

Whenever tensor or fiber products are formed over $\mathbb{F}_q$, the subscript $\mathbb{F}_q$ at $\otimes$ and $\times$ will be omitted.

By $\mathrm{pr}_1 : X \times \mathrm{Spec}\, B \to X$ the projection onto the first factor is denoted.

4.1 τ-sheaves

Definition 4. *A τ-sheaf over B on a scheme X is a pair $\underline{\mathcal{F}} := (\mathcal{F}, \tau_\mathcal{F})$ consisting of a coherent sheaf $\mathcal{F}$ on $X \times \mathrm{Spec}\, B$ and an $\mathcal{O}_{X \times \mathrm{Spec}\, B}$-linear homomorphism*

$$(\sigma \times \mathrm{id})^* \mathcal{F} \xrightarrow{\ \tau\ } \mathcal{F}.$$

A homomorphism *of τ-sheaves $\underline{\mathcal{F}} \to \underline{\mathcal{G}}$ over B on X is a homomorphism of the underlying sheaves $\phi \colon \mathcal{F} \to \mathcal{G}$ which is compatible with the action of τ.*

We often simply speak of τ-*sheaves on X*. The sheaf underlying a τ-sheaf $\underline{\mathcal{F}}$ will always be denoted by $\mathcal{F}$. When the need arises to indicate on which sheaf τ acts, we write $\tau = \tau_{\mathcal{F}}$.

The category of all τ-sheaves over A on X is denoted by $\mathbf{Coh}_\tau(X, A)$. It is an abelian A-linear category, and all constructions like kernel, cokernel, etc. are the usual ones on the underlying coherent sheaves, with the respective τ added by functoriality.

In this survey we focus on coherent objects. To alleviate our notation, we deviate from the terminology in [BP04]. What is called a τ-sheaf here is a coherent τ-sheaf in [BP04].

On any affine open $\operatorname{Spec} R \subset X$ a τ-sheaf over B is given by a finitely generated $R \otimes B$-module M together with an $R \otimes B$-linear homomorphism $R^\sigma \otimes_R M \to M$. The latter homomorphism corresponds bijectively to a $\sigma \otimes \mathrm{id}$-linear morphism $\tau \colon M \to M$. The pair (M, τ) is also called a τ-*module*.

Example 2. Due to properties 1 and 2 in the definition of a motive, any A-motive on K is a τ-module, and thus yields a τ-sheaf over A on $\operatorname{Spec} K$. However, the notion of τ-sheaf is less restrictive than that of A-motive, since we impose no local freeness conditions, or conditions on the kernel or cokernel of τ.

4.2 Examples

We now describe some further examples of τ-sheaves. All but the first and last of these correspond to families of A-motives (of fixed rank).

Most of the examples are *locally free τ-sheaves*; by this we mean τ-sheaves whose underlying sheaf is locally free. The *rank* of a locally free τ-sheaf is that of its underlying sheaf. Locally free τ-sheaves had been considered already in [TW96], where they were called ϕ-sheaves.

I. Any (coherent) sheaf $\mathcal{F}$ on $X \times \operatorname{Spec} B$ can be made into a τ-sheaf by setting $\tau = 0$. As we will see shortly, these τ-sheaves are not interesting to us.

II. The unit τ-sheaf, denoted by $\underline{\mathbb{1}}_{X,B}$, is the free τ-sheaf defined by the pair consisting of the sheaf $\mathcal{O}_{X \times \operatorname{Spec} B}$ together with the isomorphism

$$\tau \colon (\sigma \times \mathrm{id})^* \mathcal{O}_{X \times \operatorname{Spec} B} \longrightarrow \mathcal{O}_{X \times \operatorname{Spec} B},$$

which via adjunction arises from $\sigma \otimes \mathrm{id} \colon \mathcal{O}_{X \times \operatorname{Spec} B} \longrightarrow (\sigma \times \mathrm{id})_* \mathcal{O}_{X \times \operatorname{Spec} B}$.

III. The construction $\phi \mapsto M(\phi)$ from Drinfeld A-modules to A-motives generalizes in an obvious way to a functor $(\mathcal{L}, \phi) \mapsto \underline{\mathcal{M}}(\phi)$ from Drinfeld A-modules over a general base X to τ-sheaves over A on X. The rank of the Drinfeld module becomes the rank of the locally free sheaf underlying $\underline{\mathcal{M}}(\phi)$.

Similarly, any elliptic sheaf on X gives rise to a τ-sheaf over A on X, by restricting it to $X \times \operatorname{Spec} A$. For details on this, we refer to the excellent article of Blum and Stuhler; cf. [BS97].

IV. As we have seen earlier, one can define Drinfeld A-modules over a general base. So it is natural to consider corresponding moduli problems. As in the case of elliptic curves (or abelian varieties) these are not rigid, unless one introduces some level structures. To define such, we fix a proper nonzero ideal $\mathfrak{n} \subset A$. Then for any Drinfeld A-module $\phi : A \rightarrow \mathrm{End}_{\mathbb{F}_q}(\mathcal{L})$ of rank r over a scheme X, one defines the subscheme

$$\mathcal{L}_\phi[\mathfrak{n}] := \bigcap_{a \in \mathfrak{n}} \mathrm{Ker}(\mathcal{L} \xrightarrow{\phi_a} \mathcal{L}) \subset \mathcal{L}.$$

It carries an action of $A/\mathfrak{n}$ and is finite flat over X. Its degree over X is the cardinality of $(A/\mathfrak{n})^r$. If the image of $\mathrm{char}_\phi : X \rightarrow \mathrm{Spec}\, A$ is disjoint from $\mathrm{Spec}\, A/\mathfrak{n}$, we say that *the characteristic of ϕ is prime to $\mathfrak{n}$*. In this case $\mathcal{L}_\phi[\mathfrak{n}]$ is, moreover, étale and Galois over X.

For a finite discrete group G we denote by G_X the corresponding constant group scheme on X.

Definition 5. *A naive level $\mathfrak{n}$-structure on ϕ is an isomorphism*

$$\psi : (A/\mathfrak{n})^r_X \xrightarrow{\cong} \mathcal{L}_\phi[\mathfrak{n}].$$

A naive level $\mathfrak{n}$-structure can only exist if the characteristic of ϕ is prime to $\mathfrak{n}$. In that case it always exists over a finite Galois covering $Y \rightarrow X$.

Let $A(\mathfrak{n})$ denote the ring of rational functions in F regular outside ∞ and the primes dividing $\mathfrak{n}$. Then for any fixed $r \in \mathbb{N}$ one may consider the moduli functor $M^r(\mathfrak{n})$ which to an $A(\mathfrak{n})$-scheme X assigns the set of triples $(\mathcal{L}, \phi, \psi)$ (up to isomorphism), where

1. $\phi : A \rightarrow \mathrm{End}_{\mathbb{F}_q}(\mathcal{L})$ is a Drinfeld A-module of rank r, and
2. $\psi : (A/\mathfrak{n})^r_X \xrightarrow{\cong} \mathcal{L}_\phi[\mathfrak{n}]$ is a naive level $\mathfrak{n}$-structure,

such that the composite of the structure morphism $X \rightarrow \mathrm{Spec}\, A(\mathfrak{n})$ with the canonical open immersion $\mathrm{Spec}\, A(\mathfrak{n}) \hookrightarrow \mathrm{Spec}\, A$ is equal to char_ϕ.

Theorem 2 ([Dr76]). *The moduli problem $M^r(\mathfrak{n})$ is representable by an affine (noetherian) $A(\mathfrak{n})$-scheme $\mathfrak{Y}^r(\mathfrak{n})$. The line bundle $\mathcal{L}^r(\mathfrak{n})$ in its universal triple $(\mathcal{L}^r(\mathfrak{n}), \phi^r(\mathfrak{n}), \psi^r(\mathfrak{n}))$ on $\mathfrak{Y}^r(\mathfrak{n})$ is isomorphic to $\mathbb{G}_{a,\mathfrak{Y}^r(\mathfrak{n})}$. The structure morphism $\mathfrak{Y}^r(\mathfrak{n}) \rightarrow \mathrm{Spec}\, A(\mathfrak{n})$ is smooth of relative dimension $r - 1$.*

For later use, we record the following.

Proposition 3. *The τ-sheaf $\underline{\mathcal{M}}^r(\mathfrak{n}) := \underline{\mathcal{M}}(\phi^r(\mathfrak{n}))$ is locally free of rank r.*

V. One can, in fact, define moduli spaces of more general types of A-motives (which carry some polarization and level structure). The corresponding universal A-motive then again yields interesting τ-sheaves. The investigation of some of these moduli problems is ongoing work by U. Hartl and, independently, by L. Taelman.

VI. An important notion to study the reduction of abelian varieties is their Néron model. For τ-sheaves, in [Ga03] Gardeyn introduced and investigated the following substitute.

Definition 6. *Suppose* $j : U \hookrightarrow X$ *a dense open immersion. A* τ-*sheaf* $\underline{\mathcal{G}} \in$ $\mathbf{Coh}_\tau(X, A)$ *is called a* model *of* $\underline{\mathcal{F}} \in \mathbf{Coh}_\tau(U, A)$ *with respect to* j *if* $j^*\underline{\mathcal{G}} \cong \underline{\mathcal{F}}$. *A model* $\underline{\mathcal{G}}$ *is called a* maximal model *of* $\underline{\mathcal{F}} \in \mathbf{Coh}_\tau(U, A)$ *with respect to* j *if for all* $\underline{\mathcal{H}} \in \mathbf{Coh}_\tau(X, A)$ *the canonical homomorphism*

$$\mathrm{Hom}_{\mathbf{Coh}_\tau(X,A)}(\underline{\mathcal{H}}, \underline{\mathcal{G}}) \longrightarrow \mathrm{Hom}_{\mathbf{Coh}_\tau(U,B)}(\underline{\mathcal{H}}_{|U \times \mathrm{Spec}\, A}, \underline{\mathcal{F}}) \qquad (4)$$

is an isomorphism. We write $j_{\#}\underline{\mathcal{F}} := \underline{\mathcal{G}}$.

If it exists, a maximal model is always unique up to unique isomorphism. There always exists a direct limit of τ-sheaves $\underline{\mathcal{G}}$ which satisfies (4). The crucial requirement is that $\underline{\mathcal{G}}$ be coherent.

As to be expected, if $\phi : A \to \mathrm{End}_{\mathbb{F}_q}(\mathcal{L})$ is a Drinfeld A-module on X, and $j : U \hookrightarrow X$ a dense open immersion, then $j_{\#}\underline{\mathcal{M}}(j^*\phi) \cong j_{\#}(j^*\underline{\mathcal{M}}(\phi)) \cong \underline{\mathcal{M}}(\phi)$.

Theorem 3 ([Ga03]). *Suppose* X *is a smooth projective curve over* $\mathbb{F}_q$, *and* $j : U \hookrightarrow X$ *is open and dense. Then for any locally free* $\underline{\mathcal{F}} \in \mathbf{Coh}_\tau(U, A)$ *with* $\tau_{\mathcal{F}}$ *injective, a maximal model* $j_{\#}\underline{\mathcal{F}}$ *exists. It is again locally free.*

Remark 1. Suppose now that $\phi : A \to K\{\tau\}$ is a Drinfeld A-module of rank r over a complete discretely valued field K with ring of integers V, residue field k and $j : \mathrm{Spec}\, K \hookrightarrow \mathrm{Spec}\, V$. Suppose also that ϕ has reduction of rank $0 < r' < r$ over k. (The latter means that over K the Drinfeld module ϕ is isomorphic to some Drinfeld module ϕ' whose image lies in $V\{\tau\}$, and such that the reduction of ϕ' to k is a Drinfeld module of rank r' over k.)

Drinfeld (cf. [Dr76, Section 7]) has shown that in this situation, after possibly passing to a finite extension of K, there exist a Drinfeld module ϕ' of rank r' over K, a discrete A-sublattice $\Lambda \subset K$ (where A acts on K via ϕ'), and an "analytic" A-homomorphism $e_\Lambda : K \to K$, where on the left A acts via ϕ' and on the right via ϕ, such that there is a short exact sequence of A-modules

$$0 \longrightarrow \Lambda \longrightarrow K^{\mathrm{alg}} \xrightarrow{e_\Lambda} K^{\mathrm{alg}} \longrightarrow 0.$$

The transition from Drinfeld modules to τ-sheaves is contravariant, so the morphism e_Λ turns into an "analytic" morphism between A-motives $M^{\mathrm{an}}(\phi) \longrightarrow$ $M^{\mathrm{an}}(\phi')$ over K. It is again surjective, and its kernel is the analytification of $\underline{\mathbb{1}}_{\mathrm{Spec}\, K, A}^{r-r'}$. So there is a short exact sequence

$$0 \longrightarrow (\underline{\mathbb{1}}_{\mathrm{Spec}\, K, A}^{r-r'})^{\mathrm{an}} \longrightarrow M^{\mathrm{an}}(\phi) \longrightarrow M^{\mathrm{an}}(\phi') \longrightarrow 0.$$

Gardeyn shows that: This short exact sequence can be extended to a left exact sequence over a formal scheme attached to $V \otimes A$. The left two terms of the extended sequence arise via analytification from

$$0 \longrightarrow \mathbb{1}_{\operatorname{Spec} V, A}^{t-r'} \longrightarrow j_{\#}\underline{\mathcal{M}}(\phi).$$

The induced morphism $\mathbb{1}_{\operatorname{Spec} k, A}^{r-r'} \longrightarrow i^* j_{\#}\underline{\mathcal{M}}(\phi)$ on the special fiber over $\operatorname{Spec} k$ is injective and some iterate of τ vanishes on the cokernel.

The upshot of this longwinded explanation is that in the present situation the maximal extension and the Drinfeld module ϕ' shed light on opposite aspects of the given bad reduction situation. Also while ϕ' can be obtained from ϕ only via an analytic morphism, the relation between ϕ and $j_{\#}\underline{\mathcal{M}}(\phi)$ is algebraic. For more on this interesting topic, we refer to [Ga03]. In this survey maximal extensions will reappear in Remark 2 and Subsection 6.8.

4.3 Crystals

There is a large class of homomorphism in $\mathbf{Coh}_\tau(X, B)$ which one would like to regard as isomorphisms. Categorically, the correct way to deal with this is to localize at this class; cf. [We94, Section 10.3]. The reason in [BP04] to pass to the localized theory was a practical one: only there we were able to construct an "extension by zero functor" as sketched below Theorem 4. First, we describe the localization procedure:

For a τ-sheaf $\underline{\mathcal{F}}$, we define the iterates τ^n of τ by setting inductively $\tau^0 := \operatorname{id}$ and $\tau^{n+1} := \tau \circ (\sigma \times \operatorname{id})^* \tau^n : (\sigma^{n+1} \times \operatorname{id})^* \mathcal{F} \longrightarrow \mathcal{F}$. A τ-sheaf $\underline{\mathcal{F}}$ is called *nilpotent* if and only if $\tau_{\underline{\mathcal{F}}}^n$ vanishes for some $n > 0$. A corresponding notion for homomorphisms is the following.

Definition 7. *A morphism of τ-sheaves is called a* nil-isomorphism *if and only if both its kernel and cokernel are nilpotent.*

It is shown in [BP04, Chapter 2] that the nil-isomorphisms in $\mathbf{Coh}_\tau(X, B)$ form a saturated multiplicative system, and so one defines the following.

Definition 8. *The category* $\mathbf{Crys}(X, B)$ *of B-crystals on X is the localization of* $\mathbf{Coh}_\tau(X, B)$ *with respect to nil-isomorphisms.*

The category $\mathbf{Crys}(X, B)$ is again a B-linear abelian category with the induced notions of kernel, image, cokernel, and coimage. Its objects are the same as those in $\mathbf{Coh}_\tau(X, A)$. However, the homomorphisms are different. Any homomorphism $\underline{\mathcal{F}} \longrightarrow \underline{\mathcal{G}}$ in $\mathbf{Crys}(X, A)$ is represented by a diagram $\underline{\mathcal{F}} \longleftarrow \underline{\mathcal{H}} \longrightarrow \underline{\mathcal{G}}$ in $\mathbf{Coh}_\tau(X, A)$, for some $\mathcal{H} \in \mathbf{Coh}_\tau(X, A)$, and where the homomorphism $\underline{\mathcal{H}} \longrightarrow \underline{\mathcal{F}}$ is a nil-isomorphism.

Let us conclude this subsection with the simplest functor that is based on nilpotency. For a τ-sheaf $\underline{\mathcal{F}}$ on X over B, define

$$\underline{\mathcal{F}}^\tau := \Gamma(X \times \operatorname{Spec} B, \underline{\mathcal{F}})^\tau \tag{5}$$

as the B-module of global τ-invariant sections of $\underline{\mathcal{F}}$. The following result is an immediate consequence of Definition 8.

Proposition 4. *The functor $\underline{\mathcal{F}} \mapsto \underline{\mathcal{F}}^\tau$ from $\mathbf{Coh}_\tau(X, B)$ to B-modules is invariant under nil-isomorphisms. It therefore passes to a functor on the category* $\mathbf{Crys}(X, B)$, *which is again denoted $\underline{\mathcal{F}} \mapsto \underline{\mathcal{F}}^\tau$.*

4.4 Functors

On τ-sheaves, we now indicate the construction of four functors:

(a) $f^*: \mathbf{Coh}_\tau(X, B) \longrightarrow \mathbf{Coh}_\tau(Y, B)$ (*pullback*):

For $\underline{\mathcal{F}} \in \mathbf{Coh}_\tau(X, B)$, we define $f^*\underline{\mathcal{F}} \in \mathbf{Coh}_\tau(Y, B)$ as the pair consisting of the coherent sheaf $(f \times \mathrm{id})^*\mathcal{F}$ on $Y \times \operatorname{Spec} B$ and the $\sigma \times \mathrm{id}$-linear endomorphism on $(f \times \mathrm{id})^*\mathcal{F}$ induced by functoriality from τ. This defines a B-linear functor f^*.

In an analogous way we define B-linear (bi-)functors:

(b) $R^i f_*: \mathbf{Coh}_\tau(Y, B) \longrightarrow \mathbf{Coh}_\tau(X, B)$ (*push-forward*) if $f: Y \to X$ is proper, and $i \geq 0$.

(c) $- \otimes -: \mathbf{Coh}_\tau(X, B) \times \mathbf{Coh}_\tau(X, B) \longrightarrow \mathbf{Coh}_\tau(X, B)$ (*tensor product*).

(d) $- \otimes_B B': \mathbf{Coh}_\tau(X, B) \longrightarrow \mathbf{Coh}_\tau(X, B')$ (*change of coefficients*) for any homomorphism $B \to B'$.

Despite the notation, at this point it is not at all clear that the $R^i f_*$ are derived functors.

In the same way as the tensor product is defined, following [Ha77, Example II.5.16], for the construction on the underlying sheaf, one also obtains higher tensor, symmetric and exterior powers of a τ-sheaf $\underline{\mathcal{F}}$. We denote them by $\otimes^n \underline{\mathcal{F}}$, $\operatorname{Sym}^n \underline{\mathcal{F}}$ and $\bigwedge^n \underline{\mathcal{F}}$. The latter two are quotients of $\otimes^n \underline{\mathcal{F}}$.

The functors defined in (a)–(d) as well as $\otimes^n$, Sym^n and $\bigwedge^n$ all preserve nil-isomorphisms. Hence they pass to functors between the corresponding categories of crystals.

Theorem 4. *The functors f^*, $R^i f_*$, $\otimes$ and $\otimes_B B'$ on τ-sheaves induce B-linear functors:*

(a) $f^*: \mathbf{Crys}(X, B) \to \mathbf{Crys}(Y, B)$.
(b) $R^i f_*: \mathbf{Crys}(Y, B) \to \mathbf{Crys}(X, B)$ *for $i \geq 0$, if f is proper.*
(c) $- \otimes -: \mathbf{Crys}(X, B) \times \mathbf{Crys}(X, B) \to \mathbf{Crys}(X, B)$.
(d) $- \otimes_B B': \mathbf{Crys}(X, B) \to \mathbf{Crys}(X, B')$.

If f is proper, then f_* and f^* form an adjoint functor pair on crystals. Moreover, $\otimes^n$, Sym^n and $\bigwedge^n$ induce functors on $\mathbf{Crys}(X, B)$.

Now we come to a main point, namely the construction of an extension by zero in the theory of crystals. Such a functor is not present on τ-sheaves. We shall see how localization affords this functor on crystals.

To explain the construction, let $j: U \hookrightarrow X$ be an open embedding with $i: Z \hookrightarrow X$ a closed complement, and denote by $\mathcal{I}_0 \subset \mathcal{O}_X$ the ideal sheaf of Z. Then $\mathcal{I} := \operatorname{pr}_1^* \mathcal{I}_0$ is the ideal sheaf for $Z \times \operatorname{Spec} B \subset X \times \operatorname{Spec} B$. We wish to extend any τ-sheaf $\underline{\mathcal{F}}$ on U "by zero" to X.

By local considerations on X, one can always construct some coherent extension $\widetilde{\mathcal{F}}$ on $X \times \operatorname{Spec} B$ of $\mathcal{F}$. Such an extension is by no means unique, since any sheaf $\mathcal{I}^m \widetilde{\mathcal{F}}, m \in \mathbb{N}$, is also an extension of $\mathcal{F}$. To extend τ, we observe that for some $\widetilde{m} \in \mathbb{N}$ it extends to a homomorphism $(\sigma \times \mathrm{id})^* \mathcal{I}^{\widetilde{m}} \widetilde{\mathcal{F}} \to \widetilde{\mathcal{F}}$. The identity $(\sigma \times \mathrm{id})^* \mathcal{I} = \mathcal{I}^q$ inside $\mathcal{O}_{X \times \operatorname{Spec} B}$ implies that for any $m \gg 0$ one has an extension

$$\tau_m : (\sigma \times \mathrm{id})^* \mathcal{I}^m \widetilde{\mathcal{F}} \to \mathcal{I}^{m+1} \widetilde{\mathcal{F}} \overset{(*)}{\underset{\sim}{\subset}} \mathcal{I}^m \widetilde{\mathcal{F}}$$

of τ. By its construction, the pair $\widetilde{\mathcal{F}}_m := (\mathcal{I}^m \widetilde{\mathcal{F}}, \tau_m)$ is a τ-sheaf on X which extends $\mathcal{F}$. The inclusion $(*)$ implies that $i^* \widetilde{\mathcal{F}}_m$ is nilpotent—in fact, the induced τ is zero—and so $\widetilde{\mathcal{F}}_m$ has the properties of an extension by zero, except the assignment $\mathcal{F} \mapsto \widetilde{\mathcal{F}}_m$ is in no way functorial since there are many possible choices. The key observation is that all such pairs are nil-isomorphic. Passing to crystals yields the following result.

Theorem 5. *There is an exact B-linear functor*

$$j_! : \mathbf{Crys}(U, B) \to \mathbf{Crys}(X, B) : \underline{\mathcal{F}} \mapsto j_! \underline{\mathcal{F}},$$

uniquely characterized by the properties $j^* j_! = \mathrm{id}_{\mathbf{Crys}(U,B)}$ *and* $i^* j_! = 0$.

Having an extension by zero, it is well known how to define cohomology with compact supports.

Definition 9 (Cohomology with compact support). *Say f is compactifiable, i.e., $f = \bar{f} \bar{j}$ for some open immersion $\bar{j} : Y \hookrightarrow \overline{Y}$ and some proper morphism $\bar{f} : \overline{Y} \to X$. Then one defines*

$$R^i f_! := R^i \bar{f}_* \circ \bar{j}_! : \mathbf{Crys}(Y, B) \longrightarrow \mathbf{Crys}(X, B).$$

Standard arguments show that the definition is independent of the chosen factorization, e.g., [Mi80, Chapter VI, Section 3]. Furthermore, due to a result of Nagata, any morphism $f : Y \to X$ between schemes of finite type over $\mathbb{F}_q$ is compactifiable, and so in this situation the $R^i f_!$ exist; cf. [Lü93] or [CoDe].

4.5 Sheaf-theoretic properties

Since $\mathbf{Crys}(X, B)$ is an abelian category, one has the notion of exactness in short sequences. There is also a good notion of *stalk* at a point $x \in X$, where $i_x : x \hookrightarrow X$ denotes the corresponding immersion: The *stalk of a crystal $\underline{\mathcal{F}}$ (on X) at x* is defined as $i_x^* \underline{\mathcal{F}}$. The sheaf underlying the stalk $i_x^* \underline{\mathcal{F}}$ is in general *not* the stalk at x of the sheaf $\mathcal{F}$ underlying $\underline{\mathcal{F}}$. The following result justifies the definition of $i_x^* \underline{\mathcal{F}}$.

Theorem 6.

1. *For any morphism $f : Y \to X$, pullback along f is an exact functor on crystals.*
2. *A sequence of crystals is exact if and only if it is exact at all stalks.*
3. *The support of a crystal $\underline{\mathcal{F}}$, i.e., the set of $x \in X$ for which $i_x^* \underline{\mathcal{F}}$ is nonzero, is constructible.*

We note that part 1 is established by showing that the higher right derived functors of f^* on τ-sheaves are all nilpotent.

Moreover, crystals enjoy a rigidity property that is not shared by τ-sheaves, but reminiscent of properties of étale sheaves.

Theorem 7. *If f is finite radical and surjective, the functors*

$$\mathbf{Crys}(X, B) \xrightarrow[\;f_*\;]{\;f^*\;} \mathbf{Crys}(Y, B)$$

are mutually quasi-inverse equivalences of categories.

In particular, the closed immersion $f : X_{\mathrm{red}} \hookrightarrow X$ yields an isomorphism $\mathbf{Crys}(X_{\mathrm{red}}, B) \cong \mathbf{Crys}(X, B)$. This rigidity property motivates the name "crystal": crystals extend in a unique way under infinitesimal extensions.

4.6 Derived categories and functors

A major part of [BP04] is to extend the functors (a)–(d) and extension by zero to derived functors between suitable derived categories of crystals. This yields derived functors f^*, $\otimes$, $Rf_!$, and change of coefficients.

There are various good reasons to do so. For instance, only there can one properly understand derived functors such as $Rf_!$. It can be shown that the ad hoc defined functors $R^i f_*$ for proper f are indeed ith cohomologies of a right derived functor Rf_*. Moreover, derived categories is the correct setting to discuss the theory of L-functions needed in Section 5.

The objects introduced so far do not suffice to define derived functors. The reader may recall that to properly define the cohomological functors on coherent sheaves, one needs the ambient larger category of quasi-coherent sheaves. Only there can one dispose of Čech resolutions and resolutions by injectives. Similarly, in [BP04], two auxiliary categories of τ-sheaves are introduced, namely those of quasi-coherent τ-sheaves, and of inductive limits of coherent τ-sheaves. In the presence of the endomorphism τ, these two and $\mathbf{Crys}(X, B)$ are pairwise distinct. It is, in fact, an important result of [BP04] that the corresponding derived categories of bounded complexes with coherent cohomology are all equivalent. Having good comparison results of the categories, the treatment of the derived functors follows the usual path.

4.7 Flatness

An important prerequisite to discussing L-functions of crystals in Section 5 is flatness. Only to flat B-crystals can we hope to attach an L-function which takes values in $1 + tB[[t]]$. Since flatness can only be fully understood in derived categories, which we mainly avoid in this survey, we also introduce the notion of crystal of *pullback type*, which is less natural, but technically easier to handle.

Definition 10. *A crystal $\underline{\mathcal{F}}$ is flat if the functor $\underline{\mathcal{F}} \otimes - : \mathbf{Crys}(X, B) \to \mathbf{Crys}(X, B)$ is exact.*

A crystal is called *(locally) free*, if it can be represented by a (locally) free τ-sheaf. A locally free τ-sheaf is acyclic for $\otimes$ on τ-sheaves, and so it represents a flat crystal. There are other flat crystals which are easy to describe:

A τ-sheaf $\underline{\mathcal{F}} = (\mathcal{F}, \tau)$ on X is of *pullback type*, if there exists a coherent sheaf $\mathcal{F}_0$ on X such that $\mathcal{F} \cong \mathrm{pr}_1^* \mathcal{F}_0$. For an affine scheme $X = \mathrm{Spec}\, R$, a τ-sheaf is of pullback type if its underlying sheaf is of the form $M_0 \otimes B$ for some finitely generated R-module M_0. A crystal on X is called of *pullback type*, if it can be represented by a τ-sheaf of pullback type. This notion derives its importance from the following proposition.

Proposition 5. *Any crystal of pullback type is flat.*

Being of pullback type is preserved under all the functors defined so far, i.e., under pullback, tensor product (of two crystals of pullback type), change of coefficients, the functors $R^i f_!$, and $\otimes^n$, Sym^n, and $\bigwedge^n$.

Let us briefly explain the first assertion of the proposition: Since by Theorem 6 exactness can be verified on stalks, a crystal is flat if and only if all its stalks are flat. From the definition of stalk for a crystal, given directly before Theorem 6, it follows that if the crystal is represented by a τ-sheaf of pullback type, then its stalk at any point x of X is represented by a τ-sheaf of pullback type over A on $\mathrm{Spec}\, k_x$. Over the field k_x any module is flat, and hence the pullback of such a module to $k_x \otimes A$ is flat as well. This shows that all the stalks of a crystal of pullback type are flat, which we needed to verify.

Similarly, one has the following important properties of flatness.

Theorem 8.

1. *Flatness of a crystal is preserved under pullback, tensor product (of two flat crystals), change of coefficients and extension by zero.*
2. *If f is compactifiable and $\underline{\mathcal{F}}^\bullet$ is a bounded complex of flat crystals on Y, then $Rf_!\underline{\mathcal{F}}^\bullet$ is represented by a bounded complex of flat crystals.*
3. *A crystal is flat if and only if all its stalks are flat.*

We observed that any locally free crystal is flat. It is, in fact, not a simple matter to understand precisely when, or in what sense, a flat crystal may be represented by a locally free τ-sheaf. This is relevant to us since we want to attach an L-function at x to a flat crystal and a point $x \in X$ with finite residue field. Therefore, it would be desirable that its stalk at x have a representing τ-sheaf whose underlying sheaf is free and finitely generated over A. Unfortunately this is not true in general. However, we have the following important special case.

Theorem 9. *Suppose that $x = \mathrm{Spec}\, k_x$ for some finite field extension k_x of $\mathbb{F}_q$, that A is artinian and that $\underline{\mathcal{F}}$ is an A-crystal on x. Then we have the following:*

1. *The crystal $\underline{\mathcal{F}}$ has a representing τ-sheaf whose endomorphism τ is an isomorphism.*
2. *The representative from 1 is unique up to unique isomorphism; we write $\underline{\mathcal{F}}_{ss}$ for it and call it the semisimple part of $\underline{\mathcal{F}}$.*
3. *The assignment $\underline{\mathcal{F}} \mapsto \underline{\mathcal{F}}_{ss}$ is functorial.*
4. *If $\underline{\mathcal{F}}$ is flat, then the module underlying $\underline{\mathcal{F}}_{ss}$ is free over $k_x \otimes A$.*

For later use, we also record the following.

Proposition 6. *Suppose that X is the spectrum of a field, and B is regular of dimension ≤ 1 or finite. Then every flat B-crystal on X can be represented by a locally free τ-sheaf on which τ is injective.*

If $\mathcal{F}$ is represented by a torsion free τ-sheaf, then the proof in [BP04] shows that $\operatorname{Im}(\tau^n_{\mathcal{F}})$ has the asserted property for $n \gg 0$.

The most general representability result for flat crystals, shown in [BP04], is that for any flat $\mathcal{F}$ and any reduced scheme X there is a finite cover by locally closed regular subschemes X_i, so that on each X_i some "jth iterate" of $\mathcal{F}$ is representable by a "τ^j-sheaf" whose underlying sheaf is locally free. Since we will not need this, we will not give the details. Note, however, that this is reminiscent of the definition of constructibility of étale sheaves.

4.8 A test case

Throughout this section, we assume that B is finite (and an $\mathbb{F}_q$-algebra). Let $\mathbf{\acute{E}t}(X, B)$ be the category of étale sheaves of B-modules and $\mathbf{\acute{E}t}_c(X, B)$ its full subcategory of constructible sheaves. By $\mathrm{pr}_1 : X \times \operatorname{Spec} B \to X$ we denote the projection onto the first factor.

It has long been known that for such B there is a correspondence between τ-sheaves on which τ is an isomorphism and lisse étale constructible sheaves of B-modules; cf., for instance, [Ka73, Theorem 4.1.1], which we recall in Theorem 10 below. In [BP04], this correspondence was extended to an equivalence of categories with, on the one hand, $\mathbf{Crys}(X, B)$ and, on the other, $\mathbf{\acute{E}t}_c(X, B)$. In this section, we describe the functor which provides this equivalence.

We would like to remark that—under the name of "Riemann–Hilbert correspondence"—M. Emerton and M. Kisin have, for regular X, constructed another equivalence of categories between on the one hand $\mathbf{\acute{E}t}_c(X, B)$ and on the other again a category whose objects carry a σ-semilinear operation, cf. [EK04]. It turns out, and this is currently under investigation by M. Blickle and the author, that their category is dual to the category of crystals, with the duality given by the duality of sheaves, as described by Hartshorne in [Ha66]. Emerton and Kisin have extended their correspondence to formal $\mathbb{Z}_p$-coefficients. It would be interesting to see whether in some way one can extend the concept of τ-sheaf to incorporate $\mathbb{Z}/(p^n)$-coefficients.

Let $\mathcal{F}$ be a τ-sheaf over B on X. To any étale morphism $u : U \to X$ we assign the B-module $(u^*\mathcal{F})^\tau$ of τ-invariants of $u^*\mathcal{F}$. This construction is functorial in u and hence defines a sheaf of B-modules on the small étale site over X, which we denote by $\mathcal{F}_{\text{ét}}$.

The construction is also functorial in $\mathcal{F}$, that is, to any homomorphism $\phi : \mathcal{F} \to \mathcal{G}$ it associates a homomorphism $\phi_{\text{ét}} : \mathcal{F}_{\text{ét}} \to \mathcal{G}_{\text{ét}}$. Therefore, it defines a B-linear functor

$$\epsilon : \mathbf{Coh}_\tau(X, B) \to \mathbf{\acute{E}t}(X, B) : \mathcal{F} \mapsto \epsilon(\mathcal{F}) := \mathcal{F}_{\text{ét}}. \tag{6}$$

Let us mention some obvious consequences of the definition of ϵ. For any τ-sheaf $\mathcal{F}$ and étale $u : U \to X$, one has a left exact sequence

$$0 \longrightarrow \underline{\mathcal{F}}_{\text{ét}}(U) \longrightarrow (u^*\mathcal{F})(U \times \operatorname{Spec} B) \overset{1-\tau}{\longrightarrow} (u^*\mathcal{F})(U \times \operatorname{Spec} B), \qquad (7)$$

which induces a short exact sequence of étale sheaves. In the particular case $\mathcal{F} = \underline{\mathbb{1}}_{X, \mathbb{F}_q}$, sequence (7) specializes to the usual Artin–Schreier sequence

$$0 \longrightarrow (\underline{\mathbb{1}}_{X, B})_{\text{ét}} \longrightarrow \mathcal{O}_X \overset{1-\tau}{\longrightarrow} \mathcal{O}_X.$$

Thus $(\underline{\mathbb{1}}_{X, B})_{\text{ét}}$ is the constant étale sheaf with fiber B.

As another example, suppose $(M(\phi), \tau)$ is the A-motive on K attached to a Drinfeld A-module on K. Then we may apply the functor ϵ to $(M(\phi), \tau) \otimes_A A/\mathfrak{n}$ for any nonzero ideal $\mathfrak{n}$ of A. Essentially by specializing Proposition 2 to finite levels, one obtains the isomorphism

$$\phi[\mathfrak{n}] \cong \operatorname{Hom}_{A/\mathfrak{n}}(\epsilon((M(\phi), \tau) \otimes_A A/\mathfrak{n})(K^{\text{sep}}), \mathfrak{n}\Omega_A/\Omega_A).$$

In particular, ϵ can be used to define étale realizations of A-crystals.

Generalizing Artin–Schreier theory, the following result is proved by Katz in [Ka73, Theorem 4.1.1].

Theorem 10. *For a normal domain X the functor $\underline{\mathcal{F}} \mapsto \underline{\mathcal{F}}_{ét}$ defines an equivalence between the categories $\{\underline{\mathcal{F}} \in \mathbf{Coh}_\tau(X, B) : \tau_{\mathcal{F}} \text{ is an isomorphism}\}$ and $\{\mathsf{F} \in \acute{\mathbf{E}}\mathbf{t}_c(X, B) : \mathsf{F} \text{ is lisse}\}$.*

Since for nilpotent τ-sheaves $\underline{\mathcal{F}}$ one has $\underline{\mathcal{F}}_{\text{ét}} = 0$, one easily deduces that ϵ passes to a functor $\mathbf{Crys}(X, B) \to \acute{\mathbf{E}}\mathbf{t}(X, B)$. Using this, [BP04, Chapter 9] refines Theorem 10 to the following.

Theorem 11. *The functor $\mathbf{Crys}(X, B) \to \acute{\mathbf{E}}\mathbf{t}(X, B) : \underline{\mathcal{F}} \mapsto \underline{\mathcal{F}}_{ét}$ takes its image in $\acute{\mathbf{E}}\mathbf{t}_c(X, B)$. The induced functor*

$$\epsilon : \mathbf{Crys}(X, B) \to \acute{\mathbf{E}}\mathbf{t}_c(X, B)$$

is an equivalence of categories. It is compatible with all of the functors f^*, $\otimes$, $\otimes_B B'$, Sym^n, $\bigwedge^n$ and $R^i f_!$, and preserves flatness.

Remark 2. The category $\acute{\mathbf{E}}\mathbf{t}_c(X, B)$ possesses no duality. Therefore, it can neither exist for $\mathbf{Crys}(X, B)$. Also only the functors f^*, $\otimes$ and $f_!$, which we have constructed on crystals, are well behaved on $\acute{\mathbf{E}}\mathbf{t}_c(\dots)$. Therefore, one should not expect that all the functors f^*, $\otimes$, $f_!$, f_*, Hom, $f^!$, postulated by Grothendieck for a good cohomological theory, exist for $\mathbf{Crys}(\dots)$.

In special cases, one may construct some further functors. For instance, in [Bö04] it is shown that for B finite or $B = A$ and any open immersion $j : U \hookrightarrow X$ there exists a meaningful functor

$$j_\# : \mathbf{Crys}(U, A) \longrightarrow \mathbf{Crys}(X, A) \qquad (8)$$

in the sense of Definition 6. It corresponds to j_* in the étale theory. It is not called j_*, since in [BP04] this name was reserved for a different functor.

5 First applications

As explained in the introduction, one of the initial motivations to introduce the category of crystals was to give a cohomological proof of the rationality of L-functions of τ-sheaves. The rationality had been conjectured—at least in the case of families of Drinfeld modules—by Goss [Go91a], and a first proof had been given by Taguchi and Wan in [TW96] using analytical tools. Subsection 5.1 describes the cohomological proof from [BP04].

In addition to the rationality, the algebraic approach yields some extra information. Namely, the L-function is expressible in terms of cohomology with compact support. These cohomology modules are in principle computable. The main result described in Subsection 5.2, makes crucial use of this.

Subsection 5.2 is concerned with a conjecture of Goss on analytic L-functions attached to families of Drinfeld modules. It asserts that these L-functions (with values in $\mathbb{C}_\infty$) have a meromorphic continuation to a function field analogue of the complex plane; cf. [Go91a]. In the case $A = \mathbb{F}_q[t]$, this is the second main result in [TW96]. Later in [Bö02] Goss's conjecture was completely established for arbitrary A.

In the present section, we want to formulate the notions necessary to state the precise results and indicate some important steps in their proofs. Since for general A, Goss's conjecture on meromorphy is rather technical to formulate, we will only do this in the case $A = \mathbb{F}_q[t]$. A detailed treatment of Subsection 5.1 can be found in [BP04], and of Subsection 5.2 in [Bö02].

5.1 L-functions of crystals

In this subsection we assume for simplicity of exposition that B is either a finite ring or a normal domain. We write $Q(B)$ for the ring of fractions of B, so that in the latter case $Q(B)$ is a field and in the former it is simply B again. All schemes will be of finite type over $\mathbb{F}_q$. By $|X|$ we denote the closed points of a scheme X. For $x \in |X|$ we denote its residue field by k_x and its degree by $d_x := [k_x : \mathbb{F}_q]$. Moreover, $\mathcal{F}$ will denote a flat B-crystal on X.

The aim is to explain how to attach L-functions to flat crystals on X, and state their main properties, i.e., the rationality, the invariance under $Rf_!$, and the invariance under change of coefficients.

Ultimately, these L-functions must be defined in terms of their underlying τ-sheaves, and at the same time invariant under nil-isomorphisms. For artinian B, we will use Theorem 9 to choose a good canonical representative.

When B is reduced, these L-functions satisfy all the usual cohomological formulas precisely (except duality). When B possesses nonzero nilpotent elements, however, these formulas hold only up to "unipotent" factors. In some sense these factors correspond to nilpotent τ-sheaves and can therefore not be detected by our theory. So this defect is built into our theory of crystals by its very construction.

As a preparation we briefly recall the theory of the dual characteristic polynomial for endomorphisms of projective modules. Suppose M is a finitely generated projective B-module and $\phi\colon M \to M$ is a B-linear endomorphism.

Lemma 1. *Let M' be any finitely generated projective B-module such that $M \oplus M'$ is free over B. Let $\phi': M' \to M'$ be the zero endomorphism and t a new indeterminate.*

1. *The expression $\det_B(\mathrm{id} - t(\phi \oplus \phi') \mid M \oplus M') \in 1 + tB[t]$ is independent of the choice of M'. It is called the* dual characteristic polynomial of (M, ϕ) *and denoted $\det_B(\mathrm{id} - t\phi \mid M)$.*
2. *The assignment $(M, \phi) \mapsto \det_B(\mathrm{id} - t\phi \mid M)$ is multiplicative in exact sequences.*

For any $x \in |X|$, the stalk $\underline{\mathcal{F}}_x$ is flat, and so is $\underline{\mathcal{F}}_x \otimes_B Q(B)$. By Theorem 9 the latter is canonically represented by the locally free τ-sheaf $(\underline{\mathcal{F}}_x \otimes_B Q(B))_{ss}$. The endomorphism τ^{d_x} is $k_x \otimes Q(B)$-linear. By Lemma 1, part 1, the following definition makes sense.

Definition 11. *The L-function of $\underline{\mathcal{F}}$ at x is*

$$L(x, \underline{\mathcal{F}}, t) := \det_{k_x \otimes Q(B)}(\mathrm{id} - t^{d_x}\tau^{d_x} \mid (\underline{\mathcal{F}}_x \otimes_B Q(B))_{ss})^{-1} \in k_x \otimes Q(B)[[t^{d_x}]].$$

Lemma 2.

1. *The power series $L(x, \underline{\mathcal{F}}, t)$ lies in $1 + t^{d_x} B[[t^{d_x}]]$.*
2. *The assignment $\underline{\mathcal{F}} \mapsto L(x, \underline{\mathcal{F}}, t)$ is multiplicative in short exact sequences.*

The proof of part 1 needs our assumption on B.

As the number of points in $|X|$ of any given degree d_x is finite, we can form the product over the L-functions at all points $x \in |X|$ within $1 + tB[[t]]$.

Definition 12. *The L-function of $\underline{\mathcal{F}}$ is*

$$L(X, \underline{\mathcal{F}}, t) := \prod_{x \in |X|} L(x, \underline{\mathcal{F}}, t) \in 1 + tB[[t]].$$

To state the main results, we need an equivalence relation on $1 + tB[[t]]$.

Definition 13. *By $\mathfrak{n}_B$ we denote the nilradical of B, i.e., the ideal of B of nilpotent elements.*

For $P, Q \in 1 + tB[[t]]$, we write $P \sim Q$ if and only if there exists $H \in 1 + t\mathfrak{n}_B[t]$, such that $P = QH$.

If B is reduced, and so, for instance, if B is a normal domain, then $\mathfrak{n}_B = (0)$, and hence $P \sim Q$ is equivalent to $P = Q$.

Finally, if $h: B \to B'$ denotes a change of coefficients homomorphism, then its induced homomorphism $B[[t]] \to B'[[t]]$ is also denoted by h.

Theorem 12. *Suppose $f: Y \to X$ is any morphism between schemes of finite type, $h: B \to B'$ is any ring homomorphism and $\underline{\mathcal{G}}$ is a B-crystal of pullback type on Y. Then*

1. *$L(Y, \underline{\mathcal{G}}, t) \sim \prod_i L(X, R^i f_! \underline{\mathcal{G}}, t)^{(-1)^i}$;*

2. $L(Y, \underline{\mathcal{G}} \otimes_B B', t) \sim h(L(Y, \underline{\mathcal{G}}, t))$, with equality if B is artinian.

Working in the context of derived categories, both parts can be proved more generally for any bounded complex $\mathcal{F}^\bullet$ of flat crystals on Y, and with the complex $Rf_* \mathcal{F}^\bullet$ instead of the crystals $R^i f_* \underline{\mathcal{F}}$. We alert the reader that the complex $Rf_* \mathcal{F}^\bullet$ carries more information than the individual $R^i f_* \underline{\mathcal{F}}$.

A *rational function* (*over* B) is an element in $1 + tB[[t]]$ of the form P/Q for suitable polynomials $P, Q \in 1 + tB[t]$. The rationality of L-functions follows by applying Theorem 12, part 1, to the structure morphism $Y \to \mathrm{Spec}\, k$.

Corollary 1. *With Y, $\underline{\mathcal{G}}$ as in Theorem* 12, *the function $L(Y, \underline{\mathcal{G}}, t)$ is rational.*

Proof of Theorem 12 (*sketch*). The proof of the two parts are independent. We shall omit the proof of part 2. Instead, we indicate two alternative proofs of the trace formula in part 1.

A *conventional proof* might go as follows, where first we assume B to be reduced: Standard fibering techniques reduce one to the proof in the case where $Y = \mathbb{A}^1$ and $X = \mathrm{Spec}\, \mathbb{F}_q$. The formula to be proved can then as in [SGA4$\frac{1}{2}$, "Rapport" and "Fonction L mod ℓ^n et mod p"] be reduced to a trace formula over the symmetric powers of $\mathbb{A}^1_A$ for the corresponding exterior symmetric product of $\underline{\mathcal{G}}$. Again by induction on dimension, it suffices to prove this formula over $\mathbb{A}^1_A$. Its proof can be obtained from the Woodshole fixed point formula for coherent sheaves: see [SGA5, Exp. III, Corollary 6.12].

Let now B be finite and nonreduced. The assertion can be reduced to affine Y over $\mathbb{F}_q$ and to τ-sheaves which are free over B on Y. Fixing a surjection $B' := \mathbb{F}_q[x_1, \ldots, x_k] \twoheadrightarrow B$, the τ-sheaf may be lifted to one over the reduced ring B'. For B' the trace formula has been proved already. Using part 2, it follows for B.

The proof of part 1, as given in [BP04] is significantly different and based on some ideas of Anderson; cf. [An00]. To sketch it, let us fix the following situation. Let $Y = \mathrm{Spec}\, R$ be smooth and affine over $X := \mathrm{Spec}\, \mathbb{F}_q$ of dimension e with f as the structure morphism, let B be a field, and suppose that $\underline{\mathcal{G}}$ is a locally free τ-sheaf.

Using coherent duality and the Cartier operator on $\omega_{Y/\mathbb{F}_q} := \bigwedge^e \Omega_{Y/\mathbb{F}_q}$, Anderson sets $\mathcal{G}^\vee := \mathcal{H}om(\mathcal{G}, \omega_{Y/\mathbb{F}_q})$ and obtains an $\mathcal{O}_Y \otimes B$-linear homomorphism $\kappa : \mathcal{G}^\vee \longrightarrow (\sigma \times \mathrm{id})^* \mathcal{G}^\vee$. Let M be the $R \otimes B$-module underlying $\mathcal{G}$ and $M^\vee$ that underlying $\mathcal{G}^\vee$. Anderson observes that κ is strongly contracting on $M^\vee$ in the following sense:

There exists a finite dimensional B-subvector space W of $M^\vee$ such that $\kappa(W) \subset W$ and such that $M^\vee = \bigcup_{i=1}^\infty \{m \in M^\vee \mid \kappa^i(m) \in W\}$. Such a subspace is termed a *nucleus* for $(M^\vee, \kappa)$. Using elementary means, Anderson proves that

$$L(Y, \underline{\mathcal{G}}, t) = \det_B(\mathrm{id} - t\kappa \mid W)^{(-1)^{e-1}}.$$

In [BP04] it is shown that for locally free $\underline{\mathcal{G}}$ the only nonvanishing cohomology $R^i f_! \underline{\mathcal{G}}$ occurs in degree $i = e$, and that furthermore the dual of (W, κ) is via Serre duality nil-isomorphic to a suitable τ-sheaf representing $R^e f_! \underline{\mathcal{G}}$. Combining the above pieces, the proof is "complete."

The method just sketched has two main advantages. First, Anderson's proof is elementary. Second, our interpretation gives a cohomological interpretation for the nucleus and thus for his trace formula.   □

For finite B we have seen in the previous section that there is an equivalence of categories $\epsilon \colon \mathbf{Crys}(X, B) \longrightarrow \text{Ét}_c(X, B)$ which preserves, in particular, the notion of flatness. As is well known, to any constructible étale sheaf F of flat B-modules, one can also attach an L-function $L_{\text{ét}}(X, \mathsf{F}, t)$, e.g., [SGA4$\frac{1}{2}$, "Rapport" and "Fonction L mod ℓ^n"]. In [BP04] the following is shown, which except for the comparison is also proved in [SGA4$\frac{1}{2}$].

Theorem 13. *Suppose B is finite and $\underline{\mathcal{F}}$ is a flat B-crystal on X. Then*

$$L(X, \underline{\mathcal{F}}, t) = L_{\text{ét}}(X, \epsilon(\underline{\mathcal{F}}), t).$$

Hence $L_{\text{ét}}$ has a trace formula and is compatible with change of coefficients.

5.2 Goss's L-functions of crystals over A

To a scheme $\mathfrak{X}$ which is flat and of finite type over $\mathbb{Z}$, one can associate its ζ-function which is an analytic function that is convergent on a right half plane. A similar construction of global L-functions over function fields has been carried out by Goss for families of Drinfeld modules. It can easily be extended to τ-sheaves, and we now sketch this for $A = \mathbb{F}_q[t]$. For further details, for the more general case and for the case of v-adic L-functions, we refer the reader to [Go96, Chapter 8] and [Bö02].

Throughout this subsection, we fix a morphism $g \colon X \to \operatorname{Spec} A$ of finite type, and assume $B = A$. By $\underline{\mathcal{F}}$ we denote a flat A-crystal on X. The example considered originally was that of a Drinfeld A-module ϕ of rank r on $\mathcal{L}$ over some scheme X of finite type over $\mathbb{F}_q$. It yields a locally free τ-sheaf $\underline{\mathcal{M}}(\phi) \ (= \underline{\mathcal{F}})$ of rank r, and for g one takes $\operatorname{char}_\phi \colon X \to \operatorname{Spec} A$.

Exponentiation of ideals

We begin by defining a substitute for the classically used expression p^{-s}, where $s \in \mathbb{C}$ and p is a prime number. Following Goss, one sets $S_\infty := \mathbb{C}_\infty^* \times \mathbb{Z}_p$, which will replace the usual complex plane as the domain of L-functions. An element $s \in S_\infty$ will have components (z, w). One defines an addition by $(z_1, w_1) + (z_2, w_2) = (z_1 \cdot z_2, w_1 + w_2)$. As a uniformizer for $F_\infty \cong \mathbb{F}_q((1/t))$ we take $\pi_\infty := 1/t$.

Since any fractional ideal $\mathfrak{a}$ of A is principal, we may write it in the form (a) for some rational function $0 \neq a \in F$. The element $\langle a \rangle := a \pi_\infty^{-\operatorname{val}_\infty(a)}$ is a unit in $A_\infty \cong \mathbb{F}_q[[\pi_\infty]]$. We choose for a the *unique* generator of $\mathfrak{a}$ for which $\langle a \rangle$ is a 1-unit, and set $\langle \mathfrak{a} \rangle := \langle a \rangle$ as well as $\deg \mathfrak{a} := -\operatorname{val}_\infty(a)$. The exponentiation of ideals with elements in S_∞ is now defined as follows.

Definition 14.

$$\{\textit{fractional ideals of } A\} \times S_\infty \to \mathbb{C}_\infty^* : (\mathfrak{a}, (z, w)) \mapsto \mathfrak{a}^s := z^{\deg \mathfrak{a}} \langle \mathfrak{a} \rangle^w.$$

The exponentiation is bilinear for multiplication on ideals, addition on S_∞ and multiplication on $\mathbb{C}_\infty^*$. Note that the exponentiation of any 1-unit with an element of $\mathbb{Z}_p$ is well defined. The so-defined exponentiation *depends* on the choice of the uniformizing parameter $\pi_\infty = 1/t$.

One defines the embedding $\mathbb{Z} \hookrightarrow S_\infty : i \mapsto s_i := (\pi_\infty^{-i}, i)$, so that the element $\mathfrak{a}^{s_i} \in F$ is the unique generator of the ideal $\mathfrak{a}^i$ such that $\langle \mathfrak{a}^{s_i} \rangle$ is a 1-unit.

The definition of global L-functions

Let x be a closed point of X. Since X is of finite type over $\operatorname{Spec} A$, the point x lies above a unique closed point $\mathfrak{p} = \mathfrak{p}_x$ of $\operatorname{Spec} A$, and one has $d_\mathfrak{p} | d_x$ for their degrees over $\mathbb{F}_q$. Hence $L(x, \underline{\mathcal{F}}, t)^{-1} \in 1 + t^{d_x} B[t^{d_x}] \subset 1 + t^{d_\mathfrak{p}} B[t^{d_\mathfrak{p}}]$, and so $L(x, \underline{\mathcal{F}}, t)\big|_{t^{d_\mathfrak{p}} = \mathfrak{p}^{-s}}$ is well defined. In [Bö02] or in many cases in [Go96, Section 8], it is shown that there exists $c > 0$ such that the following Euler product converges for all $s \in \mathbb{H}_c := \{(z, w) \in \mathbb{C}_\infty^* \times \mathbb{Z}_p \mid |z| > c\}$.

Definition 15. *The* global L-function of $\underline{\mathcal{F}}$ at $s \in \mathbb{H}_c$ is

$$L^{\mathrm{an}}(X, \underline{\mathcal{F}}, s) := \prod_{\substack{\mathfrak{p} \in \operatorname{Max}(A)}} \prod_{\substack{x \in |X| \\ x \text{ above } \mathfrak{p}}} L(x, \underline{\mathcal{F}}, t)\big|_{t^{d_\mathfrak{p}} = \mathfrak{p}^{-s}} = \prod_{x \in |X|} L(x, \underline{\mathcal{F}}, t)\big|_{t^{d_{\mathfrak{p}_x}} = \mathfrak{p}_x^{-s}} .$$

Thus $L^{\mathrm{an}}(X, \underline{\mathcal{F}}, -) : \mathbb{H}_c \to \mathbb{C}_\infty$. The subset $\mathbb{H}_c \subset S_\infty$ is called a *half space (of convergence) of* S_∞ in analogy with the usual right half plane of $\mathbb{C}$.

Obviously, this definition depends on the morphism $g : X \to \operatorname{Spec} A$. If $\underline{\mathcal{F}} = \mathcal{M}(\phi)$ for some Drinfeld A-module ϕ on $\mathcal{L}$ over X, then Definition 15 agrees with the one originally given by Goss.

Meromorphy

For $c \in \mathbb{R}_{\geq 0}$ we define $D_c^* := \{z \in \mathbb{C}_\infty \mid |z| > c\} \subset \mathbb{P}^1(\mathbb{C}_\infty)$ as the *punctured "open" disc around ∞ of radius* c, and $\overline{D}_c^* := \{z \in \mathbb{C}_\infty \mid |z| \geq c\}$ as the corresponding "closed" disc. In particular, $\mathbb{H}_c = D_c^* \times \mathbb{Z}_p$.

To give a precise meaning to "entire," respectively, "meromorphic extension" of a global L-function to S_∞, one now changes one's viewpoint. Namely, for any fixed $w \in \mathbb{Z}_p$ one regards an L-function as a function $D_c^* \to \mathbb{C}_\infty$. With respect to a suitable topology on the resulting functions $D_c^* \to \mathbb{C}_\infty$, the variation over $w \in \mathbb{Z}_p$ will be continuous. We now describe this topology.

For $D = \overline{D}_c^*$ and $c > 0$, or $D = D_c^*$ and $c \geq 0$, we define

$$\mathrm{C}^{\mathrm{an}}(D) := \left\{ f = \sum_{n \geq 0} a_n z^{-n} \mid a_n \in \mathbb{C}_\infty, \ f \text{ converges on } D \right\}.$$

If $D = \overline{D}_c^*$, then $\mathrm{C}^{\mathrm{an}}(D)$ is isomorphic to the usual Tate algebra over $\mathbb{C}_\infty$, and one can define a Banach space structure on it by defining for any $f \in \mathrm{C}^{\mathrm{an}}(D)$ the norm

$||f||_c := \sup_{z \in \overline{D}_c^*} |f(z)|$ (which is also multiplicative). In the case $D = D_c^*$ one can only obtain a Fréchet space. The procedure is slightly more involved. Namely, let $(c_n) \subset \mathbb{R}$ be a strictly decreasing sequence with limit c. For any two elements $f, g \in C^{an}(D)$, we define their distance as

$$\text{dist}(f, g) := \sum_{m=1}^{\infty} 2^{-m} \frac{||f - g||_{c_m}}{1 + ||f - g||_{c_m}}.$$

One can show (a), that with respect to dist the $\mathbb{C}_\infty$-linear space $C^{an}(D)$ is a complete linear metric space, and (b), that the topology on this space does not depend on the choice of the sequence (c_n) (provided that $c_n \to c$ and $c_n > c$ for all n). In the following, $\mathbb{Z}_p$ is equipped with its usual locally compact topology.

Proposition 7. *If $L(X, \mathcal{F}, (z, w)))$ converges on $\mathbb{H}_c$, then the function $w \mapsto (z \mapsto L(X, \mathcal{F}, (z, w)))$ defines a continuous function $\mathbb{Z}_p \to C^{an}(D_c^*)$.*

This viewpoint bears some similarities to that of p-adic L-functions, where the complete, algebraically closed field $\mathbb{C}_p$ takes the role of $C^{an}(D_c^*)$.

For $c < c'$ one has the obvious inclusion $C^{an}(D_c^*) \subset C^{an}(D_{c'}^*)$, given as the identity on power series. Also since a power series is uniquely determined by its coefficients, an element in $C^{an}(D_{c'}^*)$ has at most one extension to an element in $C^{an}(D_c^*)$. Having this in mind, one introduces the following notions.

Definition 16 (Goss). *A continuous function $\mathbb{Z}_p \to C^{an}(D_0^*)$ is called* entire. *The quotient of two entire functions which are units on D_c^* for some $c \gg 0$ is called* meromorphic.

A global L-function $L^{an}(X, \mathcal{F}, s)$ is called entire, *respectively,* meromorphic *on S_∞ if there exists an entire, respectively, meromorphic, function h whose restriction $h: \mathbb{Z}_p \to C^{an}(D_c^*)$ agrees with $L^{an}(X, \mathcal{F}, s)$ for $c \gg 0$.*

By the remark preceding the definition, there is at most one entire function h which extends a function $L^{an}(X, \mathcal{F}, s)$. The same holds for meromorphic functions.

For a meromorphic function h, the values $h(i), i \in \mathbb{Z}$, are its *special values*. In the examples of interest, the special values at $-\mathbb{N}_0$ will typically lie in $\mathbb{C}_\infty(z)$. Since both $\mathbb{Z}$ and $\mathbb{N}_0$ are dense in $\mathbb{Z}_p$, the special values completely determine a meromorphic function. One has the following criterion in terms of special values for $L^{an}(X, \mathcal{F}, s)$ to be entire.

Proposition 8. *Let $\mathbb{H}_c$ denote a half plane of convergence for $L^{an}(X, \mathcal{F}, s)$ and write h for the corresponding continuous function $\mathbb{Z}_p \to C^{an}(D_c^*)$. Suppose there exists $\varepsilon \in \{\pm 1\}$ such that*

1. *$h(i)^\varepsilon$ is a polynomial in z^{-1} over $\mathbb{C}_\infty$ for all $i \in -\mathbb{N}_0$, and*
2. *the degrees of the polynomials $h(i)^\varepsilon$, $i \in -\mathbb{N}_0$, grow like $\mathcal{O}(\log |i|)$.*

Then $L^{an}(X, \mathcal{F}, s)^\varepsilon$ is entire.

The assumptions of the theorem can typically be achieved for X smooth over $\mathbb{F}_q$ of dimension e, for $\mathcal{F}$ locally free on $X \times \operatorname{Spec} A$, and $\varepsilon = (-1)^{e-1}$.

There are two ways to prove Proposition 8. The path taken in the proof of [Bö02, Theorem 4.15] uses directly p-adic interpolation properties. In [Go04a], Goss takes an alternative measure-theoretic approach.

Special values

The *Carlitz τ-sheaf* $\underline{C}$ over A on $\operatorname{Spec} A$ is the τ-sheaf corresponding to

$$(\mathbb{F}_q[t] \otimes \mathbb{F}_q[t], (1 \otimes t - t \otimes 1)(\sigma \otimes \operatorname{id})).$$

The following result, first observed by Taguchi and Wan, provides the following key link between algebraic and analytic L-functions.

Proposition 9. *Suppose $X = \operatorname{Spec} A \cong \mathbb{A}^1$ and $g \colon X \to \operatorname{Spec} A$ is the identity. Then for $i \in \mathbb{N}_0$ one has*

$$L^{\mathrm{an}}(X, \mathcal{F}, (z, 0) + s_{-i}) = L(X, \mathcal{F} \otimes \underline{C}^{\otimes i}, t)_{|t = z^{-1}},$$

where we recall $s \colon \mathbb{Z} \hookrightarrow S_\infty \colon i \mapsto s_i = (\pi_\infty^{-i}, i)$.

In light of Proposition 8, one would like to bound the degree of $L(X, \mathcal{F} \otimes \underline{C}^{\otimes i}, t)$ while i varies. For this we shall compute $Rg_!(\mathcal{F} \otimes \underline{C}^{\otimes i})$. If one has a τ-sheaf on $\operatorname{Spec} k$ representing it, its rank will bound the degree of the algebraic L-function. Let us denote by j the open immersion $\mathbb{A}^1 \hookrightarrow \mathbb{P}^1$ and $\bar{f} \colon \mathbb{P}^1 \to \operatorname{Spec} \mathbb{F}_q$ the structure morphism.

We proceed as follows: First, one changes the coefficients from A to its fraction field F. Next, one extends the τ-sheaves $\mathcal{F}$ and $\underline{C}$ (over F) to τ-sheaves $\widetilde{\mathcal{F}}$ and $\widetilde{\underline{C}}$ on $\mathbb{P}^1$ which represent the crystals $j_!\mathcal{F}$ and $j_!\underline{C}$, respectively. Then, one computes $R^i \bar{f}_*(\widetilde{\mathcal{F}} \otimes \widetilde{\underline{C}}^{\otimes i})$ as a τ-sheaf over F on $\operatorname{Spec} \mathbb{F}_q$. In the case at hand, the τ-sheaves $R^i \bar{f}_*, i \neq 1$, will all be nilpotent. By the trace formula, Theorem 12, we thus have

$$L(X, \mathcal{F} \otimes \underline{C}^{\otimes i}, t) = \det_F(1 - t\tau \mid R^1 \bar{f}_* \widetilde{\mathcal{F}} \otimes \widetilde{\underline{C}}^{\otimes i}).$$

Hence $L(X, \mathcal{F} \otimes \underline{C}^{\otimes i}, t)$ is a polynomial and its degree is bounded by the dimension of the F-vector space $H^1(\mathbb{P}^1_F, \widetilde{\mathcal{F}} \otimes \widetilde{\underline{C}}^{\otimes i})$.

The sheaf underlying $\widetilde{\underline{C}}$ can be taken as $\mathcal{O}_{\mathbb{P}^1_F}(-2)$. If we follow the above recipe, then by the Riemann–Roch theorem, the dimension of $H^1(\mathbb{P}^1_F, \widetilde{\mathcal{F}} \otimes \widetilde{\underline{C}}^{\otimes i})$ will grow linearly in i. The degree of $L(X, \mathcal{F} \otimes \underline{C}^{\otimes i}, t) \in \mathbb{C}_\infty[t]$ can thus grow at most linearly. But we can do much better. Namely, the sheaf $\underline{C}^{\otimes p^i}$ is nil-isomorphic to $\underline{C}^{(i)}$, which is defined by

$$(\mathbb{F}_q[t] \otimes \mathbb{F}_q[t], (1 \otimes t^{p^i} - t \otimes 1)(\sigma \otimes \operatorname{id})).$$

The latter (considered over F) has an extension $\widetilde{\underline{C}}^{(i)}$ whose underlying sheaf is $\mathcal{O}_{\mathbb{P}^1_F}(-2)$. So if we write $i = a_0 + a_1 p + a_2 p^2 + \cdots$ in its p-adic expansion with $a_i \in \{0, 1, \ldots, p - 1\}$, then

$$(\widetilde{\underline{C}}^{(0)})^{\otimes a_0} \otimes (\widetilde{\underline{C}}^{(1)})^{\otimes a_1} \otimes (\widetilde{\underline{C}}^{(2)})^{\otimes a_2} \otimes \cdots$$

also represents the crystal $j_! \underline{C}^i$, but its underlying sheaf is $\mathcal{O}_{\mathbb{P}^1_F}(-2(a_0+a_1+a_2+\cdots))$. It follows that the degree of $L(X, \underline{\mathcal{F}} \otimes \underline{C}^{\otimes i}, t)$ grows at most logarithmically in i. This combined with Propositions 8 and 9 proves the following result, which in the given form with a different proof is due to Taguchi and Wan; cf. [TW96].

Theorem 14. *Suppose $X = \operatorname{Spec} A$, $g \colon X \to \operatorname{Spec} A$ is the identity, and $\underline{\mathcal{F}}$ is locally free over A on X. Then $L^{\mathrm{an}}(X, \underline{\mathcal{F}}, s)$ is entire.*

Refining the above methods, in [Bö02] the following is shown.

Theorem 15. *Suppose X is Cohen–Macaulay and equidimensional of dimension e. Then for any locally free τ-sheaf $\underline{\mathcal{F}}$ over A on X the function $L^{\mathrm{an}}(X, \underline{\mathcal{F}}, s)^{(-1)^{e-1}}$ is entire.*

Using the representability results for flat crystals, and the theory of iterates of characteristic polynomials, as developed in [BP04], one obtains the following consequence by decomposing X into a suitable finite union of regular locally closed subschemes.

Corollary 2. *Suppose $X \to \operatorname{Spec} A$ is a scheme of finite type and $\underline{\mathcal{F}}$ is a flat A-crystal on X. Then $L^{\mathrm{an}}(X, \underline{\mathcal{F}}, s)$ is meromorphic.*

5.3 Open questions

While on the one hand, we have seen that under some reasonable set of hypotheses the analytic functions $L^{\mathrm{an}}(X, \underline{\mathcal{F}}, s)$ are meromorphic, there remain many mysteries concerning these functions. The interpolation procedure seems to identify them as a kind of p-adic L-function interpolating special values at the negative integers. At the same time, these functions also have Euler products. We pose some open problems.

Question 1. What is the arithmetic meaning of the special values?

For an analogue of the Riemann ζ-function, already in the work of Carlitz there appeared identities that are reminiscent of the formulas $\zeta(n) = \pi^n r_n$ for even $n \in \mathbb{N}$ and rational r_n. So Carlitz's formulas have some arithmetic meaning. For more on this, we refer to [Go96, Section 8.18].

Question 2. Is there a conjecture à la Birch and Swinnerton–Dyer (BSD) for A-motives?

A naive analogue of BSD cannot hold, since it is known due to a result of Poonen (cf. [Po95]) that the naive analogue of the Mordell–Weil group for a Drinfeld A-module over a field L as in Section 2 is of infinite A-rank. In [An96] in certain cases, a finite rank A-module has been constructed, that could serve as a starting point to investigate such a conjecture.

Question 3. Is there a Riemann hypothesis or a (substitute for a) functional equation?

There is no duality to be expected for crystals or τ-sheaves, as explained in Remark 2. Nevertheless, there are very intriguing calculations along these lines which are very suggestive although definitive conjectures cannot now be made; cf. [Go00] and [Go04].

6 Motives for Drinfeld cusp forms

Using geometric means, Scholl in [Sch90] has constructed for the space of cusp forms (over $\mathbb{Q}$) for each fixed weight and level a motive in the sense of Grothendieck. In this section, we want to describe a similar construction for Drinfeld modular forms in the function field case. It attaches a motive in the sense of Anderson to each space of Drinfeld cusp forms of fixed weight and level. Again for the sake of exposition, we only consider the simplest case $A = \mathbb{F}_q[t]$. Details of this appear in [Bö04].

6.1 Moduli spaces

Let $\mathfrak{n}$ be a proper nonzero ideal of A. In Subsection 4.2, in particular in Theorem 2, we recalled the definition and existence of a fine moduli space $\mathfrak{Y}^r(\mathfrak{n})$ for Drinfeld A-modules of rank r and characteristic prime to $\mathfrak{n}$ that carry a level $\mathfrak{n}$-structure. From now on, we only consider the case $r = 2$, and therefore omit the superscript r whenever $r = 2$. As in Subsection 4.2, by $\mathcal{M}(\mathfrak{n})$ we denote the τ-sheaf corresponding to the universal Drinfeld A-module on $\mathfrak{Y}(\mathfrak{n})$, and by $g_{\mathfrak{n}} : \mathfrak{Y}(\mathfrak{n}) \to \mathrm{Spec}\, A(\mathfrak{n})$ its characteristic.

The first observation we will need in the following is due to Drinfeld.

Theorem 16. *The morphism $g_{\mathfrak{n}}$ has a (canonical) smooth compactification*

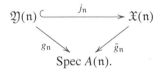

The completion of $\mathfrak{X}(\mathfrak{n})$ along the complement $\mathfrak{X}(\mathfrak{n}) \setminus \mathfrak{Y}(\mathfrak{n})$ (considered as a reduced scheme) is (formally) smooth over $\mathrm{Spec}\, A(\mathfrak{n})$, and may be considered as a disjoint union of what one might call Drinfeld–Tate curves. They describe the degeneration of rank 2 to rank 1 Drinfeld modules and have properties analogous to the usual Tate curve. For details of this construction, cf. [Bö04, vdH03, Le01].

To describe the connectivity properties of $\mathfrak{Y}(\mathfrak{n})$ and the cusps, recall that in the case of elliptic curves the existence of the Weil-pairing yields a morphism of the corresponding (compactified) moduli space to $\mathrm{Spec}\,\mathbb{Z}[\zeta_N, \frac{1}{N}]$. Over this base the moduli space is geometrically connected, and so not over $\mathbb{Z}$ itself.

Similarly one has a pairing on rank 2 Drinfeld A-modules. It induces a smooth morphism $w_{\mathfrak{n}} : \mathfrak{Y}(\mathfrak{n}) \to \mathfrak{Y}^1(\mathfrak{n})$ to the moduli space of rank 1-Drinfeld modules with a level $\mathfrak{n}$-structure, which may be extended to a smooth proper morphism $\bar{w}_{\mathfrak{n}} : \mathfrak{X}(\mathfrak{n}) \to \mathfrak{Y}^1(\mathfrak{n})$; cf. [vdH03]. The situation together with the canonical morphism $\mathfrak{Y}^1(\mathfrak{n}) \to \operatorname{Spec} A(\mathfrak{n})$ is displayed in

$$
\begin{array}{ccc}
\mathfrak{Y}(\mathfrak{n}) \hookrightarrow \mathfrak{X}(\mathfrak{n}) \longleftarrow \mathfrak{X}(\mathfrak{n}) \setminus \mathfrak{Y}(\mathfrak{n}) \\
\searrow w_{\mathfrak{n}} \quad \downarrow \bar{w}_{\mathfrak{n}} \quad \swarrow \partial w_{\mathfrak{n}} \\
\mathfrak{Y}^1(\mathfrak{n}) \\
g^1_{\mathfrak{n}} \downarrow \\
\operatorname{Spec} A(\mathfrak{n}).
\end{array}
\tag{9}
$$

The morphism $g^1_{\mathfrak{n}}$ is finite étale, of degree $d(\mathfrak{n})$; the morphism w_n is geometrically connected. Thus if we pass from $A(\mathfrak{n})$ to $\mathbb{C}_\infty$, the space $\mathfrak{Y}^1(\mathfrak{n})$ will decompose into $d(\mathfrak{n})$ copies of $\mathbb{C}_\infty$. Correspondingly, $\mathfrak{X}(\mathfrak{n})$ breaks up into $d(\mathfrak{n})$ components. Finally, under $\partial w_{\mathfrak{n}}$, the scheme $\mathfrak{X}(\mathfrak{n}) \setminus \mathfrak{Y}(\mathfrak{n})$ is isomorphic to a disjoint union of copies of $\mathfrak{Y}^1(\mathfrak{n})$. Their number is denoted by $c(\mathfrak{n})$.

6.2 Rigid analytic uniformization

Over $\mathbb{Z}[1/n]$, one has a compactification similar to $\mathfrak{X}(\mathfrak{n}) \to \operatorname{Spec} A(\mathfrak{n})$ for the arithmetic surfaces that arise as the moduli space of elliptic curves with a level N-structure. This is described in detail in [KM85]. If instead one works over the complex numbers and the finer complex topology, the situation becomes considerably simpler. The resulting curves admit a uniformization by the upper half plane, and can be realized as quotients by congruence subgroups.

The analogous procedure in the function field setting is to base change $\mathfrak{Y}(\mathfrak{n})$ via $\iota : A(\mathfrak{n}) \hookrightarrow \mathbb{C}_\infty$ to a curve over $\operatorname{Spec} \mathbb{C}_\infty$. Now one regards the curve over $\mathbb{C}_\infty$ as a rigid analytic space—we write $\mathfrak{Y}(\mathfrak{n})^{\mathrm{rig}}$—which again yields a finer (Grothendieck) topology than the Zariski topology. For details on the rigidification functor $X \mapsto X^{\mathrm{rig}}$, we refer the reader to [BGR84].

As observed by Drinfeld, there is an analogue of the upper half plane, usually denoted Ω. The rigidified moduli space $\mathfrak{Y}(\mathfrak{n})^{\mathrm{rig}}$ is, in fact, isomorphic to $\bigsqcup_{i=1}^{d(\mathfrak{n})} \Gamma(\mathfrak{n}) \backslash \Omega$ for a suitable quotient of Ω, which we now describe:

The points of Ω over $\mathbb{C}_\infty$ are given as $\Omega(\mathbb{C}_\infty) := \mathbb{P}^1(\mathbb{C}_\infty) \setminus \mathbb{P}^1(F_\infty)$. They are acted on by the group $\mathrm{GL}_2(F_\infty)$ via

$$
\mathrm{GL}_2(F_\infty) \times \Omega(\mathbb{C}_\infty) \longrightarrow \Omega(\mathbb{C}_\infty) : \left(\begin{pmatrix} a\,b \\ c\,d \end{pmatrix}, z \right) \mapsto \frac{az+b}{cz+d}.
$$

To give Ω the structure of a rigid analytic space, we describe an admissible cover (which is the analogue in rigid geometry of an atlas in differential geometry). Its construction can be best understood, if one introduces the reduction map of $\Omega(\mathbb{C}_\infty)$ to the corresponding Bruhat–Tits tree. The atlas we describe arises via pullback from a simple combinatorial Čech covering of the tree. Not wanting to introduce these notions, we now directly define the cover: Let $\mathbb{F}_\infty$ be the residue field of F_∞ and q_∞ its cardinality, and define

$$\mathfrak{U}_0(\mathbb{C}_\infty) := \{z \in \mathbb{C}_\infty \mid |z - \beta|_\infty \geq q_\infty^{-2/3} \text{ for all } \beta \in \mathbb{F}_\infty \text{ and } |z|_\infty \leq q_\infty^{2/3}\}.$$

The set $\mathfrak{U}_0(\mathbb{C}_\infty)$ is rigid analytically equivalent to the unit disc with q_∞ smaller discs removed. One can show the following:

1. If for $\gamma \in \mathrm{GL}_2(F_\infty)$ the intersection $\mathfrak{U}_0(\mathbb{C}_\infty) \cap \gamma(\mathfrak{U}_0(\mathbb{C}_\infty))$ is nonempty, then one has precisely $q_\infty + 2$ possibilities: Either the intersection is $\mathfrak{U}_0(\mathbb{C}_\infty)$, or it is

$$\mathfrak{U}_1(\mathbb{C}_\infty) = \{z \in \mathbb{C}_\infty \mid q_\infty^{1/3} \leq |z|_\infty \leq q_\infty^{2/3}\},$$

or there exists some $\beta \in \mathbb{F}_\infty$ such that it is of the form

$$\{z \in \mathbb{C}_\infty \mid q_\infty^{-1/3} \geq |z - \beta|_\infty \geq q_\infty^{-2/3}\}.$$

2. The sets in 1 different from $\mathfrak{U}_0(\mathbb{C}_\infty)$ are all translates of $\mathfrak{U}_1(\mathbb{C}_\infty)$.
3. The sets $\gamma\mathfrak{U}_0(\mathbb{C}_\infty)$, $\gamma \in \mathrm{GL}_2(F_\infty)$, form an admissible covering of $\Omega(\mathbb{C}_\infty)$.

To define a geometry (in this case a rigid analytic structure), one also has to describe a set of functions on the atlas given. A function $f : \mathfrak{U}_0(\mathbb{C}_\infty) \to \mathbb{C}_\infty$ is *rigid analytic on* $\mathfrak{U}_0(\mathbb{C}_\infty)$, if and only if it can be written as a series

$$\sum_{n \in \mathbb{N}_0} a_n z^n + \sum_{\beta \in \mathbb{F}_\infty} \sum_{n \in \mathbb{N}} b_{n,\beta}(z - \beta)^{-n}$$

which converges on all of $\mathfrak{U}_0(\mathbb{C}_\infty)$. The latter simply means that the sequences $(|a_n| q^{2/3n})$ and $(|b_{n,\beta}| q^{-2/3n})$, for all $\beta \in \mathbb{F}_\infty$, tend to zero.

Definition 17. *A function* $f : \Omega(\mathbb{C}_\infty) \to \mathbb{C}_\infty$ *is rigid analytic on* $\Omega(\mathbb{C}_\infty)$*, if for all* $\gamma \in \mathrm{GL}_2(F_\infty)$ *the restriction of* $f \circ \gamma$ *to* $\mathfrak{U}_0(\mathbb{C}_\infty)$ *is rigid analytic.*

To describe $\Gamma(\mathfrak{n}) \backslash \Omega$, we recall that one defines

$$\Gamma(\mathfrak{n}) := \{\gamma \in \mathrm{GL}_2(A) \mid \gamma \equiv \mathrm{id} \pmod{\mathfrak{n}}\}.$$

It is a discrete subgroup of $\mathrm{GL}_2(F_\infty)$, and thus acts on $\Omega(\mathbb{C}_\infty)$. Say we fix $\gamma \in \mathrm{GL}_2(F_\infty)$ and abbreviate $\mathfrak{U} := \gamma\mathfrak{U}_0$. Then for $\gamma_0 \in \Gamma(\mathfrak{n})$ one either has $\gamma_0\mathfrak{U} = \mathfrak{U}$ or $\gamma_0\mathfrak{U} \cap \mathfrak{U} = \emptyset$. The former case only occurs a finite number of times, so that the stabilizer $\mathrm{Stab}_{\Gamma(\mathfrak{n})}(\mathfrak{U})$ of $\mathfrak{U}$ in $\Gamma(\mathfrak{n})$ is finite. One may define rigid analytic quotients $\mathrm{Stab}_{\Gamma(\mathfrak{n})}(\mathfrak{U}) \backslash \mathfrak{U}$, and these can be glued to define a rigid space $\Gamma(\mathfrak{n}) \backslash \Omega$.

6.3 Cusp forms

Following the case of elliptic modular forms over number fields, one can define Drinfeld modular functions (and cusp forms) over function fields as follows.

Definition 18. *A rigid analytic function* $f : \Omega(\mathbb{C}_\infty) \to \mathbb{C}_\infty$ *is called a* modular function *of weight* $k \in \mathbb{N}$ *for* $\Gamma(\mathfrak{n})$*, if it satisfies the identity*

$$f(\gamma z) = (cz + d)^k f(z) \quad \text{for all } \gamma = \begin{pmatrix} a & b \\ c & d \end{pmatrix} \in \Gamma(\mathfrak{n}) \quad \text{and} \quad z \in \Omega(\mathbb{C}_\infty).$$

The modular functions of level $\mathfrak{n}$ are invariant under the operation of

$$\Gamma_\infty(\mathfrak{n}) := \left\{ \gamma = \begin{pmatrix} 1 & b \\ 0 & 1 \end{pmatrix} \mid b \in \mathfrak{n} \right\},$$

i.e., under translation by all $b \in \mathfrak{n}$. Another such function is

$$e_{\mathfrak{n}}(z) := z \prod_{\lambda \in \mathfrak{n} \setminus \{0\}} \left(1 - \frac{z}{\lambda} \right).$$

The function $t_{\mathfrak{n}} := e_{\mathfrak{n}}^{-1}$ is, in a suitable sense, a uniformizing parameter of $\mathfrak{n}$-invariant functions near ∞. Therefore, any modular function has a Laurent expansion

$$f(z) = \sum_{n \in \mathbb{Z}} a_n t_{\mathfrak{n}}^n(z)$$

convergent for $|z|_i \gg 0$, where $|z|_i := \inf \{|z - a| \mid a \in F_\infty\}$.

If f is a modular function of weight k for $\Gamma(\mathfrak{n})$, then so is $f \circ \gamma$ for all $\gamma \in GL_2(A)$, because $\Gamma(\mathfrak{n})$ is normal in $GL_2(A)$. Therefore, one can equally consider the expansion $\sum_{n \in \mathbb{Z}} a_{n,\gamma} t_{\mathfrak{n}}^n(z)$ of $f \circ \gamma$. Since $GL_2(A)$ acts transitively on the "cusps" of $\Gamma \backslash \Omega$, following the case of elliptic modular forms, one defines the following.

Definition 19. *A modular function f of weight k for $\Gamma(\mathfrak{n})$ is called a*

$$\left\{ \begin{array}{c} \text{modular form} \\ \text{cusp form} \\ \text{double-cusp form} \end{array} \right\} \Longleftrightarrow \forall \gamma \in GL_2(A) \ \forall n \in \mathbb{Z} \ \text{with} \ \left\{ \begin{array}{c} n < 0 \\ n < 1 \\ n < 2 \end{array} \right\} a_{n,\gamma} = 0.$$

By $M_k(\Gamma(\mathfrak{n}), \mathbb{C}_\infty)$, we denote the $\mathbb{C}_\infty$-vector space of modular forms for $\Gamma(\mathfrak{n})$ of weight k and by $S_k(\Gamma(\mathfrak{n}), \mathbb{C}_\infty)$ and $S_k^{dc}(\Gamma(\mathfrak{n}), \mathbb{C}_\infty)$ its subspaces of cusp and double-cusp forms.

The number of conditions imposed in Definition 19 is finite, since it suffices to require these conditions for matrices γ which form a set of representatives of $\Gamma(\mathfrak{n}) \backslash GL_2(A)$.

Elliptic double-cusp forms are usually not considered, since they have no meaningful interpretation. One reason to introduce them in the Drinfeld modular setting is given below in Theorem 17.

To define modular forms on $\mathfrak{Y}(\mathfrak{n})^{\mathrm{rig}}$ one uses its identification with $\bigsqcup_{i=1}^{d(\mathfrak{n})} \Gamma(\mathfrak{n}) \backslash \Omega$. Formally we set

$$M_k(\mathfrak{n}, \mathbb{C}_\infty) := M_k(\Gamma(\mathfrak{n}), \mathbb{C}_\infty)^{d(\mathfrak{n})}, \tag{10}$$

and similarly for cusp and double-cusp forms.

As in the classical situation, one may define Hecke operators (e.g., as correspondences) which act on Drinfeld modular forms. These preserve the subspaces of cusp and double-cusp forms. (Depending on their normalization, Hecke operators may also "permute" the components of $\mathfrak{Y}(\mathfrak{n})/\mathbb{C}_\infty$.)

We conclude this subsection with two examples. For the first, recall that $\mathfrak{Y}(\mathfrak{n})$ is an affine scheme, equal to $\mathrm{Spec}\, R_{\mathfrak{n}}$ for some smooth $A(\mathfrak{n})$-algebra $R_{\mathfrak{n}}$. As remarked in Theorem 2, the universal rank 2 Drinfeld A-module is a homomorphism $A \to \mathrm{End}_{\mathbb{F}_q}(\mathbb{G}_{a,\mathfrak{Y}(\mathfrak{n})})$, i.e., a homomorphism

$$\phi_{\mathfrak{n}} \colon A \to R_{\mathfrak{n}}\{\tau\}.$$

There are morphisms from $R_{\mathfrak{n}}$ to its rigidification, and then from the latter to the ring of global rigid analytic sections $\Gamma(\Omega, \mathcal{O}_{\Omega}^{\mathrm{rig}})^{d(\mathfrak{n})}$ of $\bigsqcup_{i=1}^{d(\mathfrak{n})} \Omega(\mathbb{C}_{\infty})$. This yields a homomorphism

$$\phi(\mathfrak{n}) \colon A \to \Gamma(\Omega, \mathcal{O}_{\Omega}^{\mathrm{rig}})^{d(\mathfrak{n})}\{\tau\}.$$

Proposition 10. *For any $a \in A$ consider $\phi(\mathfrak{n})_a$. The coefficient of its leading term $\tau^{2\deg a}$ lies in $S_{q^{2\deg a}-1}(\mathfrak{n}, \mathbb{C}_{\infty})$, the coefficients of the remaining terms τ^i lie in $M_{q^i-1}(\mathfrak{n}, \mathbb{C}_{\infty})$.*

As for elliptic modular forms one has an interpretation of the global sections of the sheaf of differentials on $\mathfrak{X}(\mathfrak{n})$ in terms of modular forms.

Theorem 17. *There is a canonical isomorphism*

$$H^0(\mathfrak{X}(\mathfrak{n})^{\mathrm{rig}}, \Omega_{\mathfrak{X}(\mathfrak{n})^{\mathrm{rig}}/\mathbb{C}_{\infty}}) \cong S_2^{\mathrm{dc}}(\mathfrak{n}, \mathbb{C}_{\infty}).$$

Without going into any details, the reason for the occurrence of double-cusp forms is the following. Suppose $f(z)$ is a Drinfeld modular form of weight 2 for $\Gamma(\mathfrak{n})$. Then $f(z)dz$ is a global $\Gamma(\mathfrak{n})$-invariant differential form on $\Omega(\mathbb{C}_{\infty})$. To investigate its behavior near the cusp described by $t_{\mathfrak{n}}$, observe that one has $de_{\mathfrak{n}} = dz$ from the definition of $e_{\mathfrak{n}}$, and hence

$$dt_{\mathfrak{n}} = d(e_{\mathfrak{n}}^{-1}) = -e_{\mathfrak{n}}^{-2}dz = -t_{\mathfrak{n}}^2 dz.$$

Thus near the cusp for $t_{\mathfrak{n}}$, the function $f(z)dz$ is a power series in $t_{\mathfrak{n}}$ times $-\frac{1}{t_{\mathfrak{n}}^2}dt_{\mathfrak{n}}$. Hence if we start with a double-cusp form, we obtain a global differential on $\mathfrak{X}(\mathfrak{n})$, and vice versa.

6.4 The motive

Following the guide by Eichler–Shimura and Deligne, we now define for any $k \geq 2$ the locally free τ-sheaf $\underline{\mathcal{M}}(\mathfrak{n})^{(k-2)} := \mathrm{Sym}^{k-2}\underline{\mathcal{M}}(\mathfrak{n})$, and the A-crystal

$$\underline{\mathcal{S}}^{(k)}(\mathfrak{n}) := R^1 g_{\mathfrak{n}!}\underline{\mathcal{M}}(\mathfrak{n})^{(k-2)} \tag{11}$$

on $\mathrm{Spec}\,A(\mathfrak{n})$. By Proposition 3 the τ-sheaf $\underline{\mathcal{M}}(\mathfrak{n})^{(k-2)}$ is locally free of rank $k-1$. Because $\mathfrak{Y}(\mathfrak{n})$ is affine, the corresponding module is projective and finitely generated. If we extend it to a free module and choose $\tau = 0$ on the complement, we find that $\underline{\mathcal{M}}(\mathfrak{n})^{(k-2)}$ is of pullback type. Also it is not difficult to see that the crystal $R^0 g_{\mathfrak{n}!}\underline{\mathcal{M}}(\mathfrak{n})^{(k-2)}$ is zero. Since, moreover, $\bar{g}_{\mathfrak{n}}$ is smooth and proper of relative dimension 1, Proposition 5 yields the following.

Proposition 11. *The crystal $R^i g_{\mathfrak{n}!}\underline{\mathcal{M}}(\mathfrak{n})^{(k-2)}$ is zero for $i \neq 1$. The crystal $\underline{\mathcal{S}}^{(k)}(\mathfrak{n}) = R^1 g_{\mathfrak{n}!}\underline{\mathcal{M}}(\mathfrak{n})^{(k-2)}$ is of pullback type and hence flat.*

In [Bö04], jointly with R. Pink we computed some explicit examples of such motives for $A = \mathbb{F}_q[t]$ and $\mathfrak{n} = (t)$. Other explicit examples are given in Corollaries 4, 5 and 6.

Considering geometric correspondences, one can define Hecke operators T_v for all places of A prime to $\mathfrak{n}$. They naturally act on the crystals $\underline{S}^{(k)}(\mathfrak{n})$. Let $\mathfrak{p}_v$ be the maximal ideal corresponding to v. The Hecke operators will still act as Hecke operators on the reduction of $\underline{S}^{(k)}(\mathfrak{n})$ to the fiber at v, i.e., its pullback along Spec $A/\mathfrak{p}_v \hookrightarrow$ Spec $A(\mathfrak{n})$. As basically already observed by Drinfeld, one has an Eichler–Shimura relation.

Theorem 18. *The action of T_v on the fiber of $\underline{S}^{(k)}(\mathfrak{n})$ at v is given by the action of the geometric Frobenius* Frob$_v$ *at v on this fiber.*

6.5 Its analytic realization

We now follow the guide of A-motives to define an analytic realization of the crystals $\underline{S}^{(k)}(\mathfrak{n})$. First, we pull back this crystal via $A(\mathfrak{n}) \hookrightarrow \mathbb{C}_\infty$ to a crystal over A on $\mathbb{C}_\infty$. Because it is flat, we may by using Proposition 6 represent it by a τ-module $(S^{(k)}(\mathfrak{n}), \tau)$ over A on $\mathbb{C}_\infty$ whose underlying sheaf is finitely generated projective, and on which τ is injective. In [Bö04] the following is shown.

Theorem 19. *The τ-module $(S^{(k)}(\mathfrak{n}), \tau)$ is uniformizable in the sense of Definition 3.*

Therefore, we call

$$(S^{(k)}(\mathfrak{n})\langle t \rangle)^\tau$$

the *analytic realization of the motive* $\underline{S}^{(k)}(\mathfrak{n})$. This realization carries an induced Hecke action. The following main result is shown in [Bö04].

Theorem 20. *There is a canonical Hecke-equivariant isomorphism*

$$\mathrm{Hom}_A((S^{(k)}(\mathfrak{n})\langle t \rangle)^\tau, \Omega_A) \otimes_A \mathbb{C}_\infty \cong S_k(\mathfrak{n}, \mathbb{C}_\infty).$$

The result should be compared with equation (3). The proof uses rigid analytic tools and an explicit combinatorial Čech covering of $\mathfrak{X}(\mathfrak{n})^{\mathrm{rig}}$. It would go beyond the scope of this article to give details.

Using the Hecke action one may define a Hecke-invariant filtration on $\underline{S}^{(k)}(\mathfrak{n})$ whose subquotients are flat crystals $\underline{S}_f$ corresponding to (generalized) cuspidal Drinfeld Hecke eigenforms f. Neither the filtration, nor the crystal $\underline{S}_f$ are canonical, and more precisely the crystal $\underline{S}_f$ corresponds to the Galois orbit of f, and one has to be aware that the Hecke action on $\underline{S}^{(k)}(\mathfrak{n})$ may not be semisimple. Nevertheless, these subquotients are useful.

6.6 Its étale realizations

Let us fix a place v of A. Then $\underline{S}^{(k)}(\mathfrak{n}) \otimes_A A/\mathfrak{p}_v^n$ defines a flat crystal over $A/\mathfrak{p}_v^n$ on Spec $A(\mathfrak{n})$. Via the functor ϵ from Subsection 4.8, we obtain a lisse étale sheaf of $A/\mathfrak{p}_v^n$-modules on Spec $A(\mathfrak{n})$, which say we denote by $S^{\mathrm{ét}}(\mathfrak{n}, A/\mathfrak{p}_v^n)$. Varying n, these sheaves form a compatible system, and thus a v-adic étale sheaf on Spec $A(\mathfrak{n})$. Since all sheaves in this systems are lisse, we obtain a Galois representation

$$\rho_{\mathfrak{n},k,v} \colon G_F \to \mathrm{GL}_{d_{\mathfrak{n},k}}(A_v),$$

where $d_{\mathfrak{n},k}$ is the dimension of $S_k(\mathfrak{n}, \mathbb{C}_\infty)$.

The filtration on $\underline{S}^{(k)}(\mathfrak{n})$ described at the end of the previous subsection induces also a filtration on the compatible system $S^{\mathrm{ét}}(\mathfrak{n}, A/\mathfrak{p}_v^n)$. The subquotients yield the compatible systems $\epsilon(\underline{S}_f \otimes_A A/\mathfrak{p}_v^n)$ corresponding to (Galois orbits of generalized) cuspidal Drinfeld Hecke eigenforms f. The correspondence can be made precise by using the Eichler–Shimura relation from Theorem 18. One obtains the following.

Theorem 21. *Let f be a cuspidal Drinfeld Hecke eigenform (over $\mathbb{C}_\infty$) and denote by F_f the field generated over F by the Hecke eigenvalues a_w of f, where w runs through all places of A prime to $\mathfrak{n}$. Then $[F_f : F]$ is finite and for any place v of A there exists a place v' of F_f above v and a representation*

$$\rho_{f,v} \colon \mathrm{Gal}(F^{\mathrm{sep}}/F) \longrightarrow \mathrm{GL}_1((F_f)_{v'})$$

uniquely characterized by

$$\rho_{f,v}(\mathrm{Frob}_w) = a_w \tag{12}$$

for all places w of A which are prime to $\mathfrak{n}\mathfrak{p}_v$.

This result is strikingly different from the analogous one for elliptic modular forms since there the representations are 2-dimensional. To explain this, we recall the following observation on Hecke operators which dates back to Gekeler and Goss. Namely, one can define Hecke operator $T_{\mathfrak{m}}$ for any nonzero ideal $\mathfrak{m}$ prime to $\mathfrak{n}$. In characteristic p they satisfy $T_{\mathfrak{m}\mathfrak{m}'} = T_{\mathfrak{m}} T_{\mathfrak{m}'}$ for any ideals $\mathfrak{m}$, $\mathfrak{m}'$, and in particular for $\mathfrak{m}$ a power of some prime ideal $\mathfrak{p}$. This is different from the case of characteristic zero, *but* it simply follows from the usual relation by reduction modulo p. Another reason why one should expect abelian representations is given following Corollary 6.

The characterizing property (12) is basically the same in all places v of A. This means that the representations $\rho_{f,v}$ form a compatible system of v-adic abelian representations of G_F. (The same holds for the semisimplification of the representations $\rho_{\mathfrak{n},k,v}$.) Thus extending the results of [Kh04] to the function field case, it seems natural to expect that to any Drinfeld cusp form one can attach a Grössencharacter χ_f of type A_0 such that the compatible family $\rho_{f,v}$ arises from χ_f in the way described in [Kh03, Section 4] and [Go92]. Therefore, the following natural question arises.

Question 4. Which Grössencharacters of F of type A_0 arise from Drinfeld modular forms?

Can any Grössencharacter of F of type A_0 be twisted by a power of the Grössencharacter arising from the Carlitz-module (as in [Go92]), such that it arises from a Drinfeld modular form?

Recall that $\underline{C}$ is the Carlitz-module defined above Proposition 9. Then Question 4 is a generalization of the following problem raised in [Go02].

Question 5. Can one find for any Drinfeld $\mathbb{F}_q[t]$-module ϕ on $\operatorname{Spec} \mathbb{F}_q[t]$ of rank 1 an $n \in \mathbb{N}_0$ such that $\underline{\mathcal{M}}(\phi) \otimes \underline{\mathcal{C}}^{\otimes n}$ is the motive of a modular form?

Namely, any such Drinfeld module determines a Hecke-character of type A_0 and is, moreover, uniquely determined by this Hecke-character.

6.7 *L*-functions

Having the (noncanonically defined) crystal $\underline{\mathcal{S}}_f$ attached to any cuspidal Drinfeld Hecke eigenform f of level n, using the formalism described in Subsection 5.2 one can attach an analytic *L*-function to it. It is independent of the choice of $\underline{\mathcal{S}}_f$. This yields nontrivial factors in the Euler product at all primes not dividing n. However, the function f may be an old form, i.e., defined over some smaller level n'. Thus it would be desirable to also have Euler factors at primes in $\operatorname{Spec} A/n \setminus \operatorname{Spec} A/n'$. One way to achieve this is to assign to f the analytic *L*-function of the maximal model of $\underline{\mathcal{S}}_f$ in the sense of Gardeyn; cf. Definition 6. This assignment is now also independent of the level in which f was found. Using, in particular, Theorem 14 and Theorem 21, one shows the following.

Corollary 3. *For f a cuspidal Drinfeld Hecke eigenform of minimal level n and with Hecke eigenvalues a_v, one has*

$$L_f^{\text{an}}(s) = \prod_{v \in \operatorname{Max}(A(n))} \left(1 - \frac{a_v}{\mathfrak{p}^s} \right)^{-1}$$

for $s = (z, w) \in S_\infty$ with $|z| \gg 0$. The function $L_f^{\text{an}}(s)$ is entire in the sense of Definition 16.

In [Go91] two further analytic *L*-functions are attached to a cuspidal Drinfeld modular form. These are known to be different from the one in Corollary 3.

Question 6. What is the relation between these *L*-functions, if any?

The assignment $f \mapsto L_f^{\text{an}}(s)$ described in Corollary 3, attaches to a Hecke eigenform an *L*-function. However, unlike in the situation for elliptic modular forms, there is no normalization for such forms in the Drinfeld modular setting. In the elliptic modular setting the typical requirement is that the first Fourier coefficient of f is 1. As the examples of doubly cuspidal Drinfeld Hecke eigenforms show, this coefficient may be zero in our setting.

Question 7 (*Goss*). Is there a canonical normalization of a cuspidal Drinfeld Hecke eigenform?

If the answer would be yes, then by superposition one could attach an *L*-function to any Drinfeld cusp form.

Question 8. Can the assignment $f \mapsto L_f(s)$ be realized by an analogue of the usual Mellin transform?

6.8 Double-cusp forms

We saw in Theorem 17, that double-cusp forms do play an important role in the theory of Drinfeld modular forms. So one may wonder whether there is also a motive describing them. The short answer is yes, and we will explain some of it, since it leads to another interesting question.

Let $k \geq 2$. In Subsection 6.4, we defined $\underline{S}^{(k)}(\mathfrak{n}) = R^1 \bar{g}_\mathfrak{n} j_{\mathfrak{n}!} \underline{\mathcal{M}}(\mathfrak{n})^{(k-2)}$, i.e., we first extended the crystal $\underline{\mathcal{M}}(\mathfrak{n})^{(k-2)}$ to $\mathfrak{X}(\mathfrak{n})$ by zero, and then we computed its first cohomology. In Remark 2, we noted that Gardeyn's notion of maximal model leads to a functor $j_\#$ on A-crystals provided the base X was of finite type over some field. Therefore, we may define

$$\underline{S}^{(k)}_{dc}(\mathfrak{n}) = R^1 \bar{g}_\mathfrak{n} j_{\mathfrak{n}\#} \underline{\mathcal{M}}(\mathfrak{n})^{(k-2)}.$$

One can now formulate (by adding the subscript $_{dc}$) and prove the precise analogue of Theorems 19 and 20 for double-cusp forms; cf. [Bö04].

It is, in fact, possible to completely determine the discrepancy between cusp and double-cusp forms. Namely, for any $k \geq 2$ one has a short exact sequence

$$0 \longrightarrow j_! \underline{\mathcal{M}}(\mathfrak{n})^{(k-2)} \longrightarrow j_\# \underline{\mathcal{M}}(\mathfrak{n})^{(k-2)} \longrightarrow \mathbb{1}_{\mathfrak{X}(\mathfrak{n}) \backslash \mathfrak{Y}(\mathfrak{n}), A} \longrightarrow 0$$

of crystals. The long exact sequence of cohomology then yields a four-term exact sequence of crystals

$$0 \to \bar{g}_{\mathfrak{n}*} j_\# \underline{\mathcal{M}}(\mathfrak{n})^{(k-2)} \to \bar{g}_{\mathfrak{n}*} \mathbb{1}_{\mathfrak{X}(\mathfrak{n}) \backslash \mathfrak{Y}(\mathfrak{n}), A} \to \underline{S}^{(k)}(\mathfrak{n}) \to \underline{S}^{(k)}_{dc}(\mathfrak{n}) \to 0. \quad (13)$$

The properties of diagram (9) yield $\bar{g}_{\mathfrak{n}*} \mathbb{1}_{\mathfrak{X}(\mathfrak{n}) \backslash \mathfrak{Y}(\mathfrak{n}), A} \cong (g^1_{\mathfrak{n}!} \mathbb{1}_{\mathfrak{Y}^1(\mathfrak{n}), A})^{c(\mathfrak{n})}$. Moreover, for $k \neq 2$ the left-hand term vanishes, for $k = 2$ it is isomorphic to $g^1_{\mathfrak{n}*} \mathbb{1}_{\mathfrak{Y}^1(\mathfrak{n}), A}$.

We define $\overline{S}_k(\mathfrak{n}, \mathbb{C}_\infty) := S_k(\mathfrak{n}, \mathbb{C}_\infty) / S^{dc}_k(\mathfrak{n}, \mathbb{C}_\infty)$, and set $\delta_k := 0$ for $k \geq 3$ and $\delta_k := 1$ for $k = 2$. Sequence (13) with the above identifications, the duality in Theorem 20, and the analogous duality for double-cusp forms, prove the following.

Corollary 4. *There is a fixed Hecke-module of dimension $d(\mathfrak{n})$ depending on $\mathfrak{n}$ but not on k, such that $\overline{S}_k(\mathfrak{n}, \mathbb{C}_\infty)$ is the direct sum of $c(\mathfrak{n}) - \delta_k$ copies of it.*

The Hecke-module in question arises from the arithmetic of $\mathfrak{Y}^1(\mathfrak{n})$.

Question 9. Can one give an explicit basis for the cuspidal Drinfeld Hecke eigenforms which are not double-cusp forms, in a way similarly explicit to the the description one has for Eisenstein series?

It is equally interesting to consider the consequences of (13) for the étale realization of our motives. With δ_k as above, one obtains the following.

Corollary 5. *The v-adic étale realization corresponding to $\overline{S}_k(\mathfrak{n}, \mathbb{C}_\infty)$ consists of $(c(\mathfrak{n}) - \delta_k)$ copies of $H^1(\mathfrak{Y}(\mathfrak{n})^1 / F^{\text{sep}}, \mathbb{F}_q) \otimes F_v$, considered as a Galois representation of G_F (which is unramified outside $\mathfrak{n}$).*

In particular, this shows that the étale realizations corresponding to the Hecke eigenforms in $\overline{S}_k(\mathfrak{n}, \mathbb{C}_\infty)$ give rise to Galois representations over $\mathbb{F}_q$, and thus with finite image. The proof of the corollary uses, among other things, the compatibility of the functor ϵ of Theorem 11 with coefficient change and with proper push-forward.

Specializing (13) to $k = 2$, and passing to étale realizations one finds the following.

Corollary 6. *The v-adic étale realization of $\underline{S}_{dc}^{(2)}(\mathfrak{n})$ is the Galois representation given of G_F on $H^1(\mathfrak{X}(\mathfrak{n})/F^{sep}, \mathbb{F}_q) \otimes F_v$.*

Again this describes Galois representations over $\mathbb{F}_q$ and thus with finite image. It is known that the curve $\mathfrak{X}(\mathfrak{n})/F$ is ordinary, and hence so is its Jacobian, which, say, we denote by $J(\mathfrak{n})$. One therefore has

$$H^1(\mathfrak{X}(\mathfrak{n})/F^{sep}, \mathbb{F}_q) \cong J(\mathfrak{n})[p](F^{sep}) \cong \mathbb{Z}/(p)^{\dim J(\mathfrak{n})},$$

where the first isomorphism is an isomorphism of Galois modules. Since, in particular, the semisimplification of the module in the middle is abelian, this gives another indication for the abelianess of the representations $\rho_{f,v}$.

The examples in [Bö04] show that the image of $\rho_{f,v}$ is typically infinite for $f \in S_k^{dc}(\mathfrak{n}, \mathbb{C}_\infty)$ and $k > 2$. This is related to a notion of weight that one can attach to (pure) motives. Again this notion goes back to Anderson. Its definition is similar to the notion of weight for ℓ-adic sheaves due to Deligne. Now for an elliptic cuspidal Hecke eigenform of weight k one knows that the weight of its ℓ-adic Galois representation is $k - 1$, which in turn yields the Ramanujan–Peterson conjecture. By considering examples we expect that for cuspidal Drinfeld Hecke eigenforms of weight k the following holds: The weight of their v-Galois representation is well defined and an integer in $[0, \ldots, \frac{k-1}{2}]$. A proof of this is still missing.

Acknowledgments. Much of the work presented here, even that which did not involve a direct collaboration with R. Pink, nevertheless was influenced by his comments, ideas and interest, and it is a great pleasure to thank him for this. There is a large number of people whose work was foundational, inspirational, or directly related to the contents of this survey, and that I would like to mention here in the introduction, namely, in alphabetical order, G. W. Anderson, V.-G. Drinfeld, F. Gardeyn, E.-U. Gekeler, D. Goss, U. Hartl, Y. Taguchi, D. Wan.

Particular thanks go to G. van der Geer and B. Moonen for organizing the Texel conference in 2004, and for giving me the opportunity to write this survey article. Also many thanks to B. Buth, D. Goss (again!), and N. Stalder for many helpful comments on earlier versions of this manuscript. Finally I would like to thank the referee for a number of helpful suggestions.

References

[Al96]   E.-U. Gekeler, M. van der Put, M. Reversat, and J. Van Geel, eds., *Drinfeld Modules, Modular Schemes and Applications: Proceedings of the Workshop held in Alden-Biesen, September* 9–14, 1996, World Scientific, River Edge, NJ, 1997.

[An86]   G. W. Anderson, *t*-motives, *Duke Math. J.*, **53** (1986), 457–502.

[An96]   G. W. Anderson, Log-algebraicity of twisted *A*-harmonic series and special values of *L*-series in characteristic *p*, *J. Number Theory*, **60**-1 (1996), 165–209.

[An00]   G. W. Anderson, An elementary approach to *L*-functions mod *p*, *J. Number Theory*, **80**-2 (2000), 291–303.

[BS97]   A. Blum and U. Stuhler, Drinfeld modules and elliptic sheaves, in *Vector Bundles on Curves: New Directions* (*Cetraro*, 1995), Lecture Notes in Mathematics 1649, Springer-Verlag, Berlin, New York, 1997, 110–188.

[Bö02]   G. Böckle, Global *L*-functions over function fields, *Math. Ann.*, **323** (2002), 737–795.

[Bö04]   G. Böckle, *An Eichler-Shimura Isomorphism over Function Fields between Drinfeld Modular Forms and Cohomology Classes of Crystals*, preprint, 2004; available online from www.exp-math.uni-essen.de/~boeckle.

[BP04]   G. Böckle and R. Pink, *Cohomological Theory of Crystals over Function Fields*, in preparation.

[BGR84]   S. Bosch, U. Güntzer, and R. Remmert, *Non-Archimedean Analysis*, Springer-Verlag, Berlin, Heidelberg, 1984.

[CoDe]   B. Conrad, Nagata's compactification theorem (via schemes), private notes of P. Deligne; available online from www.math.lsa.umich.edu/~bdconrad/papers.

[SGA4$\frac{1}{2}$]   P. Deligne, et al., *Cohomologie Étale: Séminaire de Géométrie Algébrique du Bois Marie, SGA4$\frac{1}{2}$*, Lecture Notes in Mathematics 569, Springer-Verlag, Berlin, New York, 1977.

[Dr76]   V. G. Drinfeld, Elliptic modules, *Math. USSR-Sb.*, **23**-4 (1976), 561–592.

[EK04]   M. Emerton and M. Kisin, *The Riemann-Hilbert correspondence for unit F-crystals*, Astérisque 293, Société Mathématique de France, Paris, 2004.

[Ga03]   F. Gardeyn, The structure of analytic τ-sheaves, *J. Number Theory*, **100**-2 (2003), 332–362.

[Ge86]   E.-U. Gekeler, *Drinfeld Modular Curves*, Lecture Notes in Mathematics 1231, Springer-Verlag, Berlin, New York, 1986.

[Go91]   D. Goss, Some integrals attached to modular forms in the theory of function fields, in *The Arithmetic of Function Fields* (*Columbus, OH*, 1991), de Gruyter, Berlin, 1992.

[Go91a]   D. Goss, *L*-series of *t*-motives and Drinfel'd modules, in *The Arithmetic of Function Fields* (*Columbus, OH*, 1991), de Gruyter, Berlin, 1992.

[Go92]   D. Goss, *L*-series of Grössencharakters of type A_0 for function fields, in *p-Adic Methods in Number Theory and Algebraic Geometry*, Contemporary Mathematics 133, American Mathematical Society, Providence, 1992.

[Go96]   D. Goss, *Basic Structures of Function Field Arithmetic*, Ergebnisse der Mathematik und ihrer Grenzgebiete 35, Springer-Verlag, Berlin, New York, 1996.

[Go00]   D. Goss, A Riemann hypothesis for characteristic *p* *L*-functions, *J. Number Theory*, **82**-2 (2000), 299–322.

[Go02]   D. Goss, Can a Drinfeld module be modular?, *J. Ramanujan Math. Soc.*, **17**-4 (2002), 221–260.

[Go04]   D. Goss, The impact of the infinite primes on the Riemann hypothesis for characteristic *p* valued *L*-series, in *Algebra, Arithmetic and Geometry with Applications* (*West Lafayette, IN*, 2000), Springer-Verlag, Berlin, 2004.

[Go04a]   D. Goss, Applications of non-Archimedean integration to the *L*-series of τ-sheaves, 2003; available online from arXiv.org/abs/math/0307376.

[SGA5]   A. Grothendieck, J. F. Boutot, A. Grothendieck, L. Illusie, and J. L. Verdier, *Cohomologie ℓ-adique et fonctions L: Séminaire de Géométrie Algébrique du Bois Marie, SGA5*, Lecture Notes in Mathematics 589, Springer-Verlag, Berlin, New York, 1977.

[Ha66]   R. Hartshorne, *Residues and Duality*, Lecture Notes in Mathematics 20, Springer-Verlag, Berlin, New York, 1966.

[Ha77]   R. Hartshorne, *Algebraic Geometry*, Graduate Texts in Mathematics 52, Springer-Verlag, Berlin, New York, 1977.

[vdH03]   G.-J. van der Heiden, *Weil Pairing and the Drinfeld Modular Curve*, Ph.D. thesis, University of Groningen, Groningen, The Netherlands, 2003; available online from www.ub.rug.nl/eldoc/dis/science/.

[Ka73]   N. Katz, *p*-adic properties of modular schemes and modular varieties, in *Modular Functions of One Variable* III, Lecture Notes in Mathematics 350, Springer, Berlin, New York, 1973.

[KM85]   N. Katz and B. Mazur, *Arithmetic Moduli of Elliptic Curves*, Princeton University Press, Princeton, NJ, 1985.

[Kh03]   C. Khare, Compatible systems of mod *p* Galois representations and Hecke characters, *Math. Res. Lett.*, 10-1 (2003), 71–84.

[Kh04]   C. Khare, Reciprocity law for compatible systems of abelian mod *p* Galois representations, 2003; available online from arXiv.org/abs/math/0309292.

[Le01]   T. Lehmkuhl, *Compactification of the Drinfeld Modular Surfaces*, Habilitationsschrift, Universität Göttingen, Göttingen, Germany, 2000.

[Lü93]   W. Lütkebohmert, On compactification of schemes, *Manuscripta Math.*, 80-1 (1993), 95–111.

[Mi80]   J. S. Milne, *Étale Cohomology*, Princeton University Press, Princeton, NJ, 1980.

[Po95]   B. Poonen, Local height functions and the Mordell-Weil theorem for Drinfel'd modules, *Compositio Math.*, 97-3 (1995), 349–368.

[Sch90]   A. J. Scholl, Motives for modular forms, *Invent. Math.*, 100-2 (1990), 419–430.

[TW96]   Y. Taguchi and D. Wan, *L*-functions of ϕ-sheaves and Drinfeld modules, *J. Amer. Math. Soc.*, 9-3 (1996), 755–781.

[We94]   C. Weibel, *An Introduction to Homological Algebra*, Cambridge University Press, Cambridge, UK, 1994.

Algebraic Stacks Whose Number of Points over Finite Fields is a Polynomial

Theo van den Bogaart and Bas Edixhoven

Mathematical Institute
University of Leiden
P. O. Box 9512
2300 RA Leiden
The Netherlands
bogaart@math.leidenuniv.nl, edix@math.leidenuniv.nl

1 Introduction

The aim of this article is to investigate the cohomology (l-adic as well as Betti) of schemes, and more generally of certain algebraic stacks, $\mathcal{X}$, that are proper and smooth over $\mathbb{Z}$ and have the property that there exists a polynomial P with coefficients in $\mathbb{Q}$ such that for every finite field $\mathbb{F}_q$ we have $\#\mathcal{X}(\mathbb{F}_q) = P(q)$. For the precise definitions and conditions the reader is invited to read the rest of the article, at least up to the statement of Theorem 2.1. Under those conditions, we prove that for all prime numbers l the étale cohomology $H(\mathcal{X}_{\overline{\mathbb{Q}}, \text{ét}}, \mathbb{Q}_l)$, considered as a representation of the absolute Galois group of $\mathbb{Q}$, is as expected: zero in odd degrees, and a direct sum of $\mathbb{Q}_l(-i)$ in degree $2i$, with the number of terms equal to the coefficient P_i of P. Our main tools here are Behrend's Lefschetz trace formula in [Beh93] and l-adic Hodge theory combined with the fact that $\mathbb{Z}$ has no nontrivial unramified extensions. Finally, using comparison theorems from l-adic Hodge theory, we obtain a corollary which says that under the extra assumption that the coarse moduli space of $\mathcal{X}$ is a quotient by a finite group, the Betti cohomology $H(\mathcal{X}(\mathbb{C}), \mathbb{Q})$ with its Hodge structure is as expected: zero in odd degree, and $\mathbb{Q}(-i)^{P_i}$ in degree $2i$.

The results in this article are motivated by a question by Carel Faber on potential applications to some moduli stacks $\mathcal{M}_{g,n}$ of stable n-pointed curves of genus g. These stacks are proper and smooth over $\mathbb{Z}$, and they also satisfy the extra hypotheses of the corollary by results of Pikaart and Boggi [PB00]. We are told that $\#\mathcal{M}_{g,n}(\mathbb{F}_q)$ is a polynomial in q for all pairs of the form $(0, n)$ with $n \geq 3$, $(1, n)$ with $1 \leq n \leq 10$, $(2, n)$ with $0 \leq n \leq 5$ (probably even up to $n = 9$), and $(3, n)$ with $0 \leq n \leq 3$ (and probably more). For genus 2 and 3 these results are due to Jonas Bergström.

As this article is motivated by its application to certain $\mathcal{M}_{g,n}$, we have not made an effort to make our results as general as possible. In particular, we have not tried to generalize comparison theorems from l-adic Hodge theory from schemes to stacks.

We hope that this article will be of help to those computing the rational Hodge structure on the cohomology of certain $\mathcal{M}_{g,n}$. Counting points, using a suitable stratification, could be easier than having to compute the cohomology, using the same stratification. We apologize for our lack of expertise in the fields of algebraic stacks and l-adic Hodge theory. Readers with more competence in these areas will probably find the contents of this article rather straightforward and the proofs too elaborate. But there seems to be a lack of "well-known facts" in the literature and we have tried hard to give precise references and proofs understandable also to the nonexpert.

Terminology, conventions

Concerning stacks, our terminology is that of [LMB00]. In particular, algebraic stacks are by definition quasi-separated.

Let k be a finite field. If $\mathcal{X}$ is a Deligne–Mumford stack of finite type over $\mathbb{Z}$, define its *number of points over k* to be

$$\#\mathcal{X}(k) = \sum_\xi \frac{1}{|\operatorname{Aut}(\xi)|},$$

where the sum is over representatives of isomorphism classes of objects in $\mathcal{X}(k)$. Here $\operatorname{Aut}(\xi)$ denotes the finite group of automorphisms of ξ.

If G is a topological group, by a G-representation (over $\mathbb{Q}_l$) we shall mean a continuous representation of G onto a finite-dimensional $\mathbb{Q}_l$-vector space equipped with the l-adic topology. We use the same notation for a representation and its underlying vector space. For any G and $n \geq 0$, the symbol $\mathbb{Q}_l^n$ denotes the trivial n-dimensional G-representation.

2 Results

Theorem 2.1. *Let $\mathcal{X}$ be a Deligne–Mumford stack over $\mathbb{Z}$ which is proper, smooth and of pure relative dimension d for some $d \geq 0$. Let S be a set of primes of Dirichlet density 1. Assume the following:*

There exists a polynomial $P(t) = \sum_{i \geq 0} P_i t^i$, with $P_i \in \mathbb{Q}$, such that

(∗) $$\#\mathcal{X}(\mathbb{F}_{p^n}) = P(p^n) + o(p^{nd/2}) \quad (n \to \infty)$$

for all $p \in S$.

Then the degree of $P(t)$ is d, and there exists a unique such polynomial satisfying $P_i = P_{d-i}$ for all $0 \leq i \leq d$. Suppose $P(t)$ is of this form. Then it has nonnegative integer coefficients and satisfies $\#\mathcal{X}(\mathbb{F}_{p^n}) = P(p^n)$ for all primes p and all $n \geq 1$. Furthermore, for all primes l and all $i \geq 0$ there is an isomorphism of $\operatorname{Gal}(\overline{\mathbb{Q}}/\mathbb{Q})$-representations

$$\mathrm{H}^i(\mathcal{X}_{\overline{\mathbb{Q}},\text{\'et}}, \mathbb{Q}_l) \simeq \begin{cases} 0 & \textit{if } i \textit{ is odd}; \\ \mathbb{Q}_l(-i/2)^{P_{i/2}} & \textit{if } i \textit{ is even}. \end{cases}$$

We remark that part of the theorem can also be stated in terms of the coarse moduli space associated to $\mathcal{X}$. Indeed, the number of points over a finite field of a Deligne–Mumford stack equals the number of points of its coarse moduli space; and furthermore, the cohomologies of both spaces with coefficients in a $\mathbb{Q}$-algebra are the same.

3 Some results on stacks

Let us make two technical remarks.

In [LMB00, Section 18] the theory of constructible sheaves of $\mathbb{Z}/l^n\mathbb{Z}$-modules over the smooth-étale site of an algebraic S-stack is developed, where S is a scheme. There is a straightforward extension of this theory to constructible l-adic sheaves, e.g., by working with projective systems of $\mathbb{Z}/l^n\mathbb{Z}$-modules modulo torsion in the usual way. We will use this without further comments.

Associated to a Deligne–Mumford stack $\mathcal{X}$ are its étale topos (denoted $\mathcal{X}_{\text{ét}}$) and its smooth-étale topos. They however give the same cohomology theory of constructible sheaves (see [LMB00, Section 12, especially Proposition 12.10.1]). This justifies the fact that we will only work with the étale topos of a Deligne–Mumford stack, but freely cite results stated in terms of the other topos.

For schemes the following proposition is classical. By lack of a precise reference, we have included a proof for the case of stacks.

Proposition 3.1. *Let $\mathcal{X}$ be a Deligne–Mumford stack which is smooth and proper over $\mathbb{Z}_p$. For every prime $l \neq p$ and every $i \geq 0$, the canonical map of $\mathrm{Gal}(\overline{\mathbb{Q}}_p/\mathbb{Q}_p)$-representations*

$$\mathrm{H}^i(\mathcal{X}_{\overline{\mathbb{F}}_p,\text{\'et}}, \mathbb{Q}_l) \to \mathrm{H}^i(\mathcal{X}_{\overline{\mathbb{Q}}_p,\text{\'et}}, \mathbb{Q}_l) \tag{1}$$

is an isomorphism. In particular, $\mathrm{H}^i(\mathcal{X}_{\overline{\mathbb{Q}}_p,\text{\'et}}, \mathbb{Q}_l)$ is unramified.

Proof. Denote by $\mathbb{Q}_p^{\mathrm{nr}}$ the maximal unramified extension of $\mathbb{Q}_p$ in $\overline{\mathbb{Q}}_p$ and let $\mathbb{Z}_p^{\mathrm{nr}}$ be its ring of integers. Set $S = \mathrm{Spec}(\mathbb{Z}_p^{\mathrm{nr}})$ and denote by s, respectively, η, its closed, respectively, generic, point. Let $\overline{\eta} \to \eta$ correspond to $\mathbb{Q}_p^{\mathrm{nr}} \to \overline{\mathbb{Q}}_p$. Consider the natural morphisms

$$\mathcal{X}_{\overline{\eta}} \xrightarrow{j} \mathcal{X}_S \xleftarrow{i} \mathcal{X}_s.$$

These maps induce continuous morphisms between the associated étale sites.

Let (U, u) be an étale neighborhood of $\mathcal{X}_S$ and let j_U be the pull-back of j along u. By [LMB00, 18.2.1(i)], for every q we have $(R^q j_* \mathbb{Q}_l)_{U,u} = R^q (j_U)_* \mathbb{Q}_l$. As U is smooth, it follows that $j_* \mathbb{Q}_l = \mathbb{Q}_l$ and $R^q j_* \mathbb{Q}_l = 0$ if $q \neq 0$. Hence the Leray spectral sequence gives an isomorphism $\mathrm{H}^i(\mathcal{X}_S, \mathbb{Q}_l) \to \mathrm{H}^i(\mathcal{X}_{\overline{\eta}}, \mathbb{Q}_l)$. But on the other hand, $\mathrm{H}^i(\mathcal{X}_S, \mathbb{Q}_l)$ is naturally isomorphic to $\mathrm{H}^i(\mathcal{X}_s, \mathbb{Q}_l)$; this follows from the proper base change theorem [LMB00, 18.5.1] for $\mathcal{X}_S$ over S and the fact that S is strictly local. $\qquad\square$

The next topic is Poincaré duality for the l-adic cohomology of certain stacks (see Proposition 3.3 below). We will obtain this by considering the cohomologies of their associated coarse moduli spaces.

Let $\mathcal{X}$ be a separated Deligne–Mumford stack of finite type over an algebraically closed field of characteristic zero. We will denote by $\overline{\mathcal{X}}$ its coarse moduli space and by $q\colon \mathcal{X} \to \overline{\mathcal{X}}$ the corresponding mapping. Note that we can cover $\overline{\mathcal{X}}$ by étale charts U such that the pull-back of U in $\mathcal{X}$ is the quotient stack of an algebraic space by a finite group [LMB00, Remark 6.2.1].

Lemma 3.2. *For every i the pull-back map*

$$q^*\colon \mathrm{H}^i(\overline{\mathcal{X}}_{\acute{e}t}, \mathbb{Q}_l) \longrightarrow \mathrm{H}^i(\mathcal{X}_{\acute{e}t}, \mathbb{Q}_l)$$

is an isomorphism.

Proof. The lemma follows from the Leray spectral sequence once we have shown that the canonical map $\mathbb{Q}_l \to Rq_*\mathbb{Q}_l$ is an isomorphism. This question is étale local on $\overline{\mathcal{X}}$ and therefore we may assume that $\mathcal{X} = [V/G]$ for some algebraic space V equipped with an action by a finite group G. Denote by $p\colon V \to \mathcal{X}$ the canonical morphism. Note that $\mathbb{Q}_l \simeq (p_*\mathbb{Q}_l)^G$. As p and qp are finite and $\mathbb{Q}[G]$ is a semisimple $\mathbb{Q}$-algebra, we obtain

$$Rq_*\mathbb{Q}_l \simeq Rq_*(p_*\mathbb{Q}_l)^G \simeq ((qp)_*\mathbb{Q}_l)^G \simeq \mathbb{Q}_l. \qquad \square$$

Now suppose that $\mathcal{X}$ is defined over $\mathbb{C}$ and smooth. Consider the complex analytic space $\overline{\mathcal{X}}^{\mathrm{an}}$ associated to $\overline{\mathcal{X}}$. It can naturally be equipped with the structure of a V-manifold, i.e., locally $\overline{\mathcal{X}}^{\mathrm{an}}$ is the quotient of a connected manifold by a finite group; cf. [Ste77].

Proposition 3.3. *Suppose $\mathcal{X}$ is a Deligne–Mumford stack which is smooth and proper over $\overline{\mathbb{Q}}$ of pure dimension d for some $d \geq 0$.*

(i) *Suppose $\mathcal{X}$ is integral. For an integer i, consider the cup product mapping*

$$\mathrm{H}^i(\mathcal{X}_{\acute{e}t}, \mathbb{Q}_l) \otimes_{\mathbb{Q}_l} \mathrm{H}^{2d-i}(\mathcal{X}_{\acute{e}t}, \mathbb{Q}_l) \longrightarrow \mathrm{H}^{2d}(\mathcal{X}_{\acute{e}t}, \mathbb{Q}_l).$$

Then the right-hand side is one dimensional and the pairing thus obtained is perfect.

(ii) *Suppose X is a smooth and proper $\overline{\mathbb{Q}}$-scheme of pure dimension d and let $f\colon X \to \mathcal{X}$ be a $\overline{\mathbb{Q}}$-morphism which is surjective and generically finite. Then for all i, the induced map*

$$f^*\colon \mathrm{H}^i(\mathcal{X}_{\acute{e}t}, \mathbb{Q}_l) \longrightarrow \mathrm{H}^i(X_{\acute{e}t}, \mathbb{Q}_l)$$

is injective.

Proof. By Lemma 3.2 and the comparison theorem between Betti and étale cohomology, it suffices to show that in case (i),

$$\mathrm{H}^i(\overline{\mathcal{X}}^{\mathrm{an}}, \mathbb{Q}) \otimes \mathrm{H}^{2d-i}(\overline{\mathcal{X}}^{\mathrm{an}}, \mathbb{Q}) \to \mathrm{H}^{2d}(\overline{\mathcal{X}}^{\mathrm{an}}, \mathbb{Q})$$

is a perfect pairing; and in case (ii) that

$$H^i(\overline{\mathcal{X}}_{\text{ét}}, \mathbb{Q}_l) \longrightarrow H^i(X_{\text{ét}}, \mathbb{Q}_l)$$

is injective. Now the singular cohomology of a V-manifold satisfies Poincaré duality ([Ste77]), from which these statements follow.   □

4 Proof of the main theorem

We will now prove Theorem 2.1. In this section, all cohomology is with respect to the étale sites. We begin with an analytic lemma used in the course of the proof.

Lemma 4.1. *Let us be given the following integers: $d \geq 0$, $d \leq r \leq 2d$, and for $0 \leq i \leq r$ also $d_i \geq 0$. Furthermore, let $p > 1$ be a real number, let $P_i \in \mathbb{Q}$ for all $i \geq 0$, with $P_i = 0$ for i large, and let $\alpha_{i,j} \in \mathbb{C}$ for $0 \leq i \leq r$ and $1 \leq j \leq d_i$. Assume $|\alpha_{i,j}| = p^{i/2}$ and*

$$\sum_{i=0}^{r}(-1)^i \sum_{1 \leq j \leq d_i} \alpha_{i,j}^n = \sum_{i \geq 0} P_i p^{ni} + o(p^{nd/2}) \quad (n \to \infty). \tag{2}$$

Then $d_i = 0$ for $i \geq d$ odd, $P_i = 0$ for $i > r/2$, while for $d/2 \leq i \leq r/2$ we have $P_i = d_{2i}$; for these i also $\alpha_{2i,j} = p^i$.

Proof. The lemma follows by induction on r. Indeed, assume that either $r = d$ or that the lemma holds for $r - 1$. As $|\sum_{1 \leq j \leq d_i} \alpha_{i,j}^n| \leq d_i p^{in/2}$, we have

$$\left| \sum_{i=0}^{r}(-1)^i \sum_{1 \leq j \leq d_i} \alpha_{i,j}^n \right| = \left| \sum_{1 \leq j \leq d_r} \alpha_{r,j}^n \right| + o(p^{nr/2}) \quad (n \to \infty)$$

and also, using (2), that $P_i = 0$ for $i > r/2$.

Note that if z is an element of a finite product $(S^1)^s$ of complex unit circles, then the closure of $\{z^n \mid n \geq 1\}$ contains the unit element. Hence for every $\epsilon > 0$ there exists an infinite subset $N \subset \mathbb{N}$ such that for all $n \in N$ and for all i and j we have $|(\alpha_{i,j} p^{-i/2})^n - 1| < \epsilon$ and, in particular,

$$\left| \sum_{1 \leq j \leq d_r} (\alpha_{r,j} p^{-r/2})^n - d_r \right| < \epsilon', \tag{3}$$

with $\epsilon' = d_r \epsilon$.

Now first suppose r is odd. Then

$$\left| \sum_{1 \leq i \leq d_r} \alpha_{r,j}^n \right| = o(p^{nr/2}) \quad (n \to \infty),$$

which by (3) implies $d_r = 0$. Therefore, if $r = d$ we are done, while if $r > d$, we can apply the induction hypotheses.

So from now on suppose r is even. From (2) it follows that

$$\left| \sum_{1 \le j \le d_r} \alpha_{r,j}^n - P_{r/2} p^{nr/2} \right| = o(p^{nr/2}) \quad (n \to \infty),$$

or, equivalently,

$$\lim_{n \to \infty} \left| \sum_{1 \le j \le d_r} (\alpha_{r,j} p^{-r/2})^n - P_{r/2} \right| = 0.$$

Together with (3), this implies $d_r = P_{r/2}$. In turn this easily leads to $\alpha_{r,j} = p^{r/2}$.

Now subtract $P_{r/2} p^{rn/2} = \sum_{1 \le j \le d_r} \alpha_{r,j}^n$ from (2) and if $r > d$ apply the induction hypotheses. $\qquad \square$

Let $\mathcal{X}$, d, and S be as in Theorem 2.1 and let $P(t) = \sum_{i \ge 0} P_i t^i$ be a polynomial satisfying ($*$). Without loss of generality we assume $P_i = P_{d-i}$ for all $0 \le i \le d$. We also fix a prime l.

By [LMB00, Theorem 16.6] and resolution of singularities, there exists a smooth and proper $\mathbb{Q}$-scheme X of pure dimension d and a surjective and generically finite map $f \colon X \to \mathcal{X}_{\mathbb{Q}}$. By removing a finite number of primes from S if necessary, we may assume that $X_{\mathbb{Q}_p}$ extends to a smooth scheme over $\mathbb{Z}_p$ for all $p \in S$. As a consequence, for all $p \ne l$ in S the representation $\mathrm{H}^i(X_{\overline{\mathbb{Q}}_p}, \mathbb{Q}_l)$ is unramified and we can consider the action of Frobenius. By [Del74i], the eigenvalues of Frobenius have complex absolute value $p^{i/2}$. Using Proposition 3.3, we obtain the same conclusions for the subrepresentation $\mathrm{H}^i(\mathcal{X}_{\overline{\mathbb{Q}}_p}, \mathbb{Q}_l)$.

Fix a prime $p \ne l$ in S. By Behrend's Lefschetz trace formula (see [Beh93] or [LMB00, Theorem 19.3.4]),

$$\sum_{i \ge 0} (-1)^i \operatorname{Tr}(\operatorname{Frob}^n, \mathrm{H}^i(\mathcal{X}_{\overline{\mathbb{F}}_p}, \mathbb{Q}_l)) = \#\mathcal{X}(\mathbb{F}_{p^n}) \tag{4}$$

for all $n \ge 1$. Let $\alpha_{i,1}, \ldots, \alpha_{i,d_i}$ be the complex roots of the characteristic polynomial of Frobenius. Applying ($*$) and Proposition 3.1, formula (4) becomes

$$\sum_{i=0}^{2d} (-1)^i \sum_{1 \le j \le d_i} \alpha_{i,j}^n = P(p^n) + o(p^{nd/2}) \quad (n \to \infty).$$

From Lemma 4.1 we now obtain that $P(t)$ has degree d and for all $d \le i \le 2d$ we have that for i even $P_{i/2} = d_i$ and $\alpha_{i,j} = p^{i/2}$, while $d_i = 0$ for i odd. Using Poincaré duality (Proposition 3.3) we obtain the same conclusions for all i. (Note that $P(t)$ is defined in such a way that $P_i = P_{d-i}$ for $0 \le i \le d/2$.)

Hence $\mathrm{H}^i(\mathcal{X}_{\overline{\mathbb{Q}}_p}, \mathbb{Q}_l)$ vanishes for any odd i, while for all even i it has dimension $P_{i/2}$—in particular, the coefficients of $P(t)$ are nonnegative integers—and furthermore,

$$\text{Tr}(\text{Frob}, \text{H}^{2i}(\mathcal{X}_{\overline{\mathbb{Q}}_p}, \mathbb{Q}_l)) = \text{Tr}(\text{Frob}, \mathbb{Q}_l(-i)^{P_i}). \tag{5}$$

This holds for all primes p in the set $S \backslash \{l\}$. But a semisimple, almost everywhere unramified, $\text{Gal}(\overline{\mathbb{Q}}/\mathbb{Q})$-representation over $\mathbb{Q}_l$ is determined by the trace of Frobenius on a set of primes of Dirichlet density 1. (A proof of this is outlined in [DDT97, Proposition 2.6].) We conclude that the semisimplification of $\text{H}^{2i}(\mathcal{X}_{\overline{\mathbb{Q}}}, \mathbb{Q}_l)$ is isomorphic to $\mathbb{Q}_l(-i)^{P_i}$.

By Proposition 3.1, $\text{H}^{2i}(\mathcal{X}_{\overline{\mathbb{Q}}_p}, \mathbb{Q}_l)$ is unramified for every prime $p \neq l$. As (5) then holds for every prime $p \neq l$, Behrend's Lefschetz trace formula gives $\#\mathcal{X}(\mathbb{F}_{p^n}) = P(p^n)$ for all $p \neq l$. Changing l, we see that this formula is valid for every prime p.

All that remains to be proved is that $\text{H}^{2i}(\mathcal{X}_{\overline{\mathbb{Q}}}, \mathbb{Q}_l)$ is semisimple, or equivalently, that its ith Tate twist $H = \text{H}^{2i}(\mathcal{X}_{\overline{\mathbb{Q}}}, \mathbb{Q}_l)(i)$ is semisimple. Note that H is unramified outside l and that the semisimplification of H is isomorphic to the trivial representation $\mathbb{Q}_l^{P_i}$.

By [LMB00, 16.6] and [DeJ96], there exists a finite extension K of $\mathbb{Q}_l$ inside $\overline{\mathbb{Q}}_l$, a proper, semistable scheme X of pure relative dimension d over the ring of integers of K and a surjective and generically finite K-morphism $f : X_K \to \mathcal{X}_K$. By [Tsu02, Theorem 1.1], $\text{H}^{2i}(X_{\overline{\mathbb{Q}}_l}, \mathbb{Q}_l)$ is a semistable representation of $\text{Gal}(\overline{\mathbb{Q}}_l/K)$. So it follows from Proposition 3.3 that $\text{H}^{2i}(\mathcal{X}_{\overline{\mathbb{Q}}_l}, \mathbb{Q}_l)$ and hence also H are potentially semistable.

Lemma 4.2. *Let* $n \geq 1$ *be an integer. Consider a short exact sequence of* $\text{Gal}(\overline{\mathbb{Q}}_l/\mathbb{Q}_l)$-*representations*

$$0 \to \mathbb{Q}_l^n \to V \to \mathbb{Q}_l \to 0. \tag{6}$$

If V *is potentially semistable, then* V *is unramified.*

Proof. By assumption, there is a finite extension K of $\mathbb{Q}_l$ inside $\overline{\mathbb{Q}}_l$, such that the restriction of V to $G_K := \text{Gal}(\overline{\mathbb{Q}}_l/K)$ is semistable. Fix such a K and denote by K_0 its maximal unramified subfield relative to $\mathbb{Q}_l$. Denote by σ the automorphism of K_0 obtained by lifting the automorphism $x \mapsto x^l$ of the residue field of K_0.

We will briefly recall some theory about semistable representations and filtered (φ, N)-modules; for more details, see [CF00].

Denote by $\underline{\text{Rep}}_{\text{st}}(G_K)$ the category of semistable representations of G_K over $\mathbb{Q}_l$ and denote by $\underline{\text{Rep}}_{\text{unr}}(G_K)$ its full subcategory of unramified representations. Let $\underline{\text{MF}}_K^f(\varphi, N)$ be the category of (weakly) admissible filtered (φ, N)-modules over K. An object of $\underline{\text{MF}}_K^f(\varphi, N)$ is a finite-dimensional K_0-vector space E equipped with a σ-semilinear bijection $\varphi : E \to E$, a nilpotent endomorphism N of E and an exhaustive and separating descending filtration $\text{Fil}^{\cdot} E_K$ on $E_K := K \otimes_{K_0} E$. We must have $N\varphi = l\varphi N$ and furthermore there is a certain admissibility condition to be satisfied (cf. [CF00, Section 3]).

All the above categories are Tannakian (so, in particular, they are all abelian $\mathbb{Q}_l$-linear $\otimes$-categories). Fontaine has constructed a functor $\text{D}_{\text{st},K}$ from $\underline{\text{Rep}}_{\text{st}}(G_K)$ to $\underline{\text{MF}}_K^f(\varphi, N)$ and the main result of [CF00] is that this is an equivalence of Tannakian categories.

To the trivial one-dimensional representation $\mathbb{Q}_l$ corresponds $D_{st,K}(\mathbb{Q}_l)$, which is just K_0 equipped with the trivial maps $\varphi = \sigma$, $N = 0$ and filtration determined by $\text{Fil}^0 K = K$ and $\text{Fil}^1 K = 0$. By abuse of notation we will denote $D_{st,K}(\mathbb{Q}_l)$ also by K_0.

We obtain natural maps of $\mathbb{Q}_l$-vector spaces

$$\text{Ext}^1_{\underline{\text{Rep}}_{\text{unr}}(G_K)}(\mathbb{Q}_l, \mathbb{Q}_l^n) \overset{i}{\hookrightarrow} \text{Ext}^1_{\underline{\text{Rep}}_{\text{st}}(G_K)}(\mathbb{Q}_l, \mathbb{Q}_l^n) \overset{\sim}{\to} \text{Ext}^1_{\underline{\text{MF}}^f_K(\varphi,N)}(K_0, K_0^n),$$

where the second map is the isomorphism induced by $D_{st,K}$. We need to show that i is an isomorphism. Since Ext^1 is additive in the second variable, it suffices to treat the case $n = 1$. We will show that the dimension of $\text{Ext}^1_{\underline{\text{Rep}}_{\text{unr}}(G_K)}(\mathbb{Q}_l, \mathbb{Q}_l)$ is at least as big as the dimension of the $\mathbb{Q}_l$-vector space $\text{Ext}^1_{\underline{\text{MF}}^f_K(\varphi,N)}(K_0, K_0)$.

First, we consider the extensions in $\underline{\text{MF}}^f_K(\varphi, N)$. Let

$$0 \to K_0 \to E \to K_0 \to 0 \tag{7}$$

be a short exact sequence in $\underline{\text{MF}}^f_K(\varphi, N)$. Choosing a splitting of the short exact sequence of vector spaces underlying (7), we write $E = K_0 \oplus K_0$ with $K_0 \to E$ being $x \mapsto (x, 0)$ and $E \to K_0$ being the projection onto the second coordinate.

As the induced sequence of filtered K-vector spaces is exact, $\text{Fil}^0 E_K = E_K$ and $\text{Fil}^1 E_K = 0$. Secondly, N has the form $\left(\begin{smallmatrix} 0 & \lambda \\ 0 & 0 \end{smallmatrix}\right)$ for some $\lambda \in K_0$. Then $N\varphi = l\varphi N$ implies $\lambda = l\sigma(\lambda)$, which is only possible if $\lambda = 0$. Hence $N = 0$ on E. Finally, we necessarily have $\varphi(1, 0) = (1, 0)$ and $\varphi(0, 1) = (\alpha, 1)$ for some $\alpha \in K_0$ and conversely giving $\alpha \in K_0$ uniquely determines φ.

Denote by K_0^v the $\mathbb{Q}_l$-vector space underlying K_0. To $\alpha \in K_0^v$ associate the unique extension (7) with $E = K_0 \oplus K_0$ and $\varphi(0, 1) = (\alpha, 1)$. This determines a surjective map

$$K_0^v \overset{j}{\longrightarrow} \text{Ext}^1_{\underline{\text{MF}}^f_K(\varphi,N)}(K_0, K_0),$$

and one checks that it is in fact $\mathbb{Q}_l$-linear. Take $x \in K_0$ and let L be the automorphism of $K_0 \oplus K_0$ given by $\left(\begin{smallmatrix} 1 & x \\ 0 & 1 \end{smallmatrix}\right)$. If we equip the source with φ and the target with φ', then L induces an equivalence of the associated extensions if $\varphi' = L^{-1}\varphi L$. It follows that $j(\alpha) = j(\alpha + \sigma(x) - x)$, so the kernel of j is a sub-$\mathbb{Q}_l$-vector space of K_0^v of codimension 1; this implies that $\dim_{\mathbb{Q}_l} \text{Ext}^1_{\underline{\text{MF}}^f_K(\varphi,N)}(K_0, K_0) \leq 1$.

But on the other hand, $\dim_{\mathbb{Q}_l} \text{Ext}^1_{\underline{\text{Rep}}_{\text{unr}}(G_K)}(\mathbb{Q}_l, \mathbb{Q}_l) = 1$. To see this, suppose V is unramified and sits in an extension of $\mathbb{Q}_l$ by $\mathbb{Q}_l$. Taking a suitable basis for V, the action of the Galois group is given by $\left(\begin{smallmatrix} 1 & \eta \\ 0 & 1 \end{smallmatrix}\right)$, where η is an unramified character $\text{Gal}(\overline{\mathbb{Q}}_l/K) \to \mathbb{Q}_l$. In other words, η is a morphism of groups $\hat{\mathbb{Z}} \to \mathbb{Q}_l$ which, being continuous, must factor through $\mathbb{Z}_l$. But $\text{Hom}(\mathbb{Z}_l, \mathbb{Q}_l)$ is one dimensional.

Thus we obtain that V in (6) is unramified when restricted to $\text{Gal}(\overline{\mathbb{Q}}_l/K)$. Then if g is an element of the inertia subgroup of $\text{Gal}(\overline{\mathbb{Q}}_l/\mathbb{Q}_l)$, the $[K : \mathbb{Q}_l]$th power of g must act trivially. But on the other hand g must act unipotently, as V sits in the short exact sequence (6). Hence g itself must act trivially and V is an unramified representation of $\text{Gal}(\overline{\mathbb{Q}}_l/\mathbb{Q}_l)$. $\square$

Now let

$$0 = H_0 \subset H_1 \subset \cdots \subset H_{P_i} = H$$

be a Jordan–Hölder filtration of H. Hence $H_j/H_{j-1} \simeq \mathbb{Q}_l$ and each H_j is unramified outside l and potentially semistable at l. Clearly, $H_1 \simeq \mathbb{Q}_l$; assume that $H_j \simeq \mathbb{Q}_l^j$ for some j $(1 \leq j \leq P_i - 1)$. Then H_{j+1} is everywhere unramified by the above lemma. However, as Minkowski's theorem says that $\mathbb{Q}$ has no nontrivial unramified extensions, we conclude that H_{j+1} is isomorphic to $\mathbb{Q}_l^{j+1}$. So by induction, $H \simeq \mathbb{Q}_l^{P_i}$. This finishes the proof of Theorem 2.1.

5 The Hodge structure

Let V be an integral, regular, quasi-projective $\mathbb{Q}$-scheme. Consider a finite group G acting on V from the right. Denote by $f : V \to Q$ the canonical projection to the quotient scheme $Q = V/G$. Note that there is a natural G-action on the module $f_* \Omega^1_{V/\mathbb{Q}}$.

We are interested in the cohomology of the quotient space Q and, in particular, in the Hodge structure of the singular cohomology of its associated analytic space. For this we will use [Ste77]. In order to be able to apply some results of [Ste77], we will need the following result.

Proposition 5.1. *Let $\Sigma \subset Q$ be a closed subset of codimension ≥ 2 containing all singular points of Q. Let $j : U \hookrightarrow Q$ be the open subscheme complementary to Σ. Then, for all p, there is a canonical isomorphism*

$$j_* \Omega^p_{U/\mathbb{Q}} \xrightarrow{\sim} (f_* \Omega^p_{V/\mathbb{Q}})^G \tag{8}$$

of $\mathcal{O}_Q$-modules.

Proof. For any $\mathbb{Q}$-scheme Z, we abbreviate $\Omega_Z := \Omega^p_{Z/\mathbb{Q}}$. Put $W = f^{-1}(U)$ and let $g : W \to U$ be the restriction of f. Note that G acts on W with quotient U. The canonical map $g^* \Omega_U \to \Omega_W$ induces a morphism

$$\alpha : \Omega_U \xrightarrow{\sim} (g_* g^* \Omega_U)^G \longrightarrow (g_* \Omega_W)^G.$$

This map is an isomorphism in the stalk at the generic point. To define (8), we apply j_* to α to obtain a map from $j_* \Omega_U$ to the sheaf $j_*(f_* \Omega_V)^G$. This last sheaf is isomorphic to $(f_* \Omega_V)^G$, as a consequence of the fact that if Z is a regular $\mathbb{Q}$-variety and $z : Z' \hookrightarrow Z$ is a dense open subset with complement of codimension ≥ 2, then $z_* \Omega_{Z'} \simeq \Omega_Z$. Another consequence of this fact is that the map (8) thus defined is independent of the choice of Σ. In particular, we have reduced the problem to showing that for any point $\eta \in U$ whose closure has codimension 1, the map α is an isomorphism in the stalk at η.

The question being local for the étale topology on U and V, it suffices to consider, for $d \geq 1$ and $n \geq 0$, the action of μ_d on $\mathbb{A}^n_{\mathbb{Q}}$, with the quotient map $\mathbb{A}^n_{\mathbb{Q}} \to \mathbb{A}^n_{\mathbb{Q}}$ mapping $(a_1, \ldots, a_n)$ to $(a_1^d, a_2, \ldots, a_n)$. The result follows from an easy calculation. $\square$

Let $j\colon U \hookrightarrow Q$ be the smooth locus of Q. As a consequence of Proposition 5.1, there is a canonical isomorphism between the De Rham complexes

$$j_* \Omega^\bullet_{U/\mathbb{Q}} \xrightarrow{\sim} (f_* \Omega^\bullet_{V/\mathbb{Q}})^G.$$

Therefore, for each i we obtain an isomorphism

$$\mathrm{H}^i(Q, j_* \Omega^\bullet_{U/\mathbb{Q}}) \xrightarrow{\sim} \mathrm{H}^i(Q, (f_* \Omega^\bullet_{V/\mathbb{Q}})^G) \xrightarrow{\sim} \mathrm{H}^i_{\mathrm{DR}}(V/\mathbb{Q})^G. \tag{9}$$

This last isomorphism follows from the fact that f is finite and that taking cohomology with $\mathbb{Q}$-coefficients commutes with taking invariants under a finite group action. The Hodge filtration on the De Rham complex induces filtrations on the vector spaces in (9) and the isomorphisms in (9) respect those filtrations.

Fix a prime l. Denote by D_{DR} the Fontaine functor (see e.g., [Tsu02]) from the category of $\mathrm{Gal}(\overline{\mathbb{Q}}_l/\mathbb{Q}_l)$-representations over $\mathbb{Q}_l$ to the category of finite-dimensional filtered $\mathbb{Q}_l$-vector spaces.

Proposition 5.2. *In the above situation, suppose furthermore that Q is proper over $\mathbb{Q}$. For every i, there is an isomorphism of filtered vector spaces*

$$\mathrm{D}_{\mathrm{DR}}(\mathrm{H}^i(Q_{\overline{\mathbb{Q}}_l, \text{ét}}, \mathbb{Q}_l)) \xrightarrow{\sim} \mathrm{H}^i(Q, j_* \Omega^\bullet_{U/\mathbb{Q}}) \otimes_{\mathbb{Q}} \mathbb{Q}_l.$$

Proof. As V is proper and smooth, the comparison theorem (see e.g., [Tsu02, Theorem A1]) states that $\mathrm{D}_{\mathrm{DR}}(\mathrm{H}^i(V_{\overline{\mathbb{Q}}_l, \text{ét}}, \mathbb{Q}_l))$ and $\mathrm{H}^i_{\mathrm{DR}}(V/\mathbb{Q}_l)$ are isomorphic. This isomorphism preserves the G-invariants. Using (9) we obtain the isomorphism of the proposition. □

With these prerequisites, we are finally ready to prove the corollary to the main theorem.

Corollary 5.3. *In the situation of Theorem 2.1, suppose furthermore that the coarse moduli space $\overline{\mathcal{X}}$ of $\mathcal{X}_{\mathbb{Q}}$ is the quotient of a smooth projective $\mathbb{Q}$-scheme by a finite group.*

Then for each i, there is an isomorphism of $\mathbb{Q}$-Hodge structures

$$\mathrm{H}^i(\overline{\mathcal{X}}(\mathbb{C}), \mathbb{Q}) \simeq \begin{cases} 0 & \text{if } i \text{ is odd,} \\ \mathbb{Q}(-i/2)^{P_{i/2}} & \text{if } i \text{ is even,} \end{cases}$$

where the left hand side is equipped with the canonical Hodge structure of [Del74ii].

Proof. We may assume that $\overline{\mathcal{X}}$ is integral. Note that $\mathrm{H}^i(\mathcal{X}_{\overline{\mathbb{Q}}_l, \text{ét}}, \mathbb{Q}_l)$ is isomorphic to $\mathrm{H}^i(\overline{\mathcal{X}}_{\overline{\mathbb{Q}}_l, \text{ét}}, \mathbb{Q}_l)$ by Lemma 3.2. Therefore, combining Theorem 2.1 and Proposition 5.2 it suffices to exhibit an isomorphism of filtered vector spaces

$$\mathrm{H}^i(\overline{\mathcal{X}}(\mathbb{C}), \mathbb{C}) \xrightarrow{\sim} \mathrm{H}^i(\overline{\mathcal{X}}, j_* \Omega^\bullet_{U/\mathbb{Q}}) \otimes_{\mathbb{Q}} \mathbb{C},$$

where $j\colon U \to \overline{\mathcal{X}}$ is the smooth locus of $\overline{\mathcal{X}}$. This is done in [Ste77, Theorem 1.12]. □

Acknowledgments. We thank Carel Faber for asking us his question, and the organizers of the 2004 Texel Island conference where this happened. We also thank Laurent Moret-Bailly for his advice on stacks and Gerard van der Geer for his comments.

References

[Beh93]   K. Behrend, The Lefschetz trace formula for algebraic stacks, *Invent. Math.*, **112** (1993), 127–149.

[CF00]   P. Colmez and J.-M. Fontaine, Construction des représentations p-adiques semi-stables, *Invent. Math.*, **140** (2000), 1–43.

[DDT97]   H. Darmon, F. Diamond, and R. Taylor, Fermat's last theorem, in *Elliptic Curves, Modular Forms and Fermat's Last Theorem* (*Hong Kong* 1993), International Press, Cambridge, UK, 1997, 2–140.

[Del74i]   P. Deligne, La conjecture de Weil I, *Publ. Math. IHES*, **43** (1974), 273–307.

[Del74ii]   P. Deligne, Théorie de Hodge III, *Publ. Math. IHES*, **44** (1974), 5–77.

[DeJ96]   A. J. de Jong, Smoothness, semi-stability and alterations, *Publ. Math. IHES*, **83** (1996), 51–93.

[LMB00]   G. Laumon and L. Moret-Bailly, *Champs algébriques*, Springer-Verlag, Berlin, 2000.

[PB00]   M. Boggi and M. Pikaart, Galois covers of moduli of curves, *Compositio Math.*, **120** (2000), 171–191.

[Ste77]   J. H. M. Steenbrink, Mixed Hodge structure on the vanishing cohomology, in *Real and Complex Singularities* (*Oslo* 1976), Sijthoff and Noordhoff, Alphen aan den Rijn, The Netherlands, 1977, 525–563.

[Tsu02]   T. Tsuji, Semi-stable conjecture of Fontaine-Jannsen: A survey, in *Cohomologies p-adiques et applications arithmétiques* (*II*), Astérisque 279, Société Mathématique de France, Paris, 2002, 323–370.

On a Problem of Miyaoka

Holger Brenner

Department of Pure Mathematics
University of Sheffield
Hicks Building
Hounsfield Road
Sheffield S3 7RH
UK
h.brenner@sheffield.ac.uk

Summary. We give an example of a vector bundle $\mathcal{E}$ on a relative curve $C \to \operatorname{Spec} \mathbb{Z}$ such that the restriction to the generic fiber in characteristic zero is semistable but such that the restriction to positive characteristic p is not strongly semistable for infinitely many prime numbers p. Moreover, under the hypothesis that there exist infinitely many Sophie Germain primes, there are also examples such that the density of primes with nonstrongly semistable reduction is arbitrarily close to one.

Introduction

In this paper, we deal with the following problem of Miyaoka [5, Problem 5.4]:

"Let C be an irreducible smooth projective curve over a noetherian integral domain R of characteristic 0. Assume that a locally free sheaf $\mathcal{E}$ on C is $\mathfrak{A}$-semistable on the generic fiber C_*. Let S be the set of primes of char. > 0 on $\operatorname{Spec} R$ such that $\mathcal{E}$ is strongly semistable. Is S a dense subset of $\operatorname{Spec} R$?"

Here a locally free sheaf $\mathcal{E}$ on a smooth projective curve over a field is called semistable, if for every coherent subsheaf $\mathcal{F} \subseteq \mathcal{E}$ the inequality $\deg(\mathcal{F})/\operatorname{rk}(\mathcal{F}) \leq \deg(\mathcal{E})/\operatorname{rk}(\mathcal{E})$ holds true. In positive characteristic, $\mathcal{E}$ is called strongly semistable if every Frobenius pullback of $\mathcal{E}$ stays semistable. Since semistability is an open property and since semistable bundles on the projective line and on an elliptic curve are strongly semistable, this problem has a positive answer for genus $g = 0, 1$.

Shepherd-Barron "rephrases" the question asking "is it true that the set $(\dots)$ of primes p modulo which $\mathcal{E}$ is not strongly semistable $(\dots)$ is finite, or at least of density 0?" [7]. He considers also higher dimensional varieties V, and one of his main results is that for $\dim V \geq 2$ and $\operatorname{rk}(\mathcal{E}) = 2$ the set Σ of prime numbers with nonstrongly semistable reduction is finite under the condition that either the Picard number of V is 1, or that the canonical bundle K_V is numerically trivial, or that the variety V is algebraically simply connected [7, Corollary 6].

Coming back to curves of genus ≥ 2, say over Spec $\mathbb{Z}$, nothing seems to be known about the following questions on the set S of prime numbers with strongly semistable reduction: Does S contain almost all prime numbers (as in the results of Shepherd-Barron)? Is S always an infinite set? Is it even possible that S is empty? Can we say anything about the density of S in the sense of analytic number theory?

In this paper we give examples of vector bundles of rank 2, which are semistable in characteristic zero, but not strongly semistable for infinitely many prime numbers. We also provide examples where the density of primes with nonstrongly semistable reduction is very high, and in fact arbitrarily close to 1 under the hypothesis that there exist infinitely many Sophie Germain prime numbers.

The example is just the syzygy bundle $\mathrm{Syz}(X^2, Y^2, Z^2)$ on the plane projective curve given by $Z^d = X^d + Y^d$ for $d \geq 5$. These bundles are semistable in characteristic zero. The point is that in positive characteristic p fulfilling certain congruence conditions modulo d, some Frobenius pullbacks of these bundles have global sections which contradict the strong semistability. It is also possible that in these examples the reduction is not strongly semistable for all prime numbers, but this we do not know.

This type of examples is motivated by the theory of tight closure. It was already used in [2] to show that there is no Bogomolov type restriction theorem for strong semistability in positive characteristic. We will come back to the impact of these examples on tight closure and on Hilbert–Kunz theory somewhere else.

1 Main results

In the following we will consider syzygy bundles of rank 2 on a smooth projective curve $C = \mathrm{Proj}\, A$ over a field K, where A is a two-dimensional normal standard-graded K-domain. Such a bundle is given by three homogeneous, A_+-primary elements $f_1, f_2, f_3 \in A$ of degree d_1, d_2, d_3 by the short exact sequence

$$0 \to \mathrm{Syz}(f_1, f_2, f_3)(m) \to \mathcal{O}(m-d_1) \oplus \mathcal{O}(m-d_2) \oplus \mathcal{O}(m-d_3) \stackrel{f_1, f_2, f_3}{\longrightarrow} \mathcal{O}(m) \to 0.$$

A global section of $\mathrm{Syz}(f_1, f_2, f_3)(m)$ is a triple (s_1, s_2, s_3) of homogeneous elements such that $\deg(s_i) + d_i = m$, $i = 1, 2, 3$, and $s_1 f_1 + s_2 f_2 + s_3 f_3 = 0$. We call m the total degree of the syzygy (s_1, s_2, s_3). The degree of such a syzygy bundle is by the short exact sequence $\deg(\mathrm{Syz}(f_1, f_2, f_3)(m)) = (2m - d_1 - d_2 - d_3)\deg(C)$. If m is such that this degree is negative and such that there exist global nontrivial sections, then this bundle is not semistable.

It is in general not easy to control the global syzygies; in the following lemma however we take advantage of the existence of a noetherian normalization of a very special type. Let $\delta(f_1, \ldots, f_n)$ denote the minimal total degree of a nontrivial syzygy for $f_1, \ldots, f_n$.

Lemma 1. *Let $P(X, Y)$ denote a homogeneous polynomial in $K[X, Y]$ of degree d and consider the projective curve C given by $Z^d - P(X, Y)$, so that $C = \mathrm{Proj}\, K[X, Y, Z]/(Z^d - P(X, Y))$. Suppose that C is smooth. Let $a_1, a_2, a_3 \in \mathbb{N}$*

and consider the syzygy bundle $\mathrm{Syz}(X^{a_1}, Y^{a_2}, Z^{a_3})(m)$ *on* C. *Write* $a_3 = dk + t$, *where* $0 \leq t < d$. *Then*

$$\delta(X^{a_1}, Y^{a_2}, Z^{a_3})$$
$$= \min\{\delta(X^{a_1}, Y^{a_2}, P(X, Y)^k) + t, \delta(X^{a_1}, Y^{a_2}, P(X, Y)^{k+1})\}.$$

Proof. A global section of $\mathrm{Syz}(X^{a_1}, Y^{a_2}, Z^{a_3})(m)$ is the same as homogeneous polynomials $F, G, H \in K[X, Y, Z]/(Z^d - P(X, Y))$ such that $FX^{a_1} + GY^{a_2} + HZ^{a_3} = 0$ and $\deg(F) + a_1 = \deg(G) + a_2 = \deg(H) + a_3 = m$. We may write $F = F_0 + F_1 Z + F_2 Z^2 + \cdots + F_{d-1} Z^{d-1}$ with $F_i \in K[X, Y]$ and similarly for G and H. We have $Z^{a_3} = Z^{dk+t} = (Z^d)^k Z^t = P(X, Y)^k Z^t$. The polynomials (F, G, H) (fulfilling the degree condition) are a syzygy if and only if for $i = 0, \ldots, d - 1$, we have

$$F_i Z^i X^{a_1} + G_i Z^i Y^{a_2} + H_j Z^j Z^{a_3} = 0, \quad \text{where } j = i - t \bmod d.$$

Now let (F, G, H) denote a nontrivial syzygy of minimal degree for $X^{a_1}, Y^{a_2}, Z^{a_3}$ and let i denote the minimal number such that $F_i \neq 0$ or $G_i \neq 0$. Since the degree is minimal we may assume by dividing through $Z^{\min(i,j)}$ that either $i = 0$ or $j = 0$, which means $i = t$.

In the first case, we can read the zero-component of the syzygy directly as a nontrivial syzygy for $X^{a_1}, Y^{a_2}, P(X, Y)^{k+1}$. In the second case the $i = t$th component of the syzygy is

$$F_t Z^t X^{a_1} + G_t Z^t Y^{a_2} + H_0 Z^{a_3} = 0.$$

Replacing Z^{a_3} through $P(X, Y)^k Z^t$ and dividing through Z^t we get a nontrivial syzygy for $X^{a_1}, Y^{a_2}, P(X, Y)^k$ of the same degree $-t$.

Suppose that we have a nontrivial syzygy $\tilde{F} X^{a_1} + \tilde{G} Y^{a_2} + \tilde{H} P(X, Y)^{k+1} = 0$ in $K[X, Y]$. Then $F = F_0 = \tilde{F}, G = G_0 = \tilde{G}$ and $H = H_{d-t} Z^{d-t} = \tilde{H} Z^{d-t}$ gives a syzygy for $X^{a_1}, Y^{a_2}, Z^{a_3}$ of the same degree.

Suppose that we have a nontrivial syzygy $\tilde{F} X^{a_1} + \tilde{G} Y^{a_2} + \tilde{H} P(X, Y)^k = 0$ in $K[X, Y]$. Multiplying with Z^t, we see that $F = F_t Z^t = \tilde{F} Z^t, G = G_t Z^t = \tilde{G} Z^t$ and $H = \tilde{H}$ gives a syzygy for $X^{a_1}, Y^{a_2}, Z^{a_3}$ of the same degree $+t$.   □

Proposition 1. *Let d and b denote natural numbers, write $b = dk + t$ with $0 \leq t < d$. Suppose that k is even and that $t > 2d/3$. Then the syzygy bundle $\mathrm{Syz}(X^b, Y^b, Z^b)$ is not semistable on the Fermat curve given by $Z^d = X^d + Y^d$.*

Proof. We will look for syzygies for X^b, Y^b and $(X^d + Y^d)^{k+1}$ of total degree $d(k + 1) + d\lfloor k/2 \rfloor = d(k + 1 + \lfloor k/2 \rfloor)$, which yields syzygies for X^b, Y^b, Z^b of the same degree by Lemma 1. To find such syzygies we have to look for multiples $H(X^d + Y^d)^{k+1} \in (X^b, Y^b)$, where $\deg(H) = d\lfloor k/2 \rfloor$. We consider for H only monomials in X^d and Y^d, so these are the $\lfloor k/2 \rfloor + 1$ monomials

$$X^{d\lfloor k/2 \rfloor}, X^{d(\lfloor k/2 \rfloor - 1)} Y^d, X^{d(\lfloor k/2 \rfloor - 2)} Y^{d2}, \ldots, Y^{d\lfloor k/2 \rfloor}.$$

The resulting monomials in the products, which do not belong to the ideal (X^b, Y^b), have the form $X^{du} Y^{dv}$ with $du + dv = d(k + 1 + \lfloor k/2 \rfloor)$ and $u, v < d(k + 1)$. So these are the monomials

$$X^{d(\lfloor k/2 \rfloor+1)} Y^{dk}, X^{d(\lfloor k/2 \rfloor+2)} Y^{d(k-1)}, \ldots, X^{dk} Y^{d(\lfloor k/2 \rfloor+1)}.$$

These are $k - (\lfloor k/2 \rfloor + 1) + 1 = k - \lfloor k/2 \rfloor$ monomials. Since k is supposed to be even, we have $\lfloor k/2 \rfloor + 1 > k - \lfloor k/2 \rfloor$, and therefore we must have a nontrivial linear relation

$$\sum_{i+j=\lfloor k/2 \rfloor} \lambda_{ij} X^{di} Y^{dj} (X^d + Y^d)^{k+1} = 0,$$

modulo (X^b, Y^b). The total degree of this nontrivial global syzygy is $d(k+1+\lfloor k/2 \rfloor)$ and the degree of the bundle is

$$\deg(\mathrm{Syz}(X^b, Y^b, Z^b)(d(k + 1 + \lfloor k/2 \rfloor))) = (2d(k + 1 + \lfloor k/2 \rfloor) - 3b) \deg(C).$$

Due to our assumptions, we have

$$2d(k + 1 + \lfloor k/2 \rfloor) = 3dk + 2d < 3dk + 3t = 3b;$$

hence the degree of the bundle is negative, but it has a nontrivial section, so it is not semistable. □

Corollary 1. *Let d and q denote natural numbers; write $q = d\ell + s$ with $0 \le s < d$. Suppose that $2s < d < 3s$. Then the syzygy bundle $\mathrm{Syz}(X^{2q}, Y^{2q}, Z^{2q})$ is not semistable on the Fermat curve given by $Z^d = X^d + Y^d$.*

Proof. We apply Proposition 1 to $b = 2q = d(2\ell) + 2s = dk + t$. Note that $t < d$ and $2d < 6s = 3t$. □

Corollary 2. *Consider the syzygy bundle $\mathcal{E} = \mathrm{Syz}(X^2, Y^2, Z^2)$ on the Fermat curve $C_K = \mathrm{Proj}\, K[X, Y, Z]/(X^d + Y^d - Z^d)$, K a field. Then $\mathcal{E}_K$ is semistable in characteristic zero for $d \ge 5$, but $\mathcal{E}_K$ is not strongly semistable in positive characteristic $p = r \bmod d$ such that some power $s = r^e$ fulfills $2s < d < 3s$. In particular, for prime numbers $d \ge 5$, $\mathcal{E}_K$ is not strongly semistable for infinitely many prime numbers p.*

Proof. Suppose first that K has characteristic zero. Then $\mathcal{E}_K$ is semistable due to [1, Proposition 6.2]; this follows for $d \ge 7$ also from the restriction theorem of Bogomolov (see [3, Theorem 7.3.5]) since the bundle is clearly stable on the projective plane.

Suppose now that K has positive characteristic p fulfilling the assumption. Then we look at $q = p^e$ so that $q = d\ell + s$ with $2s < d < 3s$, and Corollary 1 yields that $\mathrm{Syz}(X^{2q}, Y^{2q}, Z^{2q})$ is not semistable. Since this bundle is the pullback under the eth Frobenius of $\mathrm{Syz}(X^2, Y^2, Z^2)$, as follows from the short exact sequence mentioned at the beginning of this section, we infer that $\mathrm{Syz}(X^2, Y^2, Z^2)$ is not strongly semistable. For prime numbers $d \ge 5$ there are natural numbers s such that $2s < d < 3s$ and such that s is coprime to d. Due to the theorem of Dirichlet about primes in an arithmetic progression [6, Chapitre VI, Section 4, Théorème and Corollaire], there exists infinitely many prime numbers p with remainder $= s \bmod d$. □

Remark 1. The condition in Corollary 2 that d is a prime number is necessary, since for $d = 6$ and $d = 10$ there does not exist such a coprime reminder s in the range $d/3 < s < d/2$. Are these the only exceptions?

Remark 2. For $p = 1$ or $-1 \bmod d$ we have $q = 1$ or $-1 \bmod d$ for all $q = p^e$ and so Corollary 2 does not apply. It is open whether for these prime numbers the bundle is strongly semistable or not.

2 The example on the Fermat quintic

In this section we take a closer look at the example $\mathcal{E} = \mathrm{Syz}(X^2, Y^2, Z^2)$ on the Fermat quintic $Z^5 = X^5 + Y^5$ for various characteristics. Then $\mathcal{E}_K$ is semistable in characteristic zero, but $\mathcal{E}_K$ is not strongly semistable in characteristic $p \equiv 2$ or $\equiv 3 \bmod 5$. In characteristic $p \equiv 1$ or $p \equiv 4 \bmod 5$ this is not known.

Corollary 3. *Consider the syzygy bundle* $\mathcal{E} = \mathrm{Syz}(X^2, Y^2, Z^2)$ *on the Fermat quintic* $C_K = \mathrm{Proj}\, K[X, Y, Z]/(X^5 + Y^5 - Z^5)$, K *a field. Then* $\mathcal{E}_K$ *is semistable in characteristic zero, but* $\mathcal{E}_K$ *is not strongly semistable in characteristic* $p = 2$ *or* $= 3$ *mod 5. For* $p = 2 \bmod 5$ *the first Frobenius pullback of* $\mathcal{E}_K$ *is not semistable, and for* $p = 3 \bmod 5$ *the third Frobenius pullback of* $\mathcal{E}_K$ *is not semistable.*

Proof. For $p = 3 \bmod 5$ we have $q = p^3 = 2 \bmod 5$, and for $p = 2 \bmod 5$ we take $q = p$. So in both cases we get a situation treated in Corollary 2; hence $\mathrm{Syz}(X^2, Y^2, Z^2)$ is not strongly semistable. $\qquad\square$

Remark 3. In the case $p = 3 \bmod 5$, Corollary 3 shows that the third Frobenius pullback of the syzygy bundle is not semistable anymore. We show now that already the first Frobenius pullback is not semistable. Write $p = 5u + 3$ so that $2p = 5(2u + 1) + 1$ (and $k = 2u + 1$, $t = 1$ in the notation of Lemma 1). We consider the syzygies for

$$X^{5(2u+1)+1},\ Y^{5(2u+1)+1},\ (X^5 + Y^5)^{2u+1}.$$

We multiply the last term by the $u + 1$ monomials

$$XY(X^{5u}Y^0),\ XY(X^{5(u-1)}Y^5),\ \ldots,\ XY(X^0Y^{5u}).$$

The resulting polynomials are modulo the first two terms expressible in the monomials $X^{5i+1}Y^{5(3u+1-i)+1}$, where $i \leq 2u$ and $u+1 \leq i$, so these are only u many. Hence there exists a syzygy of these polynomials of degree $5(3u+1)+2$, and therefore there exists a global nontrivial syzygy for X^{2p}, Y^{2p}, Z^{2p} of degree $5(3u + 1) + 3$ by Lemma 1. This contradicts semistability, since $2(5(3u + 1) + 3) - 3(5(2u + 1) + 1) = -2$.

For example, for $p = 3$, we find for $X^6, Y^6, (X^5 + Y^5)$ the syzygy $(-Y, -X, XY)$ of total degree 7, which yields the syzygy $(-YZ, -XZ, XY)$ for X^6, Y^6, Z^6 of total degree 8 on the Fermat quintic.

Example 1. We consider the bundle $\mathrm{Syz}(X^2, Y^2, Z^2)$ on the Fermat quintic for $q = p = 7$. Then $Z^{14} = Z^{10}Z^4 = (X^5 + Y^5)^2 Z^4$ and we look for syzygies in $K[X, Y]$ for

$$X^{14}, Y^{14}, (X^5 + Y^5)^3.$$

We multiply $(X^5 + Y^5)^3$ by the monomials X^5 and Y^5. The only monomial in the products which remains modulo (X^{14}, Y^{14}) is $X^{10}Y^{10}$. Therefore, we must have a nontrivial syzygy of total degree 20, and indeed we have

$$-(X^6 + 2XY^5)X^{14} + (2X^5Y + Y^6)Y^{14} + (X^5 - Y^5)(X^5 + Y^5)^3 = 0.$$

Going back to our original setting on the Fermat curve, we get the syzygy

$$-(X^6 + 2XY^5)X^{14} + (2X^5Y + Y^6)Y^{14} + (X^5 - Y^5)ZZ^{14} = 0.$$

This shows that $\mathrm{Syz}(X^{14}, Y^{14}, Z^{14})(20)$ has a nontrivial global section, but its degree is $(2 \cdot 20 - 3 \cdot 14) \deg(C) = -2 \deg(C)$ negative. So $\mathrm{Syz}(X^2, Y^2, Z^2)$ is not strongly semistable for $p = 7$. It is easy to see that the syzygy $(-X^6 - 2XY^5, +2X^5Y + Y^6, (X^5 - Y^5)Z)$ does not have a common zero on the curve; hence we get the short exact sequence

$$0 \longrightarrow \mathcal{O} \longrightarrow \mathrm{Syz}(X^{14}, Y^{14}, Z^{14})(20) \longrightarrow \mathcal{O}(-2) \longrightarrow 0,$$

which is the Harder–Narasimhan filtration.

Example 2. We consider the example for $p = 11$, so the remainder mod 5 is 1 and we cannot expect a syzygy for X^{22}, Y^{22}, Z^{22} contradicting the semistability. We have $Z^{22} = (X^5 + Y^5)^4 Z^2$ and we look first for syzygies for $X^{22}, Y^{22}, (X^5 + Y^5)^4$. We have $(X^5 + Y^5)^4 = X^{20} + 4X^{15}Y^5 + 6X^{10}Y^{10} + 4X^5Y^{15} + Y^{20}$ and multiplication by X^{10}, X^5Y^5, Y^{10} yields modulo (X^{22}, Y^{22}) the three polynomials

$$6X^{20}Y^{10} + 4X^{15}Y^{15} + X^{10}Y^{20},$$
$$4X^{20}Y^{10} + 6X^{15}Y^{15} + 4X^{10}Y^{20},$$
$$X^{20}Y^{10} + 4X^{15}Y^{15} + 6X^{10}Y^{20}.$$

The determinant of the corresponding matrix is $50 = 6 \bmod 11$ and so these polynomials are linearly independent.

We look now at syzygies for $X^{22}, Y^{22}, (X^5 + Y^5)^5$. If we consider only powers of 5, we multiply only by X^5 and Y^5, which yields modulo (X^{22}, Y^{22}) the monomials $10X^{20}Y^{10} + 10X^{15}Y^{15} + 5X^{10}Y^{20}$ and $5X^{20}Y^{10} + 10X^{15}Y^{15} + 10X^{10}Y^{20}$, which are again linearly independent.

Remark 4. For Fermat curves of degree $d < 5$, the situation is as follows: for $d = 1$ the restriction of $\mathrm{Syz}(X^2, Y^2, Z^2)(3) = \mathrm{Syz}(X^2, Y^2, X^2 + 2XY + Y^2)(3)$ is $\mathcal{O} \oplus \mathcal{O}$, hence (strongly) semistable (characteristic $\neq 2$). For $d = 2$ the restriction of $\mathrm{Syz}(X^2, Y^2, Z^2)(2) = \mathrm{Syz}(X^2, Y^2, X^2 + Y^2)(2)$ has a nontrivial section; hence $\mathrm{Syz}(X^2, Y^2, Z^2)(3) \cong \mathcal{O}(-1) \oplus \mathcal{O}(1)$, so this is not semistable in any characteristic. For $d = 3$ the Fermat equation $Z^3 = X^3 + Y^3$ yields at once a global section $\mathcal{O} \to \mathrm{Syz}(X^2, Y^2, Z^2)(3)$ without a zero. This shows that the bundle is strongly semistable, but not stable, independent of the characteristic. For $d = 4$ we have shown in [1, Example 7.4] that for $\mathrm{char}(K) \neq 2$ the restriction is strongly semistable.

3 Using Sophie Germain primes

Do there exist examples of vector bundles that are semistable in characteristic zero and where the density of prime numbers for which the bundle is not strongly semistable is 1 or arbitrarily close to 1? The density of prime numbers might be the analytic (or Dirichlet) density or the natural density (see [6, Chapitre 5, Sections 4.1 and 4.5]). Since we will only use the fact that the set of prime numbers p such that $p = r \bmod d$, $(r, d$ coprime), has the density $1/\varphi(d)$, we will not say much about this point.

If we want to attack this question with the method of the first section, we need to know for which and for how many remainders $r \in \mathbb{Z}_d^\times$ there exists a power

$$r^e \in M = \{s : d/3 < s < d/2\}.$$

For $r = 1$ or $= -1 \bmod d$, this is not possible; on the other hand, it is always true for primitive elements if M is not empty. For a remainder r there exists some power inside M if and only if the (multiplicative) group generated by r intersects M. Therefore, we only have to count the number of generators of all the subgroups of $\mathbb{Z}_d^\times$ which contain an element of M; hence

$$\sharp\{r \in \mathbb{Z}_d^\times : \exists e \text{ such that } r^e \in M\} = \sum_{H \subseteq \mathbb{Z}_d^\times, H \cap M \neq \emptyset} \varphi(\mathrm{ord}(H)),$$

where φ denotes the Euler φ-function. Good candidates to obtain here a big cardinality are degrees d of type $d = 2h + 1$, where both d and h are prime. The numbers h with this property are called Sophie Germain primes. It is still not known whether there exist infinitely many such numbers.

Proposition 2. *Suppose that $h > 5$ is a Sophie Germain prime; set $d = 2h + 1$. Then the primes for which $\mathrm{Syz}(X^2, Y^2, Z^2)$ is not strongly semistable on the curve given by $Z^d = X^d + Y^d$ have density at least $(2h - 2)/2h = 1 - 1/h$.*

Proof. We will show that for every remainder $r \neq 1, -1 \bmod d$ there exists a power such that $r^e \in M = \{s : d/3 < s < d/2\}$. The residue class ring $\mathbb{Z}_d$ has $2h$ units; therefore every element has order $1, 2, h$ or $2h$. We only have to show that M contains primitive remainders as well as nonprimitive remainders. Since then there exist $\varphi(2h) + \varphi(h) = 2h - 2$ remainders such that some power of them belongs to M.

We first look for nonprimitive remainders. For $d > 75$ there exists always an integer n between $\sqrt{d}/\sqrt{3} < n < \sqrt{d}/\sqrt{2}$, since the length of the interval is then > 1. Thus n^2 is a square in M and hence nonprimitive. It is also true that there exists a square in M for the smaller Sophie Germain prime numbers $h = 5, 11, 23, 29$.

Now to find primitive remainders note first that $d = 3 \bmod 4$. Hence -1 is not a square in $\mathbb{Z}_d^\times$. There exist again squares between $d/2$ and $2d/3$ (check directly for $d \leq 59$). If b is such a square, then $-b = d - b$ is a nonsquare inside M, and so M contains squares as well as nonsquares (for $h = 3 \bmod 4$ one can also show by quadratic reciprocity law that $h \in M$ is a nonsquare). $\qquad\square$

Remark 5. If we would know that there exist infinitely many Sophie Germain primes, then we could conclude that the density of primes for which the bundle $\text{Syz}(X^2, Y^2, Z^2)$ is not strongly semistable on a Fermat curve can be arbitrarily high. The biggest Sophie Germain prime which I have found in the literature (see [4]) is $h = 2375063906985 \cdot 2^{19380} - 1$. There should be known results in analytic number theory which imply that the density of primes with nonstrongly semistable behavior is arbitrarily close to 1.

Example 3. Let $d = 11 = 2 \cdot 5 + 1$. The set M consists only of 4 and 5. We have $2^2 = 4$ and $4^2 = 5 \bmod 11$; hence both numbers are squares and not primitive. Thus the density of primes p such that $\text{Syz}(X^2, Y^2, Z^2)$ is not strongly semistable on $Z^{11} = X^{11} + Y^{11}$ for char $K = p$ is only $\geq 5/11$.

Example 4. Let $d = 167 = 2 \cdot 83 + 1$. Here we have $M = \{56, \ldots, 83\}$. The numbers $s = 64$ and $s = 81$ are squares, hence nonprimitive elements, and 83 is a nonsquare by Proposition 2. Thus the density of primes p such that $\text{Syz}(X^2, Y^2, Z^2)$ is not strongly semistable on $Z^{167} = X^{167} + Y^{167}$ over char $K = p$ is $\geq 165/167$.

Example 5. We look now at $h = 29$, so $d = 59$. We have $M = \{20, \ldots, 29\}$. 2 is a primitive element in $\mathbb{Z}_{59}$; hence computing 2^u, u odd (or by quadratic reciprocity law), we see that the only primitive remainders in M are 23 and 24. So in this range we have eight quadratic remainders but only two nonquadratic remainders.

We close with an example of a degree which does not come from a Sophie Germain prime.

Example 6. Let $d = 31$, which does not come from a Sophie Germain prime. The remainders s for which we know that $\text{Syz}(X^{2q}, Y^{2q}, Z^{2q})$ is not semistable for $q = s \bmod d$, are $s \in M = \{11, \ldots, 15\}$. Which remainders $p = r \bmod d$ have the property that some power $q = p^e = r^e = s = 11, \ldots, 15$? The number 3 is a primitive element modulo 31, and we have $11 = 3^{23}$, $12 = 3^{19}$, $13 = 3^{11}$, $14 = 3^{22}$, $15 = 3^{21}$. So 11, 12 and 13 are primitive, 14 generates a subgroup with 15 elements and 15 generates a subgroup with 10 elements. The number of generators of these groups are eight, eight, and four, so we have altogether 20 remainders for which some power fulfills the condition in Corollary 2. So the density of primes p for which $\mathcal{E}_K$ is not strongly semistable in characteristic char $K = p$ is at least $\geq 2/3$ (the remainders for which we do not know the answer are 1, 2, 4, 5, 8, 16, 25, 27, 29, 30).

Acknowledgment. I thank Neil Dummigan (University of Sheffield) for a useful remark concerning Sophie Germain primes.

References

[1] H. Brenner, Computing the tight closure in dimension two, *Math. Comput.*, **74**-251 (2005), 1495–1518.

[2] H. Brenner, There is no Bogomolov type restriction theorem for strong semistability in positive characteristic, *Proc. Amer. Math. Soc.*, **133** (2005), 1941–1947.

[3] D. Huybrechts and M. Lehn, *The Geometry of Moduli Spaces of Sheaves*, Vieweg, Braunschweig, Germany, 1997.

[4] K.-H. Indlekofer and A. Járai, Largest known twin primes and Sophie Germain primes, *Mathematics of Computation*, **68**-227 (1999), 1317–1324.

[5] Y. Miyaoka, The Chern class and Kodaira dimension of a minimal variety, in *Algebraic Geometry, Sendai*, 1985, Advanced Studies in Pure Mathematics 10, North-Holland, Amsterdam, 1987, 449–476.

[6] J. P. Serre, *Cours d'arithmétique*, Presses Universitaires de France, Paris, 1970.

[7] N. I. Shepherd-Barron, Semi-stability and reduction mod p, *Topology*, **37**-3 (1997), 659–664.

Monodromy Groups Associated to Non-Isotrivial Drinfeld Modules in Generic Characteristic

Florian Breuer[1] and Richard Pink[2]

[1] Department of Mathematics
University of Stellenbosch
Stellenbosch 7602
South Africa
fbreuer@sun.ac.za

[2] Department of Mathematics
ETH Zentrum
8092 Zürich
Switzerland
pink@math.ethz.ch

Summary. Let φ be a non-isotrivial family of Drinfeld A-modules of rank r in generic characteristic with a suitable level structure over a connected smooth algebraic variety X. Suppose that the endomorphism ring of φ is equal to A. Then we show that the closure of the analytic monodromy group of X in $\mathrm{SL}_r(\mathbb{A}_F^f)$ is open, where $\mathbb{A}_F^f$ denotes the ring of finite adèles of the quotient field F of A.

From this we deduce two further results: (1) If X is defined over a finitely generated field extension of F, the image of the arithmetic étale fundamental group of X on the adèlic Tate module of φ is open in $\mathrm{GL}_r(\mathbb{A}_F^f)$. (2) Let ψ be a Drinfeld A-module of rank r defined over a finitely generated field extension of F, and suppose that ψ cannot be defined over a finite extension of F. Suppose again that the endomorphism ring of ψ is A. Then the image of the Galois representation on the adèlic Tate module of ψ is open in $\mathrm{GL}_r(\mathbb{A}_F^f)$.

Finally, we extend the above results to the case of arbitrary endomorphism rings.

Key words: Drinfeld modules, Drinfeld moduli spaces, Fundamental groups, Galois representations

Subject Classifications: 11F80, 11G09, 14D05

1 Analytic monodromy groups

Let $\mathbb{F}_p$ be the finite prime field with p elements. Let F be a finitely generated field of transcendence degree 1 over $\mathbb{F}_p$. Let A be the ring of elements of F which are regular outside a fixed place ∞ of F. Let M be the fine moduli space over F of Drinfeld A-modules of rank r with some sufficiently high level structure. This is a smooth affine scheme of dimension $r - 1$ over F.

Let F_∞ denote the completion of F at ∞, and $\mathbb{C}$ the completion of an algebraic closure of F_∞. Then the rigid analytic variety $M_{\mathbb{C}}^{\mathrm{an}}$ is a finite disjoint union of spaces of the form $\Delta \backslash \Omega$, where $\Omega \subset (\mathbb{P}_{\mathbb{C}}^{r-1})^{\mathrm{an}}$ is Drinfeld's upper half-space and Δ is a congruence subgroup of $\mathrm{SL}_r(F)$ commensurable with $\mathrm{SL}_r(A)$.

Let $X_{\mathbb{C}}$ be a smooth irreducible locally closed algebraic subvariety of $M_{\mathbb{C}}$. Then $X_{\mathbb{C}}^{\mathrm{an}}$ lies in one of the components $\Delta \backslash \Omega$ of $M_{\mathbb{C}}^{\mathrm{an}}$. Fix an irreducible component $\Xi \subset \Omega$ of the preimage of $X_{\mathbb{C}}^{\mathrm{an}}$. Then $\Xi \to X_{\mathbb{C}}^{\mathrm{an}}$ is an unramified Galois covering whose Galois group $\Delta_\Xi := \mathrm{Stab}_\Delta(\Xi)$ is a quotient of the analytic fundamental group of $X_{\mathbb{C}}^{\mathrm{an}}$.

Let φ denote the family of Drinfeld modules over $X_{\mathbb{C}}$ determined by the embedding $X_{\mathbb{C}} \subset M_{\mathbb{C}}$. We assume that $\dim X_{\mathbb{C}} \geq 1$. Since M is a fine moduli space, this means that φ is non-isotrivial. It also implies that $r \geq 2$. Let $\eta_{\mathbb{C}}$ be the generic point of $X_{\mathbb{C}}$ and $\bar\eta_{\mathbb{C}}$ a geometric point above it. Let $\varphi_{\bar\eta_{\mathbb{C}}}$ denote the pullback of φ to $\bar\eta_{\mathbb{C}}$. Let $\mathbb{A}_F^f$ denote the ring of finite adèles of F. The main result of this article is the following:

Theorem 1. *In the above situation, if* $\mathrm{End}_{\bar\eta_{\mathbb{C}}}(\varphi_{\bar\eta_{\mathbb{C}}}) = A$, *then the closure of* Δ_Ξ *in* $\mathrm{SL}_r(\mathbb{A}_F^f)$ *is an open subgroup of* $\mathrm{SL}_r(\mathbb{A}_F^f)$.

The proof uses known results on the $\mathfrak{p}$-adic Galois representations associated to Drinfeld modules [Pi97] and on strong approximation [Pi00].

Theorem 1 leaves open the following natural question:

Question. *If* $\mathrm{End}_{\bar\eta_{\mathbb{C}}}(\varphi_{\bar\eta_{\mathbb{C}}}) = A$, *is* Δ_Ξ *an arithmetic subgroup of* $\mathrm{SL}_r(F)$?

Theorem 1 has applications to the analogue of the André–Oort conjecture for Drinfeld moduli spaces; see [Br]. Consequences for étale monodromy groups and for Galois representations are explained in Sections 2 and 3. The proof of Theorem 1 will be given in Sections 4 through 7. Finally, in Section 8 we outline the case of arbitrary endomorphism rings.

For any variety Y over a field k and any extension field L of k we will abbreviate $Y_L := Y \times_k L$.

2 Étale monodromy groups

We retain the notation from Section 1. Let $k \subset \mathbb{C}$ be a subfield that is finitely generated over F, such that $X_{\mathbb{C}} = X \times_k \mathbb{C}$ for a subvariety $X \subset M_k$. Let K denote the function field of X and K^{sep} a separable closure of K. Then $\eta := \mathrm{Spec}\, K$ is the generic point of X and $\bar\eta := \mathrm{Spec}\, K^{\mathrm{sep}}$ a geometric point above η. Let k^{sep} be the separable closure of k in K^{sep}. Then we have a short exact sequence of étale fundamental groups

$$1 \longrightarrow \pi_1(X_{k^{\mathrm{sep}}}, \bar\eta) \longrightarrow \pi_1(X, \bar\eta) \longrightarrow \mathrm{Gal}(k^{\mathrm{sep}}/k) \to 1.$$

Let $\hat{A} \cong \prod_{\mathfrak{p} \neq \infty} A_\mathfrak{p}$ denote the profinite completion of A. Recall that $\mathbb{A}_F^f \cong F \otimes_A \hat{A}$ and contains $\hat{A}$ as an open subring. Let φ_η denote the Drinfeld module over K corresponding to η. Its adèlic Tate module $\hat{T}(\varphi_\eta)$ is a free module of rank r over $\hat{A}$. Choose a basis and let

$$\rho \colon \pi_1(X, \bar{\eta}) \longrightarrow \mathrm{GL}_r(\hat{A}) \subset \mathrm{GL}_r(\mathbb{A}_F^f)$$

denote the associated monodromy representation. Let $\Gamma^{\mathrm{geom}} \subset \Gamma \subset \mathrm{GL}_r(\hat{A})$ denote the images of $\pi_1(X_{k^{\mathrm{sep}}}, \bar{\eta}) \subset \pi_1(X, \bar{\eta})$ under ρ.

Lemma 1. Γ^{geom} *is the closure of* $g^{-1}\Delta_\Xi g$ *in* $\mathrm{SL}_r(\hat{A})$ *for some element* $g \in \mathrm{GL}_r(\mathbb{A}_F^f)$.

Proof. Choose an embedding $K^{\mathrm{sep}} \hookrightarrow \mathbb{C}$ and a point $\xi \in \Xi$ above $\bar{\eta}$. Let $\Lambda \subset F^r$ be the lattice corresponding to the Drinfeld module at ξ. This is a finitely generated projective A-module of rank r. The choice of a basis of $\hat{T}(\varphi_\eta)$ yields a composite embedding

$$\hat{A}^r \cong \hat{T}(\varphi_\eta) \cong \Lambda \otimes_A \hat{A} \hookrightarrow F^r \otimes_A \hat{A} \cong (\mathbb{A}_F^f)^r,$$

which is given by left multiplication with some element $g \in \mathrm{GL}_r(\mathbb{A}_F^f)$. Since the discrete group $\Delta \subset \mathrm{SL}_r(F)$ preserves Λ, we have $g^{-1}\Delta g \subset \mathrm{SL}_r(\hat{A})$.

For any nonzero ideal $\mathfrak{a} \subset A$ let $M(\mathfrak{a})$ denote the moduli space obtained from M by adjoining a full level $\mathfrak{a}$ structure. Then $\pi_\mathfrak{a} \colon M(\mathfrak{a}) \twoheadrightarrow M$ is an étale Galois covering with group contained in $\mathrm{GL}_r(A/\mathfrak{a})$, and one of the connected components of $M(\mathfrak{a})_\mathbb{C}^{\mathrm{an}}$ above the connected component $\Delta \backslash \Omega$ of $M_\mathbb{C}^{\mathrm{an}}$ has the form $\Delta(\mathfrak{a}) \backslash \Omega$ for

$$\Delta(\mathfrak{a}) := \{\delta \in \Delta \mid |g^{-1}\delta g \equiv \mathrm{id} \mod \mathfrak{a}\hat{A}\}.$$

Let $X(\mathfrak{a})_{k^{\mathrm{sep}}}$ be any connected component of the inverse image $\pi_\mathfrak{a}^{-1}(X_{k^{\mathrm{sep}}}) \subset M(\mathfrak{a})_{k^{\mathrm{sep}}}$. Since k^{sep} is separably closed, the variety $X(\mathfrak{a})_\mathbb{C}$ over $\mathbb{C}$ obtained by base change is again connected. The associated rigid analytic variety $X(\mathfrak{a})_\mathbb{C}^{\mathrm{an}}$ is then also connected (cf. [Lü74, Korollar 3.5]) and therefore a connected component of $\pi_\mathfrak{a}^{-1}(X_\mathbb{C}^{\mathrm{an}})$. But one of these connected components is $(\Delta_\Xi \cap \Delta(\mathfrak{a})) \backslash \Xi$, whose Galois group over $X_\mathbb{C}^{\mathrm{an}} \cong \Delta_\Xi \backslash \Xi$ is $\Delta_\Xi / (\Delta_\Xi \cap \Delta(\mathfrak{a}))$. This implies that $g^{-1}\Delta_\Xi g$ and $\pi_1(X_{k^{\mathrm{sep}}}, \bar{\eta})$ have the same images in $\mathrm{GL}_r(A/\mathfrak{a}) = \mathrm{GL}_r(\hat{A}/\mathfrak{a}\hat{A})$. By taking the inverse limit over the ideal $\mathfrak{a}$ we deduce that the closure of $g^{-1}\Delta_\Xi g$ in $\mathrm{SL}_r(\hat{A})$ is Γ^{geom}, as desired. $\square$

Lemma 2. $\mathrm{End}_{K^{\mathrm{sep}}}(\varphi_\eta) = \mathrm{End}_{\bar{\eta}_\mathbb{C}}(\varphi_{\bar{\eta}_\mathbb{C}})$.

Proof. By construction $\bar{\eta}_\mathbb{C}$ is a geometric point above η, and $\varphi_{\bar{\eta}_\mathbb{C}}$ is the pullback of φ_η. Any embedding of K^{sep} into the residue field of $\bar{\eta}_\mathbb{C}$ induces a morphism $\bar{\eta}_\mathbb{C} \to \bar{\eta}$. Thus the assertion follows from the fact that for every Drinfeld module over a field, any endomorphism defined over any field extension is already defined over a finite separable extension. $\square$

Theorem 2. *In the above situation, suppose that* $\mathrm{End}_{K^{\mathrm{sep}}}(\varphi_\eta) = A$. *Then*

(a) Γ^{geom} *is an open subgroup of* $\mathrm{SL}_r(\mathbb{A}_F^f)$, *and*

(b) Γ *is an open subgroup of* $\mathrm{GL}_r(\mathbb{A}_F^f)$.

Proof. By Lemma 2 the assumption implies that $\text{End}_{\bar{\eta}_{\mathbb{C}}}(\varphi_{\bar{\eta}_{\mathbb{C}}}) = A$. Thus part (a) follows at once from Theorem 1 and Lemma 1. Part (b) follows from (a) and the fact that $\det(\Gamma)$ is open in $\text{GL}_1(\mathbb{A}_F^f)$. This fact is a consequence of work of Drinfeld [Dr74, Section 8, Theorem 1] and Hayes [Ha79, Theorem 9.2] on the abelian class field theory of F, and of Anderson [An86] on the determinant Drinfeld module. Note that Anderson's paper only treats the case $A = \mathbb{F}_q[T]$; the general case has been worked out by van der Heiden [He03, Chapter 4]. Compare also [Pi97, Theorem 1.8]. □

3 Galois groups

Let F and A be as in Section 1. Let K be a finitely generated extension field of F of arbitrary transcendence degree, and let $\psi : A \to K\{\tau\}$ be a Drinfeld A-module of rank r over K. Let K^{sep} denote a separable closure of K and

$$\sigma : \text{Gal}(K^{\text{sep}}/K) \longrightarrow \text{GL}_r(\mathbb{A}_F^f)$$

the natural representation on the adèlic Tate module of ψ. Let $\Gamma \subset \text{GL}_r(\mathbb{A}_F^f)$ denote its image.

Theorem 3. *In the above situation, suppose that* $\text{End}_{K^{\text{sep}}}(\psi) = A$ *and that* ψ *cannot be defined over a finite extension of F inside K^{sep}. Then Γ is an open subgroup of* $\text{GL}_r(\mathbb{A}_F^f)$.

Proof. The assertion is invariant under replacing K by a finite extension. We may therefore assume that ψ possesses a sufficiently high level structure over K. Then ψ corresponds to a K-valued point on the moduli space M from Section 1. Let η denote the underlying point on the scheme M, and let $L \subset K$ be its residue field. Then ψ is already defined over L, and σ factors through the natural homomorphism $\text{Gal}(K^{\text{sep}}/K) \to \text{Gal}(L^{\text{sep}}/L)$, where L^{sep} is the separable closure of L in K^{sep}. Since K is finitely generated over L, the intersection $K \cap L^{\text{sep}}$ is finite over L; hence the image of this homomorphism is open. To prove the theorem we may thus replace K by L, after which K is the residue field of η.

The assumption on ψ implies that even after this reduction, K is not a finite extension of F. Therefore its transcendence degree over F is ≥ 1. Let k denote the algebraic closure of F in K. Then η can be viewed as the generic point of a geometrically irreducible and reduced locally closed algebraic subvariety $X \subset M_k$ of dimension ≥ 1. After shrinking X we may assume that X is smooth. We are then precisely in the situation of the preceding section, with $\psi = \varphi_\eta$. The homomorphism σ above is then the composite

$$\text{Gal}(K^{\text{sep}}/K) \cong \pi_1(\eta, \bar{\eta}) \twoheadrightarrow \pi_1(X, \bar{\eta}) \xrightarrow{\rho} \text{GL}_r(\mathbb{A}_F^f)$$

with ρ as in Section 2. It follows that the groups called Γ in this section and the last coincide. The desired openness is now equivalent to Theorem 2 (b). □

Note. The adèlic openness for a Drinfeld module ψ as in Theorem 3, but defined over a *finite* extension of F, is conjectured yet still unproved.

4 p-Adic openness

This section and the next three are devoted to proving Theorem 1. Throughout we retain the notation from Sections 1 and 2 and the assumptions $\dim X \geq 1$ and $\text{End}_{\bar{\eta}_{\mathbb{C}}}(\varphi_{\bar{\eta}_{\mathbb{C}}}) = A$. In this section we recall a known result on p-adic openness. For any place $\mathfrak{p} \neq \infty$ of F let $\Gamma_{\mathfrak{p}}$ denote the image of Γ under the projection $\text{GL}_r(\mathbb{A}_F^f) \twoheadrightarrow \text{GL}_r(F_{\mathfrak{p}})$.

Theorem 4. $\Gamma_{\mathfrak{p}}$ *is open in* $\text{GL}_r(F_{\mathfrak{p}})$.

Proof. By construction $\Gamma_{\mathfrak{p}}$ is the image of the monodromy representation

$$\rho_{\mathfrak{p}} \colon \pi_1(X, \bar{\eta}) \longrightarrow \text{GL}_r(F_{\mathfrak{p}})$$

on the rational p-adic Tate module of φ_{η}. This is the same as the image of the composite homomorphism

$$\text{Gal}(K^{\text{sep}}/K) \cong \pi_1(\eta, \bar{\eta}) \twoheadrightarrow \pi_1(X, \bar{\eta}) \xrightarrow{\rho_{\mathfrak{p}}} \text{GL}_r(F_{\mathfrak{p}}).$$

Since K is a finitely generated extension of F, and $\text{End}_{K^{\text{sep}}}(\varphi_{\eta}) = A$ by the assumption and Lemma 2, the desired openness is a special case of [Pi97, Theorem 0.1]. $\qquad\square$

Next let $\Gamma_{\mathfrak{p}}^{\text{geom}}$ denote the image of Γ^{geom} under the projection $\text{GL}_r(\mathbb{A}_F^f) \twoheadrightarrow \text{GL}_r(F_{\mathfrak{p}})$. Note that this is a normal subgroup of $\Gamma_{\mathfrak{p}}$. Lemma 1 immediately implies the following.

Lemma 3. $\Gamma_{\mathfrak{p}}^{\text{geom}}$ *is the closure of* $g^{-1}\Delta_{\Xi}g$ *in* $\text{SL}_r(F_{\mathfrak{p}})$ *for some element* $g \in \text{GL}_r(F_{\mathfrak{p}})$.

5 Zariski density

Lemma 4. *The Zariski closure* H *of* Δ_{Ξ} *in* $\text{GL}_{r,F}$ *is a normal subgroup of* $\text{GL}_{r,F}$.

Proof. Choose a place $\mathfrak{p} \neq \infty$ of F. Then by base extension $H_{F_{\mathfrak{p}}}$ is the Zariski closure of Δ_{Ξ} in $\text{GL}_{r,F_{\mathfrak{p}}}$. Thus Lemma 3 implies that $g^{-1}H_{F_{\mathfrak{p}}}g$ is the Zariski closure of $\Gamma_{\mathfrak{p}}^{\text{geom}}$ in $\text{GL}_{r,F_{\mathfrak{p}}}$. Since $\Gamma_{\mathfrak{p}}$ normalizes $\Gamma_{\mathfrak{p}}^{\text{geom}}$, it therefore normalizes $g^{-1}H_{F_{\mathfrak{p}}}g$. But $\Gamma_{\mathfrak{p}}$ is open in $\text{GL}_r(F_{\mathfrak{p}})$ by Theorem 4 and therefore Zariski dense in $\text{GL}_{r,F_{\mathfrak{p}}}$. Thus $\text{GL}_{r,F_{\mathfrak{p}}}$ normalizes $g^{-1}H_{F_{\mathfrak{p}}}g$ and hence $H_{F_{\mathfrak{p}}}$, and the result follows. $\qquad\square$

Lemma 5. Δ_{Ξ} *is infinite.*

Proof. Let X, K, k and φ_{η} be as in Section 2. Then, as M_k is affine and $\dim X \geq 1$, there exists a valuation v of K, corresponding to a point on the boundary of X not on M_k, at which φ_{η} does not have potential good reduction. Denote by $I_v \subset \text{Gal}(K^{\text{sep}}/Kk^{\text{sep}})$ the inertia group at v. By the criterion of Néron–Ogg–Shafarevich

[Go96, Section 4.10], the image of I_v in $\Gamma_{\mathfrak{p}}^{\text{gcom}}$ is infinite for any place $\mathfrak{p} \neq \infty$ of F. In particular, Δ_{Ξ} is infinite by Lemma 3, as desired.

Alternatively, we may argue as follows. Suppose that Δ_{Ξ} is finite. Then after increasing the level structure we may assume that $\Delta_{\Xi} = 1$. Then $\Gamma_{\mathfrak{p}}^{\text{geom}} = 1$ by Lemma 3, which means that $\rho_{\mathfrak{p}}$ factors as

$$\pi_1(X, \bar{\eta}) \twoheadrightarrow \text{Gal}(k^{\text{sep}}/k) \longrightarrow \text{GL}_r(F_{\mathfrak{p}}).$$

After a suitable finite extension of the constant field k we may assume that X possesses a k-rational point x. Let φ_x denote the Drinfeld module over k corresponding to x. Via the embedding $k \subset K$ we may consider it as a Drinfeld module over K and compare it with φ_η. The factorization above implies that the Galois representations on the $\mathfrak{p}$-adic Tate modules of φ_x and φ_η are isomorphic. By the Tate conjecture (see [Tag95] or [Tam95]) this implies that there exists an isogeny $\varphi_x \to \varphi_\eta$ over K. Its kernel is finite and therefore defined over some finite extension k' of k. Thus φ_η, as a quotient of φ_x by this kernel, is isomorphic to a Drinfeld module defined over k'. But the assumption $\dim X \geq 1$ implies that η is not a closed point of M_k; hence φ_η cannot be defined over a finite extension of k. This is a contradiction.   □

Proposition 1. Δ_{Ξ} *is Zariski dense in* $\text{SL}_{r,F}$.

Proof. By construction we have $H \subset \text{SL}_{r,F}$, and Lemma 5 implies that H is not contained in the center of $\text{SL}_{r,F}$. From Lemma 4 it now follows that $H = \text{SL}_{r,F}$, as desired.   □

The above results may be viewed as analogues of André's results [An92, Theorem 1, Proposition 2], comparing the monodromy group of a variation of Hodge structures with its generic Mumford–Tate group. Our analogue of the former is Δ_{Ξ}, and by [Pi97] the latter corresponds to $\text{GL}_{r,F}$. In our situation, however, we do not need the existence of a special point on X.

6 Fields of coefficients

Let $\bar{\Delta}_{\Xi}$ denote the image of Δ_{Ξ} in $\text{PGL}_r(F)$. In this section we show that the field of coefficients of $\bar{\Delta}_{\Xi}$ cannot be reduced.

Definition. Let L_1 be a subfield of a field L. We say that a subgroup $\bar{\Delta} \subset \text{PGL}_r(L)$ lies in a model of $\text{PGL}_{r,L}$ over L_1, if there exist a linear algebraic group G_1 over L_1 and an isomorphism $\lambda_1 : G_{1,L} \xrightarrow{\sim} \text{PGL}_{r,L}$, such that $\bar{\Delta} \subset \lambda_1(G_1(L_1))$.

Proposition 2. $\bar{\Delta}_{\Xi}$ *does not lie in a model of* $\text{PGL}_{r,F}$ *over a proper subfield of* F.

Proof. As before we use an arbitrary auxiliary place $\mathfrak{p} \neq \infty$ of F. Let $\bar{\Gamma}_{\mathfrak{p}}^{\text{geom}} \lhd \bar{\Gamma}_{\mathfrak{p}}$ denote the images of $\Gamma_{\mathfrak{p}}^{\text{geom}} \lhd \Gamma_{\mathfrak{p}}$ in $\text{PGL}_r(F_{\mathfrak{p}})$. Lemma 3 implies that $\bar{\Gamma}_{\mathfrak{p}}^{\text{geom}}$ is conjugate to the closure of $\bar{\Delta}_{\Xi}$ in $\text{PGL}_r(F_{\mathfrak{p}})$. By Proposition 1 it is therefore Zariski dense in $\text{PGL}_{r,F_{\mathfrak{p}}}$. On the other hand Theorem 4 implies that $\bar{\Gamma}_{\mathfrak{p}}$ is an open subgroup of $\mathrm{PGL}_r(F_\mathfrak{p})$. It therefore does not lie in a model of $\mathrm{PGL}_{r,F_\mathfrak{p}}$ over a proper subfield of $F_\mathfrak{p}$. Thus $\bar{\Gamma}_\mathfrak{p}^{\mathrm{geom}}$ is Zariski dense and normal in a subgroup that does not lie in a model over a proper subfield of $F_\mathfrak{p}$, which by [Pi98, Corollary 3.8] implies that $\bar{\Gamma}_\mathfrak{p}^{\mathrm{geom}}$, too, does not lie in a model over a proper subfield of $F_\mathfrak{p}$.

Suppose now that $\bar{\Delta}_\Xi \subset \lambda_1(G_1(F_1))$ for a subfield $F_1 \subset F$, a linear algebraic group G_1 over F_1, and an isomorphism $\lambda_1 : G_{1,F} \xrightarrow{\sim} \mathrm{PGL}_{r,F}$. Since $\bar{\Delta}_\Xi$ is Zariski dense in $\mathrm{PGL}_{r,F}$, it is in particular infinite. Therefore F_1 must be infinite. As F is finitely generated of transcendence degree 1 over $\mathbb{F}_p$, it follows that F_1 contains a transcendental element, and so F is a finite extension of F_1. Let $\mathfrak{p}_1$ denote the place of F_1 below $\mathfrak{p}$. Since $\bar{\Gamma}_\mathfrak{p}^{\mathrm{geom}}$ is the closure of $\bar{\Delta}_\Xi$ in $\mathrm{PGL}_r(F_\mathfrak{p})$, it is contained in $\lambda_1(G_1(F_{1,\mathfrak{p}_1}))$. The fact that $\bar{\Gamma}_\mathfrak{p}^{\mathrm{geom}}$ does not lie in a model over a proper subfield of $F_\mathfrak{p}$ thus implies that $F_{1,\mathfrak{p}_1} = F_\mathfrak{p}$.

But for any proper subfield $F_1 \subsetneq F$, we can choose a place $\mathfrak{p} \neq \infty$ of F above a place $\mathfrak{p}_1$ of F_1, such that the local field extension $F_{1,\mathfrak{p}_1} \subset F_\mathfrak{p}$ is nontrivial. Thus we must have $F_1 = F$, as desired. $\square$

7 Strong approximation

The remaining ingredient is the following general theorem.

Theorem 5. *For $r \geq 2$ let $\Delta \subset \mathrm{SL}_r(F)$ be a subgroup that is contained in a congruence subgroup commensurable with $\mathrm{SL}_r(A)$. Assume that Δ is Zariski dense in $\mathrm{SL}_{r,F}$ and that its image $\bar{\Delta}$ in $\mathrm{PGL}_r(F)$ does not lie in a model of $\mathrm{PGL}_{r,F}$ over a proper subfield of F. Then the closure of Δ in $\mathrm{SL}_r(\mathbb{A}_F^f)$ is open.*

Proof. If Δ is finitely generated, then this is a special case of [Pi00, Theorem 0.2]. That result concerns arbitrary finitely generated Zariski-dense subgroups of $G(F)$ for arbitrary semisimple algebraic groups G, but it uses the finite generation only to guarantee that the subgroup is integral at almost all places of F. For Δ as above the integrality at all places $\neq \infty$ is already known in advance, so the proof in [Pi00] covers this case as well.

As an alternative, we will deduce the general case by showing that every sufficiently large finitely generated subgroup $\Delta_1 \subset \Delta$ satisfies the same assumptions. Then the closure of Δ_1 in $\mathrm{SL}_r(\mathbb{A}_F^f)$ is open by [Pi00], and so the same follows for Δ, as desired.

For the Zariski density of Δ_1 note first that the trace of the adjoint representation defines a dominant morphism to the affine line $\mathrm{SL}_{r,F} \to \mathbb{A}_F^1$, $g \mapsto \mathrm{tr}(\mathrm{Ad}(g))$. Since Δ is Zariski dense, this function takes infinitely many values on Δ. As the field of constants in F is finite, we may therefore choose an element $\gamma \in \Delta$ with $\mathrm{tr}(\mathrm{Ad}(\gamma))$ transcendental. Then γ has infinite order; hence the Zariski closure $H \subset \mathrm{SL}_{r,F}$ of the abstract subgroup generated by γ has positive dimension. Let H° denote its identity component. Since Δ is Zariski dense and $\mathrm{SL}_{r,F}$ is almost simple, the Δ-conjugates of H° generate $\mathrm{SL}_{r,F}$ as an algebraic group. By noetherian induction finitely many conjugates suffice. It follows that finitely many conjugates of γ generate a Zariski-dense subgroup of $SL_{r,F}$. Thus every sufficiently large finitely generated subgroup $\Delta_1 \subset \Delta$ is Zariski dense.

Consider such Δ_1 and let $\bar{\Delta}_1$ denote its image in $PGL_r(F)$. Consider all triples (F_1, G_1, λ_1) consisting of a subfield $F_1 \subset F$, a linear algebraic group G_1 over F_1, and an isomorphism $\lambda_1 \colon G_{1,F} \xrightarrow{\sim} PGL_{r,F}$, such that $\bar{\Delta}_1 \subset \lambda_1(G_1(F_1))$. By [Pi98, Theorem 3.6] there exists such a triple with F_1 minimal, and this F_1 is unique, and G_1 and λ_1 are determined up to unique isomorphism. Consider another finitely generated subgroup $\Delta_1 \subset \Delta_2 \subset \Delta$ and let (F_2, H_2, λ_2) be the minimal triple associated to it. Then the uniqueness of (F_1, G_1, λ_1) implies that $F_1 \subset F_2$, that $G_2 \cong G_{1,F_2}$, and that λ_2 coincides with the isomorphism $G_{2,F} \cong G_{1,F} \to PGL_{r,F}$ obtained from λ_1. In other words, the minimal model (F_1, G_1, λ_1) is monotone in Δ_1.

For any increasing sequence of Zariski-dense finitely generated subgroups of Δ, we thus obtain an increasing sequence of subfields of F. This sequence must become constant, say equal to $F_1 \subset F$, and the associated model of $PGL_{r,F}$ over F_1 is the same up to isomorphism from that point onwards. Thus we have a triple (F_1, G_1, λ_1) with $\bar{\Delta}_1 \subset \lambda_1(G_1(F_1))$ for every sufficiently large finitely generated subgroup $\bar{\Delta}_1 \subset \bar{\Delta}$. But then we also have $\bar{\Delta} \subset \lambda_1(G_1(F_1))$, which by assumption implies that $F_1 = F$. Thus every sufficiently large finitely generated subgroup of Δ satisfies the same assumptions as Δ, as desired. $\qquad\square$

Proof of Theorem 1. In the situation of Theorem 1 we automatically have $r \geq 2$, so the assertion follows by combining Propositions 1 and 2 with Theorem 5 for Δ_Ξ. $\quad\square$

8 Arbitrary endomorphism rings

Set $E := End_{\bar{\eta}_\mathbb{C}}(\varphi_{\bar{\eta}_\mathbb{C}})$, which is a finite integral ring extension of A. Write $r = r' \cdot [E/A]$; then the centralizer of E in $GL_r(\mathbb{A}_F^f)$ is isomorphic to $GL_{r'}(E \otimes_A \mathbb{A}_F^f)$. Lemma 2 implies that all elements of E are defined over some fixed finite extension of K. This means that an open subgroup of $\rho(\pi_1(X, \bar{\eta}))$ is contained in $GL_{r'}(E \otimes_A \mathbb{A}_F^f)$. Thus by Lemma 1 the same holds for a subgroup of finite index of Δ_Ξ. The following results can be deduced easily from Theorems 1, 2, and 3, using the same arguments as in [Pi97, end of §2].

Theorem 6. *In the situation of before Theorem 1, for $E := End_{\bar{\eta}_\mathbb{C}}(\varphi_{\bar{\eta}_\mathbb{C}})$ arbitrary, the closure in $GL_r(\mathbb{A}_F^f)$ of some subgroup of finite index of Δ_Ξ is an open subgroup of $SL_{r'}(E \otimes_A \mathbb{A}_F^f)$.*

Theorem 7. *In the situation of before Theorem 2, for $E := End_{K^{\mathrm{sep}}}(\varphi_\eta)$ arbitrary,*

(a) *some open subgroup of $\Gamma^{\mathrm{geom}} := \rho(\pi_1(X_{K^{\mathrm{sep}}}, \bar{\eta}))$ is an open subgroup of $SL_{r'}(E \otimes_A \mathbb{A}_F^f)$, and*

(b) *some open subgroup of $\Gamma := \rho(\pi_1(X, \bar{\eta}))$ is an open subgroup of $GL_{r'}(E \otimes_A \mathbb{A}_F^f)$.*

Theorem 8. *In the situation of before Theorem 3, for* $E := \mathrm{End}_{K^{\mathrm{sep}}}(\psi)$ *arbitrary, suppose that* ψ *cannot be defined over a finite extension of* F *inside* K^{sep}. *Then some open subgroup of* $\Gamma := \sigma(\mathrm{Gal}(K^{\mathrm{sep}}/K))$ *is an open subgroup of* $\mathrm{GL}_{r'}(E \otimes_A \mathbb{A}_F^f)$.

References

[An86]   G. Anderson, *t*-motives, *Duke Math. J.*, **53** (1986), 457–502.

[An92]   Y. André, Mumford-Tate groups of mixed Hodge structures and the theorem of the fixed part, *Compositio Math.*, **82** (1992), 1–24.

[Br]   F. Breuer, Special subvarieties of Drinfeld modular varieties, in preparation, 2005.

[Dr74]   V. G. Drinfeld, Elliptic modules, *Math. Sb.*, **94** (1974), 594–627 (in Russian); *Math. USSR-Sb.*, **23** (1974), 561–592 (in English).

[Go96]   D. Goss, *Basic Structures of Function Field Arithmetic*, Ergebnisse der Mathematik und ihrer Grenzgebiete 35, Springer-Verlag, Berlin, New York, 1996.

[Ha79]   D. R. Hayes, Explicit class field theory in global function fields, in *Studies in Algebra and Number Theory*, Advances in Mathematics: Supplementary Studies 6, Academic Press, New York, 1979, 173–217.

[He03]   G.-J. van der Heiden, *Weil Pairing and the Drinfeld Modular Curve*, Ph.D. thesis, Rijksuniversiteit Groningen, Groningen, The Netherlands, 2003.

[Lü74]   W. Lütkebohmert, Der Satz von Remmert-Stein in der nichtarchimedischen Funktionentheorie, *Math. Z.*, **139** (1974), 69–84.

[Pi97]   R. Pink, The Mumford-Tate conjecture for Drinfeld modules, *Publ. RIMS Kyoto Univ.*, **33** (1997), 393–425.

[Pi98]   R. Pink, Compact subgroups of linear algebraic groups, *J. Algebra*, **206** (1998), 438–504.

[Pi00]   R. Pink, Strong approximation for Zariski dense subgroups over arbitrary global fields, *Comm. Math. Helv.*, **75**-4 (2000), 608–643.

[Tag95]   Y. Taguchi, The Tate conjecture for *t*-motives, *Proc. Amer. Math. Soc.*, **123** ,(1995) 3285–3287.

[Tam95]   A. Tamagawa, The Tate conjecture and the semisimplicity conjecture for *t*-modules, *RIMS Kokyuroku Proc. RIMS*, **925** (1995), 89–94.

Irreducible Values of Polynomials: A Non-Analogy

Keith Conrad

Department of Mathematics
University of Connecticut
Storrs, CT 06269-3009
USA
kconrad@math.uconn.edu

1 Introduction

For a polynomial $f(T) \in \mathbb{Z}[T]$, the frequency with which the values $f(n)$ are prime has been considered since at least the eighteenth century. Euler observed, in a letter to Goldbach in 1752, that the sequence $n^2 + 1$ has many prime values for $1 \leq n \leq 1500$. Legendre assumed an arithmetic progression $an + b$ with $(a, b) = 1$ contains infinitely many primes in his work on the quadratic reciprocity law. There is also the old question of twin prime pairs n and $n + 2$, but we will focus here only on a *single* polynomial (in one variable).

An asymptotic estimate for

$$\pi_f(x) := |\{1 \leq n \leq x : f(n) \text{ is prime}\}| \tag{1.1}$$

as $x \to \infty$ amounts to a higher-degree generalization of the prime number theorem and Dirichlet's theorem. Conjectural estimates for $\pi_f(x)$ have been around since the work of Hardy and Littlewood in the early twentieth century, and this will be recalled later.

Hardy and Littlewood did not pursue a characteristic p version of this topic, but the framework is simple to set up. Let $\kappa[u]$ be the polynomial ring in one variable over a finite field κ. Given $f(T) = f(u, T)$ in $\kappa[u][T]$, how often is $f(g)$ irreducible in $\kappa[u]$ as g runs over $\kappa[u]$? We will see that it is trivial to translate Hardy and Littlewood's conjectural estimate for (1.1) into the setting of $\kappa[u]$, but a completely unexpected development will unfold: the Hardy–Littlewood conjecture in characteristic p is not always true! This discovery leads to new nontrivial theorems concerning polynomials over finite fields, and with these results the Hardy–Littlewood conjecture in characteristic p can be corrected. Moreover, the new understanding we gain in characteristic p leads to an interesting family of elliptic curves over $\kappa(u)$.

2 The classical situation

Given a nonconstant $f(T) \in \mathbb{Z}[T]$, there are two necessary conditions that f must satisfy in order for $f(n)$ to be prime infinitely often:

(1) $f(T)$ is irreducible in $\mathbb{Q}[T]$.
(2) No prime p divides $f(n)$ for every $n \in \mathbb{Z}$. (That is, for no p is the function $f : \mathbb{Z} \to \mathbb{Z}/(p)$ identically 0.)

The need for (1) is obvious. The role of (2) was first noticed by Bouniakowsky in 1854; it excludes examples (such as $T^2 - T + 2$) that are irreducible as polynomials yet have all values on $\mathbb{Z}$ containing a common prime factor (such as 2). We allow negative primes, so we do not require f to have a positive leading coefficient. Whereas (2) implies that $f(T)$ is primitive (i.e., its coefficients have no common factor), (2) is a strictly stronger condition than primitivity. We call (1) and (2) the *Bouniakowsky conditions*, and we consider the failure of (2) to be a local obstruction to the growth of $\pi_f(x)$. Condition (2) is equivalent to there being at least one pair of relatively prime values $f(m)$ and $f(n)$ for $m \neq n$, and this is how (2) is checked in practice.

Conjecture 2.1 (Bouniakowsky). For nonconstant $f(T) \in \mathbb{Z}[T]$, $f(n)$ is prime for infinitely many $n \in \mathbb{Z}$ if and only if conditions (1) and (2) hold.

Bouniakowsky's conjecture is known for f of degree 1, but no instance of it has been established when $\deg f > 1$.

The Hardy–Littlewood conjecture (also called the Bateman–Horn conjecture) makes the Bouniakowsky conjecture quantitative.

Conjecture 2.2 (Hardy–Littlewood). If $f(T) \in \mathbb{Z}[T]$ satisfies both (1) and (2), then

$$\pi_f(x) \overset{?}{\sim} C(f) \sum_{n \leq x} \frac{1}{\log |f(n)|} \sim \frac{C(f)}{\deg f} \frac{x}{\log x},$$

where $C(f) = \prod_p (1 - \omega_f(p)/p)/(1 - 1/p)$ and $\omega_f(p)$ is the number of solutions to $f(n) = 0$ in $\mathbb{Z}/(p)$.

Remark 2.3. When f is irreducible, $C(f) = 0$ if and only if one of its factors is 0, which is exactly when condition (2) fails. Assuming (1) and (2), the product $C(f)$ converges, although only conditionally when $\deg f > 1$. Rapidly convergent formulas for the Hardy–Littlewood constant $C(f)$ can be obtained from L-functions by writing $\omega_f(p)$ in terms of character values on a Frobenius element at p in the splitting field of f over $\mathbb{Q}$.

3 The characteristic p (non)analogue

Let κ be a finite field. We consider polynomials $f(T) = f(u, T)$ in $\kappa[u][T]$ that have positive T-degree. Let

$$\pi_f(n) := |\{g \in \kappa[u] : \deg g = n, \ f(g) \text{ is irreducible in } \kappa[u]\}|.$$

(One might consider a count over $\deg g \leq n$, rather than over $\deg g = n$, to be more analogous to the classical setting. If so, two points are worth noting: (i) the number of g with degree n grows exponentially with n, so sampling by degree is substantive, and (ii) the new phenomenon we will see later is essentially impossible to describe if we count by $\deg g \leq n$.)

In order for $f(g)$ to be irreducible infinitely often in $\kappa[u]$, the appropriate Bouniakowsky conditions must hold:

(1) $f(T)$ is irreducible in $\kappa(u)[T]$.
(2) There are no local obstructions: No irreducible π in $\kappa[u]$ divides $f(g)$ for every $g \in \kappa[u]$.

The following conjecture is the obvious analogue of Conjecture 2.2. We call it the Naive Conjecture. In many cases it fits numerical data well, but there are cases where the conjecture is wrong, so the "Naive" label is important.

Conjecture 3.1. Let κ have size q. When $f(T) \in \kappa[u][T]$ satisfies conditions (1) and (2),

$$\pi_f(n) \overset{?}{\sim} C(f) \sum_{\deg g=n} \frac{1}{\deg f(g)} \sim \frac{C(f)}{\deg_T f} \frac{(q-1)q^n}{n}$$

as $n \to \infty$, where $C(f) = \prod_{(\pi)}(1 - \omega_f(\pi)/N\pi)/(1 - 1/N\pi)$, $\omega_f(\pi)$ is the number of solutions to $f = 0$ in $\kappa[u]/(\pi)$, and $N\pi = |\kappa[u]/(\pi)| = q^{\deg \pi}$.

Remark 3.2. The convergence of $C(f)$ is proved just as in the classical case, and in particular depends on condition (2). Analogies between number fields and function fields suggest replacing $\deg f(g)$ in the denominator with

$$\log N(f(g)) = (\log q)(\deg f(g)).$$

Then $C(f)$ should be replaced with $(\log q)C(f)$ to maintain the same overall values on the right side. From the viewpoint of base-change properties, the product $(\log q)C(f)$ is in fact a better $\kappa[u]$-analogue of the classical Hardy–Littlewood constant than $C(f)$, and it is this product with $\log q$ which goes under the label $C(f)$ in [1].

When $\deg_T f = 1$, the Naive Conjecture is a theorem (an analogue of Dirichlet's theorem) and has been known for a long time. No case has been proved when $\deg_T f > 1$.

Numerical data when $\deg_T f > 1$, at first, present evidence in favor of the Naive Conjecture. But then we meet examples like those in the four tables below, which suggest the Naive Conjecture is *not* true in general. (In each table, the choice of $f(T)$ and κ is indicated along the top. The irreducibility of f over $\kappa(u)$ is left to the reader to check. Since $f(0)$ and $f(1)$ are relatively prime in $\kappa[u]$, the second Bouniakowsky condition is satisfied.)

In the headings of the tables, "Naive Est." means the expression just to the right of the $\sim^?$ in Conjecture 3.1, and "Ratio" means the ratio of the two sides of the $\sim^?$, which should be tending to 1 if the Naive Conjecture is true. The ratios do not seem to be tending to 1 according to the data in the tables. In Table 1, the ratios seem to tend to 2 for odd n and equal 0 for even n. In Table 2, the ratios seem to be tending to the periodic values 1,2,1,0. In Table 3, the ratios appear to be tending to a number ≈ 1.33. In Table 4, it looks like $\pi_f(n) = 0$ for $n > 0$. (Clearly $\pi_f(0) = 5$.)

Table 1. $T^4 + u$ over $\mathbb{F}_2[u]$

n	$\pi_f(n)$	Naive Est.	Ratio
9	24	14.2	1.690
10	0	25.6	0
11	92	46.5	1.978
12	0	85.3	0
13	336	157.5	2.133
14	0	292.6	0
15	1076	546.1	1.970
16	0	1024.0	0

Table 2. $T^3 + u$ over $\mathbb{F}_3[u]$

n	$\pi_f(n)$	Naive Est.	Ratio
9	1404	1458.0	0.963
10	7776	3936.6	1.975
11	10746	10736.2	1.001
12	0	29524.5	0
13	82140	81760.2	1.005
14	455256	227760.4	1.999
15	637440	637729.2	1.000
16	0	1793613.4	0

Table 3. $T^{12} + (u + 1)T^6 + u^4$ over $\mathbb{F}_3[u]$

n	$\pi_f(n)$	Naive Est.	Ratio
9	1624	1168.3	1.390
10	4228	3154.5	1.340
11	11248	8603.2	1.307
12	31202	23658.7	1.319
13	87114	65516.5	1.330
14	244246	182510.2	1.338
15	683408	511028.6	1.337
16	1914254	1437268.0	1.332

Table 4. $T^{10} + u$ over $\mathbb{F}_5[u]$

n	$\pi_f(n)$	Naive Est.	Ratio
1	0	4.0	0
2	0	10.0	0
3	0	33.3	0
4	0	125.0	0
5	0	500.0	0
6	0	2083.3	0
7	0	8928.6	0
8	0	12686.5	0
9	0	173611.1	0
10	0	781250.0	0
11	0	3551136.4	0
12	0	16276041.7	0
13	0	75120192.3	0
14	0	348772321.4	0
15	0	1627604166.7	0
16	0	7629394531.3	0

Unlike the classical case over $\mathbb{Z}$, the Bouniakowsky conditions (1) and (2) over $\kappa[u]$ are apparently *not* sufficient to guarantee that $f(T)$ takes infinitely many irreducible values in $\kappa[u]$. In fact, the Bouniakowsky conditions over $\kappa[u]$ are not sufficient to guarantee that $f(T)$ takes any irreducible values. For an example in any $\kappa[u][T]$, let $f(T) = T^{4q} + u^{2q-1}$, where q is the size of κ. This polynomial is irreducible in $\kappa(u)[T]$ and $f(0)$ and $f(1)$ are relatively prime, so the Bouniakowsky conditions are satisfied. However, it can be proved that $f(g)$ is reducible for every $g \in \kappa[u]$. (We will see a proof in Example 4.3.) This example in the case $q = 2$ was found by Swan [7] over 40 years ago, but in a different context. It seems that nobody noticed the connection to a failure of the Hardy–Littlewood conjecture (and even the Bouniakowsky conjecture) in characteristic p.

Our explanation for the unexpected examples in the tables (and others that are not given here, including polynomials $f(T)$ which are not monic in T) is a new global obstruction that has no known counterpart in characteristic 0. This is the topic of the next section.

4 Möbius fluctuations

We have found many examples that appear to deviate from the Naive Conjecture. These examples have two common properties:

(a) $f(T)$ is a polynomial in $\kappa[u][T^p]$, where p is the characteristic of κ.
(b) The sequence of ratios has interlaced limiting trends for $n \gg 0$, which fall into a cycle of one, two, or four limits. (See, respectively, Tables 3, 1, and 2.)

Not all polynomials in $\kappa[u][T^p]$ disagree (numerically) with the Naive Conjecture. For instance, $T^p + u^2$ over $\mathbb{F}_3[u]$, $\mathbb{F}_5[u]$, $\mathbb{F}_7[u]$, and $\mathbb{F}_9[u]$ appears to fit the Naive

Conjecture. We expect that the Naive Conjecture is correct if $f(T) \notin \kappa[u][T^p]$, but we have not proved anything in that direction.

A closer examination of the data behind the four tables in the previous section reveals a more subtle third common property:

(c') There is a Möbius bias: the nonzero values of $\mu(f(g))$, where μ is the Möbius function on $\kappa[u]$, are not ± 1 equally often.

The Möbius function on $\kappa[u]$ is defined by analogy with its classical counterpart: it vanishes on polynomials with a multiple irreducible factor and is ± 1 on square-free polynomials in accordance with the parity of the number of (monic) irreducible factors.

Let us explain the meaning of (c') through our four examples. In Table 1, we found numerically that $\mu(f(g)) = \mu(g^4 + u)$ is -1 when $\deg g$ is odd and is 1 when $\deg g$ is positive and even. In particular, if such a pattern persists, $g^4 + u$ must be reducible when $\deg g$ is positive and even since the Möbius value is not -1. In Table 2, we found numerically that $\mu(g^3 + u)$ is ± 1 equally often in each odd degree, while $\mu(g^3 + u) = -1$ for $\deg g \equiv 2 \bmod 4$ and $\mu(g^3 + u) = 1$ for $\deg g \equiv 0 \bmod 4$. In Table 3, we found numerically that $\mu(f(g))$ is -1 twice as often as it is 1 when sampling over g with a fixed degree ≥ 2. (While $\mu(f(g))$ can also vanish, the point is the apparent bias among nonzero values.) In Table 4, we found numerically that $\mu(f(g)) = 1$ when $\deg g > 0$. We can prove these numerically observed Möbius patterns are true in all degrees, as special cases of Theorem 4.4.

That biases in irreducibility statistics of $f(g)$ are linked to biases in nonzero values of $\mu(f(g))$ was our basic numerical discovery, but this link is a bit more subtle than the data so far suggest. Consider Table 5, where the ratio of irreducibility counts to the estimate coming from the Naive Conjecture seems to be approaching the limiting values 0, 1, 2, 1. In particular, the Naive Conjecture looks good in even degrees. Consistent with this, computations suggest $\mu(f(g))$ is equally often ± 1 in even degrees. (As before, Möbius value 0 is not taken into account.) However, though the Naive Conjecture looks bad in odd degrees (bad in different ways depending on the degree modulo 4), it appears from computations that $\mu(f(g))$ is still equally often ± 1 in odd degrees. It turns out that property (c') has to be amended (which is why it is called (c') and not (c)). This will be treated later.

In both $\mathbb{Z}$ and $\kappa[u]$, the definition of the Möbius function is useless for effective computations. But unlike the case over $\mathbb{Z}$, there is another formula for the Möbius function in $\kappa[u]$, and this does not require factoring.

Lemma 4.1. *When κ is a finite field with odd characteristic and $h \in \kappa[u]$ is nonzero,*

$$\mu(h) = (-1)^{\deg h} \chi(\operatorname{disc}_\kappa h). \tag{4.1}$$

Here χ is the quadratic character on $\kappa^\times$, with $\chi(0) = 0$, and $\operatorname{disc}_\kappa h$ is the discriminant of h.

Proof. If h has a multiple irreducible factor, then the result is obvious since both sides vanish. Assume h is separable. Both sides are multiplicative functions of h,

Table 5. $T^9 + (2u^2 + u)T^6 + (2u + 2)T^3 + u^2 + 2u + 1$ over $\mathbb{F}_3[u]$

n	$\pi_f(n)$	Naive Est.	Ratio
5	0	11.0	0
6	28	27.4	1.022
7	146	70.5	2.071
8	173	185.1	0.935
9	0	493.6	0
10	1345	1332.8	1.009
11	7348	3634.9	2.022
12	10138	9996.1	1.014
13	0	27681.4	0
14	77288	77112.5	1.002
15	432417	215915.0	2.003

so it suffices to check the case when $h = \pi$ is irreducible. Now (4.1) is equivalent to $\chi(\text{disc}_\kappa \pi) = (-1)^{\deg \pi - 1}$, which is easily checked using Galois theory of finite fields: the Frobenius over κ acts as a cycle of length $\deg \pi$ on the roots of π.   □

When κ has characteristic 2, there is a Möbius formula due to Swan [7], in terms of a characteristic 0 lifting of $\kappa[u]$ to $W(\kappa)[u]$, where $W(\kappa)$ is the Witt vectors of κ. We omit this formula. The special case when $\kappa = \mathbb{F}_2$ was described by Stickelberger at the first International Congress of Mathematicians in 1897 using a lift to $\mathbb{Z}[u]$ rather than a lift to $\mathbb{Z}_2[u]$.

Remark 4.2. When $\kappa = \mathbb{F}_p$ for $p \neq 2$ and h is squarefree in $\mathbb{F}_p[u]$, (4.1) can be rewritten as $(\frac{\text{disc}\, h}{p}) = (-1)^{n-r}$, where $n = \deg h$ and h has r distinct irreducible factors in $\mathbb{F}_p[u]$. This goes back to Pellet (1878) and is related to Stickelberger's formula $(\frac{\Delta}{p}) = (-1)^{n-r}$, where Δ is the discriminant of a number field of degree n in which p is unramified with r prime ideal factors.

Example 4.3. Let q be the size of κ and $f(T) = T^{4q} + u^{2q-1}$. For $g \in \kappa[u]$, clearly $f(g)$ is reducible when $g(0) = 0$; in fact, $\mu(f(g)) = 0$. When q is odd and $g(0) \neq 0$, a calculation using (4.1) shows $\mu(f(g)) = 1$. When q is even and $g(0) \neq 0$, it can also be shown that $\mu(f(g)) = 1$. Since $\mu(f(g))$ is never -1, we see that $f(g)$ is reducible for every $g \in \kappa[u]$. This is an example where the Bouniakowsky conditions hold but no irreducible values occur.

By a substantial extension of Swan's ideas in [7], and motivated by our numerical data, we proved the surprising fact that $\mu(f(g))$ is essentially a periodic function of g if $f(T) \in \kappa[u][T]$ is a polynomial in T^p when $p \neq 2$ or is a polynomial in T^4 when $p = 2$. (When $p = 2$, the case of polynomials in T^2 which are not polynomials in T^4 still has not been completely understood.)

Theorem 4.4 ([1]). *Let κ be a finite field with characteristic p. Let $f(T)$ be squarefree in $\kappa[u][T]$ with positive T-degree. Assume, moreover, that $f(T)$ is a polynomial in*

T^p when $p \neq 2$ and is a polynomial in T^4 when $p = 2$. When $p \neq 2$, let χ be the quadratic character on $\kappa^\times$.

There is a nonzero $M = M_{f,\kappa}$ in $\kappa[u]$ such that for $g_1 = c_1 u^{n_1} + \cdots$ and $g_2 = c_2 u^{n_2} + \cdots$ in $\kappa[u]$ with sufficiently large degrees n_1 and n_2,

$$g_1 \equiv g_2 \bmod M, \qquad n_1 \equiv n_2 \bmod 4, \qquad \chi(c_1) = \chi(c_2) \Longrightarrow \mu(f(g_1)) = \mu(f(g_2))$$

when $p \neq 2$ and

$$g_1 \equiv g_2 \bmod M, \qquad n_1 \equiv n_2 \bmod 4 \Longrightarrow \mu(f(g_1)) = \mu(f(g_2))$$

when $p = 2$.

Proof (sketch). Very briefly, the proof of Theorem 4.4 requires a careful study of resultants.

According to (4.1), $\mu(f(g))$ depends on the discriminant of $f(g)$ when $p \neq 2$. The discriminant of $f(g)$ can be expressed in terms of the resultant of $f(g)$ and $(d/du)(f(g)) = (\partial f/\partial u)(g)$. (This derivative calculation indicates why $f(T) = f(u, T)$ being a polynomial in $\kappa[u, T^p]$ is useful in the proof.) In order to exploit inductive arguments, we replace the study of the resultant $R(f(g), (\partial f/\partial u)(g))$ with $R(f_1(g), f_2(g))$, where f_1 and f_2 are fairly general polynomials in $\kappa[u, T]$. There are properties of resultants which resemble the properties of greatest common divisors, and this suggests a method for computing $R(f_1(g), f_2(g))$ by a procedure analogous to the Euclidean algorithm. However, a moment's thought about the difference between, say, $R(u^2+1, u^3+u+1)$ and $R(ug^2+(u+1)g+1, g^4+u^2g+u)$ for varying g in $\kappa[u]$ indicates why a proof that $R_{\kappa[u]}(f_1(g), f_2(g))$ has a periodic structure in g does not follow right away from any kind of basic elementary property of resultants for one-variable polynomials.

To correctly handle the varying polynomial g, we view the resultant of $f_1(g)$ and $f_2(g)$ as an algebraic function of g. This requires a combination of polynomial algebra and algebraic geometry, and is the main content of [1]. We study the geometry of the zero-scheme of $R(f_1(g), f_2(g))$ on the space of polynomials g with a fixed degree in order to get a formula for this resultant function in terms of the geometry of the intersections of the plane curves $f_1 = 0$ and $f_2 = 0$. (Recall f_1 and f_2 are in $\kappa[u][T] = \kappa[u, T]$.) This geometric formula for the resultant has the asserted periodicity by inspection. (The case of characteristic 2 has its own set of complications.)

The mod 4 congruence in the conclusion of the theorem has a simple explanation. It is essentially due to the fact that the discriminant of a polynomial of degree n picks up a sign of $(-1)^{n(n-1)/2}$ when written in terms of a resultant, and this sign depends on n mod 4.   □

Remark 4.5. The Bouniakowsky conditions (1) and (2) are irrelevant for $f(T)$ in Theorem 4.4. In particular, the hypotheses on $f(T)$ in Theorem 4.4 are preserved when κ is replaced by a finite extension, but the first Bouniakowsky condition does not have to remain true under a finite extension of κ (for the same f).

The following two examples illustrate Theorem 4.4.

Example 4.6. For $f(T)$ as in Table 1 and $g \in \mathbb{F}_2[u]$ with $\deg g \geq 1$, $\mu(f(g)) = (-1)^{\deg g}$. Here $M = 1$ and the mod 4 condition in the characteristic 2 case of Theorem 4.4 can be relaxed to a mod 2 condition.

Example 4.7. For $f(T)$ as in Table 5, and $g(u) = cu^n + \cdots$ in $\mathbb{F}_3[u]$ with $n \geq 2$, the proof of Theorem 4.4 leads to the formula

$$\mu(f(g)) = (-1)^{n(n+1)/2} \left(\frac{c}{3}\right)^{n+1} \left(\frac{g(1)^2 + g(1) + 2}{3}\right) \left(\frac{g(2)}{3}\right), \qquad (4.2)$$

where $(\frac{\cdot}{3})$ is a Legendre symbol. All of the conditions from Theorem 4.4 are seen in (4.2): $M = (u - 1)(u - 2)$, there is a mod 4 dependence on $n = \deg g$, and there is a quadratic dependence on the leading coefficient c of g.

Furthermore, (4.2) lets us prove $\mu(f(g))$ takes values 1 and -1 equally often in every degree. This means that the deviations from the Naive Conjecture in Table 5, in odd degrees, are apparently not "explained" by the distribution of nonzero values of $\mu(f(g))$ in odd degrees. But a closer look at (4.2) reveals something peculiar in odd degrees: when $\deg g \equiv 1 \mod 4$, $\mu(f(g))$ is -1 only when $f(g)$ is divisible by $u - 1$ or $u - 2$. Therefore $f(g)$ will not be irreducible even in the case that $\mu(f(g)) = -1$. Similarly, when $\deg g \equiv 3 \mod 4$, $\mu(f(g))$ is 1 only when $f(g)$ is divisible by $u - 1$ or $u - 2$. In short, if $\mu(f(g))$ is nonzero and $\deg g$ is odd, the sign of $\mu(f(g))$ is fixed when $(f(g), M) = 1$, where $M = (u - 1)(u - 2)$ is the "modulus" from (4.2). Classically, one would not expect a nonconstant $f(T) \in \mathbb{Z}[T]$ to have the long-range statistics on $\mu(f(n))$ be affected by a local constraint of the form $(f(n), m) = 1$ for some $m \in \mathbb{Z}$. But this can happen in characteristic p.

We now revise the incorrect property (c') from the start of this section, by using $M_{f,\kappa}$ from Theorem 4.4:

(c) There is a Möbius bias: the nonzero values of $\mu(f(g))$ are not ± 1 equally often when $(f(g), M_{f,\kappa}) = 1$.

The idea in (c) will be used later to correct the Naive Conjecture.

Although Theorem 4.4 does not pin down a unique choice of $M_{f,\kappa}$, it turns out that all possible choices of $M_{f,\kappa}$ are multiples of a choice with least degree. Therefore the choice of $M_{f,\kappa}$ with least degree and leading coefficient 1 could be considered a "canonical" selection. However, it is important for the proof of the full version of Theorem 4.4, as stated in [1], that we can always choose $M_{f,\kappa}$ in a geometric manner, which is not always a choice with least degree. We describe this geometric construction in characteristic $\neq 2$ for simplicity, first in a special case and then in general:

- When $f(T)$ is monic as a polynomial in T, $M_{f,\kappa}$ is the radical of the resultant of f and $\partial f/\partial u$ as polynomials in T.
- In the general case, when $f(T)$ is not necessarily monic in T, view f as a polynomial in the two variables u and T, and let Z_f be the zero locus of f in $\mathbb{A}^2_\kappa$. The projection $Z_f \to \mathbb{A}^1_\kappa$ onto the T-axis has a finite nonétale locus on Z_f, and its projection onto the u-axis is a finite set. Let $M_{f,\kappa}$ be the separable (monic) polynomial in $\kappa[u]$ having this subset of the u-axis as its zero locus.

Example 4.8. For $f(T)$ as in Example 4.7, the resultant of f and $\partial f/\partial u$ as polynomials in T is $-(u-1)^6(u-2)^9$, whose radical is $(u-1)(u-2)$. This agrees with the "modulus" for $\mu(f(g))$ according to (4.2).

Definition 4.9. *For f as in Theorem* 4.4 *and satisfying the second Bouniakowsky condition, define*

$$\Lambda_\kappa(f;n) := 1 - \frac{\sum_{\deg g=n,(f(g),M_{f,\kappa})=1} \mu(f(g))}{\sum_{\deg g=n,(f(g),M_{f,\kappa})=1} |\mu(f(g))|}. \tag{4.3}$$

The denominator sum in (4.3) is the number of g with degree n such that $f(g)$ is squarefree and relatively prime to $M_{f,\kappa}$. By work of Poonen [5] on squarefree values and relatively prime values of polynomials, this denominator is positive for $n \gg 0$. Clearly $0 \le \Lambda_\kappa(f;n) \le 2$. While there is not a unique choice for $M_{f,\kappa}$ in Theorem 4.4, the choice used in (4.3) has no impact on the long-range behavior of $\Lambda_\kappa(f;n)$: two different choices of $M_{f,\kappa}$ from Theorem 4.4 provably give sequences in (4.3) that agree for $n \gg 0$ (depending on f and κ and the choice of the Ms).

Corollary 4.10 ([1]). *Let $f(T)$ satisfy the hypotheses of Theorem* 4.4 *and the second Bouniakowsky condition. For $n \gg 0$, $\Lambda_\kappa(f;n)$ is periodic in n with period* 1, 2, *or* 4.

Proof (sketch). This follows from a careful evaluation of the formula for $\mu(f(g))$ which is established in the proof of Theorem 4.4 (taking separately $p \ne 2$ and $p = 2$). It turns out that $\Lambda_\kappa(f;n)$ depends on $n \bmod 4$, so its minimal period as a function of n is 1, 2, or 4. $\qquad\square$

The periodicity in Corollary 4.10 shows $\Lambda_\kappa(f;n)$ is a much simpler function of n than its definition suggests! In any particular example, we can use the proof of Theorem 4.4 to compute the periodic part of the sequence $\Lambda_\kappa(f;n)$. As Table 6 shows, when $f(T)$ is one of the polynomials from the previous tables, the periodic part of the sequence $\Lambda_\kappa(f;n)$ appears to fit the deviations from the Naive Conjecture for $\pi_f(n)$.

Table 6. Examples of $\Lambda_\kappa(f;n)$

Table for $f(T)$	$M_{f,\kappa}$	Periodic Part of $\Lambda_\kappa(f;n)$
Table 1	1	2,0 for $n \ge 1$
Table 2	1	1,2,1,0 for $n \ge 1$
Table 3	$u(u-1)$	4/3 for $n \ge 2$
Table 4	1	0 for $n \ge 1$
Table 5	$(u-1)(u-2)$	0,1,2,1 for $n \ge 1$

We believe the following are correct $\kappa[u]$-analogues of the conjectures of Bouniakowsky and Hardy–Littlewood over $\mathbb{Z}$.

Conjecture 4.11 ([1]). Let κ have characteristic $p \neq 2$, and let $f \in \kappa[u][T]$ have positive degree in T. Then $f(g)$ is irreducible for infinitely many g in $\kappa[u]$ if and only if the following conditions hold:

(1) $f(T)$ is irreducible in $\kappa(u)[T]$.
(2) No irreducible π in $\kappa[u]$ divides $f(g)$ for every $g \in \kappa[u]$.
(3) $f(T) \notin \kappa[u][T^p]$, or if $f(T) \in \kappa[u][T^p]$, then the periodic part of the sequence $\Lambda_\kappa(f; n)$ is not identically 0.

As in the classical case, the second condition in this conjecture is checked in practice by finding a pair of relatively prime values $f(g_1)$ and $f(g_2)$. The third condition can also be checked in practice. When κ has characteristic 2, we believe Conjecture 4.11 is correct if $f(T) \notin \kappa[u][T^2]$ or if $f(T) \in \kappa[u][T^4]$. The case of polynomials in T^2 that are not polynomials in T^4 still needs further study.

Here is a quantitative refinement of Conjecture 4.11, incorporating part of the characteristic 2 case.

Conjecture 4.12 ([1]). Let $f \in \kappa[u][T]$ satisfy the two Bouniakowsky conditions. Let $p = \mathrm{char}(\kappa)$. If $f(T) \notin \kappa[u][T^p]$, then the asymptotic relation in the Naive Conjecture is true. If $f(T)$ is a polynomial in T^p when $p \neq 2$ or $f(T)$ is a polynomial in T^4 when $p = 2$, then

$$\pi_f(n) \sim \Lambda_\kappa(f; n) \frac{C(f)}{\deg_T f} \frac{(q-1)q^n}{n} \tag{4.4}$$

as $n \to \infty$.

As in Theorem 4.4, we do not yet have a complete formulation of Conjecture 4.12 in characteristic 2, since the behavior of polynomials in T^2 that are not in T^4 is still not adequately understood.

The periodicity of $\Lambda_\kappa(f; n)$ is essential for a proper understanding of (4.4). When 0 is in the period of $\Lambda_\kappa(f; n)$, the meaning of (4.4) is that $\pi_f(n)$ equals 0 for those (large) periodic n where $\Lambda_\kappa(f; n) = 0$. In fact, it is easy to prove this: if $\Lambda_\kappa(f; n) = 0$, then for all g of degree n we have either $(f(g), M_{f,\kappa}) \neq 1$ or $\mu(f(g)) \in \{0, 1\}$. Thus, for $n \gg 0$ (depending on $M_{f,\kappa}$), the vanishing of $\Lambda_\kappa(f; n)$ implies $\pi_f(n) = 0$. In this way, by making the condition "$n \gg 0$" effective in specific examples, we can prove the 0 patterns in Tables 1, 2, 4, and 5 continue for all larger n. We have not proved a relation between Möbius statistics and irreducibility statistics for those periodic (large) n where $\Lambda_\kappa(f; n) \neq 0$, but the data in these cases agree very well with (4.4).

We said at the start of this section that some polynomials in T^p appear to fit the Naive Conjecture numerically, such as $T^p + u^2$ over $\mathbb{F}_3[u]$, $\mathbb{F}_5[u]$, $\mathbb{F}_7[u]$, and $\mathbb{F}_9[u]$. If the Naive Conjecture and (4.4) are going to be compatible, then any polynomial in T^p (for $p \neq 2$) that satisfies the Bouniakowsky conditions and agrees with the Naive Conjecture must have $\Lambda_\kappa(f; n) = 1$ for all large n. This conclusion has been confirmed in several examples of polynomials in T^p (for $p \neq 2$) where the Naive Conjecture appears to look good, e.g., we can prove $\Lambda_\kappa(T^p + u^2; n) = 1$ for $n \geq 1$ and κ any finite field of characteristic $p \neq 2$.

The ring $\kappa[u]$ corresponds to the affine line over κ. Theorem 4.4 can be extended to the coordinate ring of any smooth affine curve over κ whose smooth compactification has only one geometric point "∞" at infinity. The substitute for the sampling condition "$\deg g = n$" is "$\operatorname{ord}_\infty(g) = -n$." From this point of view, Theorem 4.4 corresponds to genus zero. The proof of the higher-genus generalization uses the work in genus zero as input, and requires additional arguments of a much more elaborate geometric character. Numerical aspects of this work are still in progress.

5 An application to elliptic curves

Having found a Möbius periodicity that is a global obstruction to the Naive Conjecture in characteristic p, we ask: why does no analogous obstruction arise over $\mathbb{Z}$? The belief in the classical Hardy–Littlewood conjecture suggests that the $\mathbb{Z}$-analogue of the new characteristic p correction factor in (4.3) is 1. This suggests the following: if $f(T) \in \mathbb{Z}[T]$ is a nonconstant (irreducible) polynomial taking at least one squarefree value, then

$$\frac{\sum_{n \le x} \mu(f(n))}{\sum_{n \le x} |\mu(f(n))|} \to 0 \tag{5.1}$$

as $n \to \infty$. The denominator of (5.1) is the number of squarefree values $f(n)$ for $n \le x$. Granting the *abc*-conjecture, work of Granville [3] shows the denominator of (5.1) is proportional to x, so (5.1) should be equivalent to

$$\frac{\sum_{n \le x} \mu(f(n))}{x} \to 0. \tag{5.2}$$

(The equivalence of (5.1) and (5.2) is unconditional when $\deg f \le 3$; the *abc*-conjecture is used for $\deg f > 3$.) When $f(T) = T$, (5.2) is equivalent to the prime number theorem. For other f of degree 1, (5.2) is equivalent to Dirichlet's theorem [6]. No other case of (5.2) has been proved. Numerical evidence for (5.2) when $\deg f > 1$ looks reasonable. An example in degree 3 is provided in Table 7.

Table 7. (5.2) for $f(T) = T^3 + 2T + 1$

x	$\frac{1}{x}\sum_{n \le x} \mu(f(n))$
10^2	$-.15$
10^3	$-.015$
10^4	$-.0009$
10^5	$.00432$
10^6	$.00028$

The Ph.D. thesis of H. Helfgott [4] gives a link between a variant on (5.2) and elliptic curves. Helfgott studies the *average root number* of an elliptic curve over

$\mathbb{Q}(T)$, which is essentially the average value of the root number of the smooth spe-cializations at $T = t \in \mathbb{P}^1(\mathbb{Q})$, with t ordered by height. This average lies in $[-1, 1]$ if it exists. Assuming two conjectures from analytic number theory about values of polynomials over $\mathbb{Z}$, Helfgott shows that the average root number of any nonisotrivial elliptic curve over $\mathbb{Q}(T)$ exists and lies strictly between -1 and 1. (When the elliptic curve has at least one place of multiplicative reduction, Helfgott can in fact prove the average is 0.) One of the two conjectures Helfgott assumes is

$$\frac{1}{x^2} \sum_{m,n \leq x} \lambda(f(m, n)) \to 0, \tag{5.3}$$

where $f(X, Y) \in \mathbb{Z}[X, Y]$ is a nonconstant nonsquare homogeneous polynomial and λ is the classical Liouville function. (Recall that $\lambda(\pm p) = -1$ for prime p and λ is totally multiplicative, e.g., $\lambda(12) = -1$.) Considering the similarity of the Möbius and Liouville functions, (5.3) bears a close resemblance to (5.2).

The natural analogue of the conjectural (5.3) for polynomials with coefficients in $\kappa[u]$ rather than $\mathbb{Z}$ is false: explicit counterexamples can be constructed from certain instances of unusual Möbius statistics in characteristic p. Might this imply that some of Helfgott's results over $\mathbb{Q}(T)$ are not true over $\kappa(u)(T)$? Yes. The following theorem says nonisotrivial elliptic curves over $\kappa(u)(T)$ with average root number 1 *do* exist in odd characteristic, with an additional interesting feature.

Theorem 5.1 ([2, Theorem 1.1]). *Let κ be any finite field with characteristic $p \neq 2$. For any $c, d \in \kappa^\times$, the Weierstrass model*

$$E_{c,d,T} : y^2 = x^3 + (c(T^2 + u)^{2p} + du)x^2 - (c(T^2 + u)^{2p} + du)^3 x \tag{5.4}$$

defines a nonisotrivial elliptic curve over $\kappa(u)(T)$ such that

(a) *for every $t \in \mathbb{P}^1(\kappa(u))$, the specialization $E_{c,d,t}$ is an elliptic curve over $\kappa(u)$ having global root number 1, and for $t \neq \infty$ there is a $\kappa(u)$-rational point of infinite order;*

(b) *the Mordell–Weil group $E_{c,d,T}(\kappa(u)(T))$ has rank 1.*

The key word in Theorem 5.1 is "nonisotrivial." Helfgott's work strongly suggests that a nonisotrivial elliptic curve over $\mathbb{Q}(T)$ should not have elevated rank. (We say an elliptic curve over $\mathbb{Q}(T)$ has *elevated rank* when the rank of all but finitely many of its specializations to elliptic curves over $\mathbb{Q}$ exceeds the rank over $\mathbb{Q}(T)$.) Granting the parity conjecture for elliptic curves over $\kappa(u)$, Theorem 5.1(a) implies the rank of $E_{c,d,t}(\kappa(u))$ is positive and even for all $t \in \mathbb{P}^1(\kappa(u)) - \{\infty\}$, so each $E_{c,d,T}$ is nonisotrivial and should have elevated rank over $\kappa(u)(T)$.

Proof (sketch). An explicit calculation of the j-invariant of $E_{c,d,T}$ shows it is non-isotrivial. (Moreover, $j(E_{c,d,T}) = j(E_{c',d',T})$ if and only if $c = c'$ and $d = d'$.)

To verify part (a), the elliptic curve over $\kappa(u)$ obtained by specialization $T \mapsto t$ for any $t \in \mathbb{P}^1(\kappa(u))$ has global root number 1 based on an analysis of local reduction types and a calculation of all the local root numbers. (It is within these local root number calculations, which are carried out in detail in [2], that one sees how we found (5.4) in the first place. This Weierstrass model was not discovered by random guessing.) We use a function field variant of the Nagell–Lutz criterion to check an explicit rational point in $E_{c,d,T}(\kappa(u)(T))$ has infinite order and retains infinite order after specialization of T to any $t \in \mathbb{P}^1(\kappa(u)) - \{\infty\}$.

The proof of part (b) amounts to showing $E_{c,d,T}(\kappa(u)(T))$ has rank at most 1. (Part (a) already tells us the rank is at least 1.) The 2-torsion is $\langle(0, 0)\rangle \cong \mathbb{Z}/2\mathbb{Z}$, so

$$\dim_{\mathbb{F}_2}(E_{c,d,T}(\kappa(u, T))/2 \cdot E_{c,d,T}(\kappa(u, T))) = 1 + r,$$

with r being the rank. We show this dimension is at most 2 by a specialization in the u-direction rather than the T-direction.

Abbreviate $E_{c,d,T}$ to $\mathscr{E}$. For any closed point $u_0 \in \mathbb{P}^1_\kappa$, with residue field κ_0 (varying with u_0), consider the natural commutative diagram

$$
\begin{array}{ccc}
\mathscr{E}(\kappa(u, T))/2 \cdot \mathscr{E}(\kappa(u, T)) & \longrightarrow & \mathscr{E}(\overline{\kappa}(u, T))/2 \cdot \mathscr{E}(\overline{\kappa}(u, T)) \\
\downarrow & & \downarrow \\
\mathscr{E}_{u_0}(\kappa_0(T))/2 \cdot \mathscr{E}_{u_0}(\kappa_0(T)) & \longrightarrow & \mathscr{E}_{\overline{u}_0}(\overline{\kappa}(T))/2 \cdot \mathscr{E}_{\overline{u}_0}(\overline{\kappa}(T))
\end{array}
$$

where the elliptic curve $\mathscr{E}_{u_0/\kappa_0(T)}$ is the u-specialization at u_0, and $\overline{u}_0 \in \overline{\kappa}$ lies over u_0.

The Lang–Néron theorem tells us $\mathscr{E}(\overline{\kappa}(u, T)) = \mathscr{E}(\kappa(u, T))$ when κ is replaced by a suitable finite extension. We make this enlargement of κ at the beginning of the proof of part (b), so we may take the top map to be an isomorphism. The enlargement of κ might depend on the choice of parameters c and d. Since part (a) has already been checked in full generality (i.e., for all finite constant fields), we may apply it to the new elliptic curve under consideration.

The proof now falls into two parts. Geometric and ramification-theoretic arguments (applicable not just to $\mathscr{E}_{/\kappa(u,T)}$, but to abelian varieties over function fields of varieties fibered over $\mathbb{P}^1$) show that the map along the right column is one-to-one for all but finitely many $\overline{u}_0$. An application of the Chebotarev density theorem and a calculation in étale cohomology show that the map along the bottom side has image with dimension at most 2 for infinitely many u_0. Therefore, by a suitable choice of u_0, we get $1 + r \leq 2$. $\square$

Although Theorem 5.1 does not include characteristic 2, further work should remove this exception.

It required several months of effort to find the curve in Theorem 5.1 and confirm all of its properties, but we never would have had the intuition that such an elliptic curve could exist in characteristic p if the investigation of a function field analogue of the classical Hardy–Littlewood conjecture had not revealed the peculiarities of the Möbius function in characteristic p. Thus, while the topic of this paper is a non-analogy between $\mathbb{Q}$ and $\kappa(u)$, the analogies between these fields provided useful insights during our work.

Acknowledgments. This paper is a summary of joint work. The results pertaining to the Hardy–Littlewood conjecture in characteristic p are joint work with B. Conrad and R. Gross [1]. The application to elliptic curves is joint work with B. Conrad and H. Helfgott [2]. Full proofs and other details can be found in the references cited.

I thank the organizers of the conference on the analogy between number fields and function fields for a stimulating week on Texel Island.

References

[1] B. Conrad, K. Conrad, and R. Gross, Irreducible specialization in genus 0, submitted.

[2] B. Conrad, K. Conrad, and H. A. Helfgott, Root numbers and ranks in positive characteristic, *Adv. Math.*, to appear.

[3] A. Granville, *ABC* allows us to count squarefrees, *Internat. Math. Res. Notices*, **19** (1998), 991–1009.

[4] H. A. Helfgott, *Root Numbers and the Parity Problem*, Ph.D. thesis, Princeton University, Princeton, NJ, 2003.

[5] B. Poonen, Squarefree values of multivariable polynomials, *Duke Math. J.*, **118** (2003), 353–373.

[6] H. N. Shapiro, Some assertions equivalent to the prime number theorem for arithmetic progressions, *Comm. Pure Appl. Math.*, **2** (1949), 293–308.

[7] R. G. Swan, Factorization of polynomials over finite fields, *Pacific J. Math.*, **12** (1962), 1099–1106.

Schemes over $\mathbb{F}_1$

Anton Deitmar

Mathematisches Institut
Universität Tübingen
Auf der Morgenstelle 10
72076 Tübingen
Germany
deitmar@uni-tuebingen.de

Introduction

Jacques Tits wondered in [5] if there would exist a "field of one element" $\mathbb{F}_1$ such that for a Chevalley group G one has $G(\mathbb{F}_1) = W$, the Weyl group of G. Recall the Weyl group is defined as $W = N(T)/Z(T)$ where T is a maximal torus, $N(T)$ and $Z(T)$ are the normalizer and the centralizer of T in G. He then showed that one would be able to define a finite geometry with W as automorphism group.

In this paper we will extend the approach of N. Kurokawa, H. Ochiai, and M. Wakayama [2] to "absolute Mathematics" to define schemes over the field of one element. The basic idea of the approach of [2] is that objects over $\mathbb{Z}$ have a notion of $\mathbb{Z}$-linearity, i.e., additivity, and that the forgetful functor to $\mathbb{F}_1$-objects therefore must forget about additive structures. A ring R for example is mapped to the multiplicative monoid $(R, \times)$. On the other hand, the theory also requires a "going up" or base extension functor from objects over $\mathbb{F}_1$ to objects over $\mathbb{Z}$. Using the analogy of the finite extensions $\mathbb{F}_{1^n}$ as in [4], we are led to define the base extension of a monoid A as

$$A \otimes_{\mathbb{F}_1} \mathbb{Z} := \mathbb{Z}[A],$$

where $\mathbb{Z}[A]$ is the monoidal ring which is defined in the same way as a group ring. Based on these two constructions, here we lay the foundations of a theory of schemes over $\mathbb{F}_1$.

1 Rings over $\mathbb{F}_1$

In this paper, a ring will always be commutative with unit element. Recall that a monoid is a set A with an associative composition that has a unit element. Homomorphisms of monoids are required to preserve units. In this paper all monoids will be commutative, so $aa' = a'a$ for all $a, a' \in A$. From now on in the rest of the paper the word 'monoid' will always mean 'commutative monoid'. For a monoid A we will write $A^\times$ for the group of invertible elements.

The category of rings over $\mathbb{F}_1$ is by definition the category of monoids. For a monoid A we will also write $\mathbb{F}_A$ to emphasize that we view A as a ring over $\mathbb{F}_1$. Let $\mathbb{F}_1 := \{1\}$ be the trivial monoid.

For an $\mathbb{F}_1$-ring $\mathbb{F}_A$, we define the *base extension to* $\mathbb{Z}$ by

$$\mathbb{F}_A \otimes \mathbb{Z} = \mathbb{F}_A \otimes_{\mathbb{F}_1} \mathbb{Z} = \mathbb{Z}[A],$$

where $\mathbb{Z}[A]$ is the monoidal ring which is defined like a group ring with the monoid structure of A giving the multiplication.

In the other direction there is the forgetful-functor F which maps a ring R (commutative with 1) to its multiplicative monoid $(R, \times)$.

Theorem 1.1. *The functor of base extension* $\cdot \otimes_{\mathbb{F}_1} \mathbb{Z}$ *is left adjoint to F, i.e., for every ring R and every $\mathbb{F}_A/\mathbb{F}_1$, we have*

$$\mathrm{Hom}_{\mathrm{Rings}}(\mathbb{F}_A \otimes_{\mathbb{F}_1} \mathbb{Z}, R) \cong \mathrm{Hom}_{\mathbb{F}_1}(\mathbb{F}_A, F(R)).$$

Proof. Let φ be a ring homomorphism from $\mathbb{F}_A \otimes_{\mathbb{F}_1} \mathbb{Z} = \mathbb{Z}[A]$ to R. Restricting it to A yields a monoid morphism from A to $(R, \times)$. So we get a map as in the theorem. Since a ring homomorphism from $\mathbb{Z}[A]$ is uniquely given by the restriction to A, this map is injective. Since, on the other hand, every monoid morphism from A to $(R, \times)$ extends to a ring homomorphism on $\mathbb{Z}[A]$, the claim follows.   □

1.1 Localization

For a monoid A write $A^\times$ for the group of invertible elements. Let S be a submonoid of A.

Lemma 1.2. *There is a monoid $S^{-1}A$ and a monoid homomorphism φ from A to $S^{-1}A$, determined up to isomorphism with the following property: $\varphi(S) \subset (S^{-1}A)^\times$ and φ is universal with this property, i.e., for every monoid B, composing with φ yields an isomorphism*

$$\mathrm{Hom}_{S \to B^\times}(A, B) \cong \mathrm{Hom}(S^{-1}A, B),$$

where the left-hand side describes the set of all monoid homomorphisms φ from A to B with $\varphi(S) \subset B^\times$.

Proof. Uniqueness is clear from the universal property. We show existence. Define $S^{-1}A$ to be the set $A \times S$ modulo the equivalence relation

$$(m, s) \sim (m', s') \Leftrightarrow \exists s'' \in S : s''s'm = s''sm'.$$

The multiplication in $S^{-1}A$ is given by $(m, s)(m', s') = (mm', ss')$. We also write $\frac{m}{s}$ for the element $[(m, s)]$ in $S^{-1}A$. The map $\varphi : m \mapsto \frac{m}{1}$ has the desired property.   □

1.2 Ideals and spectra

For two subsets S, T of a monoid A we write ST for the set of all st where $s \in S$ and $t \in T$. An *ideal* is a subset $\mathfrak{a}$ such that $\mathfrak{a}A \subset \mathfrak{a}$. This then implies that $\mathbb{Z}[\mathfrak{a}]$ is an ideal in $\mathbb{Z}[A]$. If $\varphi \colon \mathbb{F}_A \to \mathbb{F}_B$ is a morphism and $\mathfrak{a}$ is an ideal of B, then its preimage $\varphi^{-1}(\mathfrak{a})$ is an ideal of A. For a given subset T of A the set TA is the smallest ideal containing T. We call it the ideal *generated by* T.

An ideal $\mathfrak{a} \neq A$ is called *prime* if $xy \in \mathfrak{a}$ implies $x \in \mathfrak{a}$ or $y \in \mathfrak{a}$. Equivalently, an ideal $\mathfrak{a}$ is prime iff $A \setminus \mathfrak{a}$ is a submonoid (compare [1]). We define the *spectrum* of $\mathbb{F}_A$ to be the set $\mathrm{Spec}\,\mathbb{F}_A$ of all prime ideals in A. Note that this set is never empty, as the set $A \setminus A^\times$ is always a prime ideal.

If $\mathfrak{p}$ is a prime ideal, then the set $S_\mathfrak{p} = A \setminus \mathfrak{p}$ is a submonoid. We define

$$A_\mathfrak{p} := S_\mathfrak{p}^{-1} A$$

to be the *localization at* $\mathfrak{p}$. Note that for the prime ideal $c = A \setminus A^\times$ the natural map $A \to A_c$ is an isomorphism.

We now introduce a topology on $\mathrm{Spec}\,\mathbb{F}_A$. The closed subsets are the empty set and all sets of the form

$$V(\mathfrak{a}) := \{\mathfrak{p} \in \mathrm{Spec}\,\mathbb{F}_A : \mathfrak{p} \supset \mathfrak{a}\},$$

where $\mathfrak{a}$ is any ideal. One checks that $V(\mathfrak{a}) \cup V(\mathfrak{b}) = V(\mathfrak{a} \cap \mathfrak{b})$, and that $\bigcap_{i \in I} V(\mathfrak{a}_i) = V(\bigcup_i \mathfrak{a}_i)$. Thus the axioms of a topology are satisfied. The point $\eta = \eta_A = \emptyset$ is open and contained in every nonempty open set. On the other hand, the point $c = c_A = A \setminus A^\times$ is closed and contained in every nonempty closed set.

Lemma 1.3. *For every* $f \in A$ *the set*

$$V(f) := \{\mathfrak{p} \in \mathrm{Spec}\,\mathbb{F}_A : f \in \mathfrak{p}\}$$

is closed.

Proof. Let $\mathfrak{a} = Af$ be the ideal generated by f. Then $V(f) = V(\mathfrak{a})$. $\square$

2 Schemes over $\mathbb{F}_1$

2.1 The structure sheaf

Let $\mathbb{F}_A$ be a ring over $\mathbb{F}_1$. On the topological space $\mathrm{Spec}\,\mathbb{F}_A$ we define a sheaf of $\mathbb{F}_1$-rings as follows. For an open set $U \subset \mathrm{Spec}\,\mathbb{F}_A$ we define $\mathcal{O}(U)$ to be the set of functions, called *sections*, $s \colon U \to \coprod_{\mathfrak{p} \in U} A_\mathfrak{p}$ such that $s(\mathfrak{p}) \in A_\mathfrak{p}$ for each $\mathfrak{p} \in U$, and such that s is locally a quotient of elements of A. This means that we require for each $\mathfrak{p} \in U$ to exist a neighborhood V of $\mathfrak{p}$, contained in U, and elements $a, f \in A$ such that for every $\mathfrak{q} \in V$ one has $f \notin \mathfrak{q}$ and $s(\mathfrak{q}) = \frac{a}{f}$ in $A_\mathfrak{q}$.

Proposition 2.1. (a) *For each* $\mathfrak{p} \in \operatorname{Spec} \mathbb{F}_A$ *the stalk* $\mathcal{O}_\mathfrak{p}$ *of the structure sheaf is isomorphic to the localization* $A_\mathfrak{p}$.

(b) $\Gamma(\operatorname{Spec} \mathbb{F}_A, \mathcal{O}) \cong A$.

Proof. For (a) define a morphism φ from $\mathcal{O}_\mathfrak{p}$ to $A_\mathfrak{p}$ by sending each element (s, U_s) of $\mathcal{O}_\mathfrak{p}$ to $s(\mathfrak{p})$. For the injectivity assume $\varphi(s) = \varphi(s')$. On some neighborhood U of $\mathfrak{p}$, we have $s(\mathfrak{q}) = \frac{a}{f}$ and $s'(\mathfrak{q}) = \frac{a'}{f'}$ for some $a, a', f, f' \in A$. This implies that there is $f'' \in A$ with $f'' \notin \mathfrak{p}$ and $f'' f' = f'' f a'$. Assume U to be small enough to be contained in the open set

$$D(f) = \{\mathfrak{p} \in \operatorname{Spec} \mathbb{F}_A : f \notin \mathfrak{p}\}.$$

Then we conclude that $\frac{a}{f} = \frac{a'}{f'}$ holds in $A_\mathfrak{q}$ for every $\mathfrak{q} \in U$ and hence $s = s'$. For the surjectivity let $\frac{a}{f} \in A_\mathfrak{p}$ with $a, f \in A$, $f \notin \mathfrak{p}$. On $U = D(f)$ define a section $s \in \mathcal{O}(U)$ by $s(\mathfrak{q}) = \frac{a}{f} \in A_\mathfrak{q}$. Then $\varphi(s) = s(\mathfrak{p}) = \frac{a}{f}$. Part (a) is proven.

For part (b) note that the natural map $A \to A_\mathfrak{p}$ for $\mathfrak{p} \in \operatorname{Spec} \mathbb{F}_A$ induces a map $\psi: A \to \Gamma(\operatorname{Spec} \mathbb{F}_A, \mathcal{O})$. We want to show that this map is bijective. For the injectivity assume $\psi(a) = \psi(a')$. Then, in particular, these sections must coincide on the closed point which implies $a = a'$. For the surjectivity let s be a global section of $\mathcal{O}$. For the closed point c, we get an element of A by $a = s(c) \in A_c = A$. We claim that $s = \psi(a)$. Since c is contained in every nonempty closed set, its only open neighborhood is $\operatorname{Spec} \mathbb{F}_A$ itself. Since $s(c) = a$ we must have $s(\mathfrak{p}) = a$ on some neighborhood of c, hence we have it on $\operatorname{Spec} \mathbb{F}_A$, so $s = \psi(a)$. □

2.2 Monoidal spaces

A *monoidal space* is a topological space X together with a sheaf $\mathcal{O}_X$ of monoids. A *morphism of monoidal spaces* $(X, \mathcal{O}_X) \to (Y, \mathcal{O}_Y)$ is a pair $(f, f^\#)$, where f is a continuous map $f: X \to Y$ and $f^\#$ is a morphism of sheaves $f^\#: \mathcal{O}_Y \to f_* \mathcal{O}_X$ of monoids on Y. Such a morphism $(f, f^\#)$ is called *local*, if for each $x \in X$ the induced morphism $f_x^\#: \mathcal{O}_{Y, f(x)} \to \mathcal{O}_{X,x}$ satisfies

$$(f_x^\#)^{-1} \left(\mathcal{O}_{X,x}^\times \right) = \mathcal{O}_{Y, f(x)}^\times.$$

A *isomorphism of monoidal spaces* is a morphism with a two-sided inverse. An isomorphism is always local.

Proposition 2.2. (a) *For a ring* $\mathbb{F}_A$ *over* $\mathbb{F}_1$ *the pair* $(\operatorname{Spec} \mathbb{F}_A, \mathcal{O}_A)$ *is a monoidal space.*

(b) *If* $\varphi: A \to B$ *is a morphism of monoids, then* φ *induces a morphism of monoidal spaces*

$$(f, f^\#): \operatorname{Spec} \mathbb{F}_B \to \operatorname{Spec} \mathbb{F}_A,$$

thus giving a functorial bijection

$$\operatorname{Hom}(A, B) \cong \operatorname{Hom}(\operatorname{Spec} \mathbb{F}_B, \operatorname{Spec} \mathbb{F}_A),$$

where on the right-hand side one only admits local morphisms.

Proof. Part (a) is clear. For (b) suppose we are given a morphism $\varphi: A \to B$. Define a map $f: \operatorname{Spec} \mathbb{F}_B \to \operatorname{Spec} \mathbb{F}_A$ that maps $\mathfrak{p}$ to the prime ideal $f(\mathfrak{p}) = \varphi^{-1}(\mathfrak{p})$. For an ideal $\mathfrak{a}$ we have $f^{-1}(V(\mathfrak{a})) = V(\varphi(\mathfrak{a}))$, where $\varphi(\mathfrak{a})$ is the ideal generated by the image $\varphi(\mathfrak{a})$. Thus the map f is continuous. For every $\mathfrak{p} \in \operatorname{Spec} \mathbb{F}_B$ we localize φ to get a morphism $\varphi_{\mathfrak{p}}: A_{\varphi^{-1}(\mathfrak{p})} \to B_{\mathfrak{p}}$. Since $\varphi_{\mathfrak{p}}$ is the localization, it satisfies $\varphi_{p}^{-1}(B_{\mathfrak{p}}^{\times}) = A_{\varphi^{-1}(\mathfrak{p})}^{\times}$. For any open set $U \subset \operatorname{Spec} \mathbb{F}_A$ we obtain a morphism

$$f^{\#}: \mathcal{O}_A(U) \to \mathcal{O}_B(f^{-1}(U))$$

by the definition of $\mathcal{O}$, composing with the maps f and φ. This gives a local morphism of local monoidal spaces $(f, f^{\#})$. We have constructed a map

$$\psi: \operatorname{Hom}(A, B) \to \operatorname{Hom}(\operatorname{Spec} \mathbb{F}_B, \operatorname{Spec} \mathbb{F}_A).$$

We have to show that ψ is bijective. For injectivity suppose $\psi(\varphi) = \psi(\varphi')$. For $\mathfrak{p} \in \operatorname{Spec} \mathbb{F}_A$ the morphism $f_{\mathfrak{p}}^{\#}: \mathcal{O}_{A, f(\mathfrak{p})} \to \mathcal{O}_{B, \mathfrak{p}}$ is the natural localization $\varphi: A_{\varphi^{-1}(\mathfrak{p})} \to B_{\mathfrak{p}}$ and this coincides with the localization of φ'. In particular, for $\mathfrak{p} = c$ the closed point, we get $\varphi = \varphi': A \to B$.

For surjectivity let $(f, f^{\#})$ be a morphism from $\operatorname{Spec} \mathbb{F}_B$ to $\operatorname{Spec} \mathbb{F}_A$. On global sections the map $f^{\#}$ gives a monoid morphism

$$\varphi: A = \mathcal{O}_A(\operatorname{Spec} \mathbb{F}_A) \to f_* \mathcal{O}_B(\operatorname{Spec} \mathbb{F}_B) = \mathcal{O}_B(\operatorname{Spec} \mathbb{F}_B) = B.$$

For every $\mathfrak{p} \in \operatorname{Spec} \mathbb{F}_B$ one has an induced morphism on the stalks $\mathcal{O}_{A, f(\mathfrak{p})} \to \mathcal{O}_{B, \mathfrak{p}}$ or $A_{f(\mathfrak{p})} \to A_{\mathfrak{p}}$ which must be compatible with φ on global sections, so we have a commutative diagram

$$
\begin{array}{ccc}
A & \xrightarrow{\ \varphi\ } & B \\
\downarrow & & \downarrow \\
A_{f(\mathfrak{p})} & \xrightarrow{\ f_{\mathfrak{p}}^{\#}\ } & B_{\mathfrak{p}}
\end{array}
$$

Since $f(\mathfrak{p}) = (f_{\mathfrak{p}}^{\#})^{-1}(\mathfrak{p})$ it follows that $f_{\mathfrak{p}}^{\#}$ is the localization of φ and hence the claim. The last part follows since the first bijection preserves isomorphisms by functoriality and an isomorphism on the spectral side preserves closed points and can thus be extended to an isomorphism of the full spectra. $\square$

2.3 Schemes

An *affine scheme over* $\mathbb{F}_1$ is a monoidal space which is isomorphic to $\operatorname{Spec} \mathbb{F}_A$ for some A.

Lemma 2.3. *Every open subset of an affine scheme is a union of affine schemes.*

Proof. Let $U \subset \operatorname{Spec} \mathbb{F}_A$ be open. Then there is an ideal $\mathfrak{a}$ such that

$$U = D(\mathfrak{a}) = \{\mathfrak{p} \in \operatorname{Spec} \mathbb{F}_A : \mathfrak{p} \not\supset \mathfrak{a}\}.$$

So U is the set of all $\mathfrak{p}$ such that there exists $f \in \mathfrak{a}$ with $f \notin \mathfrak{p}$. For any $f \in A$, let

$$D(f) := \{\mathfrak{p} \in \operatorname{Spec} \mathbb{F}_A : f \notin \mathfrak{p}\}.$$

Then we get

$$U = \bigcup_{f \in \mathfrak{a}} D(f).$$

Let $A_f = f^{-1}A = S_f^{-1}A$ where $S_f = \{1, f, f^2, f^3, \dots\}$. One checks that the open set $D(f)$ is affine, more precisely,

$$(D(f), \mathcal{O}_A|_{D(f)}) \cong \operatorname{Spec} \mathbb{F}_{A_f}.$$

The lemma follows.   □

A monoidal space X is called a *scheme over* $\mathbb{F}_1$, if for every point $x \in X$ there is an open neighborhood $U \subset X$ such that $(U, \mathcal{O}_X|_U)$ is an affine scheme over $\mathbb{F}_1$. A *morphism* of schemes over $\mathbb{F}_1$ is a local morphism of monoidal spaces.

A point η of a topological space such that $\{\eta\}$ is open and η is contained in every nonempty open set, is called a *generic point*. It it exists, it is unique.

Proposition 2.4. *Every connected scheme over $\mathbb{F}_1$ has a generic point. Morphisms on connected schemes map generic points to generic points. If for an arbitrary scheme X over $\mathbb{F}_1$ we define*

$$X(\mathbb{F}_1) := \operatorname{Hom}(\operatorname{Spec} \mathbb{F}_1, X),$$

then we get

$$X(\mathbb{F}_1) \cong \pi_0(X),$$

where the right-hand side is the set of connected components of X.

Proof. Let X be connected. Every affine subscheme U has a generic point η_U. If U and V are affine with $U \cap V \neq \emptyset$, then $\eta_V = \eta_U$ by the uniqueness. Fix an open affine subset U. By Zorn's Lemma, there is a maximal open subset $V \supset U$ with η_U as a generic point. This set V must coincide with X. The rest is clear.   □

Let X be a scheme over $\mathbb{F}_1$ and let $(f, f^\#)$ be a morphism from X to an affine scheme $\operatorname{Spec} \mathbb{F}_A$. Taking global sections the sheaf morphism $f^\# : \mathcal{O}_A \to f_*\mathcal{O}_X$ induces a morphism $\varphi : A \to \mathcal{O}_X(X)$ of monoids.

Proposition 2.5. *The map $\psi : (f, f^\#) \mapsto \varphi$ is a bijection*

$$\operatorname{Hom}(X, \operatorname{Spec} \mathbb{F}_A) \to \operatorname{Hom}(A, \Gamma(X, \mathcal{O}_X)).$$

Proof. Let $\psi(f, f^\#) = \varphi$. For each $\mathfrak{p} \in X$ one has a local morphism $f_{\mathfrak{p}}^\# : \mathcal{O}_{A, f(\mathfrak{p})} \to \mathcal{O}_{X,\mathfrak{p}}$. Via the map $\mathcal{O}_X(X) \to \mathcal{O}_{X,\mathfrak{p}}$ the point $\mathfrak{p}$ induces an ideal $\tilde{\mathfrak{p}}$ on $\mathcal{O}_X(X)$ and $f(\mathfrak{p}) = \varphi^{-1}(\tilde{\mathfrak{p}})$. So f is determined by φ. Further, since $f_{\mathfrak{p}}^\#$ factorizes over $A_{f(\mathfrak{p})} = A_{\tilde{\mathfrak{p}} \circ \varphi} \to \mathcal{O}_X(X)_{\tilde{\mathfrak{p}}}$ it follows that $f_{\mathfrak{p}}^\#(\frac{a}{s}) = \frac{\varphi(a)}{\varphi(s)}$ and so $f^\#$ also is determined by φ, so ψ is injective. For surjectivity reverse the argument.   □

The forgetful functor from Rings to Monoids mapping R to $(R, \times)$ extends to a functor

$$\text{Schemes}/\mathbb{Z} \to \text{Schemes}/\mathbb{F}_1$$

in the following way: A scheme X over $\mathbb{Z}$ can be written as a union of affine schemes $X = \bigcup_{i \in I} \operatorname{Spec} A_i$ for some rings A_i. We then map it to $\bigcup_{i \in I} \operatorname{Spec}(A_i, \times)$, where we use the gluing maps from X.

Likewise, the base change $A \mapsto A \otimes_{\mathbb{F}_1} \mathbb{Z}$ extends to a functor

$$\text{Schemes}/\mathbb{F}_1 \to \text{Schemes}/\mathbb{Z}$$

by writing a scheme X over $\mathbb{F}_1$ as a union of affine ones, $X = \bigcup_{i \in I} \operatorname{Spec} A_i$ and then mapping it to $\bigcup_{i \in I} \operatorname{Spec}(A_i \otimes_{\mathbb{F}_1} \mathbb{Z})$, which is glued via the gluing maps from X. The fact that these constructions do not depend on the choices of affine coverings follows from Proposition 2.2, Lemma 2.3 and its counterpart for schemes over $\mathbb{Z}$.

As an example of a scheme which is not affine let us construct the projective line $\mathbb{P}^1$ over $\mathbb{F}_1$. Let $C_\infty = \{\dots, \tau^{-1}, 1, \tau, \dots\}$ be the infinite cyclic group with generator τ. Let $C_{\infty,+} = \{1, \tau, \tau^2, \dots\}$ and $C_{\infty,-} = \{1, \tau^{-1}, \tau^{-2}, \dots\}$. The inclusions give maps from $U = \operatorname{Spec} C_\infty$ to $X = \operatorname{Spec} C_{\infty,+}$ and $Y = \operatorname{Spec} C_{\infty,-}$ identifying U with open subsets of the latter. We define a new space $\mathbb{P}^1$ by gluing X and Y along this common open subset. The space X has two points, c_X, η_X, one closed and one open and likewise for Y. The space $\mathbb{P}^1$ has three points, c_X, c_Y, η, two closed and one open. The structure sheaves of X, Y, U give a structure sheaf of $\mathbb{P}^1$ making it a scheme over $\mathbb{F}_1$.

3 Fiber products

Let S be a scheme over $\mathbb{F}_1$. A pair (X, f_X) consisting of an $\mathbb{F}_1$-scheme X and a morphism $f_X \colon X \to S$ is called a *scheme over S*.

Proposition 3.1. *Let X, Y be schemes over S. There exists a scheme $X \times_S Y$ over S, unique up to S-isomorphism and morphisms from $X \times_S Y$ to X and Y such that the diagram*

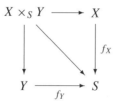

is commutative and the composition with these morphisms induces a bijection for every scheme Z over S,

$$\operatorname{Hom}_S(Z, X) \times \operatorname{Hom}_S(Z, Y) \to \operatorname{Hom}_S(Z, X \times_S Y).$$

This fiber product is compatible with $\mathbb{Z}$ extension and the usual fiber product of schemes, i.e., one has

$$(X \times_S Y) \otimes \mathbb{Z} \cong (X \otimes \mathbb{Z}) \times_{S \otimes \mathbb{Z}} (Y \otimes \mathbb{Z}).$$

Proof. This proposition follows via the gluing procedure once it has been established in the affine case. So let's assume that X, Y, S are all affine, say $X = \operatorname{Spec} A$, $Y = \operatorname{Spec} B$ and $S = \operatorname{Spec} L$. Then f_X and f_Y give monoid morphisms $\varphi_A : L \to A$ and $\varphi_B : L \to B$ and we define

$$X \times_S Y := \operatorname{Spec}(A \otimes_L B),$$

where $A \otimes_L B$ is the monoid $A \times B / \sim$ and $\sim$ is the equivalence relation generated by $(\varphi_A(l)m, n) \sim (m, \varphi_B(l)n)$ for all $m \in A$, $n \in B$, and $l \in L$. In other words, $A \otimes_L B$ is the push-out of φ_A and φ_B and so $X \times_S Y$ has the desired properties. $\square$

4 $\mathcal{O}_X$-modules

For an $\mathbb{F}_1$-ring $\mathbb{F}_A$, one defines an $\mathbb{F}_A$-*module* to be a set M together with an action $A \times M \to M$, $(a, m) \mapsto am$ such that $1s = s$ and $(aa')s = a(a's)$ for every $s \in S$ and all $a, a' \in A$. One defines $M \otimes \mathbb{Z} = \mathbb{Z}[M]$ with the obvious $\mathbb{F}_A \otimes \mathbb{Z}$-module structure. The direct sum $M \oplus N$ of A-modules is the disjoint union and the tensor product $M \otimes_A N$ is the quotient $M \times N / \sim$, where $\sim$ is the equivalence relation generated by $(am, n) \sim (m, an)$ for all $a \in A, m \in M, n \in N$. Then $M \otimes_A N$ is an A-module via $a[m, n] = [am, n]$. There are natural isomorphisms of $\mathbb{F}_A \otimes \mathbb{Z}$-modules

$$(M \oplus N) \otimes \mathbb{Z} \cong (M \otimes \mathbb{Z}) \oplus (N \otimes \mathbb{Z}),$$

$$(M \otimes_A N) \otimes \mathbb{Z} \cong (M \otimes \mathbb{Z}) \otimes_{A \otimes \mathbb{Z}} (N \otimes \mathbb{Z}).$$

Let $(X, \mathcal{O}_X)$ be a monoidal space. We define an $\mathcal{O}_X$-*module* to be a sheaf $\mathcal{F}$ of sets on X together with the structure of an $\mathcal{O}_X(U)$-module on $\mathcal{F}(U)$ for each open set $U \subset X$ such that for open sets $V \subset U \subset X$ the restriction $\mathcal{F}(U) \to \mathcal{F}(V)$ is compatible with the module structure via the map $\mathcal{O}_X(U) \to \mathcal{O}_X(V)$. If $\mathcal{F}$ is an $\mathcal{O}_X$-module and $U \subset X$ is open, then $\mathcal{F}|_U$ is an $\mathcal{O}_U = \mathcal{O}_X|_U$-module. A *morphism* of $\mathcal{O}_X$-modules $\varphi : \mathcal{F} \to \mathcal{G}$ is a morphism of sheaves such that for every open set $U \subset X$ the map $\varphi(U) : \mathcal{F}(U) \to \mathcal{G}(U)$ is a morphism of $\mathcal{O}_X(U)$-modules. The category of $\mathcal{O}_X$-modules has kernels, cokernels, images, and internal Hom's, as the presheaf $U \mapsto \operatorname{Hom}_{\mathcal{O}_X(U)}(\mathcal{F}(U), \mathcal{G}(U))$ is a sheaf called the *sheaf Hom* for any two given $\mathcal{O}_X$-modules $\mathcal{F}$ and $\mathcal{G}$. This then is naturally an $\mathcal{O}_X$-module. The *tensor product* $\mathcal{F} \otimes_{\mathcal{O}_X} \mathcal{G}$ of two $\mathcal{O}_X$-modules $\mathcal{F}, \mathcal{G}$ is the sheaf associated to the presheaf $U \mapsto \mathcal{F}(U) \otimes_{\mathcal{O}_X(U)} \mathcal{G}(U)$. Note that $\mathcal{F} \otimes \mathcal{O}_X \cong \mathcal{F}$. An $\mathcal{O}_X$-module is *free* if it is isomorphic to a direct sum of copies of $\mathcal{O}_X$. It is *locally free* if every $x \in X$ has an open neighborhood U such that $\mathcal{F}|_U$ is free. In this case the *rank* of $\mathcal{F}$ at a point x is the number of copies of $\mathcal{O}_X$ needed over any open neighborhood of x over which $\mathcal{F}$ is free. If X is connected, then the rank is the same everywhere.

For a given $\mathcal{O}_X$-module $\mathcal{F}$ let $\mathcal{F}^* := \operatorname{Hom}(\mathcal{F}, \mathcal{O}_X)$ be its *dual* module. There is a canonical morphism, called the *trace*,

$$\operatorname{tr} : \mathcal{F} \otimes \mathcal{F}^* \to \mathcal{O}_X,$$

given over an open set U by mapping $f \otimes \alpha \in \mathcal{F}(U) \otimes_{\mathcal{O}_X(U)} \mathcal{F}^*(U)$ to $\alpha(f) \in \mathcal{O}_X(U)$.

Lemma 4.1. *If $\mathcal{F}$ is locally free of rank 1, then the trace is an isomorphism. So the set of isomorphism classes of locally free $\mathcal{O}_X$-modules of rank 1 forms a group, called the* Picard *group* $\mathrm{Pic}(X)$ *of X.*

Proof. Clear. □

Proposition 4.2. *Let X be an affine scheme over $\mathbb{F}_1$, then every locally free sheaf is free, so, in particular, $\mathrm{Pic}(X)$ is trivial.*

Proof. There is a unique closed point c which is contained in every nonempty closed set, so the only open neighborhood of c is the full space X, which implies that every locally free sheaf is free. □

Proposition 4.3. *The Picard group of the projective line $\mathbb{P}^1$ is isomorphic to $\mathbb{Z}$.*

Proof. The space $\mathbb{P}^1$ has three elements, c_X, c_Y, and η. The nontrivial open sets are $U = \{\eta\}$, $X = \{\eta, c_X\}$, and $Y = \{\eta, c_y\}$. We have $\mathcal{O}(X) \cong C_{\infty,+}$, $\mathcal{O}(Y) \cong C_{\infty,-}$ and $\mathcal{O}(U) \cong C_\infty$ with the inclusions as restriction maps. Since X and Y are affine, a given invertible sheaf $\mathcal{F}$ is trivial on X and on Y. We fix isomorphisms $\alpha \colon \mathcal{F}|_X \to \mathcal{O}_X$ and $\beta \colon \mathcal{F}|_Y \to \mathcal{O}_Y$. The restriction of α and β gives two isomorphisms $\mathcal{F}(U) \to \mathcal{O}(U) = C_\infty$. These two differ by a C_∞-module automorphism of C_∞. The group of these automorphisms is isomorphic to $\mathbb{Z}$. It is easy to see that this establishes the claimed isomorphism. □

Let $f \colon X \to Y$ be a morphism of $\mathbb{F}_1$-ringed spaces. If $\mathcal{F}$ is an $\mathcal{O}_X$-module, then $f_*\mathcal{F}$ is an $f_*\mathcal{O}_X$-module. The morphism $f^\# \colon \mathcal{O}_Y \to f_*\mathcal{O}_X$ thus makes $f_*\mathcal{F}$ into an $\mathcal{O}_Y$-module, called the *direct image* of $\mathcal{F}$.

For an $\mathcal{O}_Y$-module $\mathcal{G}$ the sheaf $f^{-1}\mathcal{G}$ is an $f^{-1}\mathcal{O}_Y$-module. Recall that the functor f^{-1} is adjoint to f_*, this implies that

$$\mathrm{Hom}_X(f^{-1}\mathcal{O}_Y, \mathcal{O}_X) \cong \mathrm{Hom}_Y(\mathcal{O}_Y, f_*\mathcal{O}_X).$$

So the map $f^\# \colon \mathcal{O}_Y \to f_*\mathcal{O}_X$ gives a map $f^{-1}\mathcal{O}_Y \to \mathcal{O}_X$. We define $f^*\mathcal{G}$ to be the tensor product

$$f^{-1}\mathcal{G} \otimes_{f^{-1}\mathcal{O}_Y} \mathcal{O}_X.$$

So $f^*\mathcal{G}$ is an $\mathcal{O}_X$-module, called the *inverse image* of $\mathcal{G}$. The functors, f_* and f^* are adjoint in the sense that

$$\mathrm{Hom}_{\mathcal{O}_X}(f^*\mathcal{G}, \mathcal{F}) \cong \mathrm{Hom}_{\mathcal{O}_Y}(\mathcal{G}, f_*\mathcal{F}).$$

4.1 Localization

Let M be a module of the monoid A. Let $S \subset A$ be a submonoid. We define the localization $S^{-1}M$ to be the following module of $S^{-1}A$. As a set, $S^{-1}M$ is the set of all pairs $(m, s) = \frac{m}{s}$ with $m \in M$ and $s \in S$ modulo the equivalence relation

$$\frac{m}{s} \sim \frac{m'}{s'} \Leftrightarrow \exists s'' \in S : s''s'm = s''sm'.$$

The $S^{-1}A$-module structure is given by

$$\frac{a}{s} \cdot \frac{m}{s'} = \frac{am}{ss'}.$$

A given A-module homomorphism $\varphi \colon M \to N$ induces an $S^{-1}A$-module homomorphism $S^{-1}\varphi \colon S^{-1}M \to S^{-1}N$ by $S^{-1}\varphi(\frac{m}{s}) = \frac{\varphi(m)}{s}$. Note that $S^{-1}\varphi$ is injective/surjective if φ is.

Given an A-module M we define an $\mathcal{O}_{\mathbb{F}_A}$-sheaf $\tilde{M}$ on $X = \operatorname{Spec} \mathbb{F}_A$ as follows. For each prime ideal $\mathfrak{p}$ of A let $M_{\mathfrak{p}}$ be the localization $S_{\mathfrak{p}}^{-1}M$ at $\mathfrak{p}$. For any open set $U \subset \operatorname{Spec} \mathbb{F}_A$ we define the set $\tilde{M}(U)$ to be the set of functions $s \colon U \to \sqcup_{\mathfrak{p} \in U} M_{\mathfrak{p}}$ such that $s(\mathfrak{p}) \in M_{\mathfrak{p}}$ for each $\mathfrak{p}$ and such that s is locally a fraction, i.e., we require that for each $\mathfrak{p} \in U$ there is a neighborhood $V \subset U$ of $\mathfrak{p}$ and $m \in M$ as well as $f \in A$ with $f \notin \mathfrak{q}$ for every $\mathfrak{q} \in V$ and $s(\mathfrak{q}) = \frac{m}{f}$ in $m_{\mathfrak{q}}$. Then $\tilde{M}$ is a sheaf with the obvious restriction maps. It is an $\mathcal{O}_X$-module. For each $\mathfrak{p} \in X$ the stalk $(\tilde{M})_{\mathfrak{p}}$ coincides with the localization $M_{\mathfrak{p}}$ at $\mathfrak{p}$. For every $f \in M$ the A_f-module $\tilde{A}(D(f))$ is isomorphic to the localized module M_f. In particular, we have $\Gamma(X, \tilde{M}) \cong M$. One also has $(M \otimes_A N)^{\sim} \cong \tilde{M} \times_{\mathcal{O}_X} \tilde{N}$. For a morphism $f \colon \operatorname{Spec} B \to \operatorname{Spec} A$ one has $f_* \tilde{M} \cong (_A M)^{\sim}$, where $_A M$ is M considered as an A-module via the map $A \to B$ induced by f. Finally, we have $f^*(\tilde{N}) \cong (B \otimes_A N)^{\sim}$.

4.2 Coherent modules

Let $(X, \mathcal{O}_X)$ be a scheme over $\mathbb{F}_1$. An $\mathcal{O}_X$-module $\mathcal{F}$ is called *coherent* if every $x \in X$ has an affine neighborhood $U \cong \operatorname{Spec} \mathbb{F}_A$ such that $\mathcal{F}|_U$ is isomorphic to $\tilde{M}$ for some A-module M.

Proposition 4.4. *Let X be a scheme over $\mathbb{F}_1$. An $\mathcal{O}_X$-module $\mathcal{F}$ is coherent if and only if for every open affine subset $U = \operatorname{Spec} A$ of X there is an A-module M such that $\mathcal{F}|_U \cong \tilde{M}$.*

Proof. Let $\mathcal{F}$ be coherent and let $U = \operatorname{Spec} A$ be an affine open subset. Let $X = \bigcup_i \operatorname{Spec} A_i$ be a covering by affines such that $\mathcal{F}|_{\operatorname{Spec} A_i} \cong \tilde{M}i$ for some modules M_i. Let $f \in A_i$. Then $\mathcal{F}|_{D(f)} \cong (M_{i,f})^{\sim}$. So X has a basis of the topology consisting of affines on which $\mathcal{F}$ comes from modules. It follows that $\mathcal{F}|_U$ also is coherent, so we can reduce to the case when X is affine. Then $\mathcal{F}$ must come from a module in a neighborhood of the closed point, so $\mathcal{F}$ comes from a module. $\square$

5 Chevalley groups

5.1 GL_n

We first repeat the definition of $\mathrm{GL}_n(\mathbb{F}_1)$ as in [2]. On rings the functor GL_n, which is a representable group functor, maps R to $\mathrm{GL}_n(R) = \operatorname{Aut}_R(R^n)$. To define $\mathrm{GL}_n(\mathbb{F}_1)$

we therefore have to define $\mathbb{F}_1^n$, or more generally, modules over $\mathbb{F}_1$. Since $\mathbb{Z}$-modules are just additive abelian groups, by forgetting the additive structure one simply gets sets. So $\mathbb{F}_1$-modules are sets. For an $\mathbb{F}_1$-module S and a ring R the base extension is the free R-module generated by S: $S \otimes_{\mathbb{F}_1} R = R[S] = R^{(S)}$. In particular, $\mathbb{F}_1^n \otimes \mathbb{Z}$ should be isomorphic to $\mathbb{Z}^n$, so $\mathbb{F}_1^n$ is a set of n elements, say $\mathbb{F}_1^n = \{1, 2, \ldots, n\}$. Hence

$$
\begin{aligned}
\mathrm{GL}_n(\mathbb{F}_1) &= \mathrm{Aut}_{\mathbb{F}_1}(\mathbb{F}_1^n) \\
&= \mathrm{Aut}(1, \ldots, n) \\
&= S_n,
\end{aligned}
$$

the symmetric group in n letters, which happens to be the Weyl group of GL_n. We now extend this to rings over $\mathbb{F}_1$. One would expect that $\mathbb{F}_A^n = \mathbb{F}_1^n \otimes \mathbb{F}_A = \{1, \ldots, n\} \times A$. So we define an $\mathbb{F}_A$-module to be a set with an action of the monoid A and, in particular, $\mathbb{F}_A^n$ is a disjoint union of n-copies of A. We define

$$
\mathrm{GL}_n(\mathbb{F}_A) := \mathrm{Aut}_{\mathbb{F}_A}(\mathbb{F}_A^n).
$$

This is compatible with $\mathbb{Z}$-extension,

$$
\begin{aligned}
\mathrm{GL}_n(\mathbb{F}_A \otimes \mathbb{Z}) &= \mathrm{Aut}_{\mathbb{F}_A \otimes \mathbb{Z}}(\mathbb{F}_A^n \otimes \mathbb{Z}) \\
&= \mathrm{Aut}_{\mathbb{Z}[A]}(\mathbb{Z}[A]^n) \\
&= \mathrm{GL}_n(\mathbb{Z}[A])
\end{aligned}
$$

as required.

Note that $\mathrm{GL}_n(\mathbb{F}_A)$ can be identified with the group of all $n \times n$ matrices A over $\mathbb{Z}[A]$ with exactly one nonzero entry in each row and each column and this entry being in the group of invertible elements $A^\times$.

Proposition 5.1. *The group functor* GL_1 *on* $\mathrm{Rings}\,/\mathbb{F}_1$ *is represented by the infinite cyclic group* C_∞. *This is compatible with* $\mathbb{Z}$-extension as

$$
\mathbb{F}_{C_\infty} \otimes \mathbb{Z} = \mathbb{Z}[C_\infty] \cong \mathbb{Z}[T, T^{-1}]
$$

represents GL_1 *on rings.*

Proof. Choose a generator τ of C_∞. For any ring $\mathbb{F}_A$ over $\mathbb{F}_1$ we have an isomorphism

$$
\mathrm{Hom}(\mathbb{F}_{C_\infty}, \mathbb{F}_A) \to A^\times = \mathrm{GL}_1(\mathbb{F}_A)
$$

mapping α to $\alpha(\tau)$. $\qquad\qquad\square$

The functor GL_n on rings over $\mathbb{F}_1$ cannot be represented by a ring $\mathbb{F}_A$ over $\mathbb{F}_1$ since $\mathrm{Hom}(\mathbb{F}_A, \mathbb{F}_1)$ has only one element.

Proposition 5.2. *The functor* GL_n *on rings over* $\mathbb{F}_1$ *is represented by a scheme over* $\mathbb{F}_1$.

Proof. We will give the proof for GL_2. The general case will be clear from that. For $\mathbb{F}_A/\mathbb{F}_1$ the group $GL_2(\mathbb{F}_A)$ can be identified with the group of matrices

$$\begin{pmatrix} A^\times & \\ & A^\times \end{pmatrix} \cup \begin{pmatrix} & A^\times \\ A^\times & \end{pmatrix}.$$

We define a scheme X over $\mathbb{F}_1$ as the disjoint union $X_0 \cup X_1$ and $X_0 \cong X_1 \cong$ $\mathrm{Spec}((C_\infty \times C_\infty)$, where C_∞ is the infinite cyclic group. The group structure on $\mathrm{Hom}(\mathrm{Spec}\,\mathbb{F}_A, X)$ for every $\mathbb{F}_A/\mathbb{F}_1$ comes via a multiplication map $m: X \times_{\mathbb{F}_1} X \to X$ defined in the following way. The scheme $X \times_{\mathbb{F}_1} X$ has connected components $X_i \times_{\mathbb{F}_1} X_j$ for $i, j \in \{0, 1\}$. The multiplication map splits into components $m_{i,j}: X_i \times_{\mathbb{F}_1} X_j \to X_{i+j(2)}$. Each $m_{i,j}$ in turn is given by a monoid morphism $\mu_{i,j}: C_\infty^2 \to C_\infty^2 \times C_\infty^2$, called the comultiplication. Here $\mu_{i,j}$ maps a to $(\varepsilon^i(a), \varepsilon^j(a))$, where $\varepsilon^0(a) = a$ and $\varepsilon^1(x, y) = (y, x)$. This finishes the construction and the proof. $\qquad\square$

5.2 O_n and Sp_{2n}

The orthogonal group O_n is the subgroup of GL_n consisting of all matrices A with $Aq A^t = q$, where

$$q = \mathrm{diag}(J, \ldots, J, 1) \quad \text{with} \quad J = \begin{pmatrix} & 1 \\ 1 & \end{pmatrix},$$

the last entry 1 in q occurring only if n is odd. A computation shows that

$$O_n(\mathbb{F}_1) \cong \text{Weyl group}$$

holds here as well. Finally Sp_{2l} is the group of all A with $ASA^t = S$, where S is the $2l \times 2l$ matrix with anti-diagonal $(1, \ldots, 1, -1, \ldots, -1)$ and zero elsewhere. Likewise, $Sp_{2l}(\mathbb{F}_1)$ is the Weyl group. Both O_n and Sp_{2l} are representable by $\mathbb{F}_1$-schemes.

6 Zeta functions

Let X be a scheme over $\mathbb{F}_1$. For $\mathbb{F}_A/\mathbb{F}_1$ we write, as usual,

$$X(\mathbb{F}_A) = \mathrm{Hom}(\mathrm{Spec}\,\mathbb{F}_A, X)$$

for the set of $\mathbb{F}_A$-valued points of X. After Weil we set for a prime p,

$$Z_X(p, T) := \exp\left(\sum_{n=1}^\infty \frac{T^n}{n} \#(X(\mathbb{F}_{p^n}))\right),$$

where, of course $\mathbb{F}_{p^n}$ means the field of p^n elements and $X(\mathbb{F}_{p^n})$ stands for $X((\mathbb{F}_{p^n}, \times))$. For this expression to make sense (even as a formal power series) we must assume that the numbers $\#(X(\mathbb{F}_{p^n}))$ are all finite.

Proposition 6.1. *The formal power series $Z_X(p, T)$ defined above coincides with the Hasse–Weil zeta function of $X \otimes_{\mathbb{F}_1} \mathbb{F}_p = X_{\mathbb{F}_p}$.*

Proof. This is an immediate consequence of Theorem 1.1.   □

This type of zeta function thus does not give new insights. Recall that to get a zeta function over $\mathbb{Z}$, one considers the product

$$Z_{X \otimes \mathbb{Z}}(s) = \prod_p Z_{X_{\mathbb{F}_p}}(p^{-s}) = \prod_p Z_X(p, p^{-s}).$$

As this product takes care of the fact that the prime numbers are the prime places of $\mathbb{Z}$, over $\mathbb{F}_1$ there is only one place, so there should be only one such factor. Soulé [4], inspired by Manin [3], provided the following idea: Suppose there exists a polynomial $N(x) \in \mathbb{Z}[x]$ such that, for every p one has $\#X(\mathbb{F}_{p^n}) = N(p^n)$. Then $Z_X(p, p^{-s})$ is a rational function in p and p^{-s}. One can then ask for the value of that function at $p = 1$. The (vanishing-) order at $p = 1$ of $Z_X(p, p^{-s})^{-1}$ is $N(1)$, so the following limit exists:

$$\zeta_X(s) := \lim_{p \to 1} \frac{Z_X(p, p^{-s})^{-1}}{(p - 1)^{N(1)}}.$$

One computes that if $N(x) = a_0 + a_1 x + \cdots + a_n x^n$, then

$$\zeta_X(s) = s^{a_0}(s - 1)^{a_1} \cdots (s - n)^{a_n}.$$

For example $X = \operatorname{Spec} \mathbb{F}_1$ gives

$$\zeta_{\operatorname{Spec} \mathbb{F}_1}(s) = s.$$

For the affine line $\mathbb{A}_1 = \operatorname{Spec} C_{\infty,+}$ one gets $N(x) = x$ and thus

$$\zeta_{\mathbb{A}_1}(s) = s - 1.$$

Finally, for GL_1 one gets

$$\zeta_{\operatorname{GL}_1}(s) = \frac{s}{s - 1}.$$

In our context the question must be if we can retrieve these zeta functions from the monoidal viewpoint without regress to the finite fields $\mathbb{F}_{p^n}$? In the examples it indeed turns out that

$$N(k) = \#X(\mathbb{F}_{D_k}),$$

where D_k is the monoid $C_{k-1} \cup \{0\}$ and C_{k-1} is the cyclic group with $k - 1$ elements. Since $(\mathbb{F}_{p^n}, \times) \cong D_{p^n}$ this comes down to the following question.

Question. *Let X be a scheme over $\mathbb{F}_1$. Assume there is a polynomial $N(x)$ with integer coefficients such that $\#X(D_{p^n}) = N(p^n)$ for every prime number p and every nonnegative integer n. Is it true that $\#X(D_k) = N(k)$ for every $k \in \mathbb{N}$? Or another question: Is there a natural characterization of the class of schemes X over $\mathbb{F}_1$ for which there exists a polynomial N_X with integer coefficients such that $\#X(D_k) = N_X(k)$ for every $k \in \mathbb{N}$?*

Acknowledgments. This paper was written during a stay at Kyushu University, Japan. I thank Masato Wakayama and his students for their inspiration and warm hospitality. I also thank Zoran Skoda for bringing the paper of Kato to my attention.

References

[1] K. Kato, Toric singularities, *Amer. J. Math.*, **116**-5 (1994), 1073–1099.

[2] B. Kurokawa, H. Ochiai, and A. Wakayama, Absolute derivations and zeta functions, *Documenta Math.*, **Extra Vol.** (Kazuya Kato's Fiftieth Birthday) (2003), 565–584.

[3] Y. Manin, Lectures on zeta functions and motives (according to Deninger and Kurokawa), in *Columbia University Number Theory Seminar* (*New York,* 1992), Astérisque 228-4, Société Mathématique de France, Paris, 1995, 121–163.

[4] C. Soulé, Les variétés sur le corps à un élément, *Moscow Math. J.*, **4**-1 (2004).

[5] J. Tits, Sur les analogues algébriques des groupes semi-simples complexes, in *Colloque d'algèbre supérieure, Bruxelles du* 19 *au* 22 *décembre* 1956, Centre Belge de Recherches Mathématiques Établissements Ceuterick, Louvain, Belgium and Gauthier-Villars, Paris, 1957, 261–289.

Line Bundles and p-Adic Characters

Christopher Deninger[1] and Annette Werner[2]

[1] Mathematisches Institut
   Westfälische Wilhelms-Universität Münster
   Einsteinstraße 62
   48149 Münster
   Germany
   `deninger@math.uni-muenster.de`
[2] Fachbereich Mathematik
   Institut für Algebra und Zahlentheorie
   Universität Stuttgart
   Pfaffenwaldring 57
   70569 Stuttgart
   Germany
   `annette.werner@mathematik.uni-stuttgart.de`

1 Introduction

In the paper [De-We2], we defined isomorphisms of parallel transport along étale paths for a certain class of vector bundles on p-adic curves. In particular, these vector bundles give rise to representations of the fundamental group.

One aim of the present paper is to discuss in more detail the special case of line bundles of degree zero on a curve X with good reduction over $\overline{\mathbb{Q}}_p$. By [De-We2] we have a continuous, Galois-equivariant homomorphism

$$\alpha \colon \operatorname{Pic}^0_{X/\overline{\mathbb{Q}}_p}(\mathbb{C}_p) \longrightarrow \operatorname{Hom}_c(\pi_1^{\mathrm{ab}}(X), \mathbb{C}_p^*). \tag{1}$$

Here $\operatorname{Hom}_c(\pi_1^{\mathrm{ab}}(X), \mathbb{C}_p^*)$ is the topological group of continuous $\mathbb{C}_p^*$-valued characters of the algebraic fundamental group $\pi_1(X, x)$.

The map α can be rephrased in terms of the Albanese variety A of X as a continuous, Galois-equivariant homomorphism

$$\alpha \colon \hat{A}(\mathbb{C}_p) \longrightarrow \operatorname{Hom}_c(TA, \mathbb{C}_p^*). \tag{2}$$

Therefore, we focus in this paper on abelian varieties A over $\overline{\mathbb{Q}}_p$ with good reduction.

We also consider vector bundles of higher rank on $A_{\mathbb{C}_p} = A \otimes_{\overline{\mathbb{Q}}_p} \mathbb{C}_p$. In Section 2, we define a category $\mathfrak{B}_{A_{\mathbb{C}_p}}$ of vector bundles on $A_{\mathbb{C}_p}$ which contains all line bundles algebraically equivalent to zero. Then we define for each bundle E in $\mathfrak{B}_{A_{\mathbb{C}_p}}$ a continuous representation ρ_E of the Tate module TA on the fiber of E in the zero section. The association $E \mapsto \rho_E$ is functorial, Galois-equivariant and compatible with several natural operations on vector bundles.

Besides, we show that it is compatible with the theory for curves in [De-We2] via the Albanese morphism $X \to A$.

Every rank-2 vector bundle on $A_{\mathbb{C}_p}$ which is an extension of the trivial line bundle $\mathcal{O}$ by itself lies in $\mathfrak{B}_{A_{\mathbb{C}_p}}$. The functor ρ induces a homomorphism between $\mathrm{Ext}_{\mathfrak{B}_{A_{\mathbb{C}_p}}}(\mathcal{O}, \mathcal{O}) \simeq H^1(A, \mathcal{O}) \otimes \mathbb{C}_p$ and the group of continuous extensions

$$\mathrm{Ext}^1_{TA}(\mathbb{C}_p, \mathbb{C}_p) \simeq H^1_{\text{ét}}(X, \mathbb{Q}_p) \otimes \mathbb{C}_p.$$

Hence ρ induces a homomorphism

$$H^1(A, \mathcal{O}) \otimes \mathbb{C}_p \longrightarrow H^1_{\text{ét}}(X, \mathbb{Q}_p) \otimes \mathbb{C}_p. \tag{3}$$

In Section 3, we show that this homomorphism is the Hodge–Tate map coming from the Hodge–Tate decomposition of $H^1_{\text{ét}}(X, \mathbb{Q}_p) \otimes \mathbb{C}_p$. Here we use an explicit description of the Hodge–Tate map by Faltings and Coleman via the universal vectorial extension.

In Section 4, we consider the case of line bundles algebraically equivalent to zero on A. We prove an alternative description of the homomorphism α in (2), which shows that the restriction of α to the points of the p-divisible group of $\hat{A}$ coincides with a homomorphism defined by Tate in [Ta] using the duality of the p-divisible groups associated to A and $\hat{A}$.

In fact, the whole project started with the search for an alternative description of Tate's homomorphism for line bundles on curves which could be generalized to higher-rank bundles.

We prove that α is a p-adic analytic morphism of Lie groups whose Lie algebra map

$$\mathrm{Lie}\,\alpha : H^1(A, \mathcal{O}) \otimes \mathbb{C}_p = \mathrm{Lie}\,\hat{A}(\mathbb{C}_p) \longrightarrow \mathrm{Lie}\,\mathrm{Hom}_c(TA, \mathbb{C}_p^*) = H^1_{\text{ét}}(A, \mathbb{Q}_p) \otimes \mathbb{C}_p$$

coincides with the Hodge–Tate map (3).

On the torsion subgroups, α is an isomorphism, so that we get the following commutative diagram with exact rows:

$$
\begin{array}{ccccccccc}
0 & \longrightarrow & \hat{A}(\mathbb{C}_p)_{\text{tors}} & \longrightarrow & \hat{A}(\mathbb{C}_p) & \xrightarrow{\text{log}} & \mathrm{Lie}\,\hat{A}(\mathbb{C}_p) & = & H^1(A, \mathcal{O}) \otimes_{\overline{\mathbb{Q}}_p} \mathbb{C}_p \to 0 \\
& & \downarrow{\scriptstyle ?} & & \downarrow{\alpha} & & \downarrow{\mathrm{Lie}\,\alpha} & & \\
0 & \to & \mathrm{Hom}_c(TA, \mu) & \to & \mathrm{Hom}_c(TA, \mathbb{C}_p^*) & \to & \mathrm{Hom}_c(TA, \mathbb{C}_p) & = & H^1_{\text{ét}}(A, \mathbb{Q}_p) \otimes \mathbb{C}_p \to 0.
\end{array}
$$

Here μ is the subgroup of roots of unity in $\mathbb{C}_p^*$. If A is defined over K, the vertical maps are all $G_K = \mathrm{Gal}(\overline{\mathbb{Q}}_p/K)$-equivariant.

Besides, we determine the image of α. By $CH^\infty(TA)$ we denote the group of continuous characters $\chi : TA \to \mathbb{C}_p^*$ whose stabilizer in G_K is open. Then α induces an isomorphism of topological groups between $\hat{A}(\mathbb{C}_p)$ and the closure of $CH^\infty(TA)$ in $\mathrm{Hom}_c(TA, \mathbb{C}_p^*)$ with respect to the topology of uniform convergence.

The final section deals with a smooth and proper variety X over $\overline{\mathbb{Q}}_p$ whose H^1 has good reduction. Combining the previous results for the Albanese variety of X with Kummer theory, we obtain an injective homomorphism of p-adic Lie groups

$$\alpha^\tau : \operatorname{Pic}^\tau_{X/\overline{\mathbb{Q}}_p}(\mathbb{C}_p) \longrightarrow \operatorname{Hom}_c(\pi_1^{ab}(X), \mathbb{C}_p^*)$$

and determine its image and Lie algebra map.

There is an overlap between parts of the present paper and results independently obtained by Faltings (see [Fa2]). In [Fa2], Faltings develops a more general theory where he proves an equivalence of categories between vector bundles on a curve X over $\mathbb{C}_p$ endowed with a p-adic Higgs field and a certain category of "generalized representations" of the fundamental group of X.

The results of the present paper originally formed the first part of the preprint [De-We1]. However, this preprint will not be published since the results in its second part are contained in the much more general theory of [De-We2].

2 Vector bundles giving rise to p-adic representations

Let A be an abelian variety over $\overline{\mathbb{Q}}_p$ with good reduction, and let $\mathcal{A}$ be an abelian scheme over $\overline{\mathbb{Z}}_p$ with generic fiber A.

We denote the ring of integers in $\mathbb{C}_p = \hat{\overline{\mathbb{Q}}}_p$ by $\mathfrak{o}$, and put

$$\mathfrak{o}_n = \mathfrak{o}/p^n \mathfrak{o} = \overline{\mathbb{Z}}_p/p^n \overline{\mathbb{Z}}_p.$$

We write $\mathcal{A}_\mathfrak{o} = \mathcal{A} \otimes_{\overline{\mathbb{Z}}_p} \mathfrak{o}$ and $A_{\mathbb{C}_p} = A \otimes_{\overline{\mathbb{Q}}_p} \mathbb{C}_p$. Besides, we denote by $\mathcal{A}_n = \mathcal{A} \otimes_{\overline{\mathbb{Z}}_p} \mathfrak{o}_n$ the reduction of $\mathcal{A}$ modulo p^n.

Let $\hat{\mathcal{A}} = \operatorname{Pic}^0_{\mathcal{A}/\overline{\mathbb{Z}}_p}$ be the dual abelian scheme. Its generic fiber $\hat{A} = \operatorname{Pic}^0_{A/\overline{\mathbb{Q}}_p}$ is the dual abelian variety of A.

Definition 1. *Let $\mathfrak{B}_{\mathcal{A}_\mathfrak{o}}$ be the full subcategory of the category of vector bundles on the abelian scheme $\mathcal{A}_\mathfrak{o}$ consisting of all bundles E on $\mathcal{A}_\mathfrak{o}$ satisfying the following property: For all $n \geq 1$, there exists some $N = N(n) \geq 1$ such that the reduction $(N^*E)_n$ of N^*E modulo p^n is trivial on $\mathcal{A}_n = \mathcal{A} \otimes_{\overline{\mathbb{Z}}_p} \mathfrak{o}_n$. Here $N : \mathcal{A}_\mathfrak{o} \to \mathcal{A}_\mathfrak{o}$ denotes multiplication by N.*

Note that every vector bundle F on $A_{\mathbb{C}_p}$ can be extended to a vector bundle E on $\mathcal{A}_\mathfrak{o}$. This can be shown by induction on the rank of F. If F is a line bundle on $A_{\mathbb{C}_p}$, it corresponds to a $\mathbb{C}_p$-valued point in the Picard scheme $\operatorname{Pic}_{A/\overline{\mathbb{Q}}_p}$ of A. Since every connected component of $\operatorname{Pic}_{A/\overline{\mathbb{Q}}_p}$ contains a $\overline{\mathbb{Q}}_p$-valued point, there exists a line bundle M on A such that $F \otimes M_{\mathbb{C}_p}^{-1}$ lies in $\operatorname{Pic}^0_{A/\overline{\mathbb{Q}}_p}(\mathbb{C}_p)$. As $\operatorname{Pic}^0_{A/\overline{\mathbb{Q}}_p}(\mathbb{C}_p) = \operatorname{Pic}^0_{\mathcal{A}/\overline{\mathbb{Z}}_p}(\mathfrak{o})$, the line bundle $F \otimes M_{\mathbb{C}_p}^{-1}$ can be extended to $\mathcal{A}_\mathfrak{o}$. Therefore, it suffices to show that every line bundle M on A can be extended to a line bundle on $\mathcal{A}$. Now

A and M descend to A_K and M_K over a finite extension K of $\mathbb{Q}_p$ in $\overline{\mathbb{Q}}_p$ with ring of integers $\mathfrak{o}_K$. The Néron model $\mathcal{A}_{\mathfrak{o}_K}$ of A_K over the discrete valuation ring $\mathfrak{o}_K$ is noetherian, hence

$$\text{Pic}_{\mathcal{A}_{\mathfrak{o}_K}/\mathfrak{o}_K}(\mathfrak{o}_K) \cong CH^1(\mathcal{A}_{\mathfrak{o}_K}) \longrightarrow CH^1(A_K) \simeq \text{Pic}_{A_K/K}(K)$$

is surjective. Therefore, M can be extended to $\mathcal{A}$.

Now let F be a vector bundle on $A_{\mathbb{C}_p}$. Then there exists a short exact sequence

$$0 \longrightarrow F_1 \longrightarrow F \longrightarrow F_2 \longrightarrow 0,$$

where F_2 is a line bundle. By induction, we can assume that F_1 and F_2 can be extended to vector bundles E_1 and E_2 on $\mathcal{A}_{\mathfrak{o}}$. Flat base change gives an isomorphism

$$\text{Ext}^1(E_2, E_1) \otimes_{\mathfrak{o}} \mathbb{C}_p \xrightarrow{\sim} \text{Ext}^1(F_2, F_1),$$

hence F can also be extended to $\mathcal{A}_{\mathfrak{o}}$.

We are interested in those bundles on $A_{\mathbb{C}_p}$ which have a model in $\mathfrak{B}_{\mathcal{A}_{\mathfrak{o}}}$.

Definition 2. *Let $\mathfrak{B}_{A_{\mathbb{C}_p}}$ be the full subcategory of the category of vector bundles on $A_{\mathbb{C}_p}$ consisting of all bundles F on $A_{\mathbb{C}_p}$ which are isomorphic to the generic fiber of a vector bundle E in the category $\mathfrak{B}_{\mathcal{A}_{\mathfrak{o}}}$.*

Consider the Tate module $TA = \varprojlim A_N(\overline{\mathbb{Q}}_p)$, where $A_N(\overline{\mathbb{Q}}_p)$ denotes the group of N-torsion points in $A(\overline{\mathbb{Q}}_p)$.

By $x_{\mathfrak{o}}$, respectively, x_n, we denote the zero sections on $\mathcal{A}_{\mathfrak{o}}$, respectively, $\mathcal{A}_n$. For a vector bundle E on $\mathcal{A}_{\mathfrak{o}}$ we write $E_{x_{\mathfrak{o}}} = x_{\mathfrak{o}}^* E$ viewed as a free $\mathfrak{o}$-module of rank $r = \text{rank } E$. Similarly, we set $E_{x_n} = x_n^* E$ viewed as a free $\mathfrak{o}_n$-module of rank r. Note that

$$E_{x_{\mathfrak{o}}} = \varprojlim_n E_{x_n}$$

as topological $\mathfrak{o}$-modules, if E_{x_n} is endowed with the discrete topology.

Assume that E is contained in the category $\mathfrak{B}_{\mathcal{A}_{\mathfrak{o}}}$ and fix some $n \geq 1$. Then there exists some $N = N(n) \geq 1$, such that the reduction $(N^*E)_n$ is trivial on $\mathcal{A}_n$. The structure morphism $\lambda: \mathcal{A}_{\mathfrak{o}} \to \text{spec } \mathfrak{o}$ satisfies $\lambda_* \mathcal{O}_{\mathcal{A}_{\mathfrak{o}}} = \mathcal{O}_{\text{spec } \mathfrak{o}}$ universally. Hence $\Gamma(\mathcal{A}_n, \mathcal{O}) = \mathfrak{o}_n$, and therefore the pullback map

$$x_n^*: \Gamma(\mathcal{A}_n, (N^*E)_n) \xrightarrow{\sim} \Gamma(\text{spec } \mathfrak{o}_n, x_n^* E_n) = E_{x_n}$$

is an isomorphism of free $\mathfrak{o}_n$-modules. (Note that $N \circ x_n = x_n$.) The group $A_N(\overline{\mathbb{Q}}_p)$ acts in a natural way on $\Gamma(\mathcal{A}_n, (N^*E)_n)$ by translations. Define a representation $\rho_{E,n}: TA \to \text{Aut}_{\mathfrak{o}_n}(E_{x_n})$ as the composition:

$$\rho_{E,n}: TA \longrightarrow A_N(\overline{\mathbb{Q}}_p) \longrightarrow \text{Aut}_{\mathfrak{o}_n} \Gamma(\mathcal{A}_n, (N^*E)_n) \xrightarrow[\text{via } x_n^*]{\sim} \text{Aut}_{\mathfrak{o}_n} E_{x_n}.$$

Lemma 1. *For E in $\mathfrak{B}_{\mathcal{A}_{\mathfrak{o}}}$ the representations $\rho_{E,n}$ are independent of the choice of N and form a projective system when composed with the natural projection maps $\text{Aut}_{\mathfrak{o}_{n+1}} E_{x_{n+1}} \to \text{Aut}_{\mathfrak{o}_n} E_{x_n}$.*

Proof. If $N' = N \cdot M$ for some $M \geq 1$, and $(N^*E)_n$ is trivial, it follows from our construction that

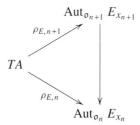

is commutative. Hence $\rho_{E,n}$ is independent of the choice of N.

Fix some $n \geq 1$ and assume that $(N^*E)_{n+1}$ is trivial. Then $(N^*E)_n$ is also trivial, and the natural action of $A_N(\overline{\mathbb{Q}}_p)$ on $\Gamma(\mathcal{A}_{n+1}, (N^*E)_{n+1})$ induces the natural action of $A_N(\overline{\mathbb{Q}}_p)$ on $\Gamma(\mathcal{A}_n, (N^*E)_n)$. Hence

$$\begin{array}{ccc}
 & \mathrm{Aut}_{\mathfrak{o}_{n+1}} E_{x_{n+1}} & \\
\overset{\rho_{E,n+1}}{\nearrow} & & \downarrow \\
TA & & \\
\underset{\rho_{E,n}}{\searrow} & & \downarrow \\
 & \mathrm{Aut}_{\mathfrak{o}_n} E_{x_n} &
\end{array}$$

is commutative, so that $(\rho_{E,n})_{n \geq 1}$ is a projective system. $\square$

By the lemma, we can define for all E in $\mathfrak{B}_{\mathcal{A}_{\mathfrak{o}}}$ an $\mathfrak{o}$-linear representation of TA by

$$\rho_E = \varprojlim_n \rho_{E,n} : TA \longrightarrow \mathrm{Aut}_{\mathfrak{o}}(E_{x_{\mathfrak{o}}}).$$

Since each $\rho_{E,n}$ factors over a finite quotient of TA, the map ρ_E is continuous, if $\mathrm{Aut}_{\mathfrak{o}}(E_{x_{\mathfrak{o}}}) \simeq \mathrm{GL}_r(\mathfrak{o})$ for $r = \mathrm{rank}\, E$ carries the topology induced by the one of $\mathfrak{o}$.

Note that for any morphism $f : E_1 \to E_2$ of vector bundles in $\mathfrak{B}_{\mathcal{A}_{\mathfrak{o}}}$ the natural $\mathfrak{o}_n$-linear map $x_n^* f : (E_1)_{x_n} \to (E_2)_{x_n}$ is TA-equivariant with respect to the actions $\rho_{E_1,n}$ and $\rho_{E_2,n}$. Hence the association $E \mapsto \rho_E$ defines a functor

$$\rho : \mathfrak{B}_{\mathcal{A}_{\mathfrak{o}}} \longrightarrow \mathbf{Rep}_{TA}(\mathfrak{o}),$$

where $\mathbf{Rep}_{TA}(\mathfrak{o})$ is the category of continuous representations of TA on free $\mathfrak{o}$-modules of finite rank.

We denote by $x = x_{\mathfrak{o}} \otimes \mathbb{C}_p$ the unit section of $A(\mathbb{C}_p)$.

Let F be a vector bundle in the category $\mathfrak{B}_{\mathcal{A}_{\mathbb{C}_p}}$. Then F can be extended to a bundle E in $\mathfrak{B}_{\mathcal{A}_{\mathfrak{o}}}$. We define a $\mathbb{C}_p$-linear representation

$$\rho_F : TA \longrightarrow \mathrm{Aut}_{\mathbb{C}_p}(F_x)$$

as $\rho_F = \rho_E \otimes_{\mathfrak{o}} \mathbb{C}_p$, where we identify F_x with $E_{x_{\mathfrak{o}}} \otimes_{\mathfrak{o}} \mathbb{C}_p$.

If E_1 and E_2 are two extensions of F lying in $\mathfrak{B}_{\mathcal{A}_{\mathfrak{o}}}$, a flat base change for H^0 of the Hom-bundle yields $\mathrm{Hom}_{\mathcal{A}_{\mathfrak{o}}}(E_1, E_2) \otimes_{\mathfrak{o}} \mathbb{C}_p \simeq \mathrm{Hom}_{\mathcal{A}_{\mathbb{C}_p}}((E_1)_{\mathbb{C}_p}, (E_2)_{\mathbb{C}_p})$.

Let $\varphi: (E_1)_{\mathbb{C}_p} \xrightarrow{\sim} F \xrightarrow{\sim} (E_2)_{\mathbb{C}_p}$ be the identifications of the generic fibers. Then there exists some $m \geq 1$ such that $m\varphi$ is the generic fiber of a morphism $\psi: E_1 \to E_2$. By functoriality, the induced map $x_0^*\psi: (E_1)_{x_0} \to (E_2)_{x_0}$ is TA-equivariant. Therefore, identifying $F_x = (E_1)_{x_0} \otimes_0 \mathbb{C}_p$ and $F_x = (E_2)_{x_0} \otimes_0 \mathbb{C}_p$, we see that $\rho_{E_1} \otimes_0 \mathbb{C}_p$ and $\rho_{E_2} \otimes_0 \mathbb{C}_p$ coincide. Hence ρ_F is well defined.

Theorem 1.

(i) *The category $\mathfrak{B}_{A_{\mathbb{C}_p}}$ is closed under direct sums, tensor products, duals, internal homs, and exterior powers. Besides, it is closed under extensions, i.e., if $0 \to F' \to F \to F'' \to 0$ is an exact sequence of vector bundles on $A_{\mathbb{C}_p}$ such that F' and F'' are in $\mathfrak{B}_{A_{\mathbb{C}_p}}$, then F is also contained in $\mathfrak{B}_{A_{\mathbb{C}_p}}$.*

(ii) *$\mathfrak{B}_{A_{\mathbb{C}_p}}$ contains all line bundles algebraically equivalent to zero. For any bundle in $\mathfrak{B}_{A_{\mathbb{C}_p}}$ the determinant line bundle is algebraically equivalent to zero.*

(iii) *The association $F \mapsto \rho_F$ defines an additive exact functor*

$$\rho: \mathfrak{B}_{A_{\mathbb{C}_p}} \longrightarrow \mathbf{Rep}_{TA}(\mathbb{C}_p),$$

where $\mathbf{Rep}_{TA}(\mathbb{C}_p)$ is the category of continuous representations of TA on finite-dimensional $\mathbb{C}_p$-vector spaces. This functor commutes with tensor products, duals, internal homs and exterior powers.

(iv) *Let $f: A \to A'$ be a homomorphism of abelian varieties over $\overline{\mathbb{Q}}_p$ with good reduction. Then pullback of vector bundles induces an additive exact functor*

$$f^*: \mathfrak{B}_{A'_{\mathbb{C}_p}} \longrightarrow \mathfrak{B}_{A_{\mathbb{C}_p}},$$

which commutes with tensor products, duals, internal homs, and exterior powers (up to canonical identifications). The following diagram is commutative:

$$
\begin{array}{ccc}
\mathfrak{B}_{A'_{\mathbb{C}_p}} & \xrightarrow{\ f^*\ } & \mathfrak{B}_{A_{\mathbb{C}_p}} \\
\rho \downarrow & & \downarrow \rho \\
\mathbf{Rep}_{TA'}(\mathbb{C}_p) & \xrightarrow{\ F\ } & \mathbf{Rep}_{TA}(\mathbb{C}_p),
\end{array}
$$

where F is the functor induced by composition with $Tf: TA \to TA'$.

Proof. (i) We only show that $\mathfrak{B}_{A_{\mathbb{C}_p}}$ is closed under extensions, the remaining assertions are straightforward. So consider an extension $0 \to F' \to F \to F'' \to 0$ with F' and F'' in $\mathfrak{B}_{A_{\mathbb{C}_p}}$. Fix some $n \geq 1$. Then we find a number N such that $(N^*F')_n$ and $(N^*F'')_n$ are trivial. Hence $(N^*F)_n$ is an extension of two trivial vector bundles. We claim that this implies the triviality of $((p^n N)^*F)_n$. It suffices to show that $(p^n)^*$ induces the zero map on $\mathrm{Ext}^1_{\mathcal{A}_n}(\mathcal{O}, \mathcal{O}) = H^1(\mathcal{A}_n, \mathcal{O})$. The diagram

$$
\begin{array}{ccc}
H^1(\mathcal{A}_n, \mathcal{O}) & \xrightarrow{(p^n)^*} & H^1(\mathcal{A}_n, \mathcal{O}) \\
\wr \uparrow & & \uparrow \wr \\
\mathrm{Lie\ Pic}^0_{\mathcal{A}_n/o_n} & \xrightarrow{\mathrm{Lie}\ p^n} & \mathrm{Lie\ Pic}^0_{\mathcal{A}_n/o_n}
\end{array}
$$

is commutative by [BLR, 8.4, Theorem 1]. Since Lie p^n is multiplication by p^n on the $\mathfrak{o}_n$-module Lie $\mathrm{Pic}^0_{\mathcal{A}_n/\mathfrak{o}_n}$, it is zero. Hence $(p^n)^*$ is indeed the zero map on $\mathrm{Ext}^1_{\mathcal{A}_n}(\mathcal{O}, \mathcal{O})$.

(ii) Since $\hat{\mathcal{A}}_{\mathfrak{o}} = \mathrm{Pic}^0_{\mathcal{A}_0/\mathfrak{o}}$ is proper, we have

$$\mathrm{Pic}^0_{\mathcal{A}_0/\mathfrak{o}}(\mathfrak{o}) = \mathrm{Pic}^0_{A_{\mathbb{C}_p}/\mathbb{C}_p}(\mathbb{C}_p),$$

so that any line bundle $L_{\mathbb{C}_p}$ on $A_{\mathbb{C}_p}$ which is algebraically equivalent to zero, can be extended to a line bundle L on $\mathcal{A}_0$ giving rise to a class in $\hat{\mathcal{A}}_{\mathfrak{o}}(\mathfrak{o})$.

$\hat{\mathcal{A}}$ descends to an abelian scheme $\hat{\mathcal{A}}_{\mathfrak{o}_K}$ over the ring of integers in some finite extension K of $\mathbb{Q}_p$ in $\overline{\mathbb{Q}}_p$. The ring $\mathfrak{o}_n = \mathfrak{o}/p^n\mathfrak{o} = \overline{\mathbb{Z}}_p/p^n\overline{\mathbb{Z}}_p$ is the union of the finite rings $\mathfrak{o}_L/p^n\mathfrak{o}_L$, where L runs through the finite extensions of K in $\overline{\mathbb{Q}}_p$. Therefore, $\hat{\mathcal{A}}_{\mathfrak{o}}(\mathfrak{o}_n) = \hat{\mathcal{A}}_{\mathfrak{o}_K}(\mathfrak{o}_n)$ is the union of all $\hat{\mathcal{A}}_{\mathfrak{o}_K}(\mathfrak{o}_L/p^n\mathfrak{o}_L)$. Since $\hat{\mathcal{A}}_{\mathfrak{o}_K}$ is of finite type over $\mathfrak{o}_K$, all these groups $\hat{\mathcal{A}}_{\mathfrak{o}_K}(\mathfrak{o}_L/p^n\mathfrak{o}_L)$ are finite. Hence $\hat{\mathcal{A}}_{\mathfrak{o}}(\mathfrak{o}_n)$ is a torsion group.

In particular, we find some N such that N annihilates the class of L_n in $\hat{\mathcal{A}}_{\mathfrak{o}}(\mathfrak{o}_n)$. Then $(N^*L)_n$ is trivial, which shows that L is contained in $\mathfrak{B}_{A_{\mathbb{C}_p}}$.

If E is a vector bundle in $\mathfrak{B}_{\mathcal{A}_0}$, then by i) its determinant line bundle L is also contained in $\mathfrak{B}_{\mathcal{A}_0}$. Hence there exists some $N \geq 1$ such that N^*L_1 is trivial on $\mathcal{A}_1$, where L_1 and $\mathcal{A}_1$ denote the reductions modulo p. If k denotes the residue field of $\mathfrak{o}$, this implies that N^*L_k is trivial on $\mathcal{A}_k$. Since the Néron–Severi group of $\mathcal{A}_k$ is torsion free, L_k lies in $\mathrm{Pic}^0_{\mathcal{A}_k/k}(k)$.

Note that $\mathcal{A}$ is projective by [Ray, Théorème XI 1.4]. Hence $\mathrm{Pic}^0_{\mathcal{A}_0/\mathfrak{o}}$ is an open subscheme of the Picard scheme $\mathrm{Pic}_{\mathcal{A}_0/\mathfrak{o}}$. Since the reduction of the point in Pic induced by L lies in Pic^0, the generic fiber of L is also contained in Pic^0, whence our claim follows.

(iii) The fact that $F \mapsto \rho_F$ is functorial on $\mathfrak{B}_{A_{\mathbb{C}_p}}$ follows from the fact that the corresponding association $E \mapsto \rho_E$ is functorial on $\mathfrak{B}_{\mathcal{A}_0}$. The remaining claims in iii) are straightforward.

(iv) It is clear that f^* induces a functor $f^*\colon \mathfrak{B}_{A'_{\mathbb{C}_p}} \to \mathfrak{B}_{A_{\mathbb{C}_p}}$ with the claimed properties. In order to show that the desired diagram commutes, it suffices to show that

$$
\begin{array}{ccc}
\mathfrak{B}_{\mathcal{A}'_0} & \xrightarrow{\ f^*\ } & \mathfrak{B}_{\mathcal{A}_0} \\
{\scriptstyle\rho}\downarrow & & \downarrow{\scriptstyle\rho} \\
\mathbf{Rep}_{TA}(\mathfrak{o}) & \xrightarrow{\ F\ } & \mathbf{Rep}_{TA}(\mathfrak{o})
\end{array}
$$

commutes, where $f\colon \mathcal{A}_0 \to \mathcal{A}'_0$ comes from the canonical extension of f to the Néron models. Let E' be an object in $\mathfrak{B}_{\mathcal{A}'_0}$. If x'_0 is the zero section of $\mathcal{A}'_0$, we have $f(x_0) = x'_0$, so that there is a canonical identification of $E'_{x'_0}$ and $(f^*E')_{x_0}$. For every $a \in A_N(\overline{\mathbb{Q}}_p)$ we have $t_{f(a)} \circ f = f \circ t_a$, where $t_{f(a)}$ and t_a are the translation maps. If we now go through the definition of the representation of TA on $(f^*E')_{x_0}$, our claim follows. $\square$

Let K be a finite extension of $\mathbb{Q}_p$ in $\overline{\mathbb{Q}}_p$ so that A is defined over K, i.e., $A = A_K \otimes_K \overline{\mathbb{Q}}_p$ for an abelian variety A_K over K. Then $G_K = \mathrm{Gal}(\overline{\mathbb{Q}}_p/K)$ acts in a natural way on TA and on $\mathbb{C}_p$ and hence on the category $\mathbf{Rep}_{TA}(\mathbb{C}_p)$ by putting both actions together. To be precise, for $\sigma \in G_K$ put $^\sigma V = V \otimes_{\mathbb{C}_p,\sigma} \mathbb{C}_p$ for every finite-dimensional $\mathbb{C}_p$-vector space V, and write $\sigma : V \to {}^\sigma V$ for the natural σ-linear map. For every representation $\varphi : TA \to \mathrm{Aut}_{\mathbb{C}_p}(V)$, we define $\sigma_* \varphi$ as the representation

$$\sigma_* \varphi : TA \xrightarrow{\sigma^{-1}} TA \xrightarrow{\varphi} \mathrm{Aut}_{\mathbb{C}_p} V \xrightarrow{c_\sigma} \mathrm{Aut}_{\mathbb{C}_p} {}^\sigma V,$$

where $c_\sigma(f) = \sigma \circ f \circ \sigma^{-1}$.

For every vector bundle F in $\mathfrak{B}_{A_{\mathbb{C}_p}}$ the vector bundle $^\sigma F = F \times_{\mathrm{spec}\, \mathbb{C}_p, \mathrm{spec}\, \sigma} \mathrm{spec}\, \mathbb{C}_p$ is also contained in the category $\mathfrak{B}_{A_{\mathbb{C}_p}}$.

Proposition 1. *For every $\sigma \in G_K$ the following diagram is commutative:*

$$
\begin{array}{ccc}
\mathfrak{B}_{A_{\mathbb{C}_p}} & \xrightarrow{\ \rho\ } & \mathbf{Rep}_{TA}(\mathbb{C}_p) \\
\Big\downarrow{g_\sigma} & & \Big\downarrow{\sigma_*} \\
\mathfrak{B}_{A_{\mathbb{C}_p}} & \xrightarrow{\ \rho\ } & \mathbf{Rep}_{TA}(\mathbb{C}_p),
\end{array}
$$

where the functor g_σ maps F to $^\sigma F$.

Proof. This can be checked directly. □

Let X be a smooth, connected, projective curve over $\overline{\mathbb{Q}}_p$. We fix a point $x \in X(\overline{\mathbb{Q}}_p)$ and denote by $\pi_1(X, x)$ the algebraic fundamental group with base point x.

In [De-We2], we define and investigate the category $\mathfrak{B}_{X_{\mathbb{C}_p}}$ of all vector bundles F on $X_{\mathbb{C}_p}$ which can be extended to a vector bundle E on $\mathfrak{X}_{\mathfrak{o}} = \mathfrak{X} \otimes_{\overline{\mathbb{Z}}_p} \mathfrak{o}$ for a finitely presented, flat and proper model $\mathfrak{X}$ of X over $\overline{\mathbb{Z}}_p$ and which have the following property: For all $n \geq 1$ there exists a finitely presented proper $\overline{\mathbb{Z}}_p$-morphism $\pi : \mathcal{Y} \to \mathfrak{X}$ with finite, étale generic fiber such that $\pi_n^* E_n$ is trivial. Here π_n and E_n denote again the reductions modulo p^n.

Besides, we define in [De-We2] a functor ρ from $\mathfrak{B}_{X_{\mathbb{C}_p}}$ to the category of continuous representations of the étale fundamental groupoid of X. In particular, every bundle F in $\mathfrak{B}_{X_{\mathbb{C}_p}}$ induces a continuous representation $\rho_F : \pi_1(X, x) \to \mathrm{Aut}_{\mathbb{C}_p}(F_x)$ of the fundamental group; cf. [De-We2, Proposition 20]. Let $\mathbf{Rep}_{\pi_1(X,x)}(\mathbb{C}_p)$ be the category of continuous representations of $\pi_1(X, x)$ on finite-dimensional $\mathbb{C}_p$-vector spaces. Then the association $F \mapsto \rho_F$ induces a functor $\rho : \mathfrak{B}_{X_{\mathbb{C}_p}} \to \mathbf{Rep}_{\pi_1(X,x)}(\mathbb{C}_p)$.

Let us now assume that X has good reduction, i.e., there exists a smooth, proper, finitely presented model $\mathfrak{X}$ of X over $\overline{\mathbb{Z}}_p$. Then $\mathcal{A} = \mathrm{Pic}^0_{\mathfrak{X}/\overline{\mathbb{Z}}_p}$ is an abelian scheme. Let $A = \mathrm{Pic}^0_{X/\overline{\mathbb{Q}}_p}$ be the Jacobian of X, and let $f : X \to A$ be the embedding mapping x to 0. After descending to a suitable finite extension of $\mathbb{Q}_p$ in $\overline{\mathbb{Q}}_p$, $\mathcal{A}$ becomes the Néron model of A. Hence f has an extension $f_0 : \mathfrak{X} \to \mathcal{A}$. Besides, f induces a homomorphism

$$f_* \colon \pi_1(X, \dot{x}) \longrightarrow \pi_1(A, 0) = TA,$$

which identifies TA with the maximal abelian quotient of $\pi_1(X, x)$.

Let F be a vector bundle in $\mathfrak{B}_{A_{\mathbb{C}_p}}$, and let E be a model of F on $\mathcal{A}_o$ such that for all $n \geq 1$ there exists some $N \geq 1$ satisfying $(N^*E)_n$ trivial. Consider the pullback of the covering $N \colon \mathcal{A} \to \mathcal{A}$ to $\mathfrak{X}$, i.e., the Cartesian diagram

$$
\begin{array}{ccc}
\mathcal{Y} & \xrightarrow{\ g\ } & \mathcal{A} \\
{\scriptstyle \pi_N}\downarrow & & \downarrow{\scriptstyle N} \\
\mathfrak{X} & \xrightarrow{\ f\ } & \mathcal{A}.
\end{array}
$$

Then $(\pi_N^* f^* E)_n$ is also trivial, and we see that $f^* F$ is contained in $\mathfrak{B}_{X_{\mathbb{C}_p}}$.

Hence pullback via f induces a functor $f^* \colon \mathfrak{B}_{A_{\mathbb{C}_p}} \to \mathfrak{B}_{X_{\mathbb{C}_p}}$.

Lemma 2. *The following diagram is commutative:*

$$
\begin{array}{ccc}
\mathfrak{B}_{A_{\mathbb{C}_p}} & \xrightarrow{\ f^*\ } & \mathfrak{B}_{X_{\mathbb{C}_p}} \\
{\scriptstyle \rho}\downarrow & & \downarrow{\scriptstyle \rho} \\
\mathbf{Rep}_{TA}(\mathbb{C}_p) & \xrightarrow{\ \tilde{f}\ } & \mathbf{Rep}_{\pi_1(X,x)}(\mathbb{C}_p),
\end{array}
$$

where $\tilde{f}$ is the functor induced by composition with the homomorphism $f_* \colon \pi_1(X, x) \to TA$.

Proof. One argues similarly as in the proof of Theorem 1(iv). $\qquad\square$

3 The Hodge–Tate map

In this section we show that the functor ρ can be used to describe the Hodge–Tate decomposition of the first étale cohomology of A, when we apply it to extensions of the trivial line bundle with itself.

Let K be a finite extension of $\mathbb{Q}_p$ in $\overline{\mathbb{Q}}_p$ such that there is an abelian variety A_K over K with $A = A_K \otimes_K \overline{\mathbb{Q}}_p$. Put $G_K = \mathrm{Gal}(\overline{\mathbb{Q}}_p/K)$.

The Hodge–Tate decomposition (originating from [Ta])

$$H_{\text{ét}}^1(A, \mathbb{Q}_p) \otimes \mathbb{C}_p \simeq (H^1(A_K, \mathcal{O}) \otimes_K \mathbb{C}_p) \oplus (H^0(A_K, \Omega^1) \otimes_K \mathbb{C}_p(-1)) \quad (4)$$

gives rise to a G_K-equivariant map

$$\theta_A^* \colon H^1(A_K, \mathcal{O}) \otimes_K \mathbb{C}_p \longrightarrow H_{\text{ét}}^1(A, \mathbb{Q}_p) \otimes \mathbb{C}_p. \quad (5)$$

As Faltings [Fa1, Theorem 4] and Coleman [Co1, p. 379], [Co3, Section 4] have shown, θ_A^* has the following elegant description.

Consider the universal vectorial extension of $\mathcal{A}$ over the ring $\overline{\mathbb{Z}}_p$,

$$0 \longrightarrow \omega_{\hat{\mathcal{A}}} \longrightarrow E \longrightarrow \mathcal{A} \longrightarrow 0.$$

Here $\omega_{\hat{\mathcal{A}}}$ is the vector group induced by the invariant differentials on $\hat{\mathcal{A}}$, i.e., $\omega_{\hat{\mathcal{A}}}(S) = H^0(S, e^* \Omega^1_{\hat{\mathcal{A}}_S/S})$ for all $\overline{\mathbb{Z}}_p$-schemes S, where e denotes the zero section.

For $\nu \geq 1$ consider the map

$$\mathcal{A}_{p^\nu}(\mathfrak{o}) \longrightarrow \omega_{\hat{\mathcal{A}}}(\mathfrak{o})/p^\nu \omega_{\hat{\mathcal{A}}}(\mathfrak{o})$$

obtained by sending a_{p^ν} to $p^\nu b_{p^\nu}$, where $b_{p^\nu} \in E(\mathfrak{o})$ is a lift of a_{p^ν}. Passing to inverse limits, we get a $\mathbb{Z}_p$-linear homomorphism

$$\theta_A : T_p A \longrightarrow \omega_{\hat{\mathcal{A}}}(\mathfrak{o}).$$

The $\mathbb{C}_p$-dual of the resulting map

$$\theta_A : T_p A \otimes \mathbb{C}_p \longrightarrow \omega_{\hat{\mathcal{A}}}(\mathfrak{o}) \otimes \mathbb{C}_p = \omega_{\hat{\mathcal{A}}}(\mathbb{C}_p)$$

is the map θ_A^* in (5).

In [Co1], Coleman proved that together with a map defined by Fontaine in [Fo],

$$H^0(A, \Omega^1) \otimes \mathbb{C}_p(-1) \longrightarrow H^1_{\text{ét}}(A, \mathbb{Q}_p) \otimes \mathbb{C}_p,$$

θ_A^* gives the Hodge–Tate decomposition.

Let us write $\text{Ext}^1_{\mathfrak{B}_{A_{\mathbb{C}_p}}}(\mathcal{O}, \mathcal{O})$ for the Yoneda group of isomorphism classes of extensions $0 \to \mathcal{O} \to \mathcal{O}(F) \to \mathcal{O} \to 0$, where F is a vector bundle in $\mathfrak{B}_{A_{\mathbb{C}_p}}$ with sheaf of sections $\mathcal{O}(F)$ and $\mathcal{O} = \mathcal{O}_{A_{\mathbb{C}_p}}$. By Theorem 1, $\mathfrak{B}_{A_{\mathbb{C}_p}}$ contains all vector bundles which are extensions of the trivial bundle by itself. Hence $\text{Ext}^1_{\mathfrak{B}_{A_{\mathbb{C}_p}}}(\mathcal{O}, \mathcal{O})$ coincides with the group $\text{Ext}^1(\mathcal{O}, \mathcal{O})$ in the category of locally free sheaves on $A_{\mathbb{C}_p}$, so that

$$\text{Ext}^1_{\mathfrak{B}_{A_{\mathbb{C}_p}}}(\mathcal{O}, \mathcal{O}) = \text{Ext}^1(\mathcal{O}, \mathcal{O}) = H^1(A_{\mathbb{C}_p}, \mathcal{O}) = H^1(A_K, \mathcal{O}) \otimes_K \mathbb{C}_p.$$

Since the functor ρ is exact by Theorem 1, it induces a homomorphism of Ext groups

$$\rho_* : \text{Ext}^1_{\mathfrak{B}_{A_{\mathbb{C}_p}}}(\mathcal{O}, \mathcal{O}) \longrightarrow \text{Ext}^1_{\textbf{Rep}_{TA}(\mathbb{C}_p)}(\mathbb{C}_p, \mathbb{C}_p).$$

There is a natural isomorphism

$$\text{Ext}^1_{\textbf{Rep}_{TA}(\mathbb{C}_p)}(\mathbb{C}_p, \mathbb{C}_p) \simeq \text{Hom}_c(TA, \mathbb{C}_p),$$

where $\text{Hom}_c(TA, \mathbb{C}_p)$ denotes the continuous homomorphisms from TA to $\mathbb{C}_p$, which is defined as follows. For every extension $0 \to \mathbb{C}_p \overset{i}{\to} V \overset{\varepsilon}{\to} \mathbb{C}_p \to 0$ in $\textbf{Rep}_{TA}(\mathbb{C}_p)$ choose any $v \in V$ with $\varepsilon(v) = 1$ and define $\psi : TA \to \mathbb{C}_p$ by $\psi(\gamma) = i^{-1}(\gamma v - v)$. Since $\text{Hom}_c(TA, \mathbb{C}_p) = \text{Hom}_{\mathbb{Z}_p}(T_p A, \mathbb{C}_p)$, we get an isomorphism

$$\text{Ext}^1_{\textbf{Rep}_{TA}(\mathbb{C}_p)}(\mathbb{C}_p, \mathbb{C}_p) \simeq H^1_{\text{ét}}(A, \mathbb{Q}_p) \otimes \mathbb{C}_p.$$

Theorem 2. *The following diagram commutes:*

$$
\begin{array}{ccc}
\mathrm{Ext}^1_{\mathfrak{B}_{A_{\mathbb{C}_p}}}(\mathcal{O}, \mathcal{O}) & \xrightarrow{\;\rho_*\;} & \mathrm{Ext}^1_{\mathbf{Rep}_{TA}(\mathbb{C}_p)}(\mathbb{C}_p, \mathbb{C}_p) \\
\| & & \downarrow{\wr} \\
H^1(A_K, \mathcal{O}) \otimes_K \mathbb{C}_p & \xrightarrow{\;\theta_A^*\;} & H^1_{\text{ét}}(A, \mathbb{Q}_p) \otimes \mathbb{C}_p,
\end{array}
\tag{6}
$$

where θ_A^* is the map (5) appearing in the Hodge–Tate decomposition.

Remark 1. Theorem 2 gives the following novel construction of the Hodge Tate map θ_A^*. Consider a class c in $H^1(\mathcal{A}_\mathfrak{o}, \mathcal{O})$. It can be viewed as an extension

$$0 \to \mathcal{O} \to \mathcal{O}(E) \to \mathcal{O} \to 0$$

of locally free sheaves on $\mathcal{A}_\mathfrak{o}$. The bundle E lies in $\mathfrak{B}_{A_\mathfrak{o}}$. Hence, for every $n \geq 1$ there is some $N \geq 1$, in fact, $N = p^n$ will do, such that $(N^* E)_n$ is the trivial rank-2 bundle on $\mathcal{A}_n$. The short exact sequence $0 \to \mathfrak{o}_n \to E_{x_n} \to \mathfrak{o}_n \to 0$ of fibers along the zero section of $\mathcal{A}_n$ becomes TA-equivariant if TA acts trivially on the $\mathfrak{o}_n$s and via the projection $TA \to A_N(\overline{K})$ and the isomorphism

$$\Gamma(\mathcal{A}_n, (N^* E)_n) \xrightarrow{\;\underset{\sim}{\text{res}}\;} E_{x_n}$$

on E_{x_n}. Passing to projective limits, we get a short exact sequence $0 \to \mathfrak{o} \xrightarrow{i} E_x \xrightarrow{q} \mathfrak{o} \to 0$ of TA-modules. Set $g_1 = i(1)$ and choose $g_2 \in E_x$ such that $q(g_2) = 1$. Then E_x is a free $\mathfrak{o}$-module on g_1 and g_2, and the action of $\gamma \in TA$ on E_x is given in terms of the basis g_1, g_2 by a matrix of the form

$$\begin{pmatrix} 1 & \beta(\gamma) \\ 0 & 1 \end{pmatrix},$$

where $\beta : TA \to \mathfrak{o}$ is a continuous homomorphism. Note that β does not depend on the choice of g_2. Viewing β as an element of $H^1_{\text{ét}}(A, \mathbb{Z}_p) \otimes \mathfrak{o}$, we have $\theta_A^*(c) = \beta$.

Proof. Since $H^1(\mathcal{A}_\mathfrak{o}, \mathcal{O}) \otimes_\mathfrak{o} \mathbb{C}_p = H^1(\mathcal{A}_{\mathbb{C}_p}, \mathcal{O})$, we find for every element in $\mathrm{Ext}^1_{\mathfrak{B}_{A_{\mathbb{C}_p}}}(\mathcal{O}, \mathcal{O})$ a p-power multiple lying in $\mathrm{Ext}^1_{\mathfrak{B}_{A_\mathfrak{o}}}(\mathcal{O}, \mathcal{O})$. Since ρ_* is a homomorphism between Yoneda Ext-groups, it suffices to show that

$$
\begin{array}{ccc}
\mathrm{Ext}^1_{\mathfrak{B}_{A_\mathfrak{o}}}(\mathcal{O}, \mathcal{O}) & \xrightarrow{\;\rho_*\;} & \mathrm{Ext}^1_{\mathbf{Rep}_{TA}(\mathfrak{o})}(\mathfrak{o}, \mathfrak{o}) \\
\| & & \downarrow{\wr} \\
H^1(\mathcal{A}_\mathfrak{o}, \mathcal{O}) & \xrightarrow{\;\theta_A^*\;} & \mathrm{Hom}_c(T_p A, \mathfrak{o})
\end{array}
$$

commutes. Consider an extension $0 \to \mathcal{O} \xrightarrow{i} \mathcal{F} \to \mathcal{O} \to 0$ on $\mathcal{A}_\mathfrak{o}$, where $\mathcal{F} = \mathcal{O}(E)$ for some E in $\mathfrak{B}_{A_\mathfrak{o}}$.

Let $\mathcal{I} = \underline{\mathrm{Iso}}_{\mathrm{Ext}}(\mathcal{O}^2, \mathcal{F})$ be the (Zariski) sheaf on $\mathcal{A}_\mathfrak{o}$ which associates to an open subset $U \subseteq \mathcal{A}_\mathfrak{o}$ the set of isomorphisms $\varphi: \mathcal{O}_U^2 \to \mathcal{F}_U$ of extensions, i.e., such that

$$\begin{array}{ccccccccc}
0 & \longrightarrow & \mathcal{O}_U & \longrightarrow & \mathcal{O}_U^2 & \overset{p}{\longrightarrow} & \mathcal{O}_U & \longrightarrow & 0 \\
 & & \| & & \downarrow{\scriptstyle \varphi} & & \| & & \\
0 & \longrightarrow & \mathcal{O}_U & \overset{i}{\longrightarrow} & \mathcal{F}_U & \longrightarrow & \mathcal{O}_U & \longrightarrow & 0
\end{array}$$

commutes. Then $c \in \mathbb{G}_a(U) = \Gamma(U, \mathcal{O})$ acts in a natural way on $\mathcal{I}(U)$ by mapping

$$\varphi \longmapsto \varphi + i \circ f_c \circ p,$$

where $f_c: \mathcal{O}_U \to \mathcal{O}_U$ is multiplication by c. Note that $\mathcal{I}$ is a $\mathbb{G}_a$-torsor on $\mathcal{A}_\mathfrak{o}$ and that the class $[\mathcal{I}]$ of $\mathcal{I}$ in $H^1(\mathcal{A}_\mathfrak{o}, \mathcal{O}) = H^1_{\mathrm{Zar}}(\mathcal{A}_\mathfrak{o}, \mathbb{G}_a)$ coincides with the image of the extension class given by $\mathcal{F}$ under the isomorphism

$$\mathrm{Ext}^1(\mathcal{O}, \mathcal{O}) \overset{\sim}{\longrightarrow} H^1(\mathcal{A}_\mathfrak{o}, \mathcal{O}).$$

The association

$$(T \overset{t}{\to} \mathcal{A}_\mathfrak{o}) \longmapsto \underline{\mathrm{Iso}}_{\mathrm{Ext}}(\mathcal{O}_T^2, t^*\mathcal{F}) \tag{7}$$

also defines a sheaf on the flat site over $\mathcal{A}_\mathfrak{o}$, which is represented by a $\mathbb{G}_a$-torsor $Z \to \mathcal{A}_\mathfrak{o}$.

We have $H^1_{\mathrm{Zar}}(\mathcal{A}_\mathfrak{o}, \mathbb{G}_a) = H^1_{\mathrm{fppf}}(\mathcal{A}_\mathfrak{o}, \mathbb{G}_a) = \mathrm{Ext}^1(\mathcal{A}_\mathfrak{o}, \mathbb{G}_a)$, so that Z can be endowed with a group structure sitting in an extension of $\mathcal{A}_\mathfrak{o}$ by $\mathbb{G}_a$:

$$0 \longrightarrow \mathbb{G}_a \overset{j}{\longrightarrow} Z \longrightarrow \mathcal{A}_\mathfrak{o} \longrightarrow 0.$$

Hence there is a homomorphism $h: \omega_{\hat{\mathcal{A}}_\mathfrak{o}} \to \mathbb{G}_a$ such that Z is the pushout of the universal vectorial extension:

$$\begin{array}{ccccccccc}
0 & \longrightarrow & \omega_{\hat{\mathcal{A}}_\mathfrak{o}} & \longrightarrow & E_\mathfrak{o} & \longrightarrow & \mathcal{A}_\mathfrak{o} & \longrightarrow & 0 \\
 & & \downarrow{\scriptstyle h} & & \downarrow & & \| & & \\
0 & \longrightarrow & \mathbb{G}_a & \longrightarrow & Z & \longrightarrow & \mathcal{A}_\mathfrak{o} & \longrightarrow & 0.
\end{array}$$

Recall that $\theta_A: T_p A \to \omega_{\hat{A}}(\mathfrak{o})$ is defined as the limit of the maps

$$\theta_{A,n}: \mathcal{A}_{p^n}(\overline{\mathbb{Z}}_p) \longrightarrow \omega_{\hat{A}}(\mathfrak{o})/p^n \omega_{\hat{A}}(\mathfrak{o})$$

associating to $a \in \mathcal{A}_{p^n}(\overline{\mathbb{Z}}_p)$ the class of $p^n b$ for an arbitrary preimage $b \in E_\mathfrak{o}(\mathfrak{o})$ of a.

By [Ma-Me, Chapter I], the natural isomorphism

$$\mathrm{Hom}_\mathfrak{o}(\omega_{\hat{A}}(\mathfrak{o}), \mathfrak{o}) \overset{\sim}{\longrightarrow} \mathrm{Lie}(\hat{\mathcal{A}}_\mathfrak{o})(\mathfrak{o}) \overset{\sim}{\longrightarrow} H^1(\mathcal{A}_\mathfrak{o}, \mathcal{O}) = \mathrm{Ext}^1(\mathcal{A}_\mathfrak{o}, \mathbb{G}_a)$$

sends a map to the corresponding pushout of E_0. Hence $\theta_A^* : H^1(\mathcal{A}_0, \mathcal{O}) \to$ $\mathrm{Hom}_c(T_p A, \mathfrak{o})$ maps the extension class of $\mathcal{F}$ to the limit of the maps

$$\mathcal{A}_{p^n}(\overline{\mathbb{Z}}_p) \xrightarrow{\theta_{A,n}} \omega_{\hat{\mathcal{A}}}(\mathfrak{o})/p^n \omega_{\hat{\mathcal{A}}}(\mathfrak{o}) \xrightarrow{h_n} \mathfrak{o}_n,$$

where h_n is induced by h.

This map can also be described as follows: For $a \in \mathcal{A}_{p^n}(\overline{\mathbb{Z}}_p) \subseteq \mathcal{A}_{p^n}(\mathfrak{o})$ choose a preimage $z \in Z(\mathfrak{o})$. Then

$$h_n \circ \theta_{A,n}(a) = \text{class of } p^n z \text{ in } \mathbb{G}_a(\mathfrak{o})/p^n \mathbb{G}_a(\mathfrak{o}) = \mathfrak{o}_n.$$

Set $Z_n = Z \otimes_{\mathfrak{o}} \mathfrak{o}_n$. Let π_{p^n} denote multiplication by p^n on $\mathcal{A}_n$. We have seen in the proof of Theorem 1(i) that $\pi_{p^n}^* \mathcal{F}_n$ is a trivial extension. Hence $\pi_{p^n}^* Z_n$ is trivial in $\mathrm{Ext}(\mathcal{A}_n, \mathbb{G}_{a,\mathfrak{o}_n})$, and there is a splitting $r : \mathcal{A}_n \to \pi_{p^n}^* Z_n$ of the extension

$$0 \longrightarrow \mathbb{G}_{a,\mathfrak{o}_n} \longrightarrow \pi_{p^n}^* Z_n \longrightarrow \mathcal{A}_n \longrightarrow 0$$

over $\mathfrak{o}_n$. Let $g : \pi_{p^n}^* Z_n \to Z_n$ denote the projection, and denote by $a_n \in \mathcal{A}_n(\mathfrak{o}_n)$ the point induced by a. Then $g(r(a_n))$ projects to zero in $\mathcal{A}_n$, hence it is equal to $j(c)$ for some $c \in \mathbb{G}_a(\mathfrak{o}_n) = \mathfrak{o}_n$. Since for any $a' \in \mathcal{A}_n(\mathfrak{o}_n)$ with $p^n a' = a_n$ the point $g(r(a'))$ is a preimage of a_n, we have $h_n \circ \theta_{A,n}(a) = c \in \mathfrak{o}_n$. Besides, Z represents the functor (7) so that the map $r : \mathcal{A}_n \to \pi_{p^n}^* Z_n$ corresponds to a trivialization

$$
\begin{array}{ccccccccc}
0 & \longrightarrow & \mathcal{O}_{\mathcal{A}_n} & \longrightarrow & \mathcal{O}_{\mathcal{A}_n}^2 & \longrightarrow & \mathcal{O}_{\mathcal{A}_n} & \longrightarrow & 0 \\
& & \| & & \downarrow{\scriptstyle \varphi} & & \| & & \\
0 & \longrightarrow & \mathcal{O}_{\mathcal{A}_n} & \longrightarrow & \pi_{p^n}^* \mathcal{F}_n & \longrightarrow & \mathcal{O}_{\mathcal{A}_n} & \longrightarrow & 0.
\end{array}
$$

The point $g(r(a_n))$ in the kernel of $Z_n(\mathfrak{o}_n) \to \mathcal{A}_n(\mathfrak{o}_n)$ corresponds to the trivialization

$$\alpha : \mathcal{O}_{\mathrm{spec}\,\mathfrak{o}_n}^2 \xrightarrow{a_n^* \varphi} a_n^* \pi_{p^n}^* \mathcal{F} \xrightarrow{\sim} 0^* \mathcal{F},$$

where 0 is the zero element in $\mathcal{A}_n(\mathfrak{o}_n)$. Besides, the trivialization

$$\beta : \mathcal{O}_{\mathrm{spec}\,\mathfrak{o}_n}^2 \xrightarrow{0^* \varphi} 0^* \pi_{p^n}^* \mathcal{F} \xrightarrow{\sim} 0^* \mathcal{F}$$

is given by the zero element in Z_n. By definition, $\alpha = \beta + i \circ f_c \circ p$. If we denote the canonical basis of $\Gamma(\mathcal{A}_n, \mathcal{O}_{\mathcal{A}_n}^2)$ by e_1, e_2, and the induced basis of $\Gamma(\mathcal{A}_n, \pi_{p^n}^* \mathcal{F})$ by f_1, f_2, it follows that $a_n^* f_2 - 0^* f_2 = i(c)$.

On the other hand, the image of E under

$$\rho_* : \mathrm{Ext}^1_{\mathfrak{B}_{\mathcal{A}_0}}(\mathcal{O}, \mathcal{O}) \longrightarrow \mathrm{Ext}^1_{\mathbf{Rep}_{T_A(\mathfrak{o})}}(\mathfrak{o}, \mathfrak{o}) = \mathrm{Hom}_c(T_p A, \mathfrak{o})$$

is the homomorphism $\gamma : T_p A \to \mathfrak{o}$, such that $\gamma \bmod p^n$ maps the p^n-torsion point $a_n \in \mathcal{A}_n(\mathfrak{o}_n)$ to the element

$$i^{-1}(0^*(\tau_{a_n}^* f_2) - 0^* f_2) = i^{-1}(a_n^* f_2 - 0^* f_2),$$

where τ_{a_n} is translation by a_n, and hence to $c = h_n \circ \theta_{A,n}(a)$. This proves our claim.   □

Corollary 1. *Let X be a smooth, connected, projective curve over $\overline{\mathbb{Q}}_p$ with good reduction and let $x \in X(\overline{\mathbb{Q}}_p)$ be a base point. The functor $\mathfrak{B}_{X_{\mathbb{C}_p}} \to \mathbf{Rep}_{\pi_1(X,x)}(\mathbb{C}_p)$ from [De-We2] induces a homomorphism*

$$\rho_* : \mathrm{Ext}^1_{\mathfrak{B}_{X_{\mathbb{C}_p}}}(\mathcal{O}, \mathcal{O}) \longrightarrow \mathrm{Ext}^1_{\mathcal{C}}(\mathbb{C}_p, \mathbb{C}_p),$$

where $\mathcal{C}$ is the category $\mathbf{Rep}_{\pi_1(X,x)}(\mathbb{C}_p)$, which makes the following diagram commutative:

$$
\begin{array}{ccc}
\mathrm{Ext}^1_{\mathfrak{B}_{X_{\mathbb{C}_p}}}(\mathcal{O}, \mathcal{O}) & \xrightarrow{\ \rho_*\ } & \mathrm{Ext}^1_{\mathcal{C}}(\mathbb{C}_p, \mathbb{C}_p) \\
\| & & \downarrow{\wr} \\
H^1(X, \mathcal{O}) \otimes_{\overline{\mathbb{Q}}_p} \mathbb{C}_p & \xrightarrow{\text{Hodge–Tate}} & H^1_{\acute{e}t}(X, \mathbb{Q}_p) \otimes_{\mathbb{Q}_p} \mathbb{C}_p.
\end{array}
\tag{8}
$$

Here the right vertical isomorphism is defined as in the case of abelian varieties, and the lower horizontal map comes from the Hodge–Tate decomposition of $H^1_{\acute{e}t}(X, \mathbb{Q}_p) \otimes \mathbb{C}_p$.

Proof. According to [De-We2, Theorem 10], $\mathfrak{B}_{X_{\mathbb{C}_p}}$ is stable under extensions, so that every vector bundle on $X_{\mathbb{C}_p}$ which is an extension of the trivial bundle by itself lies in $\mathfrak{B}_{X_{\mathbb{C}_p}}$. By [De-We2, Proposition 21], the association $F \mapsto \rho_F$, mapping a vector bundle F to the continuous representation $\rho_F : \pi_1(X, x) \to \mathrm{Aut}_{\mathbb{C}_p}(F_x)$, respects exact sequences. Hence it induces a homomorphism on Yoneda Ext groups. Denote by $f : X \to A$ the morphism of X into its Jacobian with $f(x) = 0$. By Lemma 2, the middle square in the following diagram is commutative:

$$
\begin{array}{ccc}
H^1(A_{\mathbb{C}_p}, \mathcal{O}) & \xrightarrow[\sim]{f^*} & H^1(X_{\mathbb{C}_p}, \mathcal{O}) \\
\| & & \| \\
\mathrm{Ext}^1_{\mathfrak{B}_{A_{\mathbb{C}_p}}}(\mathcal{O}, \mathcal{O}) & \xrightarrow{f^*} & \mathrm{Ext}^1_{\mathfrak{B}_{X_{\mathbb{C}_p}}}(\mathcal{O}, \mathcal{O}) \\
\downarrow{\rho_*} & & \downarrow{\rho_*} \\
\mathrm{Ext}^1_{\mathbf{Rep}_{T A}(\mathbb{C}_p)}(\mathbb{C}_p, \mathbb{C}_p) & \xrightarrow{\tilde{f}_*} & \mathrm{Ext}^1_{\mathcal{C}}(\mathbb{C}_p, \mathbb{C}_p) \\
\downarrow{\wr} & & \downarrow{\wr} \\
H^1_{\acute{e}t}(A, \mathbb{Q}_p) \otimes \mathbb{C}_p & \xrightarrow[\sim]{f^*} & H^1_{\acute{e}t}(X, \mathbb{Q}_p) \otimes \mathbb{C}_p.
\end{array}
$$

The outer squares are also commutative. Since the Hodge–Tate decomposition is functorial, our claim follows from Theorem 2.   □

4 Line bundles on abelian varieties

Let A be an abelian variety with good reduction over $\overline{\mathbb{Q}}_p$ which is defined over the finite extension K of $\mathbb{Q}_p$ in $\overline{\mathbb{Q}}_p$, i.e., $A = A_K \otimes_K \overline{\mathbb{Q}}_p$. Then for the dual abelian variety $\hat{A}$ we have $\hat{A} = \hat{A}_K \otimes_K \overline{\mathbb{Q}}_p$. Let $\mathcal{A}$ be an abelian scheme with generic fiber A and $\hat{\mathcal{A}}$ its dual abelian scheme.

By Theorem 1, all line bundles algebraically equivalent to zero on $A_{\mathbb{C}_p}$ are contained in the category $\mathfrak{B}_{A_{\mathbb{C}_p}}$. For every line bundle L on $A_{\mathbb{C}_p}$, we have $\mathrm{Aut}_{\mathbb{C}_p}(L_x) = \mathbb{C}_p^*$. Hence ρ induces a map

$$\alpha \colon \hat{A}(\mathbb{C}_p) \longrightarrow \mathrm{Hom}_c(TA, \mathbb{C}_p^*),$$

where $\mathrm{Hom}_c(TA, \mathbb{C}_p^*)$ denotes the continuous homomorphisms from TA to $\mathbb{C}_p^*$. Note that $\mathrm{Hom}_c(TA, \mathbb{C}_p^*) = \mathrm{Hom}_c(TA, \mathfrak{o}^*)$ since $\mathbb{C}_p^* = p^{\mathbb{Q}} \times \mathfrak{o}^*$. By Theorem 1 and Proposition 1, ρ is compatible with tensor products and $G_K = \mathrm{Gal}(\overline{\mathbb{Q}}_p/K)$-action. Hence α is a G_K-equivariant homomorphism.

We endow the $\mathbb{G}_m$-torsor associated to the Poincaré bundle over $\mathcal{A} \times \hat{\mathcal{A}}$ with the structure of a biextension of $\mathcal{A}$ and $\hat{\mathcal{A}}$ by $\mathbb{G}_m$, so that we have $\hat{\mathcal{A}} = \underline{\mathrm{Ext}}^1(\mathcal{A}, \mathbb{G}_m)$ in the flat topology. For all $N \geq 1$ we denote by $\mathcal{A}_N$ and $\hat{\mathcal{A}}_N$ the subschemes of N-torsion points. The long exact Hom / Ext-sequence associated to the exact sequence

$$0 \longrightarrow \mathcal{A}_N \longrightarrow \mathcal{A} \overset{N}{\longrightarrow} \mathcal{A} \longrightarrow 0$$

induces an isomorphism $\hat{\mathcal{A}}_N \overset{\sim}{\longrightarrow} \underline{\mathrm{Hom}}(\mathcal{A}_N, \mathbb{G}_m)$. Hence we get for all $n \geq 1$ a homomorphism

$$\hat{\mathcal{A}}_N(\mathfrak{o}_n) \longrightarrow \mathrm{Hom}(\mathcal{A}_N(\mathfrak{o}_n), \mathfrak{o}_n^*).$$

By composition with the reduction map

$$A_N(\overline{\mathbb{Q}}_p) \longrightarrow \mathcal{A}_N(\overline{\mathbb{Z}}_p) \longrightarrow \mathcal{A}_N(\mathfrak{o}_n)$$

and with the projection $TA \to A_N(\overline{\mathbb{Q}}_p)$, we get a homomorphism

$$\hat{\mathcal{A}}_N(\mathfrak{o}_n) \longrightarrow \mathrm{Hom}_c(TA, \mathfrak{o}_n^*).$$

Note that here $\mathfrak{o}_n$ carries the discrete topology. For $N \mid M$ the corresponding maps are compatible with the inclusion $\hat{\mathcal{A}}_N(\mathfrak{o}_n) \hookrightarrow \hat{\mathcal{A}}_M(\mathfrak{o}_n)$.

Note that the abelian group $\hat{\mathcal{A}}(\mathfrak{o}_n)$ is torsion since it is the union of the finite groups $\hat{\mathcal{A}}_{\mathfrak{o}_K}(\mathfrak{o}_L/p^n\mathfrak{o}_L)$, where L runs over the finite extensions of K in $\overline{\mathbb{Q}}_p$. Hence we get a homomorphism $\hat{\mathcal{A}}(\mathfrak{o}_n) \to \mathrm{Hom}_c(TA, \mathfrak{o}_n^*)$. Composition with the reduction map $\hat{A}(\mathbb{C}_p) = \hat{\mathcal{A}}(\mathfrak{o}) \to \hat{\mathcal{A}}(\mathfrak{o}_n)$ induces a homomorphism

$$\alpha_n \colon \hat{A}(\mathbb{C}_p) \longrightarrow \mathrm{Hom}_c(TA, \mathfrak{o}_n^*).$$

Theorem 3. *For every $n \geq 1$ and all $\hat{a} \in \hat{A}(\mathbb{C}_p)$ the $\mathfrak{o}_n^*$-valued character $\alpha_n(\hat{a})$ is the reduction of the $\mathfrak{o}^*$-valued character $\alpha(\hat{a})$ modulo p^n. Hence the homomorphism $\alpha \colon \hat{A}(\mathbb{C}_p) \to \mathrm{Hom}_c(TA, \mathfrak{o}^*)$ is the inverse limit of the α_n.*

Proof. We denote by L the line bundle on $\mathcal{A}_{\mathfrak{o}}$ corresponding to the point $\hat{a} \in \hat{A}(\mathbb{C}_p) = \hat{\mathcal{A}}(\mathfrak{o})$. If N is big enough, the reduction $\hat{a}_n$ of $\hat{a}$ modulo p^n lies in $\hat{\mathcal{A}}_N(\mathfrak{o}_n)$. By L_n we denote the reduction of L modulo p^n, i.e., L_n is the line bundle on $\mathcal{A}_n$ corresponding to $\hat{a}_n$. Then N^*L_n is trivial on $\mathcal{A}_n$. Since we identified $\hat{\mathcal{A}}$ with $\underline{\mathrm{Ext}}^1(\mathcal{A}, \mathbb{G}_m)$, the $\mathbb{G}_m$-torsor $\tilde{L}_n = L_n \setminus \{\text{zero section}\}$ on $\mathcal{A}_n$ is endowed with the structure of an extension of $\mathcal{A}_n$ by $\mathbb{G}_{m,\mathfrak{o}_n}$. Moreover, with this identification the inclusion

$$i : \underline{\mathrm{Hom}}(\mathcal{A}_N, \mathbb{G}_m) \simeq \hat{\mathcal{A}}_N \hookrightarrow \hat{\mathcal{A}}$$

is given by pushout with respect to the exact sequence $0 \to \mathcal{A}_N \to \mathcal{A} \xrightarrow{N} \mathcal{A} \to 0$. Denote by $\varphi : \mathcal{A}_{n,N} \to \mathbb{G}_{m,\mathfrak{o}_n}$ the homomorphism corresponding to $\hat{a}_n \in \hat{\mathcal{A}}_N(\mathfrak{o}_n)$. Then $\tilde{L}_n$ is given by the following pushout diagram:

$$
\begin{array}{ccccccccc}
0 & \longrightarrow & \mathcal{A}_{n,N} & \longrightarrow & \mathcal{A}_n & \xrightarrow{N} & \mathcal{A}_n & \longrightarrow & 0 \\
& & \varphi \downarrow & & s \downarrow & & \| & & \\
0 & \longrightarrow & \mathbb{G}_{m,\mathfrak{o}_n} & \longrightarrow & \tilde{L}_n & \longrightarrow & \mathcal{A}_n & \longrightarrow & 0.
\end{array}
\tag{9}
$$

By definition, $\alpha_n(\hat{a})$ is the map

$$\alpha_n(\hat{a}) : TA \longrightarrow A_N(\overline{\mathbb{Q}}_p) = \mathcal{A}_N(\mathbb{Z}_p) \longrightarrow \mathcal{A}_N(\mathfrak{o}_n) = \mathcal{A}_{n,N}(\mathfrak{o}_n) \xrightarrow{\varphi} \mathfrak{o}_n^*.$$

On the other hand, the reduction of

$$\alpha(\hat{a}) : TA \longrightarrow A_N(\overline{\mathbb{Q}}_p) \longrightarrow \mathfrak{o}^*$$

modulo p^n is obtained from the map

$$A_N(\overline{\mathbb{Q}}_p) \longrightarrow \mathfrak{o}_n^*$$

associating to $a \in A_N(\overline{\mathbb{Q}}_p) = \mathcal{A}_N(\mathbb{Z}_p)$ the element in $\mathfrak{o}_n^*$ corresponding to the natural action of $a_n \in \mathcal{A}_{n,N}(\mathfrak{o}_n)$ on $\Gamma(\mathcal{A}_n, N^*L_n) \xrightarrow{\sim} \mathfrak{o}_n$. Here we can as well regard the natural action of $a_n \in \mathcal{A}_{n,N}(\mathfrak{o}_n)$ on $\Gamma(\mathcal{A}_n, N^*\tilde{L}_n) \xrightarrow{\sim} \mathfrak{o}_n^*$ in the setting of $\mathbb{G}_m$-torsors.

Now the homomorphism $s : \mathcal{A}_n \to \tilde{L}_n$ from diagram (9) induces an element

$$s_0 = (s, \mathrm{id}) : \mathcal{A}_n \longrightarrow \tilde{L}_n \times_{\mathcal{A}_{n,N}} \mathcal{A}_n = N^*\tilde{L}_n \quad \text{in } \Gamma(\mathcal{A}_n, N^*\tilde{L}_n)$$

which is mapped to $s_0 \circ t_{a_n}$ via the action of a_n, where t_{a_n} denotes translation by a_n on $\mathcal{A}_n$. By diagram (9), the corresponding element in $\mathfrak{o}_n^*$ is equal to $\varphi(a_n)$. Therefore, $\alpha(\hat{a})$ reduces to $\alpha_n(\hat{a})$ modulo p^n. $\qquad\square$

In [Ta, Section 4], Tate considers the homomorphism

$$\alpha_T : \hat{A}(p)(\mathfrak{o}) \longrightarrow \mathrm{Hom}_c(T(A(p)), U_1)$$

defined by duality of the p-divisible groups $\mathcal{A}(p)$ and $\hat{\mathcal{A}}(p)$. Here U_1 denotes the group of units congruent to 1 in $\mathfrak{o}$. It follows from Theorem 3 that α_T coincides with the restriction of α to the open subgroup $\hat{\mathcal{A}}(p)(\mathfrak{o})$ of $\hat{\mathcal{A}}(\mathbb{C}_p)$.

We now consider the restriction α_{tors} of the map α to the torsion part of $\hat{A}(\mathbb{C}_p)$:

$$\alpha_{\mathrm{tors}} \colon \hat{A}(\mathbb{C}_p)_{\mathrm{tors}} = \hat{A}(\overline{\mathbb{Q}}_p)_{\mathrm{tors}} \longrightarrow \mathrm{Hom}_c(TA, \mathfrak{o}^*)_{\mathrm{tors}} = \mathrm{Hom}_c(TA, \mu).$$

Here $\mu = \mu(\overline{\mathbb{Q}}_p)$ is the group of roots of unity in $\mathfrak{o}^*$ or $\overline{\mathbb{Q}}_p^*$. Note that the Kummer sequence on $A_{\text{ét}}$ induces an isomorphism

$$i_A \colon H^1(A, \mu_N) \xrightarrow{\sim} H^1(A, \mathbb{G}_m)_N.$$

Proposition 2. *The map α_{tors} is an isomorphism. On $\hat{A}_N(\overline{\mathbb{Q}}_p)$ it coincides with the composition:*

$$\hat{A}_N(\overline{\mathbb{Q}}_p) = H^1(A, \mathbb{G}_m)_N \xrightarrow{i_A^{-1}} H^1(A, \mu_N) = \mathrm{Hom}_c(TA, \mu_N).$$

Proof. By Theorem 3, the restriction of α to $\hat{A}_N(\overline{\mathbb{Q}}_p)$ is the map

$$\hat{A}_N(\overline{\mathbb{Q}}_p) \longrightarrow \mathrm{Hom}(A_N(\overline{\mathbb{Q}}_p), \mu_N) = \mathrm{Hom}(TA, \mu_N)$$

coming from Cartier duality $\hat{A}_N \simeq \underline{\mathrm{Hom}}(A_N, \mathbb{G}_m)$ over $\overline{\mathbb{Q}}_p$. The canonical identification $\mathrm{Hom}_c(TA, \mu_N) = H^1(A, \mu_N)$ can be factorized by the isomorphisms

$$\mathrm{Hom}(A_N, \mu_N) \xrightarrow{\sim} \mathrm{Ext}^1(A, \mu_N) \xrightarrow{\sim} H^1(A, \mu_N),$$

where the first map is induced by the exact sequence $0 \to A_N \to A \to A \to 0$ and the second map is the forgetful map associating to an extension the corresponding μ_N-torsor. Since the diagram

$$
\begin{array}{ccccc}
\mathrm{Hom}(A_N, \mu_N) & \xrightarrow{\ \sim\ } & \mathrm{Ext}^1(A, \mu_N) & \xrightarrow{\ \sim\ } & H^1(A, \mu_N) \\
\wr\downarrow & & \downarrow & & \wr\downarrow{\scriptstyle i_A} \\
\mathrm{Hom}(A_N, \mathbb{G}_m) & \xrightarrow{\ \sim\ } & \mathrm{Ext}^1(A, \mathbb{G}_m)_N & \xrightarrow{\ \sim\ } & H^1(A, \mathbb{G}_m)_N \\
\wr\uparrow & & & & \| \\
\hat{A}_N(\overline{\mathbb{Q}}_p) & = \!\!=\!\!= & & & \hat{A}_N(\overline{\mathbb{Q}}_p)
\end{array}
$$

commutes, our claim follows. $\qquad\square$

Next, we need an elementary lemma. Consider an abelian topological group π which fits into an exact sequence of topological groups

$$0 \longrightarrow H \longrightarrow \pi \longrightarrow \hat{\mathbb{Z}}^n \longrightarrow 0$$

where H is a finite discrete group. Later π will be the abelianized fundamental group of an algebraic variety. Applying the functor $\mathrm{Hom}_c(\pi, \_)$ to the exact sequence

$$0 \longrightarrow \mu \longrightarrow \mathfrak{o}^* \xrightarrow{\ \log\ } \mathbb{C}_p \longrightarrow 0,$$

we get the sequence

$$0 \longrightarrow \mathrm{Hom}_c(\pi, \mu) \longrightarrow \mathrm{Hom}_c(\pi, \mathfrak{o}^*) \xrightarrow{\ \log_*\ } \mathrm{Hom}_c(\pi, \mathbb{C}_p) \longrightarrow 0. \qquad (10)$$

Lemma 3.

(a) *The sequence* (10) *is exact.*
(b) *In the topology of uniform convergence* $\mathrm{Hom}_c(\pi, \mathfrak{o}^*)$ *is a complete topological group. It contains* $(\mathfrak{o}^*)^n$ *as an open subgroup of finite index and hence acquires a natural structure as a Lie group over* $\mathbb{C}_p$. *Its Lie algebra is* $\mathrm{Hom}_c(\pi, \mathbb{C}_p)$ *and the logarithm map is given by* $\log_*$.

Proof. (a) Since $\mathrm{Hom}_c(\pi, \_)$ is left exact, it suffices to show that $\log_*$ is surjective. As $\mathrm{Hom}_c(H, \mathbb{C}_p) = 0$ we only have to show surjectivity of $\log_*$ for $\pi = \hat{\mathbb{Z}}^n$, hence for $\pi = \hat{\mathbb{Z}}$. We first prove that the injective evaluation map $\mathrm{ev}: \mathrm{Hom}_c(\hat{\mathbb{Z}}, \mathfrak{o}^*) \longrightarrow \mathfrak{o}^*$, $\mathrm{ev}(\varphi) = \varphi(1)$ is surjective.

Set $U_1 = \{x \in \mathfrak{o}^* \mid |x - 1| < 1\}$ and $U_0 = \{x \in \mathfrak{o}^* \mid |x - 1| < p^{-\frac{1}{p-1}}\}$. The logarithm provides an isomorphism

$$\log: U_0 \xrightarrow{\ \sim\ } V_0 = \{x \in \mathbb{C}_p \mid |x| < p^{-\frac{1}{p-1}}\},$$

whose inverse is the exponential map. Therefore, U_0 is a $\mathbb{Z}_p$-module and it follows that $\mathrm{ev}: \mathrm{Hom}_c(\mathbb{Z}_p, U_0) \xrightarrow{\ \sim\ } U_0$ is an isomorphism. We claim that $\mathrm{ev}: \mathrm{Hom}_c(\mathbb{Z}_p, U_1) \hookrightarrow U_1$ is an isomorphism as well. Fix some b in U_1. We construct a continuous map $\psi: \mathbb{Z}_p \to U_1$ with $\psi(1) = b$ as follows. There is some $N \geq 1$ such that $b^{p^N} \in U_0$. Hence there is a continuous homomorphism $\varphi: p^N \mathbb{Z}_p \longrightarrow U_1$ such that $\varphi(p^N v) = (b^{p^N})^v$ for all $v \in \mathbb{Z}$. Because of the decomposition $\mathfrak{o}^* = \mu_{(p)} \times U_1$ the group U_1 is divisible. Hence there is a homomorphism $\psi': \mathbb{Z}_p \to U_1$ whose restriction to $p^N \mathbb{Z}_p$ equals φ. It follows that ψ' is continuous as well. Because of $\psi'(1)^{p^N} = \psi'(p^N) = b^{p^N}$ there is a root of unity $\zeta \in \mu_{p^N}$ with $\psi'(1) = \zeta b$. Take the continuous homomorphism $\psi'': \mathbb{Z}_p \to \mu_{p^\infty} \subset U_1$ with $\psi''(1) = \zeta^{-1}$ and set $\psi = \psi' \cdot \psi''$.

The natural projection $\hat{\mathbb{Z}} \to \mathbb{Z}_p$ induces a commutative diagram

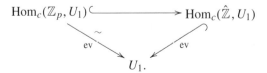

It follows that ev: $\operatorname{Hom}_c(\hat{\mathbb{Z}}, U_1) \to U_1$ is an isomorphism as well. Using the decomposition $\mathfrak{o}^* = \mu_{(p)} \times U_1$ where $\mu_{(p)}$ carries the discrete topology we conclude that ev: $\operatorname{Hom}_c(\hat{\mathbb{Z}}, \mathfrak{o}^*) \longrightarrow \mathfrak{o}^*$ is an isomorphism. Using the commutative diagram

$$
\begin{array}{ccc}
\operatorname{Hom}_c(\hat{\mathbb{Z}}, \mathfrak{o}^*) & \xrightarrow{\ \log_* \ } \operatorname{Hom}_c(\hat{\mathbb{Z}}, \mathbb{C}_p) == \operatorname{Hom}_c(\mathbb{Z}_p, \mathbb{C}_p) \\
\Big\downarrow{\wr\, ev} & \qquad\qquad\qquad\qquad\quad \Big\downarrow{\wr\, ev} \\
\mathfrak{o}^* & \xrightarrow{\qquad\ \log\ \qquad} \mathbb{C}_p,
\end{array}
$$

we see that $\log_*$ is surjective for $\pi = \hat{\mathbb{Z}}$ and hence in general.

(b) With the topology of uniform convergence, $\operatorname{Hom}_c(\pi, \mathfrak{o}^*)$ becomes a topological group. This topology comes from the inclusion of $\operatorname{Hom}_c(\pi, \mathfrak{o}^*)$ into the p-adic Banach space $C^0(\pi, \mathbb{C}_p)$ of continuous functions from π to $\mathbb{C}_p$ with the norm

$$
\|f\| = \max_{\gamma \in \pi} |f(\gamma)|.
$$

Since $\operatorname{Hom}_c(\pi, \mathfrak{o}^*)$ is closed in $C^0(\pi, \mathbb{C}_p)$ it becomes a complete metric space and hence it is a complete topological group. We now observe that the continuous evaluation map ev: $\operatorname{Hom}_c(\hat{\mathbb{Z}}, \mathfrak{o}^*) \xrightarrow{\sim} \mathfrak{o}^*$ is actually a homeomorphism. Let $x_n \to x$ be a convergent sequence in $\mathfrak{o}^*$ and let $\varphi_\nu, \varphi \colon \hat{\mathbb{Z}} \to \mathfrak{o}^*$ be the continuous homomorphisms with $\varphi_\nu(1) = x_\nu$ and $\varphi(1) = x$. Since $\mathbb{Z}^{\geq 1}$ is dense in $\hat{\mathbb{Z}}$ we get

$$
\|\varphi - \varphi_\nu\| = \max_{\gamma \in \hat{\mathbb{Z}}} |\varphi(\gamma) - \varphi_\nu(\gamma)| = \sup_{n \geq 1} |\varphi(n) - \varphi_\nu(n)| = \sup_{n \geq 1} |x^n - x_\nu^n|
$$

$$
= |x - x_\nu| \sup_{n \geq 1} |\sum_{i=0}^{n-1} x^i x_\nu^{n-i-1}| \leq |x - x_\nu|.
$$

Hence φ_ν converges uniformly to φ.

It follows that $\operatorname{Hom}_c(\hat{\mathbb{Z}}^n, \mathfrak{o}^*)$ and $(\mathfrak{o}^*)^n$ are isomorphic as topological groups. The exact sequence of topological groups

$$
0 \to \operatorname{Hom}_c(\hat{\mathbb{Z}}^n, \mathfrak{o}^*) \to \operatorname{Hom}_c(\pi, \mathfrak{o}^*) \to \operatorname{Hom}_c(H, \mathfrak{o}^*) = \operatorname{Hom}_c(H, \mu_{|H|})
$$

shows that $\operatorname{Hom}_c(\pi, \mathfrak{o}^*)$ contains $(\mathfrak{o}^*)^n$ as an open subgroup of finite index. Hence $\operatorname{Hom}_c(\pi, \mathfrak{o}^*)$ becomes a Lie group over $\mathbb{C}_p$. It is clear that the analytic structure depends only on π and not on the choice of an exact sequence $0 \to H \to \pi \to \hat{\mathbb{Z}}^n \to 0$ as above. The remaining assertions have to be checked for $\pi = \hat{\mathbb{Z}}^n$ and hence for $\pi = \hat{\mathbb{Z}}$ only where they are clear by the preceeding observations. $\square$

Remark 2. The proof shows that the topologies of uniform and pointwise convergence on $\operatorname{Hom}_c(\pi, \mathfrak{o}^*)$ coincide.

By [Bou, III, Section 7, number 6], there is a logarithm map on an open subgroup U of the p-adic Lie group $\hat{A}(\mathbb{C}_p)$, mapping $U \to \operatorname{Lie} \hat{A}(\mathbb{C}_p)$, such that the kernel consists of the torsion points in U. Since $\hat{A}(\mathbb{C}_p)/U$ is torsion (see [Co2, Theorem 4.1]), the logarithm has a unique extension to the whole Lie group $\hat{A}(\mathbb{C}_p)$. It is surjective since the $\mathbb{C}_p$-vector space Lie $\hat{A}(\mathbb{C}_p)$ is uniquely divisible. Therefore, we have the exact sequence

$$0 \longrightarrow \hat{A}(\mathbb{C}_p)_{\text{tors}} \longrightarrow \hat{A}(\mathbb{C}_p) \overset{\log}{\longrightarrow} \text{Lie } \hat{A}(\mathbb{C}_p) = H^1(A, \mathcal{O}) \otimes \mathbb{C}_p \longrightarrow 0. \quad (11)$$

Using Proposition 2 and Lemma 3, we therefore get a commutative diagram with exact rows and G_K-equivariant maps

$$
\begin{array}{ccccccccc}
0 & \longrightarrow & \hat{A}(\mathbb{C}_p)_{\text{tors}} & \longrightarrow & \hat{A}(\mathbb{C}_p) & \overset{\log}{\longrightarrow} & \text{Lie } \hat{A}(\mathbb{C}_p) & = & H^1(A,\mathcal{O}) \otimes_{\overline{\mathbb{Q}}_p} \mathbb{C}_p \longrightarrow 0 \\
& & \alpha_{\text{tors}} \downarrow \wr & & \alpha \downarrow & & \downarrow \tilde{\alpha} & & \\
0 & \rightarrow & \text{Hom}_c(TA, \mu) & \rightarrow & \text{Hom}_c(TA, \mathfrak{o}^*) & \rightarrow & \text{Hom}_c(TA, \mathbb{C}_p) & = & H^1_{\text{ét}}(A, \mathbb{Q}_p) \otimes \mathbb{C}_p \longrightarrow 0.
\end{array}
\quad (12)
$$

Here $\tilde{\alpha}$ is the map induced by α on the quotients.

We will prove next that α is a p-adically analytic map of p-adic Lie groups. It follows that $\tilde{\alpha} = \text{Lie } \alpha$.

Lemma 4. *Let* $\beta \colon \hat{A}(\mathbb{C}_p) \times TA \to \mathfrak{o}^*$ *be the pairing induced by the homomorphism* $\alpha \colon \hat{A}(\mathbb{C}_p) \to \text{Hom}_c(TA, \mathfrak{o}^*)$. *Then* β *is continuous. In particular,* α *is continuous.*

Proof. Denote by $r_n \colon \hat{A}(\mathfrak{o}) \to \hat{A}(\mathfrak{o}_n)$ the reduction map. Since the kernel of r_n is p-adically open, it contains an open neighborhood $W \subseteq \hat{A}(\mathbb{C}_p)$ of zero.

Fix $(\hat{a}, \gamma) \in \hat{A}(\mathbb{C}_p) \times TA$ mapping to $z = \beta(\hat{a}, \gamma)$. We show that the preimage of the open neighborhood $z(1 + p^n \mathfrak{o})$ is open. Let $N \geq 1$ be big enough so that $r_n(\hat{a})$ is contained in $\hat{A}_N(\mathfrak{o}_n)$. If U denotes the kernel of the projection $TA \to A_N(\overline{\mathbb{Q}}_p)$, the neighborhood $(\hat{a} + W, \gamma + U)$ of $(\hat{a}, \gamma)$ maps to $z(1 + p^n \mathfrak{o})$ under β. Since the topology on $\text{Hom}_c(TA, \mathfrak{o}^*)$ is the topology of pointwise convergence by the remark following Lemma 3, continuity of β implies continuity of α.   □

Let us fix some $\gamma \in TA$, and denote by ψ_γ the induced homomorphism

$$\psi_\gamma = \beta(-, \gamma) \colon \hat{A}(\mathbb{C}_p) \longrightarrow \mathfrak{o}^*.$$

Proposition 3. *ψ_γ is an analytic map, hence a Lie group homomorphism.*

Proof. We will briefly write $\psi = \psi_\gamma$ in this proof. It suffices to show that ψ is analytic in a neighborhood of the zero element $e_{\mathbb{C}_p} \in \hat{A}(\mathbb{C}_p)$.

Let $e \in \hat{\mathcal{A}}_\mathfrak{o}(\mathfrak{o})$ be the zero section of $\hat{\mathcal{A}}_\mathfrak{o}$. Since $\hat{\mathcal{A}}_\mathfrak{o}$ is smooth over $\mathfrak{o}$, there is a Zariski open neighborhood $U \subseteq \hat{\mathcal{A}}_\mathfrak{o}$ of e of the form

$$U = \text{spec } \mathfrak{o}[x_1, \ldots, x_{m+r}]/(f_1, \ldots, f_m)$$

such that the matrix $(\frac{\partial f_i}{\partial x_{r+j}}(e))_{i,j=1\ldots m}$ is invertible over $\mathfrak{o}$.

By the theorem of implicit functions (see, e.g., [Col, A.3.4]), $U(\mathfrak{o})$ contains an open neighborhood V of e in the p-adic topology, such that the projection map $q \colon U(\mathfrak{o}) \subseteq$

$\mathfrak{o}^{m+r} \to \mathfrak{o}^r$ given by $(x_1, \ldots, x_{m+r}) \longmapsto (x_1, \ldots, x_r)$ maps e to $(0, \ldots, 0)$ and induces a homeomorphism

$$q : V \longrightarrow V_1$$

between V and an open ball $V_1 \subseteq \mathfrak{o}^r$ around zero. This is an analytic chart around $e_{\mathbb{C}_p}$. A function f on V is locally analytic around $e_{\mathbb{C}_p}$, iff it induces a function on V_1 which coincides on a ball $V_1' \subseteq V_1$ around 0 with a power series in $x_1, \ldots, x_r$ converging pointwise on V_1'.

Since by [Bou, III, Section 7, Propositions 10 and 11] the logarithm map on $\hat{A}(\mathbb{C}_p)$ is locally around $e_{\mathbb{C}_p}$ an analytic isomorphism respecting the group structures, there exists an open subgroup H of $\hat{A}(\mathbb{C}_p)$ such that $(p^\nu H)_{\nu \geq 0}$ is a basis of open neighborhoods of $e_{\mathbb{C}_p}$. By shrinking V if necessary, we can assume that $V \subseteq H$.

For $n \geq 1$ we denote by r_n as before the reduction map

$$r_n : \hat{A}(\mathbb{C}_p) = \hat{A}(\mathfrak{o}) \longrightarrow \hat{A}(\mathfrak{o}_n) = \hat{A}_n(\mathfrak{o}_n),$$

where $\hat{A}_n = \hat{A} \otimes \mathfrak{o}_n$.

Since the kernel of r_n is an open subgroup of $\hat{A}(\mathbb{C}_p)$, it contains $p^{\nu_n} H$ for a suitable $\nu_n \geq 0$. Hence $p^{\nu_n} V$ is contained in the kernel of r_n, which implies that for all $x \in V$ the point $r_n(x)$ lies in the scheme $\hat{A}_{n, p^{\nu_n}}$ of p^{ν_n}-torsion points in $\hat{A}_n$.

The element γ in TA induces a point in $\mathcal{A}_{n, p^{\nu_n}}(\mathfrak{o}_n)$, whose image under the Cartier duality morphism

$$\mathcal{A}_{n, p^{\nu_n}}(\mathfrak{o}_n) \longrightarrow \operatorname{Hom}(\hat{\mathcal{A}}_{n, p^{\nu_n}}, \mathbb{G}_{m, \mathfrak{o}_n})$$

we denote by ψ_n. Then $\psi(x)$ for $x \in V$ is by definition the element in $\mathfrak{o}^* \subseteq \mathfrak{o}$ satisfying $\psi(x) \equiv \psi_n(r_n(x)) \bmod p^n$ for all n.

Let $U_n = U \otimes_{\mathfrak{o}} \mathfrak{o}_n = \operatorname{spec} \mathfrak{o}_n[x_1, \ldots, x_{m+r}]/(\overline{f}_1, \ldots, \overline{f}_m)$ be the reduction of the affine subscheme $U \subseteq \hat{A}_{\mathfrak{o}}$. We write $\overline{f}$ for the reduction of a polynomial f over $\mathfrak{o}$ modulo p^n.

Then $U_n \cap \hat{A}_{n, p^{\nu_n}} = \operatorname{spec} \mathfrak{o}_n[x_1, \ldots, x_{m+r}]/\mathfrak{a}$ for some ideal $\mathfrak{a}$ containing $(\overline{f}_1, \ldots, \overline{f}_m)$. Since ψ_n is an algebraic morphism, it is given on $U_n \cap \hat{A}_{n, p^{\nu_n}}$ by a unit in $\mathfrak{o}_n[x_1, \ldots, x_{m+r}]/\mathfrak{a}$, which is induced by a polynomial $\overline{g}_n \in \mathfrak{o}_n[x_1, \ldots, x_{m+r}]$. Let $g_n \in \mathfrak{o}[x_1, \ldots, x_{m+r}]$ be a lift of $\overline{g}_n$.

The implicit function theorem also implies that possibly after shrinking V and V_1, we find power series $\theta_1, \ldots, \theta_m \in \mathbb{C}_p[[x_1, \ldots, x_r]]$ converging in all points of $V_1 \subseteq \mathfrak{o}^r$, such that the map $V_1 \xrightarrow{q^{-1}} V \subseteq U(\mathfrak{o}) \subseteq \mathfrak{o}^{m+r}$ is given by

$$(x_1, \ldots, x_r) \longmapsto (x_1, \ldots, x_r, \theta_1(x_1, \ldots, x_r), \ldots, \theta_m(x_1, \ldots, x_r)).$$

For all $i = 1, \ldots, m$ and all $n \geq 1$ let $h_{i,n} \in \mathbb{C}_p[x_1, \ldots, x_r]$ be a polynomial satisfying $\theta_i(x) - h_{i,n}(x) \in p^n \mathfrak{o}$ for all $x \in V_1$. We can obtain $h_{i,n}$ by truncating θ_i suitably.

Then the map $V_1 \xrightarrow{\sim} V \xrightarrow{r_n} U_n(\mathfrak{o}_n) \cap \hat{A}_{n, p^{\nu_n}}(\mathfrak{o}_n) \xrightarrow{\psi_n} \mathfrak{o}_n^*$ maps the point $(x_1, \ldots, x_r)$ to $\overline{g}_n(\overline{x}_1, \ldots, \overline{x}_r, \overline{\theta_1(x_1, \ldots, x_r)}, \ldots, \overline{\theta_m(x_1, \ldots, x_r)})$.

Hence for all $x = (x_1, \ldots, x_r) \in V_1$ we have

$$\psi(q^{-1}(x)) - g_n(x_1, \ldots, x_r, h_{1,n}(x_1, \ldots, x_r), \ldots, h_{m,n}(x_1, \ldots, x_r)) \in p^n \mathfrak{o}.$$

Thus ψ is the uniform limit of polynomials with respect to the coordinate chart V_1. This implies our claim.   □

Corollary 2. *The homomorphism* $\alpha \colon \hat{A}(\mathbb{C}_p) \longrightarrow \mathrm{Hom}_c(TA, \mathfrak{o}^*)$ *is analytic.*

Proof. Recall from the proof of Lemma 3 that evaluation in $1 \in \hat{\mathbb{Z}}$ induces an iso-morphism $\mathrm{Hom}_c(\hat{\mathbb{Z}}, \mathfrak{o}^*) \simeq \mathfrak{o}^*$. Hence $\mathrm{Hom}_c(TA, \mathfrak{o}^*) \simeq (\mathfrak{o}^*)^{2g}$ by evaluation in a $\hat{\mathbb{Z}}$-basis $\gamma_1, \ldots, \gamma_{2g}$ of TA. This isomorphism induces the analytic structure on $\mathrm{Hom}_c(TA, \mathfrak{o}^*)$. Hence our claim follows from Proposition 3.   □

Let us now determine the Lie algebra map induced by α.

Recall from Section 3 that the map

$$\theta_A^* \colon H^1(A_K, \mathcal{O}) \otimes_K \mathbb{C}_p \longrightarrow H^1_{\text{ét}}(A, \mathbb{Q}_p) \otimes \mathbb{C}_p$$

coming from the Hodge–Tate decomposition of $H^1_{\text{ét}}(A, \mathbb{Q}_p) \otimes \mathbb{C}_p$ is the dual of a $\mathbb{Z}_p$-linear homomorphism

$$\theta_A \colon T_p A \longrightarrow \omega_{\hat{A}}(\mathfrak{o})$$

defined using the universal vectorial extension of $\mathcal{A}$.

Theorem 4. *We have* $\mathrm{Lie}\,\alpha = \theta_A^*$.

Proof. We give two proofs of this fact.

(1) According to [Ta, Section 4], the diagram

$$
\begin{array}{ccccc}
\hat{\mathcal{A}}(p)(\mathfrak{o}) & \xrightarrow{\;\log\;} & \mathrm{Lie}\,\hat{\mathcal{A}}(p) & =\!=\!= & H^1(A, \mathcal{O}) \otimes_{\overline{\mathbb{Q}}_p} \mathbb{C}_p \\[2pt]
{\scriptstyle \alpha_T}\downarrow & & \downarrow{\scriptstyle \mathrm{Lie}\,\alpha_T} & & \\[2pt]
\mathrm{Hom}_c(T(\mathcal{A}(p)), U_1) & \xrightarrow{\;\log_*\;} & \mathrm{Hom}_c(T(\mathcal{A}(p)), \mathbb{C}_p) & =\!=\!= & H^1_{\text{ét}}(A, \mathbb{Q}_p) \otimes \mathbb{C}_p
\end{array}
$$

commutes. As we have seen, α_T is the restriction of α to $\hat{\mathcal{A}}(p)(\mathfrak{o})$. Combining Cole-man's work in [Co1] and [Co3, Section 4] with Fontaine's results, specifically [Fo, Proposition 11], it follows that $\mathrm{Lie}\,\alpha_T = \theta_A^*$. Hence $\mathrm{Lie}\,\alpha = \theta_A^*$.

(2) The result can also be proved directly. Consider for $\gamma \in T_p(A)$ the analytic map $\psi_\gamma \colon \hat{A}(\mathbb{C}_p) \longrightarrow \mathfrak{o}^*$ induced by α. It induces a Lie algebra homomorphism

$$\mathrm{Lie}\,\psi_\gamma \colon \mathrm{Lie}\,\hat{A}(\mathbb{C}_p) \longrightarrow \mathrm{Lie}\,\mathfrak{o}^* \xrightarrow{\;\sim\;} \mathfrak{o},$$

where we identify $\mathrm{Lie}\,\mathfrak{o}^*$ with $\mathfrak{o}$ by means of the invariant differential $\frac{dT}{T}$ on $\mathbb{G}_{m,\mathfrak{o}} = \mathrm{spec}\,\mathfrak{o}[T, T^{-1}]$.

It suffices to show that for all $\gamma \in T_p A$ the map $\mathrm{Lie}\,\psi_\gamma$ is given by the invariant differential $\theta_A(\gamma)$, i.e., that

$$\psi_\gamma^* \frac{dT}{T} = \theta_A(\gamma).$$

We use the notation from the proof of Proposition 3. Fix some $n \geq 1$ and recall the morphism

$$\psi_n : U_n \cap \hat{A}_{n,p^{\nu n}} = \mathrm{spec}\, \mathfrak{o}_n[x_1, \ldots, x_{m+r}]/\mathfrak{a} \longrightarrow \mathbb{G}_{m,\mathfrak{o}_n}$$

induced by a polynomial $\overline{g}_n \in \mathfrak{o}_n[x_1, \ldots, x_{m+r}]$. This $\overline{g}_n$ defines a morphism

$$\varphi_n^{(n)} : U_n \longrightarrow \mathbb{G}_{a,\mathfrak{o}_n}.$$

Denote by $i : \mathbb{G}_{m,\mathfrak{o}} \to \mathbb{G}_{a,\mathfrak{o}}$ the obvious closed immersion, and by i_n the induced morphism over $\mathfrak{o}_n$. The diagram

is commutative, where κ_n is the canonical closed immersion.

Similarly, the lift g_n of $\overline{g}_n$ to $\mathfrak{o}[x_1, \ldots, x_{m+r}]$ defines a morphism $\varphi^{(n)} : U \to \mathbb{G}_{a,\mathfrak{o}}$ over $\mathfrak{o}$, which induces an analytic map $\varphi^{(n)} : U(\mathfrak{o}) \longrightarrow \mathfrak{o}$.

The space of invariant differentials $\omega_{\hat{A}}(\mathfrak{o})$ is an $\mathfrak{o}$-lattice in the $\mathbb{C}_p$-vector space $\omega_{\hat{A}}(\mathbb{C}_p)$. For any analytic map $h : V \to \mathbb{C}_p$ we denote by $(dh)_e$ the element in the cotangential space $\omega_{\hat{A}}(\mathbb{C}_p) \simeq M_{e_{\mathbb{C}_p}}/M_{e_{\mathbb{C}_p}}^2$ given by the class of $h - h(e)$ modulo $M_{e_{\mathbb{C}_p}}^2$. Here $M_{e_{\mathbb{C}_p}}$ is the ideal of germs of analytic functions vanishing at $e_{\mathbb{C}_p}$.

If $j : U \hookrightarrow \mathbb{A}_{\mathfrak{o}}^{m+r}$ denotes the obvious closed immersion, the fact that $(\frac{\partial f_i}{\partial x_{j+r}}(e))_{i,j=1\ldots m}$ is invertible implies that

$$\omega_{\hat{A}}(\mathfrak{o}) = \Gamma(\mathrm{spec}\,\mathfrak{o}, e^* \Omega_{U/\mathfrak{o}}^1)$$

is freely generated over $\mathfrak{o}$ by the differentials $j^*(dx_1)_e, \ldots, j^*(dx_r)_e$.

For all $x \in V$ we have $i \circ \psi(x) \equiv i_n \circ \psi_n(r_n(x)) \bmod p^n$, hence $i \circ \psi(x) - \varphi^{(n)}(x) \in p^n \mathfrak{o} \subseteq \mathfrak{o}$. Since $\varphi^{(n)}$ is a polynomial map, we may assume by shrinking V, if necessary, that for all n the function $(i \circ \psi - \varphi^{(n)}) \circ q^{-1}$ is given by a pointwise converging power series $\sum_{I=(i_1,\ldots,i_r)} a_I^{(n)} x_1^{i_1} \ldots x_r^{i_r}$ on the chart $V_1 \xrightarrow{q^{-1}} V$, where $V_1 = (p^t \mathfrak{o})^r$ for some $t \geq 0$. For every multiindex I this implies $p^{t(i_1+\cdots+i_r)-n} a_I^{(n)} \in \mathfrak{o}$. Then

$$d(i \circ \psi - \varphi^{(n)})_e = a_{(1,0,\ldots,0)}^{(n)}(j^* dx_1)_e + \cdots + a_{(0,\ldots,0,1)}^{(n)}(j^* dx_r)_e \in p^{n-t} \omega_{\hat{A}}(\mathfrak{o}).$$

In particular, for $n \geq t$ this implies $d(i \circ \psi)_e \in \omega_{\hat{A}}(\mathfrak{o})$. Under the isomorphism $\omega_{\hat{A}}(\mathfrak{o})/p^n \omega_{\hat{A}}(\mathfrak{o}) \to \omega_{\hat{A}}(\mathfrak{o}_n)$, the element $(d\varphi^{(n)})_e \in \omega_{\hat{A}}(\mathfrak{o})$ maps to $(d\varphi_n^{(n)})_e \in \omega_{\hat{A}}(\mathfrak{o}_n)$. Moreover, the diagram above implies that

$$\kappa_n^*(d\varphi_n^{(n)})_e = (d(i_n \circ \psi_n))_e = \psi_n^* \left(\frac{dT}{T} \right)_e .$$

Besides, we can assume that $v_n \geq n$, so that $\mathfrak{o}_n$ is annihilated by p^{v_n}. Then the exact sequence

$$\omega_{\hat{\mathcal{A}}_n} \xrightarrow{(p^{v_n})^*} \omega_{\hat{\mathcal{A}}_n} \longrightarrow \omega_{\hat{\mathcal{A}}_{n,p^{v_n}}} \longrightarrow 0$$

induces an isomorphism $\kappa_n^* \colon \omega_{\hat{\mathcal{A}}_n} \xrightarrow{\sim} \omega_{\hat{\mathcal{A}}_{n,p^{v_n}}}$. This implies

$$\psi^* \left(\frac{dT}{T} \right)_e = d(i \circ \psi)_e \equiv (d\varphi^{(n)})_e \bmod p^{n-t} \omega_{\hat{\mathcal{A}}}(\mathfrak{o})$$

$$\equiv (\kappa_n^*)^{-1} \psi_n^* \left(\frac{dT}{T} \right)_e \bmod p^{n-t} \omega_{\hat{\mathcal{A}}}(\mathfrak{o}).$$

Now we take a closer look at the map θ_A.

By definition, $\theta_A(\gamma) \equiv p^{v_n} b_{p^{v_n}} \bmod p^n \omega_{\hat{\mathcal{A}}}(\mathfrak{o})$, where $b_{p^{v_n}} \in E(\mathfrak{o})$ is an arbitrary lift of the p^{v_n}-torsion point $a_{p^{v_n}}$ in $\mathcal{A}(\mathfrak{o})$ induced by $\gamma \in T_p A$ to the universal vectorial extension E.

Cartier duality $[\,,\,] \colon \mathcal{A}_{n,p^{v_n}} \times \hat{\mathcal{A}}_{n,p^{v_n}} \to \mathbb{G}_{m,\mathfrak{o}_n}$ induces a homomorphism

$$\tau_n \colon \mathcal{A}_{n,p^{v_n}} \longrightarrow \omega_{\hat{\mathcal{A}}_{n,p^{v_n}}},$$

given by $a \mapsto [a, -]^* \frac{dT}{T}$. Now we use an argument of Crew (see [Cr, Section 1] and also [Ch, Lemma A.3]) to show that $\theta_A(\gamma) \equiv (\kappa_n^*)^{-1} \tau_n(\overline{a_{p^{v_n}}}) \bmod p^n \omega_{\hat{\mathcal{A}}}(\mathfrak{o})$, where $\overline{a_{p^{v_n}}} \in \mathcal{A}_n(\mathfrak{o}_n)$ is the reduction of $a_{p^{v_n}}$.

Namely, by [Ma-Me, Chapter I, (2.6.2)], the universal vectorial extension $E_n = E \otimes \mathfrak{o}_n$ of $\mathcal{A}_n$ is isomorphic to the pushout of the sequence

$$0 \to \mathcal{A}_{n,p^{v_n}} \to \mathcal{A}_n \to \mathcal{A}_n \to 0$$

by $(\kappa_n^*)^{-1} \circ \tau_n$. Hence we have a commutative diagram with exact rows

Let $\overline{b_{p^{v_n}}}$ be the image of $b_{p^{v_n}}$ under the reduction map $E(\mathfrak{o}) \to E_n(\mathfrak{o}_n)$. Since multiplication by p^{v_n} on $\mathcal{A}(\mathfrak{o})$ is surjective, we can find some $\overline{c} \in \mathcal{A}_n(\mathfrak{o}_n)$ with $p^{v_n}\overline{c} = \overline{a_{p^{v_n}}}$. Then $f(\overline{c})$ differs from $\overline{b_{p^{v_n}}}$ by an element in $\omega_{\hat{\mathcal{A}}_n}(\mathfrak{o}_n)$, which implies

$$p^{\nu_n}\overline{b_{p^{\nu_n}}} = f(p^{\nu_n}\overline{c})$$
$$= (\kappa_n^*)^{-1}\tau_n(\overline{a_{p^{\nu_n}}}),$$

so that indeed

$$\theta_A(\gamma) \equiv (\kappa_n^*)^{-1}\tau_n(\overline{a_{p^{\nu_n}}}) \bmod p^n\omega_{\hat{A}}(\mathfrak{o}).$$

By definition, $\tau_n(\overline{a_{p^{\nu_n}}}) = \psi_n^*(\frac{dT}{T})_e$ so that for all n,

$$\psi^*\left(\frac{dT}{T}\right)_e \equiv (\kappa_n^*)^{-1}\psi_n^*\left(\frac{dT}{T}\right)_e \bmod p^{n-t}\omega_{\hat{A}}(\mathfrak{o})$$
$$\equiv \theta_A(\gamma) \bmod p^{n-t}\omega_{\hat{A}}(\mathfrak{o}),$$

which implies our claim.   □

Corollary 3. *The map α is injective.*

Proof. This follows from Theorem 4 and diagram (12) since $\tilde{\alpha} = \mathrm{Lie}\,\alpha = \theta_A^*$ is injective.   □

By Theorem 4, the following diagram is commutative:

$$
\begin{array}{ccccccc}
\hat{A}(\mathbb{C}_p) & = & H^1(A_{\mathbb{C}_p},\mathcal{O}^*)^0 & \xrightarrow{\ \log\ } & H^1(A_{\mathbb{C}_p},\mathcal{O}) & = & \mathrm{Lie}\,\mathrm{Pic}^0_{A_K/K}(\mathbb{C}_p) \\
& & \downarrow{\scriptstyle\alpha} & & \downarrow{\scriptstyle\theta_A^*} & & \downarrow{\scriptstyle\mathrm{Lie}\,\alpha} \\
& & \mathrm{Hom}_c(TA,\mathbb{C}_p^*) & \xrightarrow{\ \log_*\ } & \mathrm{Hom}_c(TA,\mathbb{C}_p) & = & \mathrm{Lie}\,\mathrm{Hom}_c(TA,\mathbb{C}_p^*).
\end{array}
$$

This is in a certain way analogous to the following diagram for the Lie group

$$U = \left\{\begin{pmatrix} 1 & * \\ 0 & 1 \end{pmatrix}\right\} \subset \mathrm{GL}_2,$$

which we derive from Theorem 2:

$$
\begin{array}{ccccccc}
\mathrm{Ext}^1_{\mathfrak{B}_{A_{\mathbb{C}_p}}}(\mathcal{O}) & = & H^1(A_{\mathbb{C}_p},U(\mathcal{O})) & \xrightarrow[\sim]{\ \log=\mathrm{id}\ } & H^1(A_{\mathbb{C}_p},\mathrm{Lie}\,U(\mathcal{O})) & = & H^1(A_{\mathbb{C}_p},\mathcal{O}) \\
& & \downarrow{\scriptstyle\rho_*} & & & & \downarrow{\scriptstyle\theta_A^*} \\
& & \mathrm{Hom}_c(TA,U(\mathbb{C}_p)) & \xrightarrow{\ \log_*=\mathrm{id}\ } & \mathrm{Hom}_c(TA,\mathrm{Lie}\,U(\mathbb{C}_p)) & = & H^1_{\text{ét}}(A,\mathbb{Q}_p)\otimes\mathbb{C}_p.
\end{array}
$$

In the first diagram, the underlying group is $\mathbb{G}_m$, in the second it is $U \simeq \mathbb{G}_a$.

The next corollary was already observed by Tate in his context of p-divisible groups [Ta, Theorem 3].

Corollary 4. *The map α induces an isomorphism of abelian groups*

$$\alpha\colon \hat{A}_K(K) \xrightarrow{\sim} \mathrm{Hom}_{c,G_K}(TA, \mathfrak{o}^*).$$

Proof. According to [Ta, Theorems 1 and 2], we have $H^0(G_K, \mathbb{C}_p) = K$ and $H^0(G_K, \mathbb{C}_p(-1)) = 0$. Hence the Hodge–Tate decomposition and Theorem 4 imply that $\tilde{\alpha} = \mathrm{Lie}\,\alpha$ induces an isomorphism:

$$\tilde{\alpha} \colon H^1(A_K, \mathcal{O}) \xrightarrow{\sim} H^0(G_K, H^1_{\text{ét}}(A, \mathbb{Q}_p) \otimes \mathbb{C}_p) = \mathrm{Hom}_{c,G_K}(TA, \mathbb{C}_p).$$

We have $H^0(G_K, \hat{A}(\mathbb{C}_p)) = \hat{A}_K(K)$. This follows for example by embedding $\hat{A}_K$ into some $\mathbb{P}^N$ over K and using the corresponding result for $\mathbb{P}^N$. The latter is a consequence of the decomposition $\mathbb{P}^N = \mathbb{A}^N \amalg \cdots \amalg \mathbb{A}^0$ over K and the equality $H^0(G_K, \mathbb{C}_p) = K$. The corollary follows by applying the 5-lemma to the diagram of Galois cohomology sequences obtained from (12).   □

We next describe the image of the map α on $\hat{A}(\mathbb{C}_p)$.

Definition 3. *A continuous character* $\chi \colon TA \to \mathbb{C}_p^*$ *is called smooth if its stabilizer in* G_K *is open. The group of smooth characters of* TA *is denoted by* $Ch^\infty(TA)$.

Note that we have

$$Ch^\infty(TA) = \varinjlim_{L/K} \mathrm{Hom}_{c,G_L}(TA, \mathfrak{o}^*),$$

where L runs over the finite extensions of K in $\overline{\mathbb{Q}}_p$. It is also the biggest G_K-invariant subset S of $\mathrm{Hom}_c(TA, \mathbb{C}_p^*)$ such that the G_K-action on S^δ is continuous. Here S^δ is S endowed with the discrete topology.

Replacing A_K by A_L in Corollary 4 we find that α induces an isomorphism

$$\alpha \colon \hat{A}(\overline{\mathbb{Q}}_p) \xrightarrow{\sim} Ch^\infty(TA) \subset \mathrm{Hom}_c(TA, \mathfrak{o}^*).$$

Let $Ch(TA)$ be the closure of $Ch^\infty(TA)$ in $\mathrm{Hom}_c(TA, \mathfrak{o}^*)$ or equivalently in $C^0(TA, \mathbb{C}_p)$. Then $Ch(TA)$ is also a complete topological group.

Theorem 5. *The map* α *induces an isomorphism of topological groups*

$$\alpha \colon \hat{A}(\mathbb{C}_p) \xrightarrow{\sim} Ch(TA).$$

Proof. By Lemma 4, α is continuous. Hence

$$\alpha(\hat{A}(\mathbb{C}_p)) = \alpha(\overline{\hat{A}(\overline{\mathbb{Q}}_p)}) \subset \overline{\alpha(\hat{A}(\overline{\mathbb{Q}}_p))} = Ch(TA).$$

It now suffices to show that α is a closed map. Namely, because of

$$Ch^\infty(TA) \subset \alpha(\hat{A}(\mathbb{C}_p)) \subset Ch(TA),$$

it will follow that $\alpha(\hat{A}(\mathbb{C}_p)) = Ch(TA)$ and α will be a homeomorphism onto its image.

So let $Y \subset \hat{A}(\mathbb{C}_p)$ be a closed set. Let $\alpha(y_n)$ for $y_n \in Y$ be a sequence which converges to some χ in $\mathrm{Hom}_c(TA, \mathfrak{o}^*)$. Since the map $\log_*$ in (10) is continuous in the uniform topologies it follows that $\mathrm{Lie}\,\alpha(\log y_n) = \log_* \alpha(y_n)$ converges to $\log_* \chi$. Because of the equality

$$\mathrm{Hom}_c(TA, \mathbb{C}_p) = \mathrm{Hom}_{\mathbb{Z}_p}(T_p A, \mathbb{C}_p)$$

the topology of uniform convergence on this space coincides with its topology as a finite-dimensional $\mathbb{C}_p$-vector space. The map $\mathrm{Lie}\,\alpha$ is a $\mathbb{C}_p$-linear injection by the Hodge–Tate decomposition and Theorem 4. Hence it is a closed injection and therefore the sequence $\log y_n$ converges. As log is a local homeomorphism there is a convergent sequence $\tilde{y}_n \in \hat{A}(\mathbb{C}_p)$ with $\log \tilde{y}_n = \log y_n$. Writing $\tilde{y}_n = y_n + t_n$ with $t_n \in \hat{A}(\mathbb{C}_p)_{\mathrm{tors}}$ we get $\alpha(\tilde{y}_n) = \alpha(y_n) + \alpha(t_n)$. The sequence $\alpha(y_n)$ converges by assumption and the sequence $\alpha(\tilde{y}_n)$ converges because α is continuous. Hence the sequence $\alpha(t_n) = \alpha_{\mathrm{tors}}(t_n)$ converges. The groups $\hat{A}(\mathbb{C}_p)_{\mathrm{tors}}$ and $\mathrm{Hom}_c(TA, \mu)$ are the kernels of the locally topological homomorphisms log, respectively, $\log_*$. Hence they inherit the discrete topology from the *p*-adic topologies on $\hat{A}(\mathbb{C}_p)$, respectively, $\mathrm{Hom}_c(TA, \mathfrak{o}^*)$. Therefore, the algebraic isomorphism α_{tors} is trivially a homeomorphism and hence the sequence t_n converges. It follows that the sequence y_n converges to some $y \in Y$. By continuity of α the sequence $\alpha(y_n)$ converges to $\alpha(y)$. Thus $\alpha(Y)$ is closed as was to be shown. □

The following example was prompted by a question of Damian Roessler.

Example. Fix some σ in G_K. Since α is G_K-equivariant we know that if $\hat{a} \in \hat{A}(\mathbb{C}_p)$ corresponds to the character $\chi: TA \to \mathfrak{o}^*$, then $\sigma(\hat{a})$ corresponds to $^\sigma \chi = \sigma \circ \chi \circ \sigma_*^{-1}$. Here σ_* is the action on TA induced by σ.

How about the character $\sigma \circ \chi: TA \to \mathfrak{o}^*$? Using Theorem 5 we will now show that it also corresponds to an element of $\hat{A}(\mathbb{C}_p)$ provided that A_K has complex multiplication over K. For this, we have to check that in the CM case, the subgroup $Ch(TA)$ is invariant under the homeomorphism $\chi \mapsto \sigma \circ \chi$ of $\mathrm{Hom}_c(TA, \mathfrak{o}^*)$. It suffices to show that $Ch^\infty(TA)$ is invariant. For χ in $Ch^\infty(TA)$, there is a finite normal extension N/K such that χ is G_N-invariant, i.e., $\tau^{-1}\chi\tau_* = \chi$ for all τ in G_N. It follows that

$$\tau^{-1}(\sigma\chi)\tau_* = \sigma(\sigma^{-1}\tau^{-1}\sigma\chi)\tau_* = \sigma\chi(\sigma^{-1}\tau^{-1}\sigma)_*\tau_* = \sigma\chi[\sigma, \tau]_*$$

where we define the commutator by $[\sigma, \tau] = \sigma^{-1}\tau^{-1}\sigma\tau$. By the CM assumption, the image of G_K in the automorphism group of TA is abelian. Hence $[\sigma, \tau]_*$ acts trivially on TA and we have thus shown that $\tau^{-1}(\sigma\chi)\tau_* = \sigma\chi$ for all $\tau \in G_N$. Hence $\sigma \circ \chi$ lies in $Ch^\infty(TA)$. This proves the claim.

For $\hat{a} \in \hat{A}(\mathbb{C}_p)$ let $\hat{a}_\sigma \in \hat{A}(\mathbb{C}_p)$ be the element corresponding to $\sigma \circ \chi$ via Theorem 5. By construction, the map $(\sigma, \hat{a}) \mapsto \hat{a}_\sigma$ determines a new action of G_K on $\hat{A}(\mathbb{C}_p)$. It seems to be a nice exercise in CM-theory to give an explicit description of this action.

5 Line bundles on varieties and their p-adic characters

Consider a smooth and proper variety X_K over a finite extension K of $\mathbb{Q}_p$. Varieties are supposed to be geometrically irreducible. We assume that $H^1(X_K)$ has good reduction in the sense that the inertia group I_K of G_K acts trivially on the étale cohomology group $H^1(X, \mathbb{Q}_l)$ for some prime $l \neq p$. Here $X = X_K \otimes \overline{K}$.

The abelianization of the fundamental group $\pi_1(X, x)$ is independent of the choice of a base point x and will be denoted by $\pi_1^{\mathrm{ab}}(X)$. It carries an action of the Galois group G_K even if X_K does not have a K-rational point. We will now attach a p-adic character of $\pi_1^{\mathrm{ab}}(X)$ to any line bundle L on $X_{\mathbb{C}_p}$ whose image in the Néron–Severi group of $X_{\mathbb{C}_p}$ is torsion.

It is known that $B_K := \mathrm{Pic}^0_{X_K/K}$ is an abelian variety over K. Its dual is the Albanese variety $A_K = \mathrm{Alb}_{X_K/K}$ of X_K over K. We put $B = B_K \otimes_K \overline{\mathbb{Q}}_p$ and $A = A_K \otimes_K \overline{\mathbb{Q}}_p$.

Using the Kummer sequence and divisibility of $\mathrm{Pic}^0_{X_K/K}(\overline{\mathbb{Q}}_p)$ one gets an exact sequence

$$0 \longrightarrow B_N(\overline{\mathbb{Q}}_p) \longrightarrow H^1(X, \mu_N) \longrightarrow NS(X)_N \longrightarrow 0 \qquad (13)$$

for every $N \geq 1$. Since the Néron–Severi group of X is finitely generated it follows that $T_l B = H^1(X, \mathbb{Z}_l(1))$ for every l. Thus B_K and hence also A_K have good reduction. For sufficiently large N in the sense of divisibility we have

$$NS(X)_{\mathrm{tors}} = NS(X)_N.$$

Applying $\mathrm{Hom}(\_, \mu_N)$ to the exact sequence (13) and passing to projective limits therefore gives an exact sequence of G_K-modules:

$$0 \longrightarrow \mathrm{Hom}(NS(X)_{\mathrm{tors}}, \mu) \longrightarrow \pi_1^{\mathrm{ab}}(X) \longrightarrow TA \longrightarrow 0. \qquad (14)$$

Here we have used the perfect Galois equivariant pairing coming from Cartier duality

$$A_N(\overline{\mathbb{Q}}_p) \times B_N(\overline{\mathbb{Q}}_p) \longrightarrow \mu_N.$$

For every prime number l the pro-l part of the sequence (14) splits continuously since $T_l A$ is a free $\mathbb{Z}_l$-module. Hence (14) splits continuously and applying $\mathrm{Hom}_c(\_, \mathfrak{o}^*)$ we get an exact sequence of G_K-modules

$$0 \longrightarrow \mathrm{Hom}_c(TA, \mathfrak{o}^*) \longrightarrow \mathrm{Hom}_c(\pi_1^{\mathrm{ab}}(X), \mathfrak{o}^*) \longrightarrow NS(X)_{\mathrm{tors}} \longrightarrow 0.$$

We set

$$\mathrm{Hom}_c^0(\pi_1^{\mathrm{ab}}(X), \mathfrak{o}^*) = \mathrm{Ker}(\mathrm{Hom}_c(\pi_1^{\mathrm{ab}}(X), \mathfrak{o}^*) \longrightarrow NS(X)_{\mathrm{tors}}).$$

Recall from Section 3 the continuous injective homomorphism

$$\alpha \colon \mathrm{Pic}^0_{X_K/K}(\mathbb{C}_p) = \hat{A}(\mathbb{C}_p) \hookrightarrow \mathrm{Hom}_c(TA, \mathfrak{o}^*) = \mathrm{Hom}_c^0(\pi_1^{\mathrm{ab}}(X), \mathfrak{o}^*).$$

Moreover, α is a locally analytic homomorphism of p-adic Lie groups over $\mathbb{C}_p$. Using Theorem 4, we see that α fits into a commutative diagram with exact rows:

$$
\begin{array}{ccccccccc}
0 & \to & \mathrm{Pic}^0_{X_K/K}(\mathbb{C}_p)_{\mathrm{tors}} & \to & \mathrm{Pic}^0_{X_K/K}(\mathbb{C}_p) & \xrightarrow{\log} & \mathrm{Lie}\,\mathrm{Pic}^0_{X_K/K}(\mathbb{C}_p) & = & H^1(X_K,\mathcal{O})\otimes_K\mathbb{C}_p \to 0 \\
& & \downarrow{\scriptstyle\alpha_{\mathrm{tors}}}\ {\wr} & & \uparrow{\scriptstyle\alpha} & & \downarrow{\scriptstyle\mathrm{Lie}\,\alpha} & & \\
0 & \to & \mathrm{Hom}^0_c(\pi_1^{\mathrm{ab}}(X),\mu) & \to & \mathrm{Hom}^0_c(\pi_1^{\mathrm{ab}}(X),\mathfrak{o}^*) & \xrightarrow{\log_*} & \mathrm{Hom}_c(\pi_1^{\mathrm{ab}}(X),\mathbb{C}_p) & = & H^1(X_{\text{ét}},\mathbb{Q}_p)\otimes\mathbb{C}_p \to 0.
\end{array}
$$
$$\tag{15}$$

Here we have set

$$
\mathrm{Hom}^0_c(\pi_1^{\mathrm{ab}}(X),\mu) = \mathrm{Hom}^0_c(\pi_1^{\mathrm{ab}}(X),\mathfrak{o}^*) \cap \mathrm{Hom}_c(\pi_1^{\mathrm{ab}}(X),\mu)
$$
$$
= \mathrm{Hom}^0_c(\pi_1^{\mathrm{ab}}(X),\mathfrak{o}^*)_{\mathrm{tors}}.
$$

Furthermore, note that:

$$
\mathrm{Lie}\,\mathrm{Hom}^0_c(\pi_1^{\mathrm{ab}}(X),\mathfrak{o}^*) = \mathrm{Hom}_c(\pi_1^{\mathrm{ab}}(X),\mathbb{C}_p).
$$

The map $\mathrm{Lie}\,\alpha$ coincides with the inclusion map coming from the Hodge–Tate decomposition of $H^1(X_{\text{ét}},\mathbb{Q}_p)\otimes\mathbb{C}_p$. This follows from Theorem 4 and the functoriality of this decomposition.

Set

$$
Ch^\infty(\pi_1^{\mathrm{ab}}(X))^0 = \varinjlim_{L/K} \mathrm{Hom}^0_{c,G_L}(\pi_1^{\mathrm{ab}}(X),\mathfrak{o}^*)
$$

and let $Ch(\pi_1^{\mathrm{ab}}(X))^0$ be its closure in $\mathrm{Hom}^0_c(\pi_1^{\mathrm{ab}}(X),\mathfrak{o}^*)$. We make similar definitions with the ^{0}s omitted.

It follows from Theorem 5 that α induces a topological isomorphism of complete topological groups:

$$
\alpha:\ \mathrm{Pic}^0_{X_K/K}(\mathbb{C}_p) \xrightarrow{\sim} Ch(\pi_1^{\mathrm{ab}}(X))^0.
$$

We will now extend the domain of definition of α to $\mathrm{Pic}^\tau_{X_K/K}(\mathbb{C}_p)$. This is the group of line bundles on $X_{\mathbb{C}_p}$ whose image in $NS(X_{\mathbb{C}_p}) = NS(X)$ is torsion. We thus have an exact sequence

$$
0 \longrightarrow \mathrm{Pic}^0_{X_K/K}(\mathbb{C}_p) \longrightarrow \mathrm{Pic}^\tau_{X_K/K}(\mathbb{C}_p) \longrightarrow NS(X)_{\mathrm{tors}} \longrightarrow 0. \tag{16}
$$

Theorem 6. *There is a G_K-equivariant map α^τ which makes the following diagrams with exact rows commute:*

$$
\begin{array}{ccccccccc}
0 & \longrightarrow & \mathrm{Pic}^0_{X_K/K}(\mathbb{C}_p) & \longrightarrow & \mathrm{Pic}^\tau_{X_K/K}(\mathbb{C}_p) & \longrightarrow & NS(X)_{\mathrm{tors}} & \to & 0 \\
& & \downarrow{\scriptstyle\alpha} & & \downarrow{\scriptstyle\alpha^\tau} & & \| & & \\
0 & \to & \mathrm{Hom}^0_c(\pi_1^{\mathrm{ab}}(X),\mathfrak{o}^*) & \to & \mathrm{Hom}_c(\pi_1^{\mathrm{ab}}(X),\mathfrak{o}^*) & \to & NS(X)_{\mathrm{tors}} & \to & 0
\end{array}
$$

and

$$0 \to \mathrm{Pic}^{\tau}_{X_K/K}(\mathbb{C}_p)_{\mathrm{tors}} \longrightarrow \mathrm{Pic}^{\tau}_{X_K/K}(\mathbb{C}_p) \xrightarrow{\log} \mathrm{Lie\,Pic}^{\tau}_{X_K/K}(\mathbb{C}_p) = H^1(X_K, \mathcal{O}) \otimes_K \mathbb{C}_p \to 0$$

$$\downarrow \wr \alpha^{\tau}_{\mathrm{tors}} \qquad\qquad \downarrow \alpha^{\tau} \qquad\qquad\qquad \downarrow \mathrm{Lie}\,\alpha^{\tau} = \mathrm{Lie}\,\alpha$$

$$0 \to \mathrm{Hom}_c(\pi_1^{\mathrm{ab}}(X), \mu) \to \mathrm{Hom}_c(\pi_1^{\mathrm{ab}}(X), \mathfrak{o}^*) \to \mathrm{Hom}_c(\pi_1^{\mathrm{ab}}(X), \mathbb{C}_p) = H^1(X_{\text{ét}}, \mathbb{Q}_p) \otimes \mathbb{C}_p \to 0.$$

The map α^{τ} is an injective and locally analytic homomorphism of p-adic Lie groups. Its restriction $\alpha^{\tau}_{\mathrm{tors}}$ to torsion subgroups is the inverse of the Kummer isomorphism:

$$i_X : \mathrm{Hom}_c(\pi_1^{\mathrm{ab}}(X), \mu) = \varinjlim_N H^1(X, \mu_N) \xrightarrow{\sim} \varinjlim_N H^1(X, \mathbb{G}_m)_N = \mathrm{Pic}^{\tau}_{X_K/K}(\mathbb{C}_p)_{\mathrm{tors}}.$$

The map α^{τ} induces a topological isomorphism of complete topological groups

$$\alpha^{\tau} : \mathrm{Pic}^{\tau}_{X_K/K}(\mathbb{C}_p) \xrightarrow{\sim} Ch(\pi_1^{\mathrm{ab}}(X)).$$

Proof. First, note that $\mathrm{Pic}_{X_K/K}(\overline{\mathbb{Q}}_p)_N = \mathrm{Pic}_{X_K/K}(\mathbb{C}_p)_N$ and $\mathrm{Pic}^{\tau}_{X_K/K}(\overline{\mathbb{Q}}_p)_N = \mathrm{Pic}^{\tau}_{X_K/K}(\mathbb{C}_p)_N$ because this holds for Pic^0 and because $NS(X_{\mathbb{C}_p}) = NS(X)$.

As $\mathrm{Pic}^0_{X_K/K}(\mathbb{C}_p)$ is divisible, the sequence (16) gives a short exact sequence

$$0 \longrightarrow \mathrm{Pic}^0_{X_K/K}(\mathbb{C}_p)_{\mathrm{tors}} \longrightarrow \mathrm{Pic}^{\tau}_{X_K/K}(\mathbb{C}_p)_{\mathrm{tors}} \longrightarrow NS(X)_{\mathrm{tors}} \longrightarrow 0.$$

We claim that the following diagram with exact rows commutes:

$$
\begin{array}{ccccccccc}
0 & \longrightarrow & \mathrm{Pic}^0_{X_K/K}(\mathbb{C}_p)_{\mathrm{tors}} & \longrightarrow & \mathrm{Pic}^{\tau}_{X_K/K}(\mathbb{C}_p)_{\mathrm{tors}} & \longrightarrow & NS(X)_{\mathrm{tors}} & \to & 0 \\
 & & \downarrow \alpha_{\mathrm{tors}} & & \downarrow i_X^{-1} & & \| & & \\
0 & \to & \mathrm{Hom}_c^0(\pi_1^{\mathrm{ab}}(X), \mu) & \to & \mathrm{Hom}_c(\pi_1^{\mathrm{ab}}(X), \mu) & \to & NS(X)_{\mathrm{tors}} & \to & 0.
\end{array}
\tag{17}
$$

If we make the identifications explicit which define the maps of the left square, we see that on N-torsion it is the outer rectangle of the following diagram

$$
\begin{array}{ccccccc}
\mathrm{Pic}^0_{X_K/K}(\mathbb{C}_p)_N & = & \hat{A}_N(\mathbb{C}_p) & = & B_N(\mathbb{C}_p) & \hookrightarrow & H^1(X, \mathbb{G}_m)_N \\
\downarrow \alpha_{\mathrm{tors}} \quad \boxed{1} & & \downarrow \alpha_{\mathrm{tors}} \quad \boxed{2} & & & & \quad \downarrow \wr i_X^{-1} \\
\mathrm{Hom}_c^0(\pi_1^{\mathrm{ab}}(X), \mu_N) = \mathrm{Hom}_c(TA, \mu_N) & = & \mathrm{Hom}(A_N, \mu_N) & = & B_N \subset H^1(X, \mathbb{G}_m)_N & \xrightarrow{\sim} & H^1(X, \mu_N).
\end{array}
$$

Now $\boxed{1}$ is commutative by definition, and $\boxed{2}$ commutes since the restriction of α to $\hat{A}_N(\mathbb{C}_p)$ is the map $\hat{A}_N(\mathbb{C}_p) \to \mathrm{Hom}_c(TA, \mu_N) = \mathrm{Hom}(A_N, \mu_N)$ coming from Cartier duality. Hence the outer rectangle commutes as well.

The right square in diagram (17) is commutative since the second map in the exact sequence (13) is induced by i_X.

We now define α^{τ} on

$$\mathrm{Pic}^{\tau}_{X_K/K}(\mathbb{C}_p) = \mathrm{Pic}^0_{X_K/K}(\mathbb{C}_p) + \mathrm{Pic}^{\tau}_{X_K/K}(\mathbb{C}_p)_{\mathrm{tors}}$$

by setting it equal to α on $\mathrm{Pic}^0_{X_K/K}(\mathbb{C}_p)$ and to i_X^{-1} on $\mathrm{Pic}^\tau_{X_K/K}(\mathbb{C}_p)_{\mathrm{tors}}$. This is well defined since by the commutativity of (17), the maps α and i_X^{-1} agree on

$$\mathrm{Pic}^0_{X_K/K}(\mathbb{C}_p) \cap \mathrm{Pic}^\tau_{X_K/K}(\mathbb{C}_p)_{\mathrm{tors}} = \mathrm{Pic}^0_{X_K/K}(\mathbb{C}_p)_{\mathrm{tors}}.$$

The remaining assertions follow without difficulty. Note that $\mathrm{Pic}^\tau_{X_K/K}(\mathbb{C}_p)_{\mathrm{tors}}$ carries the discrete topology as a subspace of $\mathrm{Pic}^\tau_{X_K/K}(\mathbb{C}_p)$ since it is the kernel of the locally topological log map. $\square$

Acknowledgments. It is a pleasure to thank Damian Roessler and Peter Schneider for interesting discussions and suggestions.

References

[BLR]   S. Bosch, W. Lütkebohmert, and M. Raynaud, *Néron Models*, Springer-Verlag, Berlin, 1990.

[Bou]   N. Bourbaki, *Groupes et algèbres de Lie*, Herrmann, Paris, 1972, Chapters 2 and 3.

[Ch]   J. A. Chambert-Loir, Extension universelle d'une variété abélienne et hauteurs des points de torsion, *Comp. Math.*, **103** (1996), 243–267.

[Co1]   R. Coleman, Hodge-Tate periods and p-adic abelian integrals, *Invent. Math.*, **78** (1984), 351–379.

[Co2]   R. Coleman, Reciprocity laws on curves, *Comp. Math.*, **72** (1989), 205–235.

[Co3]   R. Coleman, The universal vectorial biextension and p-adic heights, *Invent. Math.*, **103** (1991), 631–650.

[Col]   P. Colmez, *Intégration sur les variétés p-adiques*, Astérisque 248, Société Mathématique de France, Paris, 1998.

[Cr]   R. Crew, Universal extensions and p-adic periods of elliptic curves, *Comp. Math.*, **73** (1990), 107–119.

[De-We1]   C. Deninger and A. Werner, Vector bundles and p-adic representations I, preprint, 2003; arXiv:mathNT/0309273.

[De-We2]   C. Deninger and A. Werner, Vector bundles on p-adic curves and parallel transport, preprint, 2004; arXiv:mathAG/0403516.

[EGAIII]   A. Grothendieck, *Éléments de Géométrie Algébrique* III, Publications Mathématiques IHES 11, 17, Institut des Hautes Études Scientifiques, Bures-sur-Yvette, France, 1961, 1963.

[Fa1]   G. Faltings, Hodge–Tate structures and modular forms, *Math. Ann.*, **278** (1987), 133–149.

[Fa2]   G. Faltings, A p-adic Simpson correspondence, preprint, 2003.

[Fo]   J.-M. Fontaine, Formes différentielles et modules de Tate des variétés abéliennes sur les corps locaux, *Invent. Math.*, **65** (1982), 379–409.

[Ma-Me]   B. Mazur and W. Messing, *Universal Extensions and One Dimensional Crystalline Cohomology*, Lecture Notes in Mathematics 370, Springer-Verlag, Berlin, New York, Heidelberg, 1974.

[Ray]   M. Raynaud, Spécialisation du foncteurs de Picard, *Publ. Math. IHES*, **38** (1970), 27–76.

[Ta]   J. Tate, p-divisible groups, in *Proceedings of a Conference on Local Fields, Driebergen* 1966, Springer-Verlag, Berlin, 1967, 158–183.

Arithmetic Eisenstein Classes on the Siegel Space: Some Computations

Gerd Faltings

Max-Planck-Institut für Mathematik
Vivatsgasse 7
D-53111 Bonn
Germany
gerd@mpim-bonn.mpg.de

1 Introduction

The classical theory of Eisenstein series and Eisenstein cohomology is rather well developed; see [12] or [6]. For Shimura varieties it gives topological cohomology classes with rational coefficients. In the present paper we study whether these classes extend to l-adic étale classes over the integers with the prime l inverted of a number field. We show this for a certain special type of class on the Siegel moduli space, in fact, even construct the class in K-theory. However, this construction follows known techniques and is exceptionally simple, and there are many other more difficult cases left. As an example, we treat the remaining class for genus 2, where we identify the relevant L-factors and bound the denominator of Eisenstein-classes using p-adic Hodge theory. Unfortunately, even here we cannot solve all relevant problems, and the proofs use unpleasant explicit calculations.

The results were inspired by Kato's construction of an Euler system in the cohomology of modular curves. All in all, we get some insight, but there remain many open questions and problems whose solution requires new techniques.

2 Preliminaries

Suppose $f: A \to S$ is an abelian scheme over a scheme S of relative dimension g. We also choose a prime l and assume that l is invertible in S. Then the direct images $\mathbb{R}^i f_*(\mathbb{Z}_l)$ are smooth and equal to the ith exterior power of the dual of the Tate-module $T_l(A)$ (which is a smooth l-adic sheaf on S). It is an old observation of Lieberman that the $\mathbb{Q}_l$-adic direct image $\mathbb{R} f_*(\mathbb{Q}_l)$ splits as the direct sum of all $\mathbb{R}^i_*(\mathbb{Q}_l)[-i]$. This follows because multiplications by integers n define endomorphisms of A/S which induce on the ith direct images multiplication by n^i. On the individual pieces the direct image n_* (which is defined because multiplication by n is finite) acts as n^{2g-i}.

More precisely, one can find a map

$$\sum_{i=0}^{2g} \phi_i : \ \oplus_{i=0}^{2g} \mathbb{R}^i f_*(\mathbb{Z}_l)[-i] \to \mathbb{R}f_*(\mathbb{Z}_l)$$

which induces on the ith cohomology multiplication by $i!$. Namely, the zero-section $S \to A$ induces a direct sum decomposition

$$\tau_{\leq 1}(\mathbb{R}f_*(\mathbb{Z}_l)) = \mathbb{Z}_l \oplus \mathbb{R}^1 f_*(\mathbb{Z}_l[-1]).$$

The cup-product is antisymmetric, so antisymmetrizing under the symmetric group (acting on $\mathbb{R}^1 f_*(\mathbb{Z}_l)^{\otimes i}$) gives

$$\mathbb{R}^i f_*(\mathbb{Z}_l) = \wedge^i \mathbb{R}^1 f_*(\mathbb{Z}_l) \to \mathbb{R}f_*(\mathbb{Z}_l)[i],$$

which induces multiplication by $i!$ on cohomology.

The ϕ_i are multiplicative in the sense that

$$\phi_{i+j}(\alpha \cup \beta) = \binom{i+j}{i} \phi_i(\alpha) \cup \phi_j(\beta).$$

Also for any integers m prime to l the trace m_* under multiplication by m acts on the image of ϕ_i as m^{2g-i}. This holds because ϕ_i commutes with the pullback m^* and because of the projection formula

$$m_* \cdot m^* = m^{2g}.$$

Furthermore, if $\gamma_0 \in H^{2g}(A, \mathbb{Z}_l(g))$ denotes the class of the zero-section (defined by Poincaré duality) and $\bar{\gamma}_0$ its projection in $\mathbb{R}^{2g} f_*(\mathbb{Z}_l(g))$, we can find an explicit integer N such that

$$N \cdot \phi_{2g}(\bar{\gamma}_0) = N \cdot (2g)! \cdot \gamma_0.$$

Namely, choose an integer m which is a generator of $\mathbb{Z}_l^*$ if $l > 2$, and (say) $m = 3$ if $l = 2$. Then the product

$$\prod_{i=0}^{2g-1} (m_* - m^{2g-i})$$

annihilates $\tau_{\leq 2g-1}(\mathbb{R}f_*(\mathbb{Z}_l))$ and thus the difference $\phi_{2g}(\gamma_0) - (2g)! \cdot \gamma_0$. As m_* fixes γ_0 as well as $\phi_{2g}(1)$, we obtain the claim with

$$N = \prod_{i=0}^{2g-i} (1 - m^{2g-i}).$$

If $l > 2$, one checks that the maximal l-power in this product is equal to $\sum_{\mu \geq 0}[2g/(l-1)l^\mu]$. For $l = 2$, it increases by g.

If $x : S \to A$ is a section of f, the obstructions to divide x by l-powers define a class $d(x) \in H^1(S, T_l(A))$. Furthermore, translation by x defines an operator T_x^* on cohomology. These are related as follows: Interior multiplication defines a pairing

$$\wedge^i(T_l(A)) \times \mathbb{R}^{i+j} f_*(\mathbb{Z}_l) \to \mathbb{R}^j f_*(\mathbb{Z}_l);$$

thus

$$H^i(S, \wedge^i(T_l(A))) \times H^0(S, \mathbb{R}^{i+j} f_*(\mathbb{Z}_l)) \to H^i(S, \mathbb{R}^j f_*(\mathbb{Z}_l)).$$

Furthermore, $\mathbb{R}^{2g} f_*(\mathbb{Z}_l(g))$ is canonically isomorphic to $\mathbb{Z}_l$, with generator γ_0, and the ith cup-power of $d(x)$ defines an element in the first factor of the pairing. We then have the following formula which holds in

$$H^j(A, \mathbb{Q}_l(g)) = \prod H^{j-i}(S, \mathbb{R}^i f_*(\mathbb{Q}_l(g))).$$

Lemma 1. *We have*

$$T_x^*(\alpha) = \sum \langle d(x)^n, \alpha \rangle / n!.$$

Proof. We first consider the action of T_x^* on $\tau_{\leq 1}(\mathbb{R} f_*(\mathbb{Z}_l))$. This complex splits (use 0^*), and $T_x^* - T_0^*$ defines a map

$$\mathbb{R}^1 f_*(\mathbb{Z}_l)[-1] \to \mathbb{Z}_l$$

that is a class in $H^1(S, T_l(A))$. It can be described as follows.

There exists a universal element in $H^1(A, f^*(T_l(A))$, whose reduction modulo l^n classifies the $A[l^n]$-torsor given by the exact sequence

$$0 \longrightarrow A[l^n] \longrightarrow A \overset{l^n \cdot}{\longrightarrow} A \longrightarrow 0.$$

The action of $T_x^* - T_0^*$ is described by pulling back this torsor via x. But this pullback is classified by $d(x)$, and this implies the above formula (even without denominators) for the action of T_x^* on $\tau_{\leq 1}(\mathbb{R} f_*(\mathbb{Z}_l))$.

For the action on $\mathbb{R}^i f_*(\mathbb{Q}_l)$ one uses the cup product and that interior multiplication by $d(x)$ is a derivation for that cup product. In fact, this argument even allows us to control denominators:

We have defined maps

$$\phi_i : \mathbb{R}^i f_*(\mathbb{Z}_l)[-i] \to \mathbb{R} f_*(\mathbb{Z}_l),$$

by antisymmetrizing the i-fold cup product of ϕ_1. They induce multiplication by $i!$ on cohomology. Then our argument gives the formula

$$T_x^*(\phi_j(\alpha)) = \sum_{i=0}^{j} \binom{j}{i} \phi_{j-i}(\langle d(x)^i, \alpha \rangle). \qquad \square$$

3 First construction of an Eisenstein class

For $A \to S$ as before, we fix an integer n prime to l. Let

$$U = A - A[n] \subset A,$$

and let $f_U : U \to S$ be the restriction of f to U. Then $\mathbb{R}^i f_{U,*}(\mathbb{Z}_l(g))$ coincides with $\mathbb{R}^i f_*(\mathbb{Z}_l(g))$ for $i < 2g - 1$, while in degree $2g - 1$ we have an injection

$$\mathbb{R}^{2g-1} f_*(\mathbb{Z}_l(g)) \subset \mathbb{R}^{2g-1} f_{U,*}(\mathbb{Z}_l(g))$$

with cokernel equal to the kernel of

$$f_{A[n],*}(\mathbb{Z}_l) \to \mathbb{Z}_l.$$

The characteristic class of

$$A[n] - n^{2g} \cdot \{0\}$$

defines a global section of this cokernel. Furthermore, for any nonzero integer m the trace m_* acts on $\mathbb{R} f_{U,*}(\mathbb{Z}_l)$ by first restricting to the complement of $A[mn]$ and then applying the usual trace for the finite morphism $m\cdot$. The obstruction to lift our class to $\mathbb{R} f_{U,*}(\mathbb{Z}_l(g))$ lies in the hypercohomology of $\tau_{\leq 2g-1}(\mathbb{R} f_*(\mathbb{Z}_l(g)))$, thus is annihilated by

$$\prod_{i=0}^{2g-1} (m_* - m^{2g-i}),$$

for any integer m. We choose m prime to n and a generator of $\mathbb{Z}_l^*$, as before. Then m_* respects the characteristic classes of $A[n]$ and of $\{0\}$, and we obtain by applying the operator

$$\prod_{i=0}^{2g-1} (m_* - m^{2g-i})$$

a lift of N times our class. This lift may depend on the chosen nullhomotopy of our operator, but applying it again eliminates the indeterminacy. Thus after multiplication by N^2, we obtain a canonical class z_n in $H^{2g-1}(U, \mathbb{Z}_l(g))$. However, it may still depend on the choice of the integer m.

In view of the dependence on n we write $U_n = A - A[n]$. Then on $U_{n_1 n_2}$, one obtains the formula

$$z_{n_1 n_2} = n_1^*(z_{n_2}) + n_2^2 \cdot z_{n_1} = n_2^*(z_{n_1}) + n_2^1 \cdot z_{n_2}.$$

It holds because we always try to lift the same element in $H^0(A[n_1 n_2], \mathbb{Z}_l)$. Also for any r prime to n we have for the direct images that $r_*(z_n) = z_n$. Now our fundamental class is defined by evaluating z_n at torsion-points $x \in A(S)$ of order prime to n.

Definition 1. *We set* $z_n(x) = x^*(z_n) \in H^{2g-1}(S, \mathbb{Z}_l(g))$. *For any integer n such that* $n \cdot x = x$ *let* $z(x) = 1/(n^{2g} - 1) \cdot z_n(x) \in H^{2g-1}(S, \mathbb{Q}_l(g))$.

It follows from the equations above that $z(x)$ does not depend on the choice of n.

The theory generalizes to semiabelian varieties. Such a variety G is an extension of an abelian variety A by a torus T,

$$0 \longrightarrow T \longrightarrow G \longrightarrow A \longrightarrow 0.$$

As before, the direct images $\mathbb{R}^i f_*(\mathbb{Z}_l)$ are exterior powers over the dual of the Tate module $T_l(G)$, and the direct image splits rationally. Any section $x: S \to G$ defines a class $c(x) \in H^1(S, T_l(G))$, and the translation action of x is given by the exponential of interior multiplication by that class. Also we have the direct image with compact support $\mathbb{R} f_!(\mathbb{Z}_l)$ which is dual to the usual direct image (Poincaré duality), and on it the translation action is the exponential of interior multiplication as well.

Our class is an eigenvector for the Hecke operators: Choose a prime p not dividing the order of x. The p-adic Hecke algebra acts by correspondences (on the moduli space parametrizing A) which send A to a formal linear combination of A's with p-isogenies $A \to A'$, and the torsion-point x' which is the image of x. It has one generator which sends A to all its quotients $A' = A/H$, with $H \subset A[p]$ an isotropic subgroup of order p^g. In addition for each $0 < i < g$ there are operators which map A to quotients A/H with $H \subset A[p^2]$ isotropic of order p^{2g}, and $H \cap A[p]$ of order p^{2g-i}.

Now by the projection formula we have

$$z(x') = \sum_{h \in H} z(x + h),$$

so the sum of all $z(x')$ (which describes the Hecke action) is equal to a sum over $z(x+h)$, each summand occurring with multiplicity equal to the number of subgroups H (of the type prescribed by the Hecke operator) which contain h. For the first type (maximal isotropic $H \subset A[p]$) the number is

$$\prod_{j=1}^{g}(1 + p^j)$$

(for $h = 0$), respectively,

$$\prod_{j=1}^{g-1}(1 + p^j)$$

(for $h \neq 0$). Thus this Hecke operator maps $z(x)$ to

$$\prod_{j=1}^{g-1}(1 + p^j) \cdot (z(px) + p^g z(x)).$$

For the other Hecke operators, indexed by $0 < i < g$, we similarly obtain numbers which only depend on whether h vanishes, or has precise order p, or precise order p^2. Thus the Hecke operator maps $z(x)$ to a certain linear combination of $z(x)$, $z(px)$, and $z(p^2 x)$.

In fact, we have not used the polarization, and could instead divide by arbitrary subgroups $H \subset A[p]$. These generate the Hecke algebra for $GL(2g)$ instead of $GSp(2g)$, and again map $z(x)$ into linear combinations of $z(x)$ and $z(px)$.

4 Behavior under degeneration

We investigate what happens to our class at the boundary of the Siegel moduli space. More precisely, we consider rigid quotients $G = \tilde{G}/\iota(Y)$, where $\tilde{G}$ is an extension of an abelian variety B of dimension $g - 1$ by the torus $\mathbb{G}_m$, and $Y \cong \mathbb{Z}$ a group of periods, as is explained in the book [3]. The varieties B are classified by a moduli space of abelian varieties of dimension $g - 1$, the extensions $\tilde{G}$ by the dual B^t of the universal abelian variety B, and finally the period map ι lifts a certain map $\mathbb{Z} \to B$ which is determined by polarizations. For principal polarizations $B = B^t$, and the projection of ι to B classifies the extension $\tilde{G}$. After choosing one lift, ι differs from this lift by a map $n \to q^n$, where q is an element of the base. Thus the formal completion of the moduli stack $\bar{A}_g$ along this boundary stratum looks as follows:

Start with the moduli stack A_{g-1}, with its universal abelian variety $B \to A_{g-1}$. Over the fiber product $B^t \times_{A_{g-1}} B$ we have the Poincaré bundle $\mathcal{P}$ (whose total space is the universal semiabelian $\tilde{G}$), and the associated $\mathbb{A}^1$-bundle. We name the first factor B^t because for arbitrary polarizations it is the dual of the universal B. In our case $B^t = B$, but it is still useful to have different names for the factors. Denote by $M \to B^t$ the $\mathbb{G}_m$-torsor opposite to the restriction of $\mathcal{P}$ to the diagonal. The formal completion of the associated $\mathbb{A}^1$-bundle along $\{0\}$ is the desired formal completion of $\bar{A}_g$ along its boundary stratum. To keep our previous notation denote $S^\circ = M$, let S be the associated $\mathbb{A}^1$-bundle, $j : S^\circ \to S$ the inclusion, and $\hat{S}$ the formal completion of S along the zero-section $i : B^t \to S$.

Over $\hat{S}$ the universal semiabelian scheme G is the quotient (via the Mumford construction) of the universal semiabelian extension $\tilde{G}$, by a group of periods $Y = \mathbb{Z}$. That is, over each affine ring of the formal scheme we have such a G. The periods lift the map $Y \times B \to B$ which is $-\,$id on $\{1\} \times B$. This lifts canonically to the desired period map $Y \to \tilde{G}(S^\circ)$ over S°.

The Mumford construction of the quotient G can be described as follows: Locally we obtain degeneration data as above, with q denoting a local equation of $\{0\} \subset \mathbb{A}^1$. Then embed $\tilde{G}$ into a relatively complete model $\tilde{P}$ which (in this case) can be given explicitly: In the projective line $\mathbb{P}^1$ over our base, blow up the codimension-2 subscheme defined as the intersection of $\{0, \infty\}$ and $\{q = 0\}$. The result has as special fiber the join of three $\mathbb{P}^1$s. Repeat the procedure at the ends and continue, to obtain a scheme $\tilde{P}_0$ which over $\{q \neq 0\}$ is equal to $\mathbb{P}^1$, and over $\{q = 0\}$ to an infinite join of $\mathbb{P}^1$s. $\mathbb{G}_m$ operates on $\tilde{P}_0$, and the operation of q on the generic fiber extends (and shifts the infinitely many components in the special fiber). This construction is insensitive to multiplying q by a unit, thus globalizes to our formal completion of $\bar{A}_g$.

Now

$$\tilde{P} = \tilde{G} \times^{\mathbb{G}_m} \tilde{P}_0$$

is the induced $\tilde{G}$-scheme. Its special fiber is an infinite join of $\mathbb{P}^1$-bundles over $B^t \times B$, indexed by integers $r \in \mathbb{Z}$, and the rth component is the projective bundle associated to the $\mathbb{G}_m$-torsor $\mathcal{P} \otimes \mathrm{pr}_1^*(M)^{\otimes -r}$. It admits a free operation of $Y = \mathbb{Z}$, and the quotient $P = \tilde{P}/Y$ exists first as a formal scheme but is seen to be algebraic because it is relatively projective. It contains the universal semiabelian G as open subset, corresponding to the open subset of $\tilde{P}$ which is the union of all Y-translates of $\tilde{G}$.

Next, we introduce level structure. Choose an integer $N \geq 3$ which will be assumed sufficiently divisible so that all occurring torsion points are N-torsion and work over $\mathbb{Z}[1/N, \zeta_N]$. The stack $A_{g,N}$ classifying principally polarized abelian varieties with level-N-structure is actually a scheme, and its toroidal compactification $\bar{A}_{g,N}$ at least an algebraic space, and for suitable choices also a scheme. Over $\mathcal{A}_{g-1}$ we have the universal $g-1$-dimensional abelian scheme B_N, with level-N-structure. The scheme S_N° is the $\mathbb{G}_m$-torsor M_N over B_N^t opposite to the restriction of the Nth power of the Poincaré bundle to the diagonal, and S_N the associated $\mathbb{A}^1$-bundle. The universal semiabelian $\tilde{G}_N$ now corresponds to the Nth power of the Poincaré bundle, and the period map $\iota\colon Y \to \tilde{G}_N$ is equal to the tautological map multiplied by N, so that it extends canonically to $1/N \cdot \mathbb{Z}$ (the image of $1/N$ lifts $-\,\mathrm{id}\colon B_N^t \to B_N$). Then $\tilde{G}[N]$ has a canonical basis (as the extension is defined by the Nth power of the Poincaré bundle), and so has the Mumford quotient G (add $\iota(1/N)$). Furthermore, we obtain $\tilde{P}_N$ whose special fiber is an infinite join of $\mathbb{P}^1$-bundles over $B_N^t \times B_N$, indexed by integers r. The rth component is the projective completion of the $\mathbb{G}_m$-torsor $\mathcal{P}^{\otimes N} \otimes \mathrm{pr}_1^*(M_N)^{\otimes -r}$.

To compute the invariants $z(x)$ we consider its image in the cohomology of $i^* \mathbb{R} j_*(\mathbb{Z}_l)$ on the special fiber B_N. The coefficients $i^* \mathbb{R} j_*(\mathbb{Z}_l)$ depend only on the normal bundle of the embedding of the special fiber:

In general, if $Y \subset X$ is a smooth divisor in a smooth scheme, denote by X' the affine bundle associated to $\mathcal{O}_X(Y)$, and by $Y' = X$ its zero-section. We have a tautological section $s\colon X \to X'$ with $s^*(Y') = Y$, and s^* induces an isomorphism on vanishing cycles. The same holds for the pullback of affine bundles via $Y \subset X$, and the assertion follows. Similarly, we construct a class on an open subset of the special fiber of P whose boundary is the characteristic class of the (relative to S°) zero-cycle $G[n] - n^{2g} \cdot \{0\}$, as follows:

This zero-cycle lifts to a cycle $\mathcal{Z}$ on $\tilde{G}$, for example, as the sum of all translates $\iota(i/n) + \tilde{G}[n]$ ($0 \le i < n$), minus $n^{2g} \cdot \{0\}$. We can indeed find such a zero-cycle with rational coefficients such that its characteristic class in $H_!^{2g}(\tilde{G}, \mathbb{Z}_l(g))$ vanishes. Namely, we replace each point $\iota(i/n) + x$ (with $x \in \tilde{G}[n]$) by a linear combination

$$\sum_{r \in \mathbb{Z}} a_r \cdot \{\iota(r + i/n) + x\}$$

with $\sum_r a_r = 1$, $\sum_r a_r (r + i/n)^s = 0$ for $0 < s < 2g$. This is possible with r ranging from 0 to $2g - 1$, because the determinant of this system of linear equations is nonzero. As the characteristic class of a point is obtained by the action by translation on the characteristic class of $\{0\}$, as this action by translation is the exponential of the class of the point in $H^1(S, T_l(\tilde{G}))$, and as this class is up to torsion proportional to $r + i/n$, the assertion follows.

We may also assume that for any integer $m \equiv 1 \,(\mathrm{mod}\, n)$ the cycle $m_*(\mathcal{Z})$ (that is, multiply each point in $\mathcal{Z}$ by m) differs from $\mathcal{Z}$ by a cycle of the form $\iota(1) + \mathcal{Z}_1 - \mathcal{Z}_1$, where $\mathcal{Z}_1$ is itself cohomologous to zero: Namely, in the notation above we have

$$m_* \left(\sum_r a_r \cdot \{\iota(r + i/n) + x)\} - \sum_r a_r \cdot \{\iota(r + i/n) + x\} \right)$$
$$= \sum_r a_r \cdot (\{\iota(m \cdot (r + i/n)) + x\} - \{\iota(r + i/n) + x\})$$

$$= (\iota(1) + \mathcal{Z}_1) - \mathcal{Z}_1,$$

with

$$\mathcal{Z}_1 = \sum_r a_r \sum_{s=0}^{(m-1)(r+i/n)-1} \{\iota(s + r + i/n) + x\}.$$

The class of $\mathcal{Z}_1$ is then a linear combination, with coefficients that are cohomology classes, of the sums

$$\sum_r a_r \cdot \sum_{s=0}^{(m-1)(r+i/n)-1} (s + r + i/n)^t,$$

for $0 \le t < 2g$. Here the inner sum is a polynomial in $r + i/n$ without constant term and of degree $t + 1$. It vanishes if we choose the a_r such that

$$\sum_r a_r (r + i/n)^t = 0$$

for $0 < t \le 2g$ (that is, one degree higher than before).

Now fix one $m > 1$ as above. There exist cohomology classes

$$\tilde{z}, \tilde{z}_1 \in H^{2g-1}(\tilde{G} - |\mathcal{Z}|, \mathbb{Q}_l(g))$$

with boundaries the classes of $\mathcal{Z}$, $\mathcal{Z}_1$, and thus

$$(m_* - \mathrm{id})(\tilde{z}) - (\iota(1)_* - \mathrm{id})(\tilde{z}_1)$$

has trivial boundary and extends to a class in $H^{2g-1}(\tilde{G}, \mathbb{Q}_l(g))$. This class lies in the image of $(m_* - \mathrm{id})$ and correcting $\tilde{z}$ we may assume that it vanishes. That is, $m_*(\mathcal{Z}) - \mathcal{Z}$ lies in the image of $\iota(1)_* - \mathrm{id}$, in the inductive limit of cohomologies with compact support of $\tilde{G} -$ (finite union of $\iota(r + i/n) + x$). It then follows that the same holds for any other $m' \equiv 1 \pmod{n}$.

Now recall that we have defined a compactification $\tilde{P}_N$ of $\tilde{G}_N$ by first forming the associated $\mathbb{P}^1$-bundle over B_N and then repeatedly blowing up $\{0, \infty\}$ in the special fiber. It admits an action of $1/N \cdot Y$. After sufficiently (but only finitely) many of these blowups all sections in the support of $\mathcal{Z}$ extend to sections of this blowup and lie in the union of the $1/n \cdot Y$-translates of $\tilde{G}$. If we extend the derived direct image over S_N° under the inclusion $\tilde{G}_N - |\mathcal{Z}| \to \tilde{G}_N$ of $\mathbb{Q}_l(g)$ by zero to all of $\tilde{P}$, the derived direct image under j gives a sheaf on our partial blowup which is the extension by zero of the corresponding sheaf on the complement of $\{0, \infty\}$. Thus $\tilde{z}$ induces a cohomology class with coefficients in the sheaf $i^* \mathbb{R} j_*(\mathbb{Q}_l(g))$, over $\tilde{P}_N - |\mathcal{Z}|$, with support contained in finitely many of the irreducible components of the special fiber of $\tilde{P}_N$. If we represent $i^* \mathbb{R} j_*(\mathbb{Q}_l(g))$ by a complex of Y-equivariant injective sheaves (better: multiply by a common denominator and then use coefficients $\mathbb{Z}/(l^s)$ for all s) we may form the (locally finite) sum of all Y-translates of our class, and it defines a class on the quotient with boundary the projection of $\mathcal{Z}$. Equivalently, we take the trace of $\tilde{z}$ under the projection $\tilde{P} \to P$, and it defines there a class z, with boundary $\mathcal{Z}$. Also one checks that it is fixed by m_*: The closure of the graph of multiplication by m on $\tilde{G}$ defines another relatively complete model $\tilde{P}^* \subset \tilde{P} \times \tilde{P}$, and m_* on $i^* \mathbb{R} j_*(\mathbb{Q}_l(g))$ is given by pullback via one projection and push forward via the other. As $\tilde{z}$ is invariant under this operation up to classes which vanish under the projection to P, we get our claim. It implies that our projection is the class z_n. Namely, it differs from it only by the image of a class in $H^{2g-1}(P_N, i^* \mathbb{R} j^*(\mathbb{Q}_l(g)))$, and $m_* - \mathrm{id}$ is bijective on that group (reduce to Y-equivariant cohomology on $\tilde{P}_N$, and filter by the pullback of $i^* \mathbb{R} j_*(\mathbb{Q}_l(g))$ on $B_N^t \times B_N$ etc.). In other words, we have represented that class in terms of the Mumford construction.

To compute its bad reduction we have to do the same for the class $\tilde{z}_N$. For this denote by the inclusion of $\tilde{P}_N^\circ$ into $\tilde{P}_N$ by j. Then the derived direct image $\mathbb{R} j_*(\mathbb{Q}_l(g))$ has cohomology in degrees 0, 1, 2. In degree 0, this cohomology is $\mathbb{Q}_l(g)$, in degree 1 the direct sum of the direct images of $\mathbb{Q}_l(g-1)$ on the components ($\mathbb{P}^1$-bundles over $B_N^t \times B_N$) E_r, and in degree 2 the direct sum of the direct images of $\mathbb{Q}_l(g-2)$ on the intersections $E_r \cap E_{r+1}$ of two consecutive components. A cohomology-class

$$w \in H_!^m(P_N^*, \mathbb{Q}_l(g)) = H_!^m(P_N, \mathbb{R} j_*(\mathbb{Q}_l(g)))$$

has as image in $\tau_{\geq 2} \mathbb{R} j_*(\mathbb{Q}_l(g))$ a finite sum of classes in

$$H^{m-2}(E_r \cap E_{r+1}, \mathbb{Q}_l(g-2)).$$

As this sum lifts to $\tau_{\geq 1}(\mathbb{R} j_*(\mathbb{Q}_l(g))$ the relevant obstruction vanishes, which means that for each r the cohomology class in $H^m(E_r, \mathbb{Q}_l(g-1)$ which is the sum of the direct images of the two contributions from $E_r \cap E_{r-1}$ and from $E_r \cap E_{r+1}$ vanishes. It then follows by induction that all these vanish individually, starting from the fact that this holds for very small r (because of compact support). Hence we conclude that our class lifts to $H_!^m(P_N, \tau_{\leq 1} \mathbb{R} j_*(\mathbb{Q}_l(g)))$, and the lift is unique up to the boundary of a sum of classes in $H_!^{m-3}(E_r \cap E_{r+1}, \mathbb{Q}_l(g-2))$.

Projecting modulo $\tau_{\leq 0} \mathbb{R} j_*(\mathbb{Q}_l(g))$, we obtain a direct sum of classes in $H^{m-1}(E_r, \mathbb{Q}_l(g-1))$, unique up to the possibility to shift the direct image of a class in $H^{m-3}(E_r \cap E_{r+1}, \mathbb{Q}_l(g-2))$ from degree r to degree $r+1$. Finally, the sum of the direct images of this class in $H_!^{m+1}(P_N, \mathbb{Q}_l(g))$ is the obstruction to lift to $\tau_{\leq 1} \mathbb{R} j_*(\mathbb{Q}_l(g))$ and thus vanishes.

Recall that the cohomology of a $\mathbb{P}^1$-bundle with section is the direct sum of two copies of the cohomology of the base, one mapping to it via pullback π^* and the other via direct image of the section. As in our case we may shift these direct images between components, we may assume that for all $r \neq 0$ our classes are pullback of classes

$$w_r \in H_!^{m-1}(B_N^t \times B_N, \mathbb{Q}_l(g-1)),$$

while for $r = 0$ we obtain the direct sum of the pullback of w_0 and the direct image (via $E_0 \cap E_1 \subset E_0$) of a class in $H_!^{m-3}(B_N^t \times B_N, \mathbb{Q}_l(g-2))$.

Now we use the condition that the sum of the direct images of our classes vanishes. Restricting to E_r for $r \neq 0, 1$ we get the equation ($N_r = $ normal bundle)

$$i_{0,*}(w_{r-1}) + i_{\infty,*}(w_{r+1}) + c_1(N_r) \cup w_r = 0.$$

As $N_r = M(-\{0, \infty\})$, and as $i_{0,*} - i_{\infty,*}$ is multiplication by the first Chern class of $\mathcal{P}^{\otimes N} \otimes M_N^{-r}$, this reads

$$i_{\infty,*}(w_{r-1} + w_{r+1} - 2w_r) + \pi^*(c_1(M) \cup w_r$$
$$+ (Nc_1(\mathcal{P}_N) - rc_1(M_N)) \cup (w_r - w_{r-1})) = 0.$$

Thus the first component vanishes, that is, $w_{r-1} + w_{r+1} = 2w_r$ for $r \neq 0, 1$, and it follows by induction that all w_r vanish. The same then follows for the additional contribution on E_0. In other words, any section of $H_!^m(\tilde{P}_N^\circ, \mathbb{Q}_l(g))$ extends to a section of $H^m(\tilde{P}_N, \mathbb{Q}_l(g))$.

This picture has to be modified if we deal with $\tilde{z}$. Namely, for each r we have a section i_r of $\tilde{P}_N$ over the preimage in S_N of the graph of multiplication by $-r$: $B_N^t \to B_N$, and, in fact, we also have sections if we translate by N-torsion-points x of $\tilde{G}_N$. Now $\tilde{z}$ is also allowed to have poles along these sections (which give subschemes of codimension g), with constant residues in $\mathbb{Q}_l$. This means that in our analysis we have to add to the divisors E_r also the subschemes $i_r(S_N)$. As the new (constant) residues are regular at the boundary it follows as before that the projection of $\tilde{z}_N$ to the truncation $\tau_{\geq 2}$ of the direct image vanishes, so we can lift as before to $\tau_{\leq 1}$, and again get classes

$$w_r \in H^{2g-2}(B_N^t \times B_N, \mathbb{Q}_l(g-1)).$$

Furthermore, the sum of the direct images of all classes vanishes. Restricting to E_r this gives a relation between the w_r, which now, however, gets an additional term coming from the section i_r. Namely, the sum of the residues of $\tilde{z}_N$ on the $\tilde{G}_N[N]$-translates of i_r is a number a_r, which vanishes if r is not divisible by N/n. Also we know that $\sum_r a_r \cdot r^t$ vanishes for $0 \leq t \leq 2g$. Now i_r contributes to our equation the direct image $i_{r,*}$ of the graph of $-r$: $B_N^t \to B_N$, that is, $i_{\infty,*}(1, -r)_*(c_\Delta)$, with c_Δ the class of the diagonal in $H^{2g-2}(B_N^t \times B_N, \mathbb{Q}_l(g-1))$ ($i_{0,*}$ gives the same result because the bundle E_r is trivial over the graph of $-r$). Hence we have the equation (for $r \neq 0, 1$)

$$w_{r-1} + w_{r+1} - 2w_r + a_r(1, -r)_*(c_\Delta) = 0.$$

It has the unique solution

$$w_r = \sum_{s < r}(s - r)a_s(1, -s)_*(c_\Delta)$$

(the sum vanishes for big r because $(1, -s)_*(c_\Delta)$ is a polynomial in s of degree $2g - 2$). It also follows that $\tilde{z}_N$ can be lifted to $\tau_{\leq 1}$ of the direct image in such a way that its residue along each E_j is pullback of a section w_r on $B_N^t \times B_N$, with no additional local component.

For completeness we also note (although it is not needed) that from the fact that the sum of direct images vanishes on E_r we have only used half, namely, that the component given by $i_{\infty,*}$ vanishes. In addition, we have a component which is π_r^* of

$$c_1(M) \cup w_r + (Nc_1(\mathcal{P}_N) - rc_1(M_N)) \cup (w_r - w_{r-1}).$$

One checks that this indeed vanishes for our solutions w_r because

$$(sc_1(M) - Nc_1(\mathcal{P}_N) \cup (1, -s)_*(c_\Delta) = 0$$

by the projection formula, since $\mathrm{pr}_1^*(M)^{\otimes s} \cong \mathcal{P}_N^{\otimes N}$ over the graph of $(1, -s)$.

Finally, we derive that the residue of $z(x)$ at the boundary depends only on the q-component $q^{r/N}$ of the torsion-point x. It is equal to the sum

$$- 1/(n^{2g} - 1) \sum_{s < t, t \equiv r (\bmod N)} (s - t) a_s (1, -r)^*(1, -s)_*(c_\Delta)$$

$$= -1/(n^{2g} - 1) \sum_{s < r}(s - r)^{2g-1} a_s c_{\{0\}}.$$

To evaluate this sum, introduce new variables $u = s/N$, $v = t/N$, and consider the function

$$f(v) = \sum_{j \in \mathbb{Z}, u < v + t} (u - v - j)^{2g-1} a_{Nu}.$$

It is periodic in v with period 1, and its $(2g)$th derivative is $-(2g - 1)!$ times the sum of $a_{Ns} \cdot (\delta$-distribution in the projection of s to $\mathbb{R}/\mathbb{Z})$. The latter is

$$n^{2g}(-\delta_0 + \sum_{j=0}^{n-1} 1/n \cdot \delta_{j/n}).$$

If we expand $f(v)$ in a Fourier series, we can read off the coefficients from those of the δ-distributions. The result is

$$f(v) = (2g - 1)! n^{2g} \sum_{k \in \mathbb{Z} - n \cdot \mathbb{Z}} - \exp(2\pi i k v)/(2\pi i k)^{2g}.$$

If v is such that $nv - v$ is an integer, this is equal to

$$- (2g - 1)! n^{2g} \sum_{k \in \mathbb{Z} - n \cdot \mathbb{Z}} \exp(2\pi i k v)/(2\pi i k)^{2g}$$

$$= -(2g - 1)!(n^{2g} - 1) \sum_{k \neq 0} \exp(2\pi i k v)/(2\pi i k)^{2g}.$$

Thus finally the residue of $z(x)$ in a torsion-point $x \in \iota(v) + \tilde{G}$ is equal to the L-series

$$-(2g - 1)! N^{2g-1}/(2\pi i)^{2g} \sum_{k \neq 0} \exp(2\pi i k v)/k^{2g} cl_{\{0\}}.$$

The factor N^{2g-1} is the degree of the covering $S_N \to S$.

Remark 1. The fact that we get a multiple of $c_{\{0\}}$ can also be read off from its transformation under the isogeny which is multiplication by n on B^t and by m^2 on the Poincaré bundle. It corresponds to dividing G by its isotropic subgroup $\tilde{G}[n] + T[n^2]$.

5 The class in K-theory

We show that our Eisenstein class is induced by a class in K-theory, more precisely from a multiple of it which lies in K_1. We construct such a class which is invariant under pushforwards m_* but we do not know whether this property makes it unique. However, this invariance suffices to show that we obtain from it the étale class if we apply the Chern character. The construction uses a little bit of topology but not the sophisticated machinery of infinite loop-spaces.

Higher K'-theory is defined in [11]. For a noetherian scheme X we consider the exact category $\mathcal{M}(X)$ of coherent sheaves on X, form the Q-category $Q\mathcal{M}(X)$ and its classifying space $BQ\mathcal{M}(X)$. Morphisms $M \to N$ in $Q\mathcal{M}(X)$ consists of isomorphisms of coherent sheaves $M \cong N_2/N_1$, where $N_1 \subseteq N_2 \subseteq N$ is a layer of subsheaves. Then $BQ\mathcal{M}(X)$ is the geometric realization of the simplicial set whose n-simplices are sequences $N_0 \to N_1 \to \cdots \to N_n$ of maps in $Q\mathcal{M}(X)$. Finally,

$$K_i'(X) = \pi_{i+1}(BQ\mathcal{M}(X)).$$

For example, for any coherent sheaf M there are two maps $0 \to M$ in $Q\mathcal{M}(X)$ corresponding to the layers $(0) \subseteq (0)$ and $M \subseteq M$. Their composition defines a loop around (0) which represents the class of M in $K_0'(X)$. We note that $BQ\mathcal{M}(X)$ is a connected H-space with the addition defined by direct sums, and thus an abelian group in the category of CW-complexes up to homotopy. (The map $(x, y) \to (x, x + y)$ induces an isomorphism on homotopy groups of $BQ\mathcal{M}(X) \times BQ\mathcal{M}(X)$, thus is a homotopy equivalence.) Also in the definition of $\mathcal{M}(X)$ we may consider only coherent sheaves up to canonical isomorphisms, by passing to an equivalent subcategory. Thus in the following we identify sheaves which are canonically isomorphic.

For a relative abelian scheme $A \to S$, of relative dimension g, the Fourier transform

$$\mathcal{F}: BQ\mathcal{M}(A) \to BQ\mathcal{M}(A^t)$$

is defined as

$$\mathcal{F} = \mathrm{pr}_{2,*}(\mathrm{pr}_1^* \cup \mathcal{P}_A),$$

where $\mathcal{P}_A$ denotes the Poincaré bundle on $A \times_S A^t$; see [10]. The composition $\mathcal{F} \circ \mathcal{F}$ is equal to $(-1)^g \cdot [-\mathrm{id}_A]$, thus $\mathcal{F}$ induces homotopy equivalences on the $BQ\mathcal{M}$s. Finally, for direct images under multiplications, we have

$$\begin{aligned}
\mathcal{F}([m]_*(x)) &= \mathrm{pr}_{2,*}(\mathrm{pr}_1^*[m]_*(x) \cup \mathcal{P}_A) \\
&= \mathrm{pr}_{2,*}([m] \times \mathrm{id})_*(\mathrm{pr}_1^*(x) \cup \mathcal{P}^{\otimes m}) \\
&= \mathrm{pr}_{2,*}(\mathrm{id} \times [m])^*(\mathrm{pr}_1^*(x) \cup \mathcal{P}) \\
&= [m]^* \mathcal{F}(x).
\end{aligned}$$

Thus the Fourier transform intertwines $[m]_*$ and $[m]^*$.

Now assume that S is quasi-projective. We note that for a quasi-projective scheme X and any vector bundle $\mathcal{E}$ on X multiplication by $\mathcal{E} - \mathrm{rank}(\mathcal{E})$ is nilpotent (up to homotopy) on $BQ\mathcal{M}(X)$. Namely, we may tensor $\mathcal{E}$ by a suitable ample line bundle and then assume that there exist (at most $\dim(X) + 1$) maps $\mathcal{O}_X^r \to \mathcal{E}$ such that locally near any given point of X at least one of them is an isomorphism. Then the tensor product of all these complexes is acyclic and realizes the desired homotopy.

As for our abelian scheme A $[m]_*[m]^*$ is equal to multiplication by $[m]_*(\mathcal{O}_A)$, it follows that $[m]_*[m]^* - m^{2g}$ id is nilpotent on $BQ\mathcal{M}(A)$, up to homotopy. By Fourier transform this also holds for the other composition $[m]^*[m]_* - m^{2g}$ id. Especially if $x \in A(S)$ denotes an m-torsion point, then translation by x defines an automorphism T_x of $BQ\mathcal{M}(A)$ with

$$[m]_* T_x = [m]_*,$$

and thus $m^{2g}(T_x - \mathrm{id})$ is nilpotent up to homotopy. As $T_x^m = \mathrm{id}$ it follows easily that for some sufficiently high power m^r the multiple $m^r(T_x - \mathrm{id})$ is nullhomotopic (we may take $r = (2g + 1)(\dim(A) + 1)$).

Thus we obtain a homotopy between $m^r T_x$ and m^r, which can be considered as a path from $f = m^r T_x$ to $g = m^r$ in the mapping space Map of continuous maps from $BQ\mathcal{M}(A)$ into itself, with the compact open topology. We would like this homotopy to be canonical, which we can achieve after increasing r: Namely, two paths p, up to homotopy, from f to g differ by a closed loop in the fundamental group of either f or of g. As Map is an H-space itself these fundamental groups can be identified with the fundamental groups at the origin and the two actions coincide. Also the group structure is induced by the H-space structure. Finally, composition with f or g on the left acts on the fundamental group as m^r, f and g commute (not just up to homotopy), and $f^m = g^m$. Any path p from f to g induces paths $f^{m-i-1} g^i \circ p$ from $f^{m-i} g^i$ to $f^{m-i-1} g^{i+1}$, by composing on the left with $f^{m-i-1} g^i$, and thus a closed loop l around $f^m = g^m$. Adding a closed loop h to our initial homotopy changes l by $m^{r+1} h$, so we obtain a canonical path $\sum_i f^{m-i-1} g^i \circ p - h$ (independent of initial choices, up to homotopy) between $m^{r+1} f = m^{2r+1} T_x$ and $m^{r+1} g = m^{2r+1}$. This procedure commutes with compositions on the right. Especially for any other integer n we have $[n]_* T_x = T_{nx} [n]_*$ and thus composing our homotopy between $m^{r+1} T_x$ and m^{r+1} with $[n]_*$ on the left is the same as composing the canonical homotopy between $m^{r+1} T_{nx}$ and m^{r+1} with $[n]_*$ on the right. A similar statement holds for arbitrary isogenies.

As an application we note that we obtain an explicit homotopy between the two loops in $BQ\mathcal{M}(A)$ representing $m^{2r+1} \mathcal{O}_{\{0\}}$ and $m^{2r+1} \mathcal{O}_{\{x\}}$. Restricting to $A - \{0, x\}$ this gives a class in $\pi_2(BQ\mathcal{M}(A - \{0, x\}))$, invariant under all $[n]_*$ for which $nx = x$. Pulling back via torsion-points y (well defined because of finite Tor-dimension) gives an Eisenstein class in $K_1(S)_{\mathbb{Q}}$. Applying the Chern character into étale cohomology, we get an étale cohomology class in $A - \{0, x\}$ with the correct residues at the boundary and invariant under $[n]_*$. This must be the class used in the étale construction. That is, the Chern character maps the K-theory class to the étale class.

6 The class in complex analytic cohomology

The Eisenstein class in Betti cohomology lifts to Deligne cohomology, or equivalently has a Hodge-theoretic analogue. This of course already follows from the previous construction in higher K-theory, by applying a regulator. However, it is possible to give an explicit formula, by calculating with Green currents. We represent this calculation, without touching the various connections to "K-theory with metrics" or "K_1-chains in Arakelov–Chow groups."

So assume given a relative abelian scheme $A \to S$, with S a complex analytic manifold, of relative dimension g, and two sections $x, y \in A(S)$ which are torsion and everywhere nonzero. Then there exists a $(g-1, g-1)$-current (differential form valued distribution) G with

$$\partial \bar{\partial} G / \pi i = \delta_x - \delta_y.$$

As usual we consider two such Green's currents equivalent if their difference lies in the sums of the images of ∂ and $\bar{\partial}$. In fact, we shall construct such a distribution such that for any integer $m > 1$ with $mx = x, my = y$ we have

$$([m]_* - \mathrm{id})^2 (G) = G.$$

If S is projective this determines G uniquely up to equivalence, and it is fixed by $[m]^*$, up to equivalence. In general I do not know about uniqueness. Finally, G will be C^∞ outside of $\{x, y\}$, and evaluation at 0 gives a form on S. So let us first construct G.

If $H = \tilde{A}$ denotes the $\mathbb{C}^g$ bundle over S defined by the Lie-algebra $\mathrm{Lie}(A)$, we can write A as quotient $A = H/\Lambda$, with $\Lambda \subset H$ locally isomorphic to $\mathbb{Z}^{2g}$. If Ω_A denotes the dual of the Lie-algebra of A we have a natural map $\Omega_A \to \mathcal{O}_H$ whose image is the ideal defining the origin in H.

We can use this to define a Green's current for the zero-section. Endow Ω_A with the hermitian metric defined by a polarization on A, which then defines a metric connection ∇ and canonical representatives for the Chern classes (the coefficients of the characteristic polynomial for $-\nabla^2/2\pi i$). Now exterior multiplication by our tautological section defines a differential v on the pullback to H of the exterior algebra $\bigwedge^\bullet \Omega_A$, which makes the resulting Koszul complex a resolution of the ideal of the zero-section. If ∇ denotes the metric connection we can form the superconnection

$$A_t = \nabla + \sqrt{t}(v + v^{\mathrm{ad}})$$

and form the components of bidegree $(g-1, g-1)$ of the regularized integral (compare [2])

$$G_0 = (2\pi i)^{1-g} \int_{t=0}^{\infty} \mathrm{tr}_s(N_H \exp(-A_t^2)) \frac{dt}{t}.$$

Then

$$\partial \bar{\partial} G_0 / 2\pi i = \delta_0 - (-1)^g c_g(\Omega_A).$$

Here tr_s denotes the supertrace (alternating sum of traces) on endomorphisms of $\mathrm{End}(\bigwedge^\bullet \Omega_A)$, N_H the number-operator (equal to $-q$ on Ω_A^q), the integrand decays exponentially at infinity away from the zero-section, and at zero we use zeta-regularization which amounts to the following:

Suppose $f(t)$ is differentiable at the origin and decays sufficiently fast at infinity. Then

$$\int_{t=0}^{\infty} f(t) \frac{dt}{t} = \int_{t=0}^{\infty} (f(t) - f(0)e^{-t}) \frac{dt}{t}.$$

Note that for an invertible $\lambda > 0$, we then have

$$\int_{t=0}^{\infty} f(\lambda t) \frac{dt}{t} = \int_{t=0}^{\infty} f(t) \frac{dt}{t} - \log(\lambda) f(0).$$

It follows that under the action on H by a homothety λ (amounting to multiplying t by $\sqrt{|\lambda|}$), G_0 changes by

$$1/2 \log(|\lambda|) \, \mathrm{ch}_{g-1}(\wedge^{\bullet} \Omega_A),$$

which in turn is invariant under homotheties.

Translating G_0 by elements $\tilde{x}$ or $\tilde{y}$ and taking the difference gives a Green's current for $\{\tilde{x}\} - \{\tilde{y}\}$. Formally summing over all translates under elements of Λ should give a Green's current on A. However, to make this sum converge we use the following procedure:

Suppose x and y are torsion sections of A. Choose locally in S liftings $\tilde{x}_{\mu}$ and $\tilde{y}_{\nu}$ to elements of $\Lambda_{\mathbb{Q}}$, and rational numbers a_{μ}, b_{ν}, such that for any polynomial $P \in S[\Lambda^{\mathrm{dual}}]$ of degree $\leq 2g + 1$, we have

$$\sum a_{\mu} P(\tilde{x}_{\mu}) = \sum b_{\nu} P(\tilde{y}_{\nu}).$$

This is possible; for example, apply to the formal linear combination $\tilde{x} - \tilde{y}$ (using some lifts) the operator $\prod_{j=1}^{2g+1}([m] - m^j)$ and divide by $\prod_{j=1}^{2g+1}(1 - m^j)$ for some integer $m > 1$ with $mx = x, my = y$.

Now the infinite sum

$$\sum_{\lambda \in \Lambda} \left(\sum_{\mu} a_{\mu} G_0(z - \tilde{x}_{\mu} - \lambda) - \sum_{\nu} b_{\nu} G_0(z - \tilde{y}_{\nu} - \lambda) \right)$$

converges for $z \in H$, as a sum of currents, because G_0 is C^{∞} outside 0 and invariant under homotheties. Thus the inner sum decays like $\|\lambda\|^{-(2g+2)}$ by Taylor's formula. Another way to define it is to use Hecke summation:

If we multiply G_0 by $\phi(z)^{-s}$ the resulting sum converges absolutely for $\mathrm{Re}(s) > 2g - 1$ (for arbitrary choices of lifts $\tilde{x}$ and $\tilde{y}$) and has an analytic continuation to the whole complex s-plane (by our argument with suitable linear combinations of lifts $\tilde{x}, \tilde{y}$). Thus we can take its value at $s = 0$. It follows that the result is independent of our regularization procedure, that is, of the choice of the $\tilde{x}_{\mu}$ and $\tilde{y}_{\nu}$. That it is annihilated by $([m]_* - \mathrm{id})^2$ follows from the vanishing of $([m^{-1}]^* - \mathrm{id})^2(G_0)$ by a simple double-sum argument, using independence from regularizations.

Going back we compute G_0 in more detail. We have

$$A - t^2 = t\{v, v^{\mathrm{ad}}\} + \sqrt{t}\{\nabla, v + v^{\mathrm{ad}}\} + \nabla^2.$$

Forgetting t-powers the first term is multiplication by the square-norm $\phi(z) = \|s\|^2$ of the tautological section of Lie_A over H, the second sum of exterior multiplication by $\nabla'(s)$ and of its adjoint, and the third the curvature R_Ω of Ω_A which acts as a derivation on $\wedge^\bullet \Omega_A$. Furthermore, the operator $N_H \exp(-A_t^2)$ lives naturally in the Clifford-algebra $C(\Omega_A \oplus \mathrm{Lie}_A)$ which acts on $\wedge^\bullet \Omega_A$ by interior and exterior multiplication. This Clifford algebra is filtered by the degree of the Clifford polynomials, the associated graded is isomorphic to the exterior algebra, and the supertrace vanishes on F^{2g-1}. Furthermore, N_H has Clifford degree 2 (its leading term is the Kähler form ω), and for the other terms the Clifford degree matches the degree as a differential form. It follows that to get differential forms of degree $2g-2$ we only need to compute with leading terms, thus can replace the Clifford algebra by the exterior algebra. As a result,

$$
\begin{aligned}
&\mathrm{tr}_s(N_H \exp(-A_t^2)) \\
&= (-1)^{g-1}\omega \sum_{a+b=g-1} \exp(-t\phi)t^a(\nabla's)^a(\nabla''s^{\mathrm{ad}})^a R_\Omega^b/(a!a!b!).
\end{aligned}
$$

Here note that this has to be computed in the exterior algebra over $\Omega_A \oplus \mathrm{Lie}_A \oplus \Omega_S^{1,0} \oplus \Omega_S^{0,1}$, and what we really want is the coefficient of the canonical $2g$-form in $\wedge^{2g}(\Omega_A \oplus \mathrm{Lie}_A)$.

Thus the regularized integral becomes

$$
(-1)^{g-1}\omega \left(\sum_{a+b=g-1} \phi^{-a}(\nabla's)^a(\nabla''s^{\mathrm{ad}})^a R_\Omega^b/(aa!b!) - \log(\phi)R_\Omega^{g-1}/(g-1)! \right).
$$

Finally, in the regularized sum

$$
\sum_{\lambda \in \Lambda} \left(\sum_\mu a_\mu G_0(z - \tilde{x}_\mu - \lambda) - \sum_\nu b_\nu G_0(z - \tilde{y}_\nu - \lambda) \right)
$$

the terms which are purely polynomial in z disappear, which applies to the terms with $a = 0$ in the big sum in the formula for G_0.

Next, we restrict this class to the zero-section and continue computing. Obviously we need expressions for R_Ω and for the derivatives $\nabla(v)$. For this note that the bundle (on S) $\mathcal{E} = \Lambda \otimes \mathcal{O}_S$ admits an integrable connection ∇, a complex conjugation, and a symplectic form given by the polarization. Furthermore, it has a maximal isotropic subbundle Ω_A with quotient the dual Lie_A. Finally, $\mathcal{E}$ is the direct sum of Ω_A and its complex conjugate, and the symplectic product on $\mathcal{E}$ induces the hermitian norm on Ω_A. Let Π denote the projection onto Ω_A, with kernel its complex conjugate. It is a C^∞ endomorphism of $\mathcal{E}$.

The metric connections on Ω_A and Lie_A are given by $\Pi\nabla\Pi$, respectively, its complex conjugate. Also

$$
\kappa = (1 - \Pi)\nabla\Pi
$$

defines a holomorphic one-form on S with values in the symmetric homomorphisms from Ω_A to $\mathrm{Lie}(A)$, that is, an element

$$\kappa \in S^2(\text{Lie}_A) \otimes \Omega_S^1.$$

This is also called the Kodaira–Spencer class. If κ^{ad} denotes its complex conjugate adjoint (for the inner product on $\mathcal{E}$), we have

$$[\nabla, \Pi] = \kappa + \kappa^{\text{ad}}$$

and

$$\kappa^{\text{ad}}\kappa = \Pi R_\Omega \Pi.$$

Also an element $\lambda \in \Lambda$ defines a section $v_\lambda \in \text{Lie}_A$ and $\nabla(v_\lambda)$ can be computed as

$$(1 - \Pi)\nabla(1 - \Pi)(\lambda) = -(1 - \Pi)[\nabla, \Pi](\lambda) = -\kappa(\lambda).$$

Finally, we find $\nabla(v_\lambda^{\text{ad}}) = \kappa^{\text{ad}}(\lambda)$.

To continue the computation use the Hermitian form to identify Lie_A with the complex conjugate of Ω_A. Then R_Ω becomes the contraction of $\kappa\kappa^{\text{ad}}$, a $(1, 1)$-form with values in $\Omega_A^{\otimes 2} \otimes \text{Lie}_A^{\otimes 2}$ contracting to a $(1, 1)$-form with coefficients $\Omega_A \otimes \text{Lie}_A$. Similarly, $\nabla(v_\lambda)\nabla(v_\lambda^{\text{ad}})$ becomes the contraction of $-\Pi(\lambda) \otimes (1 - \Pi)(\lambda)$. Thus all terms can be expressed in terms of the Kodaira–Spencer class κ. However, I see no significant simplification in the resulting formula.

7 Classical Eisenstein cohomology

Here we try to explain what can be derived from the classical theory of Eisenstein series, as discussed in the article [6]. However, we restrict ourselves to the Siegel space classifying principally polarized abelian varieties of dimension 2 to avoid complications.

So denote by G the algebraic group $\text{GSp}(4)$, by P its parabolic corresponding to degenerations with one-dimensional torus, by Q the Siegel-parabolic (corresponding to degenerations into tori), and by B its Borel. Associated to G is a Shimura variety SH_G which is a projective limit of quasi-projective schemes, and admits an action of the group $G(\mathbb{A}_f)$ of finite adeles. In fact, SH_G is the quotient of the product of $G(\mathbb{A}_f)$ and the Siegel upper half space, under the action of $G(\mathbb{Q})^+$ (symplectic similitudes with positive multiplier. Positivity comes in because the relevant symmetric space for symplectic similitudes has two connected components). It has a "compactification" which is the projective limit of reductive Borel–Serre compactifications. Its boundary strata are first indexed by a conjugacy class of a parabolic, that is, by either P, Q, or B. Furthermore, the set of P-strata is the quotient $G(\mathbb{A}_f)/P(\mathbb{Q})^+$, the Q-strata are $G(\mathbb{A}_f)/Q(\mathbb{Q})$, and similarly for B. More precisely, we mean the projective limits of quotients under compact open subgroups of $G(\mathbb{A}_f)$.

If U denotes the unipotent radical of the relevant parabolic (P, Q, or B), then $U(\mathbb{Q})$ is dense in $U(\mathbb{A}_f)$, so in the quotient we may divide on the right by $U(\mathbb{A}_f)$ instead of $U(\mathbb{Q})$, or even by $P^0(\mathbb{A}_f)$ and $Q^0(\mathbb{A}_f)$, where $P^0 \subset P$ and $Q^0 \subset Q$ denote the derived subgroups, or the kernels of the canonical maps into $\mathbb{G}_m^2$.

The induced action of $G(\mathbb{A}_f)$ on the cohomology of SH_G is admissible, that is, each vector is fixed by an open compact subgroup K_f of $G(\mathbb{A}_f)$. We may assume that this open subgroup is the product of open compacts $K_l \subset G(\mathbb{Q}_l)$, for all primes l, and $K_l = G(\mathbb{Z}_l)$ for almost all l. The space of K_f-fixed vectors admits an action of the Hecke algebra for K_f, that is, the K_f-biinvariant functions with compact support on $G(\mathbb{A}_f)$. This Hecke algebra is the tensor product of the local Hecke algebras for each factor K_l. We say that a vector is strictly Eisenstein if there exists a finite set of primes and a character χ of finite order of $T(\mathbb{A}_f)/T(\mathbb{Q})$ such that for $l \neq S$, χ is unramified in l, our vector is fixed by K_l, and the local Hecke algebra acts on it by the same character by which it acts on the K_f-invariant vector in the produced representation $\mathrm{Pro}_{B(\mathbb{Q}_l)}^{G(\mathbb{Q}_l)}(\chi)$. "Produced" means continuous functions on $G(\mathbb{Q}_l)$ which transform via χ under $B(\mathbb{Q}_l)$, as opposed to "induced" which uses tensor products. We also call an element Eisenstein if it lies in a Hecke module which has a finite Jordan–Hölder series consisting of strictly Eisenstein modules.

We need the following argument, due to Manin and Drinfeld: If a square-integrable differential form on $\mathrm{SH}_g(\mathbb{C})$ is a Hecke eigenform and Eisenstein, it is invariant under the derived group of $G(\mathbb{R})$: Namely, we may assume that it is a Hecke eigenfunction for almost all local Hecke algebras, with eigenvalues prescribed by a character of finite order. For primes $l \equiv 1 \bmod N$, with sufficiently divisible N, the local Hecke algebra then acts like it acts on the functions on $G(\mathbb{Q}_l)/B(\mathbb{Q}_l)$. In more concrete terms we have a square-integrable differential form λ on a finite union of quotients $\mathbb{D}/\Gamma$, with D a symmetric domain and Γ a congruence subgroup of G. Then there exist finitely many elements $\gamma_1, \ldots, \gamma_r \in G(\mathbb{Q})$, and a congruence subgroup $\Gamma' \subset \Gamma$, such that $\mathrm{Ad}(\gamma_i)(\Gamma') \subset \Gamma$, and such that on $\mathbb{D}/\Gamma'$ we have for the sum

$$\sum_i \gamma_i^*(\lambda) = r \cdot \lambda.$$

By elementary Hilbert-space geometry this means that λ is fixed by each individual γ_i. However, the subgroup generated by all such elements (for various Hecke operators) is dense in the derived subgroup of $G(\mathbb{R})$.

We need a more sophisticated version of this argument: For the parabolic P or Q we can apply the Jacquet functor to the induced representation $\mathrm{Ind}_{B(\mathbb{Q}_l)}^{G(\mathbb{Q}_l)}(\chi)$, that is, form covariants under the unipotent radical $U(\mathbb{Q}_l)$. This is a representation of the corresponding Levi subgroup, and we can look for subquotients which are unitary, that is, admit an invariant positive definite hermitian product. This implies that the central character has to be unitary, and it then follows easily that only the trivial subquotient can be unitary. This implies that Eisenstein classes in L^2-cohomology lie in sums of one-dimensional subrepresentations.

The Eisenstein cohomology can be analyzed using [6]: Namely, we know the Eisenstein part of the L^2-cohomology, since this cohomology coincides with the space of square-integrable harmonic forms, and any such form is constant. We obtain powers of the first Chern class of the line bundle ω, or more precisely of the curvature form defined by its canonical metric. Furthermore, the L^2-cohomology coincides with the intersection cohomology (Zucker conjecture), and the latter can be described by weights. Finally, we can analyze the difference between intersection cohomology and usual cohomology, and thus determine its Eisenstein part. For technical reasons we use cohomology with compact support and obtain the desired results by duality. The reason for this is that only special weight profiles are used in [6] (see Remark 35.4 there), and these are better adapted to cohomology with compact support. Also we concentrate on cohomology with compact support in degrees at least 3, dual to usual cohomology in degrees at most 3.

The weights are defined by data involving the root system. In our case it is of type C_2. If one chooses an Euclidean vector space with orthonormal base $\{e_1, e_2\}$ its roots can be given as $\{\pm 2e_1, \pm 2e_2, \pm e_1 \pm e_2\}$, the positive roots as $\{2e_1, 2e_2, e_1 \pm e_2\}$, $\rho = 2e_1 + e_2$. The Weyl orbit of ρ consists of $\{\pm 2e_1 \pm e_2, \pm e_1 \pm 2e_2\}$. The parabolic P has the simple root $2e_2$ in its Levi component, whose center is spanned by $H_P = 2e_1$. For the parabolic Q these are replaced by $e_1 - e_2$ and $H_Q = e_1 + e_2$. Finally, a maximal weight $w(\rho) - \rho$ appears in the weight profile μ corresponding to the intersection cohomology of the Baily–Borel compactification [6, Theorem 23.2] if it is dominant for the relevant Levi and its inner product with H_P, respectively, H_Q, respectively, both, is strictly greater than it is for $-\rho$. The latter means that the inner product of $w(\rho)$ with H_P or H_Q is strictly positive. For P this leaves the ρ-transforms $\{2e_1 + e_2, e_1 + 2e_2\}$, for Q $\{2e_1 + e_2, 2e_1 - e_2\}$, and for the Borel their union. These correspond precisely to the elements in the Weyl group of length ≤ 1, and thus the intersection cohomology coincides with the cohomology of the truncation $\tau_{\leq 1} \mathbb{R} j_*(\mathbb{C})$ on the reductive Borel–Serre compactification.

Next, the first direct image has as stalks at the boundary the first cohomology of the unipotent radical of the parabolic corresponding to the stratum. The strata are again Shimura varieties, now associated to the group GL(2).

On the P-stratum the first direct image has rank 2 and corresponds to the irreducible two-dimensional local system on the upper half plane. For the Q-stratum we obtain the three-dimensional local system, and for the Borel the direct sum of two one-dimensional systems. These coincide with the direct images of the systems on the P-stratum, respectively, Q-stratum. All in all, the first direct image is the direct sum of the middle extensions of nontrivial local systems. Thus its cohomology coincides with the intersection cohomology, and any Eisenstein component in degree 2 comes from constant harmonic forms which must vanish. So finally the second cohomology of the first direct image is totally non-Eisenstein, and the Eisenstein part of the intersection cohomology in degree at least 3 coincides with that of the cohomology of the zeroth direct image, that is, with the Eisenstein part in the cohomology of the reductive Borel–Serre compactification.

Finally, there exists an exact triangle relating the cohomology with compact support of SH_G, the cohomology of the reductive Borel–Serre compactification, and the cohomology of the boundary. Similarly, the cohomology of the boundary sits in an exact triangle whose other terms are the direct sum of the cohomologies of the closures of the P- and Q-stratum, and the cohomology of the B-stratum. Again by the Manin–Drinfeld argument the Eisenstein parts of these cohomologies consist of constant harmonic forms, thus no contribution to H^1. Thus finally the Eisenstein part of the cohomology of the boundary consists of the constants in degree 0, of the Eisenstein part in the quotient of the produced representation $\mathrm{Pro}_{B(\mathbb{Q})}^{G(\mathbb{A}_f)}(\mathbb{C})$ under the sum of $\mathrm{Pro}_{P(\mathbb{Q})}^{G(\mathbb{A}_f)}(\mathbb{C})$ and of $\mathrm{Pro}_{Q(\mathbb{Q})}^{G(\mathbb{A}_f)}(\mathbb{C})$ in degree 1, and of the Eisenstein part in the direct sum of $\mathrm{Pro}_{P(\mathbb{Q})}^{G(\mathbb{A}_f)}(\mathbb{C})$ and $\mathrm{Pro}_{Q(\mathbb{Q})}^{G(\mathbb{A}_f)}(\mathbb{C}^+)$ in degree 2. Here "produced representation" means functions $f : G(\mathbb{A}_f) \to \mathbb{C}$, left-invariant (because we produce from the trivial representation) under $P(\mathbb{Q})$ or $Q(\mathbb{Q})$, with $G(\mathbb{A}_f)$ acting via right translation. The exponent "+" refers to the action of the quotient $\mathrm{GL}(2, \mathbb{Q})/\mathrm{GL}(2, \mathbb{Q})^+$ where the nontrivial element acts by orientation reversing maps, that is, with a negative sign on top cohomology. Finally, the Eisenstein part of the cohomology with compact support becomes

$$\mathrm{Pro}_{B(\mathbb{Q})}^{G(\mathbb{A}_f)}(\mathbb{C})/(\mathrm{Pro}_{P(\mathbb{Q})}^{G(\mathbb{A}_f)}(\mathbb{C}) + \mathrm{Pro}_{Q(\mathbb{Q})}^{G(\mathbb{A}_f)}(\mathbb{C}))$$

in degree 2, and

$$\mathrm{Pro}_{P(\mathbb{Q})}^{G(\mathbb{A}_f)}(\mathbb{C}^+)/\text{constants} \oplus \mathrm{Pro}_{Q(\mathbb{Q})}^{G(\mathbb{A}_f)}(\mathbb{C}^-)$$

in degree 3, and finally the constant multiples of powers of $c_1(\omega)$ in degrees 4 and 6. Dually the Eisenstein part of the usual cohomology has the constants in degree 0 and 2,

$$\mathrm{Ind}_{P(\mathbb{Q})}^{G(\mathbb{A}_f)}(\mathbb{C}^+)/\text{constants} \oplus \mathrm{Ind}_{Q(\mathbb{Q})}^{G(\mathbb{A}_f)}(\mathbb{C}^-)$$

in degree 3, and some induced representation in degree 4. Here "induced representations" are defined by using tensor products. For later use we note that the induced representations in degree 3 can be detected as follows: In the toroidal compactification we have strata at infinity indexed by P (corresponding to degenerations with one-dimensional torus) and Q (total degeneration). A class in degree 3 induces on the P-stratum a class in degree 2 with coefficients in the first direct image, which can be integrated over the torus bundle as we did in the previous chapter. Over the Q-stratum we obtain a section of the third direct image (which is supported in the most degenerate stratum).

Also we note that by strong approximation (which is rather elementary in our case) we may replace in the induced (and produced) representations $B(\mathbb{Q})$ by its product with $B^0(\mathbb{A}_f)$, where B^0 denotes the kernel of all characters into $\mathbb{G}_m$, and similarly for P and Q. Hence the induced representations split into sums of representations induced from characters.

Next, we show that these representations already occur in étale cohomology over the rationals $\mathbb{Q}$, with coefficients in $\mathbb{Q}_l$ (which means the inductive limit of the $\mathbb{Q}_l$-adic cohomologies of the SH_K): First of all, we get over the complex numbers cohomology-classes with coefficients $\mathbb{Q}$ instead of $\mathbb{C}$, and these give rise to $\mathbb{Q}_l$-adic classes over the algebraic closure $\bar{\mathbb{Q}}$. Next, we know that the powers of $c_1(\omega)$ are classes over $\mathbb{Q}$. So finally by the Leray spectral sequence it suffices to show that the induced representations in degree 3 are Galois-invariant. For this we use the toroidal compactification of our moduli space. It has stratum corresponding to P where the abelian variety degenerates to an extension of an elliptic curve by a torus. Considering the leading term gives an invariant linear form on the cohomology with coefficients $\mathbb{Q}_l(2)$, thus the representation $\text{Ind}_{P(\mathbb{Q})}^{G(\mathbb{A}_f)}(\mathbb{Q}_l)/$ constants occurs in the cohomology of $\mathbb{Q}_l(2)$. This we have already seen before.

For the other representation $\text{Ind}_{Q(\mathbb{Q})}^{G(\mathbb{A}_f)}(\mathbb{Q}_l)$ we consider the Q-stratum, where the abelian variety degenerates to a two-dimensional torus. The stratum corresponds to a torus embedding of dimension 3: Namely, the universal degenerating abelian variety is a quotient $G = \tilde{G}/\iota(Y)$, where $\tilde{G} \cong \mathbb{G}_m^2$ is a torus with character group X, and ι a period map on $X = Y = \mathbb{Z}^2$ which corresponds to a symmetric definite bilinear form b parametrized by the torus embedding. The possible level-n-structures on G are parametrized by first choosing a maximal isotropic $\mathbb{Z}/(n)^2$ in the symplectic $\mathbb{Z}/(n))^4$, then identifying it with

$$\tilde{G}[n] = \text{Hom}(X, \mu_n),$$

and finally extending ι to $1/n \cdot Y$ by extracting an nth root out of b. It follows that the Galois-action on $\text{Ind}_Q^G(\mathbb{Q}_l)$ is induced by the cyclotomic character $\text{Gal}(\bar{\mathbb{Q}}/\mathbb{Q}) \to \hat{\mathbb{Z}}^*$ which takes values in the $\mathbb{A}_f$-points of the center of the Levi-group GL_2 of Q, and thus acts on the induced representation.

The restriction of cohomology to the most degenerate strata defines an invariant linear form on the étale cohomology with coefficients $\mathbb{Q}_l(3)$, with values in this induced representation, and respects Galois-actions. Thus the Q-induced representation gives classes with coefficients $\mathbb{Q}_l(3)$. Thus, all in all, we have constructed Eisenstein classes in $H^3(\text{SH}_G, \mathbb{Q}_l(2))$ and $H^3(\text{SH}_G, \mathbb{Q}_l(3))$. Note that they are not quite unique because of the Eisenstein classes in degrees 0 and 2.

Finally, there is a relation between Galois and Hecke action, weaker than the classical Eichler–Shimura (see [3, Chapter 7]). Namely, the maximal torus T of G consists of diagonal matrices $\text{diag}(a, b, c, d) \in \mathbb{G}_m^4$ with $ab = cd$. For an unramified character χ of $T(\mathbb{Q}_l)$ at a prime l "of good reduction" the geometric Frobenius Frob_l at l satisfies an equation of degree 4 over the Hecke-algebra, which means that on a subspace where the Hecke-algebra acts like on the induced representation from the $B(\mathbb{Q}_l)$-representation χ (an unramified character of $T(\mathbb{Q}_l)$), Frob_l satisfies the quartic equation

$$(\text{Frob}_l - a_1)(\text{Frob}_l - a_2)(\text{Frob}_l - a_3)(\text{Frob}_l - a_4) = 0,$$

where $a_1 = \chi(p, p, 1, 1)$, $a_2 = p\chi(p, 1, 1, p)$, $a_3 = p^2\chi(1, p, p, 1)$, and $a_4 = p^3\chi(1, 1, p, p)$.

8 Interpolation

Following Hida (initially in [7] and later extended to more complicated situations) we can interpolate the Eisenstein series to a p-adic meromorphic function. Unfortunately, we obtain little information about the location of its poles. For this purpose denote by χ the character $Q \to \mathbb{G}_m$ trivial on the center, and by Q° its kernel. Also $\rho : \mathbb{G}_m \to Q$

denotes the "dual" cocharacter which is contracting on the unipotent radical of Q, and such that the centralizer of ρ is the Levi-subgroup. The composition $\chi \circ \rho$ is multiplication by $g = 2$.

Choose some level-structure at primes away from p, and denote by M^* a complex representing the projective limit of the cohomologies of the Shimura varieties associated to all open compact subgroups of $G(\mathbb{Z}_p)$. The transition-maps are traces. M^* can be represented by a perfect complex of $\mathbb{Z}_p[G(\mathbb{Z}_p)]$-modules, bounded below, and the $G(\mathbb{Z}_p)$-action extends to $G(\mathbb{Q}_p)$, up to homotopy.

Next, form the covariants of M^* under $Q^\circ(\mathbb{Z}_p)$. It is still a complex over $\mathbb{Z}_p[\mathbb{G}_m(\mathbb{Z}_p)]$, and admits an action of a Hecke operator U_p, as follows:

First apply $\rho(p)$ and then form the trace (essentially under the $\mathbb{F}_p$-points of the unipotent radical) to make the result Q°-invariant.

The U_p-operator defines a splitting of the $Q^\circ(\mathbb{Z}_p)$-invariants in two spaces on which U_p is either invertible or topologically nilpotent. Here the topology is defined by the natural topology on $\mathbb{Z}_p[G(\mathbb{Z}_p)]$ defined by the p-adic topologies on the coefficients and on the group, and the assertion holds because it is true on the covariants under suffiently many compact open subgroups of $G(\mathbb{Z}_p)$. Also the splitting of the complex may depend on the choice of U_p in its homotopy-class, but two different choices lead to homotopy-equivalent decompositions. This follows because the induced splitting on the cohomology is canonical. Also any endomorphism of the complex which commutes up to homotopy with U_p, is homotopic to an endomorphism respecting the decomposition.

Denote by N^* the subcomplex of the $Q^\circ(\mathbb{Z}_p)$-coinvariants where U_p is invertible. This complex is perfect over $\mathbb{Z}_p[\mathbb{G}_m(\mathbb{Z}_p)]$, or equivalently its cohomology is finite over $\mathbb{Z}_p$ if we form coinvariants under the 1-units $1 + p\mathbb{Z}_p$, and also divide by p. However, this turns out to be the U_p-invertible subspace of the cohomology (with coefficients $\mathbb{F}_p$) of the Shimura variety associated to the compact open subgroup of $G(\mathbb{Z}_p)$ which is the preimage of $Q(\mathbb{F}_p) \subset G(\mathbb{F}_p)$.

The cohomology groups of N^* are finite modules over the Noetherian and reduced ring $\mathcal{A} = \mathbb{Z}_p[[\mathbb{G}_m(\mathbb{Z}_p)]]$ which is the completion of the group-ring in the adic topology defined by the kernel of the surjection to $\mathbb{F}_p[\mathbb{F}_p^*]$. Thus there exists for each fixed cohomological degree a dense Zariski-open subset of the spectrum of $\mathcal{A}$, such that in this subset the cohomology is flat and commutes with base change. Furthermore, on it we have an operation of the Hecke algebra $\mathbb{T}$ consisting of Hecke operators at primes different from p. Their action on Eisenstein series defines a homomorphism $\mathbb{T} \to \mathcal{A}$, and we can consider its generalized eigenspace on the cohomology of N^*. Again this is flat and commutes with base change on a dense open subset of $\mathrm{Spec}(\mathcal{A})$. If we make base change by a nontrivial character of finite order of $\mathbb{G}_m(\mathbb{Z}_p)$ the corresponding generalized eigenspace becomes the Eisenstein cohomology. The half of it induced from P has p-adically nilpotent U_p and contributes nothing. For the other half induced from Q we obtain either nothing or a space of dimension 1, if the character is odd. It follows that our generalized $\mathbb{T}$-eigenspace has generic ranks 1 or 0. So we find in it an element which generates it in all generic points of $\mathrm{Spec}(\mathcal{A})/(1 + (-1))$. Its constant term along the parabolic Q (given by considering $\mathbb{R}^3 j_*(\mathbb{Z}_p)$) is an element of $\mathcal{A}$ which is nonzero in all generic points, as one sees by evaluating at characters of finite order. By the same argument the constant term along P vanishes. All in all, we have the following.

Theorem 1. *There exists a cohomology class with coefficients in the fraction field of $\mathcal{A}$, which interpolates the Eisenstein class at points where it is regular and which correspond to characters of finite order.*

9 The classical Eisenstein series

We have seen in the previous section that over the complex numbers there exists an Eisenstein class in degree 3 associated to the parabolic Q and to odd characters. We can exhibit an explicit modular form of weight 3 which represents this class. Namely, suppose $\chi : (\mathbb{Z}/(n))^* \to \mathbb{C}^*$ is a primitive Dirichlet-character with $\chi(-1) = -1$. Then for a symmetric 2×2 matrix $Z = X + iY$ in the Siegel-space the series

$$E(\chi, Z) = \sum_{C,D} \chi(\det(C)) \det(CZ + D)^{-3}$$

defines such a modular form. The sum is over coprime pairs of matrices C, D with CD^t symmetric, C invertible modulo n, D divisible by n, modulo the action by left multiplication of invertible matrices $U \in \mathrm{GL}(2, \mathbb{Z})$. The exponent 3 lies on the boundary of the domain of absolute convergence but one can define the series by Hecke summation (more details later). A more intrinsic definition goes as follows: In the symplectic space $\Lambda = \mathbb{Z}^4$ consider maximal isotropic sublattices L of rank 2. Then the sum is over all such sublattices such that the reduction $L/(n)$ is the subspace L_0 generated by the first two coordinate vectors. Namely, L is the image of the linear map given by the 4×2-matrix (C, D). Note that sublattices are parametrized by lines in

$$\wedge^2(\Lambda) = \mathbb{Z}^6$$

which satisfy a certain quadratic equation (the Plücker condition). Furthermore, the symplectic form gives a linear form on $\wedge^2(\Lambda)$ and isotropic lattices correspond to lines in the kernel, that is, points in a Plücker-quadric in $\mathbb{P}^4$. Each such line has a canonical generator of its determinant $\det(L)$, up to factor ± 1. The reduction of $\det(C)$ modulo n is the determinant of the induced map modulo n between L and L_0, or the value modulo n of one of the coordinates of L.

The determinant $\det(CZ + D)$ defines another linear form l_Z on $\wedge^2(\Lambda)$, and E_Z is the sum over $\chi(\det(L))l_Z(L)^{-3}$, with L running over points in the Plücker-quadric in $\mathbb{P}^4$ which reduce modulo n to the given point L_0. It converges if one sums in the order given by the norm of L because of the factor $\chi(\det(L))$ which leads to cancellations. Also we can identify Λ with the first cohomology $H^1(A(Z), \mathbb{Z})$ of the principally polarized abelian variety

$$A(Z) = \mathbb{G}_m^2 / \exp(2\pi i \mathbb{Z}^2 Z)$$

classified by Z and then l_Z corresponds to the projection onto $H^2(A, \mathcal{O}_A)$ in the Hodge-filtration. The third power of this line bundle is isomorphic to the dual of the 3-differentials on the moduli space, via the Kodaira–Spencer isomorphism, and thus the infinite sum defines indeed a three-form. Note also that we evaluate at the multiplicative cusp, that is, the maximal isotropic subgroup in $A(Z)[n]$ consists of the torsion-points of the torus $\mathbb{G}_m^2$.

The effect of Hecke operators on $E(\chi, Z)$ can be computed as follows (as in [3, Chapter 7, Proposition 4.4]): A Hecke operator acts formally by pulling back a class via all isogenies $A \to A'$ of a certain type. These correspond to certain sublattices $\Lambda' \subset \Lambda$. For example, for primes l not dividing n, one assumes that Λ/Λ' is a maximal isotropic subgroup of $\Lambda/(l^r)$ with a prescribed sequence of elementary divisors $l^a, l^b, l^{r-a}, l^{r-b}$. If one applies this to the sum defining $E(\chi, Z)$, one notes that the isotropic $L \subset \Lambda$ correspond one to one to the isotropic

$$L' = L \cap \Lambda' \subset \Lambda'$$

because the reductions modulo n are isomorphic. However, the canonical generators of the determinant differ by a factor $[L : L']$, and the Kodaira–Spencer isomorphisms differ by l^{3r}, so that the corresponding summands differ by

$$\chi([L : L'])[L : L']^{-3}l^{3r}.$$

If we sum over all Λ' the result is independent of L and given by the following rule:

There exists a homomorphism ρ_Q from the Hecke algebra for GSp(4) (symplectic similitudes) to that for its Levi-subgroup $GL(2) \times \mathbb{G}_m$ (pairs (A, D) in GL(2) with $AD^t = scalar$) by integrating over the unipotent radical of the parabolic Q. In down to earth terms we consider all $\Lambda' \subset \Lambda$ as before and count the induced lattices $L' \subset L$ and $\Lambda'/L' \subset \Lambda/L$, for any fixed maximal isotropic L. Then the Hecke action on $E(\chi, Z)$ is via the character induced from the character $\chi(\det(A)) \det(D)^3$ of the Hecke algebra of $GL(2) \times \mathbb{G}_m$.

At prime divisors p of n we can still consider the operator U_p which maps A to the sum of all A' such that the kernel of $A \to A'$ is a maximal isotropic subgroup of $A[p]$ not meeting the given one. These correspond to $\Lambda' \subset \lambda$ mapping to an isotropic complement of $L_0/(p)$. For such a Λ' we always have $L' = pL$ (if $L/(n) = L_0/(n)$), and $E(\chi, Z)$ is fixed by U_p.

Before we come to the computation of Fourier coefficients (following [5]) we check that Hecke summation applies. The isotropic $L \subset \Lambda$ reducing modulo n to the fixed L_0 generated by the first two coordinates can be identified with the set of symmetric 2×2-matrices S, with coefficients in $\mathbb{Q}$, such that S is integral and divisible by n at primes dividing n. Namely, let $L_\mathbb{Q} = \mathrm{graph}(\mathrm{id}, S)$. Furthermore, the image $C(\mathbb{Z}^2) \subset \mathbb{Z}^2$ is the subgroup of elements λ for which $S(\lambda)$ is still integral. For such an S define the denominator $D(S)$ as a diagonal 2×2-matrix with positive integral entries δ_1, δ_2 the elementary divisors of C. That is, δ_2 is the smallest common multiple of the denominators of the entries of S, and δ_1 has a similar but slightly more complicated definition. In any case $D(s)$ depends only on the reduction of S modulo $\mathbb{Z}$ and is the product, over all primes p, of factors depending only on the p-primary component of this reduction. Finally, our infinite sum becomes the sum of

$$\chi(\det(D(S))) \det(D(S))^{-3} \det(Z + S)^{-3},$$

so for the Hecke summation we have to multiply these by $(\det(D(S) \cdot |\det(Z + S)|)^{-s}$ and try to analytically continue to $s = 0$. Here we first sum over all translates by integral $S \in n\mathbb{Z}^3$, and then over residue-classes of S modulo $n\mathbb{Z}^3$. The first sum converges absolutely for $\mathrm{Re}(s) > -1$ (use Poisson summation to replace it by an integral) and defines an infinitely differentiable periodic function $\phi(Z, s)$ in Z, analytic in s. We then have to sum over its translates under $S \in n\mathbb{Z}^3$. Here such a translate gets a factor of absolute size $D(S)^{-3-\mathrm{Re}(s)}$.

$S \mod (n)$ is the sum of its p-primary components S_p, for p a prime not dividing n. If such an S_p has denominator $\mathrm{diag}(p^\alpha, p^{\alpha+\beta})$ it is of the form

$$S_p = p^{-\alpha-\beta} u x x^t + p^\alpha S_p'$$

with a unit $u \in \mathbb{Z}_p^*$, a unimodular vector x, and a symmetric matrix S_p'. If we let u run over an arbitrary element of $\mathbb{Z}_p$, and S' be an arbitrary symmetric matrix, we get in this manner $p^{3\alpha+2\beta}$ residue classes of symmetric matrices, and the fraction of them which does not have the denominator $p^\alpha, p^{\alpha+\beta}$ is $O(1/p)$. It then follows that if we form the sum over all pairs α_p, β_p (for all primes p not dividing n) and for each such pair all S_p- residues as above, we make an error which is bounded by the sum over all $p^{-2-\mathrm{Re}(s)}$ and thus converges absolutely for $\mathrm{Re}(s) > -1$. A similar argument allows us to ignore terms where some $\alpha_p > 0$ or $\sum \beta_p > 1$. Thus we are left with the infinite sum

$$\sum_{p,x,u} p^{-3-s} \chi(p) \phi(Z + (u/p) x x^t, s).$$

Here p runs over all primes not dividing n, u over all elements of $\mathbb{F}_p$, and x over all unimodular vectors in $\mathbb{F}_p^2$ modulo scalars, that is, over $\mathbb{P}^1(\mathbb{F}_p)$.

Now express $\phi(Z, s)$ as a Fourier series in $\mathrm{Re}(Z)$ (for fixed imaginary part). The Fourier coefficients $a(T, s)$ are indexed by even integral matrices T (which correspond to integral quadratic forms $T(x) = \mathrm{Tr}(x T x^t / 2)$) and they are holomorphic in s. If we sum a Fourier term $a(T, s) \exp(\pi i \, \mathrm{Tr}(\mathrm{Re}(Z)T))$ over all multiples $u/p \cdot x x^t$ the result vanishes unless $T(x)$ is divisible by p. Furthermore, for a given T this holds for at most two xs unless T is divisible by p. Thus again up to an absolutely convergent error term we need only sum over terms with T divisible by p, which amounts to forming the average of $\phi(Z, s)$ over all p-division points, which then again we may replace by the average $\bar{\phi}(s)$ over all $\mathrm{Re}(Z) \mod (1)$. Thus we end up with

$$\sum_p p^{-1-s} \chi(p) \bar{\phi}(s)$$

which is, in fact, analytic for $\mathrm{Re}(s) > -1$ because χ is odd and thus nontrivial.

Next, we want to compute Fourier coefficients in more detail but for a slightly different series. In general we can form Eisenstein series

$$E(M, \chi) = \sum_L \chi(L) l_Z(L)^{-3}$$

where we sum over all isotropic $L \subset \Lambda$ such that the intersection of L/nL with the subspace of $\Lambda/(n)$ generated by the first two basis elements is a fixed submodule $M \subseteq L_0/(n)$. They are sums transforms of our previous one $(M = L_0/(n))$ under the symplectic group and also can be defined via Hecke summation. For p a prime divisor of n the U_p-operator replaces Λ by all smaller Λ' which are preimages of an isotropic complement L_1 modulo p to $L_0/(p)$. As the intersection $L' = L \cap \Lambda'$, we have for the new intersection $M' \supseteq M$ with equality if and only if the p-torsion of M is a direct summand in the p-torsion of $L_0/(n)$ and in addition L_1 contains a complement to M/pM in $L/(p)$. The number of such L_1 is 1, p^2, p^3 depending whether M/pM has rank 0, 1, 2, and the corresponding indices $[L : L']$ are 1, p, p^2. It follows that U_p transforms $E(M, \chi)$ into a linear transformation of $E(M', \chi)$ with $M \subseteq M'$, where the coefficient of $E(M, \chi)$ is equal to 0, p^3, p^2, 1 depending on whether M is not a direct summand at p or a direct summand of rank 0, 1, 2. For example, $M = L_0/(n)$ defines, as we already know, a U_p-eigenfunction with eigenvalue 1. Also we derive that the eigenvalues of U_p on the space of $E(M, \chi)$s are 0, 1, p^2, p^3. One obtains eigenvectors with the eigenvalue 1 if one applies a sufficiently high power of $U_p(U_p - p^2)(U_p - p^3)$ to $E(M, \chi)$s with $M/(p) = (0)$. We shall need such eigenvectors and thus compute the Fourier coefficients of

$$E(0, \chi) = \sum \chi(\det(D)) \det(CZ + D)^3,$$

where the sum is over all coprime (C, D) with CD^t symmetric and C divisible by n, up to common unimodular left factors. In fact, we are really interested in the projection to the $U_p = 1$-eigenspace, for all p dividing n. This is given by applying a suitable power of $U_p(U_p - p^2)(U_p - p^3)$. However, once we know that the p-powers in the denominators are bounded, it is simpler to form the limit $\lim U_p^n(E)$ which converges p-adically. This we shall do when we compute local factors at p. Note that the T-Fourier coefficient of $U_p(E)$ is the pT-coefficient of E, by explicit computation at the multiplicative cusp, so passing to the limit of Fourier expansions is easy.

Firstly the Fourier coefficients can be evaluated termwise by integrating against $\exp(-\pi i \operatorname{Tr}(TZ)$ (avoiding Hecke regularization). Secondly they are invariant under transformations $T \mapsto UTU^t$ with unimodular U. The coefficient for $T = 0$ is evaluated by letting Z go to infinity and corresponds to the subsum where $C = 0$, so it is 1. For T of rank 1 we can assume that $T = \operatorname{diag}(2u, 0)$ with $u > 0$, and let $z_{2,2}$ approach infinity. Then only Cs annihilating the second coordinate vector contribute. Multiplying by unimodular U we may assume that $C = \operatorname{diag}(nc, 0)$ with $c > 0$. Then D is lower triangular and can still be multiplied by a unipotent matrix, and one can choose a unique representative $D = \operatorname{diag}(d, 1)$ with d prime to nc. So we have to compute the Fourier coefficients of $\sum \chi(d)(cn\tau + d)^{-3}$. We treat the coefficient of $\exp(2\pi i u\tau)$ for $u > 0$ and with no prime factors not dividing n (that is, u divides a power of n). More general us will be treated by other means.

This coefficient turns out to be

$$(-2\pi i)^3 u^2/2 \sum_{c>0, d\in\mathbb{Z}/(nc)^*} \chi(d) \exp(2\pi i ud/(nc))/(nc)^3.$$

The sum is over all rational numbers $s = d/(nc)$ with denominator divisible by n, modulo integers. Such a number is the sum over its p-primary components and this makes the sum an infinite product over all primes p. For p not dividing n we get as factor

$$\sum_{m=0}^{\infty} \chi(p^n) p^{-3m} \sum_{x \in \mathbb{Z}/(p^m)^*} \exp(2\pi i u x/p^m) = 1 - \chi(p)/p^3$$

(the inner sum vanishes for $m > 1$), while for p dividing n in exact power p^r, we have

$$\sum_{m=r}^{\infty} p^{-3m} \sum_{d \in \mathbb{Z}/(p^m)^*} \chi(d) \exp(2\pi i u d/p^m).$$

If we sum first over $d \equiv 1 \mod (p^r)$ the sum vanishes unless u is divisible by p^{m-r}. If this is so we get a linear combination of value $\chi(d)$ where the coefficients have denominators bounded by $p^{-3r-2(m-r)}$.

All in all, we get the following result: If we multiply the Fourier coefficient by the L-value $L(3, \chi)/(-2\pi i)^3$ (with trivial Euler factors at primes p dividing n) the result is a linear combination of values $\chi(d)$, with coefficients in $\mathbb{Q}(\mu_n)$ and denominators dividing $2n^3$.

So we get to the most interesting case of strictly positive definite Ts. The partial sum where C is not invertible does not contribute to this Fourier coefficient because these terms are invariant under translations by real multiples of some xx^t. So we consider the sum with invertible Cs. Then we can multiply C by a unique unimodular matrix to make it equal to the denominator of the symmetric matrix $S = C^{-1}D$, to get

$$\sum_S \chi(\det(D(S)S) \det(D(S)^{-3} \det(Z + S)^{-3}.$$

The sum is over all symmetric S with denominator divisible by n. If we use the formula in [5, Chapter 4, 7.6, p. 294], the factor π^{-nr} there seems to be superfluous) the Fourier coefficient of T in this sum becomes

$$-\pi^5/3 \cdot \det(T)^{3/2} \sum_S \chi(\det(D(S)S) \det(D(S))^{-3} \exp(\pi i \operatorname{Tr}(ST)),$$

where now the sum is over symmetric S modulo $\mathbb{Z}^3$ with denominator divisible by n. Similarly as before, the sum becomes the product, over all primes p, of the sum with S p-primary. Finally, we do not need to do this only for Ts such that for any prime the quadratic form defined by T is not proportional to a square modulo p^2, and does not vanish modulo p. The other Ts will be handled by Hecke operators.

We now first compute the local factors at primes p not dividing n. The denominator of S_p has the form $\operatorname{diag}(p^\alpha, p^{\alpha+\beta})$. The sum over denominators with $\alpha \geq 2$ vanishes because the sum over a fixed class modulo $1/p$ vanishes. Similarly, the contributions from $\alpha = 1$ can be computed by first summing over all S_p with $\alpha \leq 1$ (which vanishes) and then subtracting the contributions with $\alpha = 0$. Finally, the S_p with $\alpha = 0$ are of the form $p^\beta u x x^t$ with $u \in \mathbb{Z}/(p^\beta)^*$ and $x \in \mathbb{P}^1(\mathbb{Z}/(p^\beta))$ (or better x a unimodular element, and the pair (u, x) is determined up to the action of units). Thus we obtain as local factor

$$\sum_{\beta \geq 0} \chi(p^\beta) p^{-3\beta} \sum_{u, x \bmod p^\beta} \exp\left(\frac{2\pi i u T(x)}{p^\beta}\right)$$

$$- \chi(p^2) p^{-6} - \sum_{\beta \geq 0} \chi(p^{2+\beta}) p^{-6-3\beta} \sum_{u, x \bmod p^{\beta+1}} \exp\left(\frac{2\pi i u T(x)}{p^{\beta+1}}\right).$$

In the first term, the sums over u vanish unless $T(x)$ is divisible by $p^{\beta-1}$. If this is the case, then the sum is $-p^{\beta-1}$ if $T(x)$ is not divisible by p^β, and $p^\beta - p^{\beta-1}$ if it is. So we have to count the number of times this will happen:

Define ϵ by the rule that $\epsilon = 1$ if $T \bmod (p)$ is nonsingular isotropic, $\epsilon = -1$ if it is nonsingular anisotropic, and $\epsilon = 0$ iff $T(x) \equiv u\lambda(x)^2 \bmod (p)$ is a square. Note that T is proportional to the norm-form on an ideal in the integers of an imaginary quadratic number field, and the three values of ϵ distinguish whether p decomposes, remains inert, or ramifies in this imaginary quadratic field. That is, $\epsilon = \epsilon_T(p)$ with ϵ_T the associated quadratic odd Dirichlet character.

For $\beta > 0$ the number of x where $T(x)$ is divisible by $p^{\beta-1}$ is $p + 1$ if $\beta = 1$, $2p, p, 0$ if $\beta = 2$ and $\epsilon = 1, 0, -1$, and $2p, 0, 0$ for $\beta > 2$. Divisibility by p^β happens in 2, 1, 0 cases if $\beta = 1$, and 2, 0, 0 cases if $\beta > 1$. If $\epsilon = \pm 1$ in the first sum the subsums with $\beta > 1$ vanish, and the same holds for the last term and $\beta > 0$. Thus we obtain

$$1 + (1 - \chi(p)p^{-3})\chi(p)p^{-3}(-p - 1 + p(1 + \epsilon)) - \chi(p^2)p^{-6}$$
$$= (1 - \chi(p)p^{-3})(1 + \chi(p)\epsilon_T(p)p^{-2}).$$

For $\epsilon = 0$ the result becomes

$$1 + (1 - \chi(p)/p^3)((-p - 1 + p)\chi(p)/p^3 - p^2\chi(p^2)/p^6) - \chi(p^2)/p^6$$
$$= (1 - \chi(p)/p^3)(1 - \chi(p^2)/p^4).$$

Thus in all cases we obtain

$$(1 - \chi(p)/p^3)(1 - \chi(p)^2/p^4)/(1 - \epsilon_T(p)\chi(p)/p^2).$$

Hence if we multiply by $L(3, \chi)L(4, \chi^2)/\pi^7$ we obtain as product (over primes p not dividing n) the L-value $L(2, \chi\epsilon_T)/\pi^2$.

Finally, we consider the prime factors p of n. Suppose p^r is the exact power of p dividing n. Denote by $M(\alpha, \beta)$ the set of symmetric matrices S with denominator $\mathrm{diag}(p^\alpha, p^{\alpha+\beta})$, modulo integral matrices. We have to compute

$$\sum_{\alpha \geq r, \beta \geq 0} p^{-6\alpha-3\beta} \sum_{S \in M(\alpha, \beta)} \chi(p^{6\alpha+3\beta} \det(S)) \exp(\pi i \, \mathrm{Tr}(ST)).$$

In fact, we want to write this as a sum of values of χ_p (the p-local factor of χ) on elements of $\mathbb{Z}/(p^r)^*$, and bound the denominators of the coefficients. That is, we consider subsums where $p^{6\alpha+3\beta}\det(S)$ lies in a fixed congruence class modulo p^r. First we note that these denominators are powers of p, and they are bounded because the sums over $M(\alpha, \beta)$ vanish if $\alpha > r$ or if $\alpha + \beta$ is sufficiently big: the first because the sum over S in a fixed congruence class mod p^{-1} vanishes (as T is not divisible by p), and the second because we may act (via USU^t) with the group of 2×2-matrices $U \equiv 1 \bmod (p^r)$. If we sum over an orbit, we may instead sum over the orbit of the action on T which contains all $T' \equiv T \bmod (p^s)$ for some sufficiently big s. If $p > 2$ and p does not divide the discriminant of T, we may choose $s = r$; if it divides the discriminant only to the first power, choose $s = r + 1$, etc. This sum vanishes if the denominator of S contains a higher p-power, that is, if $\alpha + \beta > s$.

Next, as already announced, we pass to the $U_p = 1$-eigenspace by replacing T by $p^m T$ and forming the p-adic limit for $m \to \infty$. We extend the character χ to $(\mathbb{Q})_p^*$ by setting $\chi(p) = 1$, and the denominator to all nonsingular symmetric S by the rule $D(S) = \mathrm{diag}(p^\alpha, p^{\alpha+\beta})$ if S has elementary divisors $p^{-\alpha}$, $p^{-\alpha-\beta}$. Then the local factor for $p^m T$ is

$$\int_{\alpha \geq r-m} \chi(\det(S)) |\det(S)|^3 \exp(\pi i \operatorname{Tr}(ST)) dS,$$

where dS denotes the usual p-adic Haar-measure and the integral is defined as the (infinite) sum of integrals over orbits under the USU^t-action of matrices $U \equiv 1 \bmod (p^r)$. Namely, these integrals over orbits vanish unless α and $\alpha + \beta$ are bounded above by a constant depending only on T.

Finally, we pass to the p-adic limit $m \to \infty$ and get the integral over all S,

$$\int \chi(\det(S)) |\det(S)|^3 \exp(\pi i \operatorname{Tr}(ST)) dS,$$

which converges p-adically because for small (and usually negative) α and β the volume of an orbit has as denominator a p-power growing like $p^{-3\alpha-2\beta}$ while $\det(S)^3$ gives $p^{-6\alpha-3\beta}$ in the numerator. Also we have used the character χ to simplify notation, but we still really want to integrate over S with fixed determinant $(\bmod \ p^r)$. If we replace T by UTU^t with $U \in \mathrm{GL}(2, \mathbb{Q}_p)$ we get a factor $|\det(U)|^{-3}\chi(\det(U)^2$, thus if we multiply the integral by $\chi(\det(T)|\det(T)|^{3/2}$ we get a function which is invariant under such actions, that is, depends only on χ and on the isomorphism class over $\mathbb{Q}_p$ of the quadratic form T, that is, on $\det(T)$ modulo squares.

To compute these integrals we may assume that either T defines a nonsingular quadric or that it is proportional to a square modulo p but not modulo p^2, that $T(x)$ is integral but the discriminant is not divisible by p^2. We already know that the integral over $M(\alpha, \beta)$ vanishes unless $\alpha \leq r, \alpha + \beta \leq r + 1$ (or $\leq r$ if p does not divide the discriminant of T).

We first compute integrals over $M(\alpha, 0)$ $(\alpha \leq r)$. Let s denote the smallest integer $\geq r/2$. A congruence class of $S \bmod (p^{s-\alpha})$ consists of elements $S + \delta S$ with δS symmetric and divisible by $p^{s-\alpha}$, and $\operatorname{Tr}(S^{-1}\delta S)$ divisible by p^r (to get fixed determinant modulo this p-power). The sum over $\exp(\pi i \operatorname{Tr}(\delta ST))$ vanishes unless $p^{-\alpha}T$ is a scalar multiple of $(p^\alpha S)^{-1}$ modulo p^{-s} or T is a scalar multiple of

$(p^\alpha S)^{-1}$ modulo $p^{\alpha-s}$. If $\alpha > s$ this cannot happen if p divides the discriminant of T, and otherwise means that $p^\alpha S$ is a multiple of T^{-1} modulo $p^{\alpha-s}$. If this holds the integrand is constant on each congruence class and thus the integral a multiple of the volume of this class, that is, of $p^{3\alpha-3s}$, multiplied by $p^{-6\alpha}$. If $\alpha = r > 1$ is odd (so $s = (r+1)/2$), we can improve this a little: Namely, if we consider δS divisible by $p^{s-1-\alpha}$, the condition on the determinant becomes

$$\mathrm{Tr}(S^{-1}\delta S - (S^{-1}\delta S)^2/2 \equiv 0 \bmod (p^r)$$

(trace of the logarithm), so we can find a system of representatives δS divisible by $p^{s-1-\alpha}$ with $\det(p^\alpha(S+\delta S))$ fixed modulo p^r by first finding δS^0 with $\mathrm{Tr}(S^{-1}\delta S^0) = 0$ (and this element divisible by p^{s-1}), and then replacing them by

$$\delta S = \delta S^0 + S/4 \cdot \mathrm{Tr}((S^{-1}\delta S^0)^2).$$

The integral then becomes a multiple of a quadratic Gauss-sum and we win an additional factor p (and a fourth root of unity), because $\mathrm{Tr}(z^2)$ is a nondegenerate quadratic form on the space of traceless z (modulo p) with $p^\alpha Sz$ symmetric. For $\alpha = r = 1$, we can do the same:

We have to sum $\exp(\pi i/p \cdot \mathrm{Tr}(sT))$ for s ("$= pS$") a symmetric matrix with entries in $\mathbb{F}_p$ with given (invertible) determinant. If T has rank-1 modulo p this reduces to a sum $\sum_S \exp(2\pi i S(x)/p)$ over S with a given invertible determinant, with x some nonzero vector in $\mathbb{F}_p^2$. We claim that any value for $S(x)$ is taken with the same multiplicity p. Namely, if $S(x) = 0$ we are free to choose the second isotropic line among the p complements to $\mathbb{F}_p x$, and the determinant then fixes S uniquely. If $S(x) \neq 0$ we can do the same argument for the perpendicular space to the line $\mathbb{F}_p x$.

If $T \bmod (p)$ is nondegenerate we may assume that it is $\mathrm{diag}(\epsilon, -1)$. The S with $\mathrm{Tr}(ST) = \lambda$ and $\det(S) = \mu$ correspond to solutions of

$$\epsilon(s_{22} - \lambda/(2\epsilon))^2 - s_{12}^2 = \mu + \lambda^2/(4\epsilon).$$

The form on the left is either anisotropic and represents any nonzero value $p + 1$-times, and zero once, or isotropic and represents nonzero values $p - 1$-times and zero $2p - 1$-times. These numbers are congruent modulo p and it follows that the exponential sum is divisible by p.

It follows that the sum of contributions from $M(\alpha, 0)$s is an integral multiple of $p^{-9r/2}$ if r is even, of $p^{(-1-9r)/2}$ if r is odd, and of p^{-6s} if p divides the discriminant of T.

So we come to $M(\alpha, \beta)$ with $\beta > 0$. To parametrize it choose unimodular elements $x \in \mathbb{Z}_p^2$ which are a set of representatives for $\mathbb{P}^1(\mathbb{Z}_p)$, and for each x choose a $y = y(x)$ such that $\{x, y\}$ forms a basis with determinant 1. Then $M(\alpha, \beta)$ consists of all matrices $u/p^{\alpha+\beta}xx^t + v/p^\alpha yy^t$ with $u, v \in \mathbb{Z}_p^*$ and $x, y = y(x)$ as above. Moreover, the determinant of S is uv. To integrate we sum over a set of representatives modulo $p^{r+\beta}$ (for x, u), respectively, p^r (for v), and multiply by $p^{2(\alpha-r)}$. Then the contribution from $M(\alpha, \beta)$ becomes the sum (over x, u, v as specified above, with uv a fixed element modulo p^r)

$$p^{-4\alpha-3\beta-2r} \sum_{x,u,v} \exp(2\pi i(u/p^{\alpha+\beta}T(x) + v/p^{\alpha}T(y))).$$

Assume first that p does not divide the discriminant of T and that T is anisotropic. Then $T(x) \in \mathbb{Z}_p^*$ and we can choose for $y(x)$ a generator of the line perpendicular (for T) to x, so that $T(x)T(y)$ is the discriminant of T. We then can multiply u by $T(x)$ and v by $T(y)$ and obtain as a result

$$(p+1)p^{-4\alpha-2\beta-r-1} \sum_{u,v} \exp(2\pi i(u/p^{\alpha+\beta} + v/p^{\alpha})).$$

The sum over u vanishes unless $\alpha + \beta \leq r$. As $\beta > 0$ the summands only depend on v modulo p^{r-1}. Then summing over a fixed congruence class of u mod p^{r-1} gives zero unless $\alpha + \beta \leq r-1$, etc. This process stops at $\alpha + \beta \leq 1$, and we get as result an integral multiple of p^{-r-3}.

Finally, we consider the case that T is the norm form of a ramified quadratic extension of $\mathbb{Z}_p$. Here we can achieve that exactly one of $T(x), T(y)$ is divisible by p (and then only to the first power). For $T(x)$ divisible by p, we obtain the subsum

$$p^{-4\alpha-3\beta-2r} p^{r+\beta-1} \sum \exp(2\pi i(u/p^{\alpha+\beta-1} + v/p^{\alpha}));$$

for $T(y)$ divisible

$$p^{-4\alpha-3\beta-2r} p^{r+\beta} \sum \exp(2\pi i(u/p^{\alpha+\beta} + v/p^{\alpha-1})).$$

Both sums are over units u mod $p^{r+\beta}$, v mod p^r with uv a fixed value mod p^r. The second sum vanishes unless $\alpha + \beta \leq r$, in which case $\alpha \leq r-1$ so that the summands depend only on v mod p^{r-1}, so we can sum over a class of u mod p^{r-1}, so the sum vanishes unless $\alpha + \beta \leq r-1$, etc. as before. That is, the second sum is an integral multiple of $p-r-3$. So we can concentrate on the first sum.

As before it does not pose any problems if $\beta > 1$, so assume $\beta = 1$. Then we can sum over u and v mod p^{α}, that is, we get

$$p^{-5\alpha-2} \sum \exp(2\pi i(u+v)/p^{\alpha}).$$

Now the sum is over units u, v mod p^{α} with fixed product. As usual if s denotes the smallest integer $\geq \alpha$ we might first sum over u in a fixed class modulo p^s. That sum either vanishes or the summands are constant, so it is divisible by $p^{\alpha-s}$. Also if $\alpha > 2$ is odd we may win an additional factor $p^{1/2}$ from a quadratic Gauss sum. However, this does not seem to work for $\alpha = 1$. Thus finally the integral is divisible by $p^{-9r/2-2}$ if $r = \alpha > 1$, and by p^{-7} if $r = 1$.

Putting everything together, we come to the main claim of this section.

Theorem 2. *For all primes $p \geq 5$,*

$$\pi^{-7} n^9 L(3, \chi) L(4, \chi^2) \lim_{m \to \infty} U_p^m(E(0, \chi))$$

has Fourier coefficients in $\mathbb{Q}(\mu_n, \chi)$) which are p-integral. If $p \geq 7$, there exists a Fourier series with coefficients in the ideal of the group ring $\mathbb{Z}[\mathbb{Z}/(n)^*]$ generated by $1 + [-1]$, from which we derive this product via evaluation at χ, and this series has coefficients in $\mathbb{Q}(\mu_n)$ which are integral at p.

Proof. Of course there exists a unique such series with complex coefficients. This result is known for the product of L-values $n^9 (2\pi)^{-7} L(3, \chi) L(4, \chi^2)$ (one gets Bernoulli distributions; see [13, Chapter 5, Theorem 5.11] for negative values of s, and apply the functional equation in [13, p. 29]), and for the Fourier coefficients for T with squarefree discriminant away from n. After that the general case will follow by applying the Hecke algebra:

Namely, apply Hecke operators at primes l not dividing n. Our Eisenstein series is an eigenvector with eigenvalue integral at p. (It may contain an l in the denominator depending on choice of normalizations.) On the other hand, the Fourier coefficients of $T_l(E)$ can be computed from the Fourier coefficients of E, by restricting quadratic forms to sublattices and rescaling. If we know that some multiple $p^s E$ has p-integral Fourier coefficients at Ts where the discriminant of T is not divisible by l^2 (for any l not dividing n), we choose a T with smallest discriminant such that the T-Fourier coefficient is not p-integral. We then find a T' with smaller discriminant and a Hecke operator T_l such that the T'-coefficient of $T_l(E)$ is a sum (with p-adic units as coefficient) of the T-Fourier coefficient of E and other Fourier coefficients which are already known to be p-integral.

Assume first that T is divisible by l, and let $T' = T/l$. For T_l we use the Hecke operator which maps A to all quotients B under maximal isotropic subgroups in $A[l]$. To compute its effect on q-expansions write $A = \tilde{G}/\iota(X)$ where $\tilde{G} \cong \mathbb{G}_m^2$ is a torus with character-group $X \cong \mathbb{Z}^2$ and ι corresponds to a definite symmetric bilinear form b on X. Then E becomes a power-series in values of b on positive semidefinite elements $T \in S^2(X)$, and the coefficients are the Fourier coefficients. The possible isogenies $A \to B$ are first classified by their multiplicative degree, that is, B is quotient of a torus with character group $Y \subseteq X, Y \supseteq lX$, which can have index $1, l$, or l^2. Furthermore, the restriction to $lX \times Y$ of the bilinear form b' on Y must coincide with the restriction of b to $X \times Y$, if we identify lX with X. Thus the T' Fourier coefficient of $T_l(E)$ (for $T' \in S^2(X)$) is obtained by summing over all $Y \subseteq X$ the sum of all T-Fourier coefficients of E, where $T \in S^2(Y)$ and $T' = T/l$. These come with l-powers as coefficients which arise formally because we use a trace operation (from the torus with character group $S^2(X) + 1/l S^2(Y)$ to that with character group $S^2(X)$) which also annihilates contributions from T' with T'/l not integral. Note that the discriminant of T' differs from the discriminant of T by a factor $[X : Y]^2/l^2$. There is only one case where this is l^{-2}, namely, if $X = Y$ (the isogeny is étale) and $T = lT'$. So the coefficient of T' in $T_l(E)$ is equal to l^3 times the coefficient of T (this is the relevant l-power) plus a linear combination of coefficients with lower discriminant.

If the smallest "bad" T is such that T has discriminant divisible by l^2 but T itself is not divisible by l, there exists a unique $Y \subset X$ of index l such that T lies in $l^2 S^2(X)$. We then use the Hecke operator T_l which associates to A all quotients

B under a maximally isotropic subgroup of $A[l^2]$ whose intersection with $A[l]$ has order l^3. At the cusps this amounts to choosing sublattices $l^2 X \subset Y \subset X$ of index l or l^3, and a b' which coincides with b on $l^2 X \times Y$. Again the T'-Fourier coefficient of $T_l(E)$ is the sum of $T = l^2 T'$-coefficients for E for which T lies in $S^2(Y)$, multiplied with some l-power. We now can copy the previous argument.

Thus we get an Eisenstein series whose q-expansion at the multiplicative cusp (which classifies isogenies $A \to B$ with purely multiplicative kernel at p) is p-integral, and which is fixed by U_p. It then follows that as a section of $\omega_A^{\otimes 3}$ it has p-integral q-expansion also at the other cusps:

Namely, such cusps correspond to isogenies $A \to B$ whose kernel is not purely multiplicative at p. The U_p-operator maps this to the formal sum of isogenies $A' \to B' = A$ where the kernel of $A \to A'$ contains a multiplicative p-component. That is, $U_p(E)$ is obtained by pullback of a section of $\omega_{A'}^{\otimes 3}$ and forming a suitable trace. The pullback is divisible by p, and the trace preserves integrality. Finally, the multiplicative degree of the kernel of $A' \to B'$ is at least as big as the multiplicative degree of the kernel of $A \to B$. By decreasing induction over this multiplicative degree we then get the result.   □

10 Application of p-adic Hodge theory

The p-adic version of Hodge theory gives relations between p-adic coherent cohomology and p-adic étale cohomology of the generic fiber. We apply this to the moduli stack S_n which classifies abelian surfaces A with a principal polarization, an isogeny $A \to B$ whose kernel is a maximal isotropic subgroup of $A[n]$ étale locally isomorphic to $\mathbb{Z}/(n)^2$, and a generator of its determinant. Actually by using some auxiliary level-structure we can assume that S_n is a scheme. Also it admits smooth toroidal compactifications over $\mathbb{Q}$ (see [3]). However, what is missing is a toroidal model over $\mathbb{Z}_p$.

Fortunately, we can still apply p-adic Hodge theory as follows: Denote by S the moduli scheme classifying A (and perhaps some auxiliary level structure), which has a smooth model over $\mathbb{Z}_p$ as well as a smooth toroidal compactification $\bar{S}$, and choose as $\mathbb{Z}_p$-model for $\bar{S}_n$ the normalization of $\bar{S}$ in the generic fiber. Then our Eisenstein series defines a regular section of $\omega_A^{\otimes 3}$ over $\bar{S}_n$. Here ω_A extends to $\bar{S}$ as a line bundle of differentials on the universal semiabelian scheme G. Furthermore, if $\bar{V}$ denotes the integral closure of $V = \mathbb{Z}_p$ in $\bar{\mathbb{Q}}_p$ and $\rho \in \bar{V}$ an element of valuation $1/(p-1)$ (for example, $\rho = \zeta_p - 1$), and $\bar{R}$ the integral closure in the maximal extension of $R[1/p]$ unramified over $S \otimes \mathbb{Q}_p$ of the ring R of an open affine in $\bar{S}$, then we have a functorial extension of $S^2(\Omega_G) \otimes_R \hat{\bar{R}}$ by $\rho^{-1}\hat{\bar{R}}$ inducing a functorial 3-extension of

$$\Omega_R^3 \otimes_R \hat{\bar{R}} = \omega_G^{\otimes 3} \otimes_R \hat{\bar{R}}$$

by $\rho^{-3}\hat{\bar{R}}(3)$. See [4, Section 2c, p. 206]. If $\mathbb{L}$ denotes the locally constant étale sheaf on $S \otimes \mathbb{Q}$ which is the direct image of $\mathbb{Z}_p$ on S_n, then E defines a compatible system of sections of $\omega_G^{\otimes 3} \otimes_R \hat{\bar{R}} \otimes \mathbb{L}$ and thus a class in the cohomology of $\rho^{-3}\hat{\bar{R}} \otimes \mathbb{L}(3)$, over the topos described in [4, Chapter 3, p. 214]. However, it is known [4, Chapter 4, Theorem 9] that this cohomology is almost isomorphic to $H^3(S \otimes \bar{\mathbb{Q}}, \mathbb{L}(3)) \otimes \rho^{-3}\hat{\bar{V}}$ or to $H^3(S_n \otimes \bar{\mathbb{Q}}, \mathbb{Z}_p(3)) \otimes \rho^{-3}\hat{\bar{V}}$, so that we end up with an étale cohomology-class. As this construction is equivariant the class is Eisenstein, so after inverting p becomes a multiple of the known Eisenstein class. By computing residues at the multiplicative cusp (corresponding to residues of logarithmic differentials) we get that these two classes coincide after inverting p. Also we know that both classes are invariant under $\mathrm{Gal}(\bar{\mathbb{Q}}_p/\mathbb{Q}_p)$. As the Galois invariants in $\rho^{-3}\bar{V}/(p^s)$ are contained in the sum of $V/(p^s)$ and elements annihilated by ρ we see that our integral class is uniquely (as $p \geq 7$) the sum of an étale class with coefficients $\mathbb{Z}_p$ and a class annihilated by ρ. The first component then is an integral version of the Eisenstein class. Over the ordinary locus of S_n its image in the Galois cohomology of $\hat{\bar{R}}$ is given by the three-form E. All in all, we have found a class in $H^3(S_n \otimes \bar{\mathbb{Q}}_p, \mathbb{Z}_p(3))$ which is invariant under $\mathrm{Gal}(\bar{\mathbb{Q}}_p/\mathbb{Q}_p)$, and which is an eigenvector for the Hecke action. However, it is not clear how to proceed further (invariance under $\mathrm{Gal}(\bar{\mathbb{Q}}/\mathbb{Q})$, class over $\mathbb{Z}[1/p]$, etc.).

References

[1] V. Berkovich, Vanishing cycles for formal schemes II, *Invent. Math.*, **125** (1996), 367–390.

[2] J. M. Bismut, H. Gillet, and C. Soulé, Bott-Chern currents and complex immersions, *Duke Math. J.*, **60** (1990), 255–284.

[3] C. L. Chai and G. Faltings, *Degeneration of Abelian Varieties*, Springer-Verlag, Berlin, 1990.

[4] G. Faltings, P-adic Hodge theory, *J. Amer. Math. Soc.*, **1** (1988), 255–299.

[5] E. Freitag, *Siegelsche Modulformen*, Springer-Verlag, Berlin, 1983.

[6] M. Goresky, G. Harder, and R. MacPherson, Weighted cohomology, *Invent. Math.*, **116** (1994), 139–213.

[7] H. Hida, Iwasawa modules attached to congruences of cusp forms, *Ann. Sci. École Norm. Sup.*, **19** (1986), 231–273.

[8] R. Kiehl, Theorem A und Theorem B in der nichtarchimedischen Funktionentheorie, *Invent. Math.*, **2** (1967), 256–273.

[9] G. Laumon and L. Moret-Bailly, *Champs algébriques*, Springer-Verlag, Berlin, 2000.

[10] S. Mukai, Duality between $D(X)$ and $D(\hat{X})$ with its applications to Picard sheaves, *Nagoya Math. J.*, **81** (1981), 153–175.

[11] D. Quillen, Higher algebraic K-theory I, in H. Bass, ed., *Algebraic K-Theory* I: *Higher K-Theories*, Lecture Notes in Mathematics 341, Springer-Verlag, Berlin, 1973, 85–147.

[12] J. Schwermer, On Euler products and residual Eisenstein cohomology classes for Siegel modular varieties, *Forum Math.*, **7** (1995), 1–28.

[13] L. C. Washington, *Introduction to Cyclotomic Fields*, Springer-Verlag, New York, 1982.

Uniformizing the Stacks of Abelian Sheaves

Urs Hartl

Institute of Mathematics
University of Freiburg
Eckerstraße 1
D-79104 Freiburg
Germany
urs.hartl@math.uni-freiburg.de

Abstract. Elliptic sheaves (which are related to Drinfeld modules) were introduced by Drinfeld and further studied by Laumon–Rapoport–Stuhler and others. They can be viewed as function field analogues of elliptic curves and hence are objects "of dimension 1." Their higher dimensional generalizations are called abelian sheaves. In the analogy between function fields and number fields, abelian sheaves are counterparts of abelian varieties. In this article we study the moduli spaces of abelian sheaves and prove that they are algebraic stacks. We further transfer results of Čerednik–Drinfeld and Rapoport–Zink on the uniformization of Shimura varieties to the setting of abelian sheaves. Actually the analogy of the Čerednik–Drinfeld uniformization is nothing but the uniformization of the moduli schemes of Drinfeld modules by the Drinfeld upper half space. Our results generalize this uniformization. The proof closely follows the ideas of Rapoport–Zink. In particular, analogues of p-divisible groups play an important role. As a crucial intermediate step we prove that in a family of abelian sheaves with good reduction at infinity, the set of points where the abelian sheaf is uniformizable in the sense of Anderson, is formally closed.

Subject Classifications: 11G09 (11G18, 14L05)

Introduction

In arithmetic algebraic geometry the moduli spaces of abelian varieties are of great importance. For instance, they have played a major role in Faltings' proof of the Mordell conjecture [16], the proof of Fermat's Last Theorem [7], and the proof of Langlands reciprocity for GL_n over nonarchimedean local fields of characteristic zero by Harris–Taylor [22]. Therefore, their structure and especially their reduction at bad primes is intensively studied. One way to investigate their reduction is through p-adic uniformization. This was begun by Čerednik [6] and Drinfeld [11] and continued by Rapoport–Zink [36]. Čerednik–Drinfeld obtained the uniformization of certain Shimura curves of EL-type by a formal scheme whose associated rigid-analytic space is Drinfeld's p-adic upper half plane. This formal scheme can be viewed as a moduli space for p-divisible groups which are isogenous to a fixed supersingular p-divisible group. See Boutot–Carayol [5] for a detailed account. Rapoport–Zink generalized these results to the (partial) uniformization of higher dimensional Shimura varieties by more general moduli spaces for p-divisible groups.

In this article we study *abelian sheaves* as positive characteristic analogues of abelian varieties. We investigate their moduli spaces and prove that these are algebraic stacks. Then our aim is to transfer the above uniformization results to the case of positive characteristic. For the case considered by Čerednik–Drinfeld this was accomplished already by Drinfeld [10]; see below. We use this as a guide line to transfer the results of Rapoport–Zink. We also obtain a partial uniformization of the moduli stacks of abelian sheaves. In the Hilbert–Blumenthal situation a similar uniformization result was obtained by Stuhler [40] using different methods.

Let us explain what abelian sheaves are by first going back 30 years to Drinfeld's elliptic sheaves. Exploiting the analogy between number fields and function fields, Drinfeld [10, 12] invented the notions of *elliptic modules* (today called *Drinfeld modules*) and the dual notion of *elliptic sheaves*. These structures are analogues of elliptic curves for characteristic p in the following sense. Their endomorphism rings are rings of integers in global function fields of positive characteristic or orders in central division algebras over the later. On the other hand, the moduli spaces are varieties over smooth curves over a finite field. Through these two aspects in which global function fields of positive characteristic come into play, elliptic sheaves and variants of them proved to be fruitful for establishing the Langlands correspondence for GL_n over local and global function fields of positive characteristic. See the work of Drinfeld [10, 13, 14], Laumon–Rapoport–Stuhler [33], and Lafforgue [30]. Beyond this the analogy between elliptic modules and elliptic curves is abundant.

In this spirit, Anderson [1] introduced higher dimensional generalizations of Drinfeld's elliptic modules and called them *abelian t-modules*. The concept of *abelian sheaves* is a higher dimensional generalization of elliptic sheaves. Both serve as characteristic p analogues of abelian varieties. Abelian sheaves were studied in various special instances in the past. In this article we intend to give a systematic treatment. The definition of abelian sheaves is as follows. Let C be a smooth projective curve over $\mathbb{F}_q$ and let $\infty \in C(\mathbb{F}_q)$ be a fixed point. For every $\mathbb{F}_q$-scheme S we denote by σ the endomorphism of $C \times_{\mathbb{F}_q} S$ that acts as the identity on the coordinates of C and as $b \mapsto b^q$ on the sections $b \in \mathcal{O}_S$. Now an *abelian sheaf of rank r and dimension d over S* consists of the following data: a collection of locally free sheaves $\mathcal{F}_i$ of rank r on $C \times_{\mathbb{F}_q} S$ satisfying a certain periodicity condition. These sheaves are connected by two commuting sets of morphisms $\Pi_i : \mathcal{F}_i \to \mathcal{F}_{i+1}$ and $\tau_i : \sigma^* \mathcal{F}_i \to \mathcal{F}_{i+1}$ such that coker Π_i and coker τ_i are locally free $\mathcal{O}_S$-modules of rank d, supported, respectively, on $\infty \times S$ and on the graph of a morphism $c : S \to C$ called the *characteristic* of the abelian sheaf. An abelian sheaf of dimension 1 is the same as an elliptic sheaf. In this sense abelian sheaves are higher dimensional elliptic sheaves. The notion of abelian sheaf is dual to the notion of abelian t-module and related to Anderson's t-motives. In fact, if the characteristic is different from ∞, an abelian sheaf over a field is nothing but a pure t-motive equipped with additional structure at infinity (Section 2). Therefore, we like to view abelian sheaves as characteristic p analogues of polarized abelian varieties.

Abelian sheaves with appropriately defined level structure possess moduli spaces which are algebraic stacks locally of finite type over the curve C. The morphism to C is given by assigning to an abelian sheaf its characteristic. We denote the algebraic stacks of abelian sheaves of rank r and dimension d with H-level structures by $Ab\text{-}Sh_H^{r,d}$. Here $H \subset \mathrm{GL}_r(\mathbb{A}_f)$ is a compact open subgroup and $\mathbb{A}_f$ are the finite adeles of C. It should be noted that opposed to the case of elliptic sheaves, the stacks of abelian sheaves will in general not be schemes, not even if we add high level structures. This is due to the fact that for every level there are abelian sheaves having nontrivial automorphisms (see Remark 4.2). The nonrepresentability also reflects in the uniformization; see below.

Then our aim is to study the uniformization of these moduli stacks at ∞. Let z be a uniformizing parameter of C at ∞. In the case of elliptic sheaves, Drinfeld [10, 11] showed that the moduli stacks are in fact smooth affine schemes which can be uniformized by a formal scheme $\Omega^{(r)}$. This formal scheme is the characteristic p version of the one used by Čerednik and Drinfeld to uniformize Shimura curves. Correspondingly, it is a moduli space for certain formal groups on which multiplication with z is an isogeny. All this was worked out in detail by Genestier [18]. We like to call these formal groups "z-divisible groups." They play an important role also in our uniformization results.

So let us next explain some facts about z-divisible groups. Naturally these groups are of most use over schemes on which z is not a unit. Therefore, we will from now on work over schemes S in $\mathcal{Nilp}_{\mathbb{F}_q[\![z]\!]}$, the category of $\mathbb{F}_q[\![z]\!]$-schemes on which z is locally nilpotent. However, since it is important to separate the two roles played by z as a uniformizing parameter at ∞ and as an element of $\mathcal{O}_S$, we use the symbol z only for the first and we denote the image of z in $\mathcal{O}_S$ by ζ. Then S belongs to $\mathcal{Nilp}_{\mathbb{F}_q[\![\zeta]\!]}$.

Classically p-divisible groups may be studied via their Dieudonné modules. There is a corresponding notion for z-divisible groups. A *Dieudonné $\mathbb{F}_q[\![z]\!]$-module over S* is a finite locally free $\mathcal{O}_S[\![z]\!]$-module $\widehat{\mathcal{F}}$ with a σ-linear endomorphism F such that

1. coker F is locally free as an $\mathcal{O}_S$-module,
2. $(z - \zeta)$ is nilpotent on coker F,

The theory of z-divisible groups and their Dieudonné $\mathbb{F}_q[\![z]\!]$-modules resembles many facets of the theory of p-divisible groups; see [23].

Also z-divisible groups are related to abelian sheaves through their Dieudonné $\mathbb{F}_q[\![z]\!]$-modules. Namely, the completion of an abelian sheaf over S at $\infty \times S$ is a Dieudonné $\mathbb{F}_q[\![z]\!]$-module. The connection between abelian sheaves and z-divisible groups parallels the situation for abelian varieties. In particular, there is an analogue of the Serre–Tate Theorem relating the deformation theory of abelian sheaves to the deformation theory of their z-divisible groups.

Finally, we come to the uniformization of $Ab\text{-}Sh_H^{r,d}$ at infinity. We begin by describing the uniformizing spaces. Let r, d, k, ℓ be positive integers with $\frac{d}{r} = \frac{k}{\ell}$ and k and ℓ relatively prime. Let $\mathcal{O}_\Delta$ be the ring of integers in the central skew field over $\mathbb{F}_q(\!(z)\!)$ of invariant k/ℓ. A *special z-divisible $\mathcal{O}_\Delta$-module over $S \in \mathcal{Nilp}_{\mathbb{F}_{q^\ell}[\![\zeta]\!]}$* is a z-divisible group E of height $r\ell$ and dimension $d\ell$ with an action of $\mathcal{O}_\Delta$ prolonging the action of $\mathbb{F}_q[\![z]\!]$, such that the inclusion $\mathbb{F}_{q^\ell} \subset \mathcal{O}_\Delta$ makes Lie E into a locally free $\mathbb{F}_{q^\ell} \otimes_{\mathbb{F}_q} \mathcal{O}_S$-module of rank d. The z-divisible groups associated to abelian sheaves of rank r and dimension d are special z-divisible $\mathcal{O}_\Delta$-modules. Let $\mathbb{E}$ be a special z-divisible $\mathcal{O}_\Delta$-module over $\operatorname{Spec} \mathbb{F}_{q^\ell}$. Then the moduli problem of special z-divisible $\mathcal{O}_\Delta$-modules which are isogenous to $\mathbb{E}$ is solved by a formal scheme G locally formally of finite type over $\operatorname{Spf} \mathbb{F}_{q^\ell}[\![\zeta]\!]$. The latter means that the reduced closed subscheme G_{red} of G is locally of finite type over $\operatorname{Spec} \mathbb{F}_{q^\ell}$.

We fix an abelian sheaf $\overline{\mathbb{M}}$ of rank r and dimension d over $\operatorname{Spec} \mathbb{F}_{q^\ell}$ whose restriction to $C \smallsetminus \infty$ satisfies $\tau_i = \operatorname{Id}_r \cdot \sigma^*$ and we let $\mathbb{E}$ be its z-divisible group. The Newton polygon of $\mathbb{E}$ is a straight line. Let Z be the set of points s of $\mathcal{A}b\text{-}\mathcal{S}h_H^{r,d} \times_C \infty$ such that the universal abelian sheaf $\mathcal{F}_s$ over s is isogenous to $\overline{\mathbb{M}}$ over an algebraic closure of $\kappa(s)$. It is an important step to show that Z is the set of points over which the Newton polygons of $\mathbb{E}$ and of the z-divisible group associated to $\mathcal{F}_s$ coincide. This implies that Z is a closed subset. We consider the formal completion $\mathcal{A}b\text{-}\mathcal{S}h_H^{r,d}{}_{/Z}$ of $\mathcal{A}b\text{-}\mathcal{S}h_H^{r,d}$ along Z. It is no longer an algebraic stack, but it is a *formal algebraic stack over* $\operatorname{Spf} \mathbb{F}_{q^\ell}[\![\zeta]\!]$. Formal algebraic stacks are generalizations of algebraic stacks in the same sense as formal schemes generalize usual schemes. (The relevant facts on formal algebraic stacks are collected in an appendix.) Now we can uniformize $\mathcal{A}b\text{-}\mathcal{S}h_H^{r,d}{}_{/Z}$ as follows. Being isogenous to the z-divisible group $\mathbb{E}$ of $\overline{\mathbb{M}}$, the universal special z-divisible $\mathcal{O}_\Delta$-module on G gives rise to an abelian sheaf on G which we call its *algebraization*. Let $J(Q)$ be the group of quasi-isogenies of $\overline{\mathbb{M}}$. There are natural embeddings of the group $J(Q)$ into $\operatorname{GL}_r(\mathbb{A}_f)$ and into the group of quasi-isogenies of $\mathbb{E}$. The latter group acts on the formal scheme G. We let $J(Q)$ act diagonally on $G \times \operatorname{GL}_r(\mathbb{A}_f)$. Taking into account level structures we obtain the following result, which we will later formulate as Uniformization Theorem 12.6:

There is a canonical 1-isomorphism of formal algebraic $\operatorname{Spf} \mathbb{F}_{q^\ell}[\![\zeta]\!]$*-stacks*

$$\Theta : J(Q) \backslash G \times \operatorname{GL}_r(\mathbb{A}_f)/H \xrightarrow{\sim} \mathcal{A}b\text{-}\mathcal{S}h_H^{r,d}{}_{/Z} \times_{\operatorname{Spf} \mathbb{F}_q[\![\zeta]\!]} \operatorname{Spf} \mathbb{F}_{q^\ell}[\![\zeta]\!].$$

At this point, note that the nonrepresentability of the algebraic stacks $\mathcal{A}b\text{-}\mathcal{S}h_H^{r,d}$ also reflects in the uniformization. Namely, since all unipotent subgroups of $J(Q)$ are torsion, in general a discrete subgroup of $J(Q)$ cannot act fixed point free on G. So the quotients $J(Q) \backslash G \times \operatorname{GL}_r(\mathbb{A}_f)/H$ can only be formal algebraic stacks and not formal algebraic spaces. This phenomenon does not occur in the p-adic uniformization of Shimura-varieties.

We mention an interesting aspect of the proof that is also related to the uniformizability of t-motives. Namely, by work of Gardeyn [17], the t-motive associated to an abelian sheaf over a complete field extension K of $\mathbb{F}_q((\zeta))$ is *uniformizable* in the sense of Anderson [1] if and only if firstly the abelian sheaf extends to an abelian sheaf over the valuation ring R of a finite extension of K, and secondly its reduction modulo the maximal ideal of R is isogenous to $\overline{\mathbb{M}}$. We show that the second condition is closed. More precisely if $\mathcal{F}$ is an abelian sheaf of rank r and dimension d over $S \in \mathcal{N}ilp_{\mathbb{F}_q[\![\zeta]\!]}$, then the set of points in S over which $\mathcal{F}$ is isogenous to $\overline{\mathbb{M}}$ is closed. This is the key ingredient in the proof of the Uniformization Theorem. It is proved in

Section 11 where we avoid the language of stacks and proceed in more down-to-earth terms.

Let us end by explaining the relation of our Uniformization Theorem to the results of Rapoport–Zink [36] and Drinfeld [10]. In the case of elliptic sheaves we have

$$\mathcal{A}b\text{-}\mathcal{S}h_H^{r,d}/Z = \mathcal{A}b\text{-}\mathcal{S}h_H^{r,d} \times_C \operatorname{Spf} \mathbb{F}_q[\![\zeta]\!].$$

This follows from the fact that there is only one polygon between the points $(0, 0)$ and $(r, 1)$ with nonnegative slopes and integral break points, namely, the straight line. So all special z-divisible $\mathcal{O}_\Delta$-modules have the same Newton polygon as $\mathbb{E}$, and therefore Z is all of $\mathcal{A}b\text{-}\mathcal{S}h_H^{r,d} \times_C \infty$. Moreover, in this case G is the formal scheme $\Omega^{(r)}$ introduced above and $J(Q) \cong \operatorname{GL}_r(Q)$. So we recover Drinfeld's uniformization theorem

$$\operatorname{GL}_r(Q) \backslash \Omega^{(r)} \times \operatorname{GL}_r(\mathbb{A}_f)/H \xrightarrow{\sim} M_H^r \times_C \operatorname{Spf} \mathbb{F}_{q^\ell}[\![\zeta]\!]$$

where M_H^r is the moduli scheme of Drinfeld modules of rank r with H-level structure.

Compared to [36], our uniformization theorem is analogous to the uniformization of the formal completion of Shimura varieties along the most supersingular isogeny class. In this sense uniformizable abelian sheaves correspond to supersingular abelian varieties. There is no doubt that the more general uniformization in [36] of an arbitrary isogeny class of abelian varieties also carries over to the setting of abelian sheaves. Furthermore, we have only described the uniformization at ∞ in this article. But the analogous uniformization results at other places of C should likewise hold. For example in the case of $\mathcal{D}$-elliptic sheaves these were described by Hausberger [26].

Notation

Throughout this article, we will denote by

$\mathbb{F}_q$	the finite field having q elements and characteristic p,
C	a smooth projective geometrically irreducible curve over $\mathbb{F}_q$,
$\infty \in C(\mathbb{F}_q)$	a fixed point,
$C' = C \smallsetminus \infty$	
$A = \Gamma(C', \mathcal{O}_{C'})$	the ring of regular functions on C',
$Q = \mathbb{F}_q(C)$	the function field of C, viz. the field of fractions of A,
Q_∞	the completion of Q at ∞,
A_∞	the ring of integers in Q_∞,
$v \in M_f$	the finite places of C, i.e., the points of C',
A_v	the completion of A at the finite place v of C,
$\widehat{A} = \prod\limits_{v \in M_f} A_v$	
$\mathbb{A}_f = Q \otimes_A \widehat{A}$	the finite adeles of C,
d, r, k, ℓ	positive integers with $\frac{d}{r} = \frac{k}{\ell}$ and k and ℓ relatively prime,

Δ the central skew field over Q_∞ of invariant k/ℓ,

$\mathcal{O}_\Delta$ its ring of integers.

All schemes, as well as their products and morphisms between them, are supposed to be over Spec $\mathbb{F}_q$. If X is a scheme, we let $\mathcal{S}ch_X$ be the category of X-schemes. For two schemes X and Y, we write $X \times Y$ for their product over Spec $\mathbb{F}_q$. A similar notation will be employed for the tensor product over $\mathbb{F}_q$. If $i : Y \hookrightarrow X$ is a closed immersion of schemes and $\mathcal{F}$ is a quasi-coherent sheaf on X, we denote the restriction $i^*\mathcal{F}$ by $\mathcal{F}|_Y$. As is customary, we will use the term vector bundle for a locally free coherent sheaf on a scheme.

Starting with Section 6, we denote by

ζ an indeterminant over $\mathbb{F}_q$,

$\mathbb{F}_q[\![\zeta]\!]$ the ring of formal power series in ζ,

Spf $\mathbb{F}_q[\![\zeta]\!]$ the formal scheme which is the formal spectrum of $\mathbb{F}_q[\![\zeta]\!]$,

$\mathcal{N}ilp_{\mathbb{F}_q[\![\zeta]\!]}$ the category of schemes over Spf $\mathbb{F}_q[\![\zeta]\!]$, viz. the category of schemes over Spec $\mathbb{F}_q[\![\zeta]\!]$ on which ζ is locally nilpotent.

From Section 6, on all schemes will be in $\mathcal{N}ilp_{\mathbb{F}_q[\![\zeta]\!]}$.

Let S be a scheme. We denote by

$\sigma_S : S \to S$ its Frobenius endomorphism which acts as the identity on points and as the q-power map on the structure sheaf,

$C_S = C \times S$

$\sigma = \mathrm{id}_C \times \sigma_S$ the endomorphism of C_S that acts as the identity on the coordinates of C and as $b \mapsto b^q$ on the elements $b \in \mathcal{O}_S$.

For a divisor D on C, we denote by $\mathcal{O}_{C_S}(D)$ the invertible sheaf on C_S whose sections have divisor $\geq -D$. If $\mathcal{F}$ is a coherent sheaf on C_S, we set $\mathcal{F}(D) := \mathcal{F} \otimes_{\mathcal{O}_{C_S}} \mathcal{O}_{C_S}(D)$. This notation applies in particular to the divisor $D = n \cdot \infty$ for an integer n.

Part One: Abelian Sheaves

1 Definition of abelian sheaves

Let S be a scheme and fix a morphism $c : S \to C$. Let $\mathcal{J}$ be the ideal sheaf on C_S of the graph of c.

Definition 1.1. *An abelian sheaf $\underline{\mathcal{F}} = (\mathcal{F}_i, \Pi_i, \tau_i)$ of rank r, dimension d, and characteristic c over S is a ladder of vector bundles $\mathcal{F}_i$ on C_S of rank r and injective homomorphisms Π_i, τ_i of $\mathcal{O}_{C_S}$-modules ($i \in \mathbb{Z}$) of the form*

$$
\begin{array}{ccccccccc}
\cdots & \longrightarrow & \mathcal{F}_{i-1} & \xrightarrow{\Pi_{i-1}} & \mathcal{F}_i & \xrightarrow{\Pi_i} & \mathcal{F}_{i+1} & \xrightarrow{\Pi_{i+1}} & \cdots \\
 & & \uparrow{\scriptstyle \tau_{i-2}} & & \uparrow{\scriptstyle \tau_{i-1}} & & \uparrow{\scriptstyle \tau_i} & & \\
\cdots & \longrightarrow & \sigma^*\mathcal{F}_{i-2} & \xrightarrow{\sigma^*\Pi_{i-2}} & \sigma^*\mathcal{F}_{i-1} & \xrightarrow{\sigma^*\Pi_{i-1}} & \sigma^*\mathcal{F}_i & \xrightarrow{\sigma^*\Pi_i} & \cdots
\end{array}
$$

subject to the following conditions (for all $i \in \mathbb{Z}$):

1. *the above diagram is commutative;*
2. *the morphism* $\Pi_{i+\ell-1} \circ \cdots \circ \Pi_i$ *identifies* $\mathcal{F}_i$ *with the subsheaf* $\mathcal{F}_{i+\ell}(-k \cdot \infty)$ *of* $\mathcal{F}_{i+\ell}$;
3. *the cokernel of* Π_i *is a locally free* $\mathcal{O}_S$-*module of rank d;*
4. *the cokernel of* τ_i *is a locally free* $\mathcal{O}_S$-*module of rank d and annihilated by* $\mathcal{J}^d$.

A morphism between two abelian sheaves $(\mathcal{F}_i, \Pi_i, \tau_i)$ *and* $(\mathcal{F}'_i, \Pi'_i, \tau'_i)$ *is a collection of morphisms* $\mathcal{F}_i \to \mathcal{F}'_i$ *which commute with the* Πs *and the* τs.

Let us make a few remarks. By condition 2 the cokernel of Π_i is supported at ∞. Moreover, due to the periodicity condition 2 we have $\tau_{i+\ell n} = \tau_i \otimes \mathrm{id}_{\mathcal{O}_{C_S}(kn)}$ for all $n \in \mathbb{Z}$. Finally, the reader should be aware that we allow that the ideal sheaf $\mathcal{J}$ acts nontrivially on coker τ_i. In this respect our abelian sheaves are more general than the abelian sheaves studied so far in the literature. As such, we mention elliptic sheaves [12, 2], which we discuss later, and $\mathcal{D}$-elliptic sheaves [33] which are abelian sheaves equipped with an action of an order $\mathcal{D}$ in a central division algebra over Q.

Definition 1.2. *We denote by* $\mathcal{A}b\text{-}\mathcal{S}h^{r,d}(S)$ *the category whose objects are the abelian sheaves of rank r and dimension d over S and whose morphisms are the isomorphisms of abelian sheaves. If* $S' \to S$ *is a morphism of schemes, the pullback of an abelian sheaf over S is an abelian sheaf over* S'. *This defines a fibered category* $\mathcal{A}b\text{-}\mathcal{S}h^{r,d}$ *over the category of* $\mathbb{F}_q$-*schemes, which is a stack for the fppf-topology. The functor which assigns to an object of* $\mathcal{A}b\text{-}\mathcal{S}h^{r,d}(S)$ *the characteristic* $c : S \to C$ *defines a 1-morphism of stacks*

$$\mathcal{A}b\text{-}\mathcal{S}h^{r,d} \to C.$$

Next, we introduce level structures on abelian sheaves. Let $I \subset C' = C \smallsetminus \infty$ be a finite closed subscheme and let $\underline{\mathcal{F}} = (\mathcal{F}_i, \Pi_i, \tau_i)$ be an abelian sheaf of rank r over S. Then the restrictions $\mathcal{F}_i|_{I \times S}$ are all isomorphic via the morphisms Π_i. We call this restriction $\underline{\mathcal{F}}|_{I \times S}$. The same holds for the morphisms τ_i. So we obtain a morphism

$$\tau|_{I \times S} : \sigma^* \underline{\mathcal{F}}|_{I \times S} \longrightarrow \underline{\mathcal{F}}|_{I \times S}$$

which we consider as a σ-linear map of $\underline{\mathcal{F}}|_{I \times S}$ to itself. In this article we always assume that the characteristic $c(S)$ of $(\mathcal{F}_i, \Pi_i, \tau_i)$ is disjoint from I. Due to this assumption, $\tau|_{I \times S}$ is an isomorphism. We consider the *functor of* τ-*invariants of* $\underline{\mathcal{F}}|_{I \times S}$,

$$(\underline{\mathcal{F}}|_I)^\tau : \mathcal{S}ch_S \longrightarrow \qquad \mathcal{O}_I\text{-modules}$$
$$T/S \longmapsto \ker \mathrm{H}^0(I \times T, \tau|_{I \times T} - \mathrm{id}_{\underline{\mathcal{F}}|_{I \times T}}).$$

In Böckle–Hartl [3, Theorem 2.5] the following fact is proved.

Proposition 1.3. *The functor* $(\underline{\mathcal{F}}|_I)^\tau$ *is representable by a finite étale scheme over S which is a* $\mathrm{GL}_r(\mathcal{O}_I)$-*torsor.*

Definition 1.4. *An I*-*level structure on* $(\mathcal{F}_i, \Pi_i, \tau_i)$ *over S is an isomorphism*

$$\bar{\eta} : (\underline{\mathcal{F}}|_{I \times S}, \tau|_{I \times S}) \overset{\sim}{\longrightarrow} (\mathcal{O}^r_{I \times S}, \mathrm{Id}_r \cdot \sigma^*)$$

from $\underline{\mathcal{F}}|_{I \times S}$ *to* $\mathcal{O}^r_{I \times S}$ *that commutes with the* σ-*linear endomorphisms* $\tau|_{I \times S}$ *on one and* $\mathrm{Id}_r \cdot \sigma^*$ *on the other side.*

If the characteristic $c(S)$ meets I, then $\underline{\mathcal{F}}$ does not possess any I-level structure.

Proposition 1.5. *Let $\underline{\mathcal{F}}$ be an abelian sheaf over a field. Then the automorphism group of $\underline{\mathcal{F}}$ is finite.*

Proof. We may assume that the base field is algebraically closed. Let $\bar{\eta}$ be an I-level structure on $\underline{\mathcal{F}} = (\mathcal{F}_i, \Pi_i, \tau_i)$. It induces a group homomorphism

$$\alpha_I : \operatorname{Aut}(\underline{\mathcal{F}}) \to \operatorname{Aut}(\mathcal{O}^r_{I \times S}, \operatorname{Id}_r \cdot \sigma^*) = \operatorname{GL}_r(\mathcal{O}_I).$$

The latter group is finite. We claim that α_I is injective for some sufficiently large finite subscheme $I \subset C'$. From this the proposition will follow. To establish the claim let $(f_i : \mathcal{F}_i \to \mathcal{F}_i)_i$ be an automorphism of $\underline{\mathcal{F}}$. Note that if f_0 is the identity then f_i must also be the identity for all i. We now consider a finite flat morphism $\pi : C \to \mathbb{P}^1_{\mathbb{F}_q}$. Since the map $\pi_* : \operatorname{Aut}_C(\mathcal{F}_0) \to \operatorname{Aut}_{\mathbb{P}^1}(\pi_*\mathcal{F}_0)$ is injective we may assume $C = \mathbb{P}^1$. Then the vector bundle $\pi_*\mathcal{F}_0$ decomposes

$$\pi_*\mathcal{F}_0 = \bigoplus_{i=1}^{s} \mathcal{O}_{\mathbb{P}^1}(n_i)$$

for uniquely determined integers $n_1 \geq \cdots \geq n_s$. We let $P \in \mathbb{P}^1$ be a point and set $I = (n_1 - n_s + 1) \cdot P$. Then α_I is injective. $\square$

Next, we want to give a different definition of I-level structures. Note that via the natural isomorphism $(\underline{\mathcal{F}}|_I)^\tau \otimes_{\mathcal{O}_I} \mathcal{O}_{I \times S} \xrightarrow{\sim} \underline{\mathcal{F}}|_{I \times S}$ the I-level structures $\bar{\eta}$ on $\underline{\mathcal{F}}$ over a connected S correspond bijectively to the isomorphisms of $\mathcal{O}_I$-modules

$$\bar{\eta}' : (\underline{\mathcal{F}}|_I)^\tau(S) \xrightarrow{\sim} \mathcal{O}^r_I.$$

We use this observation to define more general level structures by introducing the adelic point of view. Let $\underline{\mathcal{F}}$ be an abelian sheaf of rank r over S. We define the functor

$$(\underline{\mathcal{F}}|_{\widehat{A}})^\tau : \mathcal{S}ch_S \longrightarrow \widehat{A}\text{-modules}$$
$$T/S \longmapsto \varprojlim_I (\underline{\mathcal{F}}|_I)^\tau(T),$$

where the limit is taken over all finite closed subschemes $I \subset C'$. Assume that S is connected and choose an algebraically closed base point $\iota : s \to S$. Due to Proposition 1.3 we may view $(\underline{\mathcal{F}}|_{\widehat{A}})^\tau$ as the $\widehat{A}[\pi_1(S, s)]$-module $(\iota^*\underline{\mathcal{F}}|_{\widehat{A}})^\tau(s)$.

We consider the set $\operatorname{Isom}_{\widehat{A}}((\underline{\mathcal{F}}|_{\widehat{A}})^\tau, \widehat{A}^r) := \operatorname{Isom}_{\widehat{A}}((\iota^*\underline{\mathcal{F}}|_{\widehat{A}})^\tau(s), \widehat{A}^r)$ of isomorphisms of $\widehat{A}$-modules. Via its natural action on $\widehat{A}^r$ the group $\operatorname{GL}_r(\widehat{A})$ acts on this set from the left. Via its action on $(\iota^*\underline{\mathcal{F}}|_{\widehat{A}})^\tau(s)$ the group $\pi_1(S, s)$ acts on it from the right.

Definition 1.6. *Let $H \subset \operatorname{GL}_r(\widehat{A})$ be a compact open subgroup. An H-level structure on $\underline{\mathcal{F}}$ over S is an H-orbit in $\operatorname{Isom}_{\widehat{A}}((\underline{\mathcal{F}}|_{\widehat{A}})^\tau, \widehat{A}^r)$ which is fixed by $\pi_1(S, s)$. (Because of this latter condition, the notion of level structure is independent of the chosen base point.)*

In particular, if $H = H_I = \ker(\mathrm{GL}_r(\widehat{A}) \to \mathrm{GL}_r(\mathcal{O}_I))$, an H-level structure is nothing else than an I-level structure. Note that as before an H-level structure can only exist if the characteristic $c(S)$ does not meet the set of places v of C for which $H_v \neq \mathrm{GL}_r(A_v)$.

Definition 1.7. *Let $\mathcal{A}b\text{-}\mathcal{S}h_H^{r,d}(S)$ be the category whose objects are the abelian sheaves of rank r and dimension d together with an H-level structure over S and whose morphisms are the isomorphisms of abelian sheaves which respect the level structures. Analogous to Definition 1.2, this defines a stack $\mathcal{A}b\text{-}\mathcal{S}h_H^{r,d}$ over C.*

For $H = H_\emptyset = \mathrm{GL}_r(\widehat{A})$ the definition of H-level structure is vacuous. Therefore, we have $\mathcal{A}b\text{-}\mathcal{S}h_{H_\emptyset}^{r,d} = \mathcal{A}b\text{-}\mathcal{S}h^{r,d}$. We will show in Section 3 that these stacks are algebraic over C.

There is a free action of the group $\mathbb{Z}$ on these stacks given on objects by the map

$$[n] : (\mathcal{F}_i, \Pi_i, \tau_i) \mapsto (\mathcal{F}_{i+n}, \Pi_{i+n}, \tau_{i+n}).$$

Example 1.8 (Drinfeld modules and elliptic sheaves). Let $d = 1$ and let $H = H_I$. Then an abelian sheaf is what was called an *elliptic sheaf* in Blum–Stuhler [2]. We consider the open substack

$$\mathcal{A}b\text{-}\mathcal{S}h_H^{r,1} \times_C C'$$

of abelian sheaves with characteristic disjoint from ∞. It is shown in [2, Theorem 3.2.1] that there is a 1-isomorphism between the stack $\mathcal{D}r\text{-}\mathcal{M}od_I^r$ of Drinfeld A-modules of rank r with I-level structure and the open and closed substack of $\mathcal{A}b\text{-}\mathcal{S}h_H^{r,1} \times_C C'$ consisting of those $(\mathcal{F}_i, \Pi_i, \tau_i)$ with $\deg(\mathcal{F}_0|_{C_s}) = 1 - r$ for each algebraically closed point s of S.

On the other hand, in $\mathcal{A}b\text{-}\mathcal{S}h_H^{r,1}$ we always have $\deg(\mathcal{F}_i|_{C_s}) = \deg(\mathcal{F}_0|_{C_s}) + i$. Using the free action of $\mathbb{Z}$ on $\mathcal{A}b\text{-}\mathcal{S}h_H^{r,1}$ we thus obtain a 1-isomorphism of stacks

$$\mathcal{A}b\text{-}\mathcal{S}h_H^{r,1} \times_C C' \cong \mathbb{Z} \times \mathcal{D}r\text{-}\mathcal{M}od_I^r .$$

The first factor gives the degree of $\mathcal{F}_0$.

In fact if $I \neq \emptyset$ the stack $\mathcal{D}r\text{-}\mathcal{M}od_I^r$ is a smooth affine scheme of finite type over $\mathbb{F}_q$. See for example [2, Theorem 2.3.8]. So $\mathcal{A}b\text{-}\mathcal{S}h_H^{r,1} \times_C C'$ is a smooth scheme locally of finite type in this case.

We will give another example in Section 4 which shows that for $d > 1$ one can neither expect that $\mathcal{A}b\text{-}\mathcal{S}h_H^{r,d}$ is a scheme, nor that it is smooth over C'.

2 Relation to Anderson's t-motives

We will show that abelian sheaves with characteristic disjoint from ∞ are the same as polarized A-motives, a variant of Anderson's [1] t-motives. Let S be a scheme and fix a characteristic morphism $c : S \to C'$ disjoint from ∞. Let $\Gamma(c) \subset C_S$ be the graph of c and let $\mathcal{J}$ be the ideal sheaf defining $\Gamma(c)$.

Definition 2.1. *A (pure) polarized A-motive* $(\mathcal{F}, \tau)$ *of rank* r, *dimension* d, *and characteristic* c *over* S *consists of a vector bundle* $\mathcal{F}$ *on* C_S *of rank* r *and a morphism of coherent sheaves* $\tau : \sigma^* \mathcal{F} \to \mathcal{F}(k \cdot \infty)$ *such that*

1. *the cokernel of* τ *is supported on* $\Gamma(c) \cup \infty$, *the part supported on* $\Gamma(c)$ *is annihilated by* $\mathcal{J}^d$,
2. *for every* $i = 1, \ldots, \ell$ *the image of* $\tau^i : \sigma^{i*} \mathcal{F} \to \mathcal{F}(ik \cdot \infty)$ *lies in* $\mathcal{F}(k \cdot \infty)$ *and* $\tau^\ell : \sigma^{\ell*} \mathcal{F} \to \mathcal{F}(k \cdot \infty)$ *is a local isomorphism at* ∞,
3. *locally at* ∞, $\mathcal{F}$ *is contained in the image* $\tau \mathcal{F}$.

Here we denote by τ^i the composition $(\tau \otimes \mathrm{id}_{\mathcal{O}((i-1)k \cdot \infty)}) \circ \cdots \circ \sigma^{(i-1)*} \tau$ mapping

$$\sigma^{i*} \mathcal{F} \longrightarrow \sigma^{(i-1)*} \mathcal{F}(k \cdot \infty) \longrightarrow \cdots \longrightarrow \sigma^* \mathcal{F}((i-1)k \cdot \infty) \longrightarrow \mathcal{F}(ik \cdot \infty).$$

Lemma 2.2. *Conditions 1 and 2 imply that* τ *is injective and that the part of* coker τ *which is supported on* $\Gamma(c)$ (*respectively, on* ∞) *is a locally free* $\mathcal{O}_S$-*module of rank* d (*respectively*, $(\ell - 1)d$).

Proof. Consider the exact sequences of $\mathcal{O}_S$-modules induced by τ

$$0 \to \ker \tau \longrightarrow \sigma^* \mathcal{F} \longrightarrow \mathrm{im}\, \tau \to 0$$
$$\downarrow \tau$$
$$0 \to \mathrm{im}\, \tau \to \mathcal{F}(k \cdot \infty) \to \mathrm{coker}\, \tau \to 0.$$

For a point $s \in S$ we tensor them with the residue field $\kappa(s)$ at s to obtain

$$0 \longrightarrow \mathrm{Tor}_1^{\mathcal{O}_S}(\mathrm{im}\, \tau, \kappa(s)) \longrightarrow (\ker \tau)_s \longrightarrow (\sigma^* \mathcal{F})_s \xrightarrow{\ \beta\ } (\mathrm{im}\, \tau)_s \longrightarrow 0$$
$$\downarrow \tau \otimes \mathrm{id}$$
$$0 \longrightarrow \mathrm{Tor}_1^{\mathcal{O}_S}(\mathrm{coker}\, \tau, \kappa(s)) \longrightarrow (\mathrm{im}\, \tau)_s \xrightarrow{\ \alpha\ } \mathcal{F}(k \cdot \infty)_s \longrightarrow (\mathrm{coker}\, \tau)_s \longrightarrow 0.$$

We claim that $\tau \otimes \mathrm{id}$ is injective. Indeed, consider a point on C_s and its local ring which is a PID. Then over this PID $\tau \otimes \mathrm{id}$ is a morphism between finite free modules of the same rank with torsion cokernel. Hence by the elementary divisor theorem $\tau \otimes \mathrm{id} = \alpha\beta$ must be injective.

Now the surjectivity of β shows that α is injective. Therefore, $\mathrm{Tor}_1^{\mathcal{O}_S}(\mathrm{coker}\, \tau, \kappa(s))$ vanishes and by the local criterion for flatness [15, 6.8], coker τ is a flat $\mathcal{O}_S$-module. This in turn implies the flatness of im τ. Hence $(\ker \tau)_s$ is identified with the kernel of β. However, β is an isomorphism and thus $(\ker \tau)_s$ is zero. By Nakayama's lemma ker τ is zero and τ is injective. Finally let $\mathcal{G}$ be the part of coker τ which is supported on $\Gamma(c)$. Since the characteristic is disjoint from ∞, $\mathcal{G}$ is a direct summand of coker τ. Then condition 2 implies that coker $\tau^\ell = \bigoplus_{i=0}^{\ell-1} \sigma^{i*} \mathcal{G}$ and the lemma follows. $\square$

If $S = \mathrm{Spec}\, K$ is the spectrum of a field, $A = \mathbb{F}_q[t]$, and $(\mathcal{F}, \tau)$ is a polarized A-motive over S, then the $K[t, \tau]$-module $\Gamma(C'_S, \mathcal{F})$ is a *pure t-motive of weight* d/r as defined by Anderson [1]. Compared to Anderson's definition, however, our *polarized* A-motive contains additional data at infinity which rigidifies the structure of the A-motive. Since Anderson's t-motives serve as characteristic p analogues of abelian varieties this may justify our terminology.

As in Section 1, we can define H-level structures on polarized A-motives for compact open subgroups $H \subset \mathrm{GL}_r(\widehat{A})$. Correspondingly, we obtain the stack $\mathcal{P}ol_H^{r,d}$ of polarized A-motives with H-level structure. This is a stack over C'.

Theorem 2.3. *There is a 1-isomorphism of stacks*

$$\mathit{Ab}\text{-}\mathcal{S}h_H^{r,d} \times_C C' \cong \mathcal{P}ol_H^{r,d}.$$

Proof. We construct two mutually 2-inverse 1-morphisms T and T' between the stacks in question. First consider the 1-morphism $T : \mathit{Ab}\text{-}\mathcal{S}h_H^{r,d} \times_C C' \to \mathcal{P}ol_H^{r,d}$ which assigns to an abelian sheaf $(\mathcal{F}_i, \Pi_i, \tau_i)$ over S with characteristic disjoint from ∞ the polarized A-motive of rank r and dimension d consisting of

$$\mathcal{F} := \mathcal{F}_0, \qquad \tau := \Pi_{\ell-1} \circ \cdots \circ \Pi_1 \circ \tau_0 : \sigma^* \mathcal{F} \to \mathcal{F}(k \cdot \infty).$$

Clearly, this construction is also compatible with level structures.

Conversely, let $(\mathcal{F}, \tau)$ be a polarized A-motive of rank r and dimension d over S. For $0 \le i \le \ell$ we set

$$\mathcal{F}_i := \mathcal{F} + \cdots + \tau^i \mathcal{F} \subset \mathcal{F}(k \cdot \infty).$$

Then $\mathcal{F}_i$ is equal to $\mathcal{F}$ outside ∞ due to the definition of τ. And locally at ∞ it is isomorphic to $\sigma^{i*}\mathcal{F}$ due to condition 3 of Definition 2.1. Therefore, $\mathcal{F}_i$ is locally free on C_S. Let $\Pi_i : \mathcal{F}_i \to \mathcal{F}_{i+1}$ be the inclusion and let $\tau_i : \sigma^*\mathcal{F}_i \to \mathcal{F}_{i+1}$ be the morphism τ. Again by condition 3, coker Π_i is supported at ∞ and coker τ_i is supported on $\Gamma(c)$ and annihilated by $\mathcal{J}^d$. By Lemma 2.2 these cokernels are locally free $\mathcal{O}_S$-modules of rank d. Furthermore, by condition 2 we have $\mathcal{F}_\ell = \mathcal{F}(k \cdot \infty)$. For arbitrary $i \in \mathbb{Z}$ we take $n \in \mathbb{Z}$ such that $0 \le i - n\ell < \ell$ and define

$$\mathcal{F}_i := \mathcal{F}_{i-n\ell} \otimes_{\mathcal{O}_{C_S}} \mathcal{O}_{C_S}(nk \cdot \infty), \qquad \Pi_i := \Pi_{i-n\ell} \otimes \mathrm{id}, \qquad \tau_i := \tau_{i-n\ell} \otimes \mathrm{id}.$$

Then $(\mathcal{F}_i, \Pi_i, \tau_i)$ is an abelian sheaf of rank r and dimension d over S. This construction is also compatible with level structures and defines a 1-morphism $T' : \mathcal{P}ol_H^{r,d} \to \mathit{Ab}\text{-}\mathcal{S}h_H^{r,d} \times_C C'$. One easily proves that T and T' are mutually 2-inverse. $\qquad\square$

Remark 2.4. Example 1.8 together with Theorem 2.3 shows that every Drinfeld module carries a canonical polarization. This parallels the situation for elliptic curves.

3 Algebraicity of the stacks of abelian sheaves

In this section we will prove the following theorem.

Theorem 3.1. *Ab-$\mathcal{S}h_H^{r,d}$ is an algebraic stack in the sense of Deligne–Mumford [9], locally of finite type over C.*

Note that the example in Section 4 below shows that it will in general not be smooth over C. Nevertheless, if the characteristic is disjoint from ∞ and one stratifies the stack according to the isomorphy type of the $\mathcal{O}_C$-modules coker Π_i and coker τ_i, then each stratum will be smooth over C'.

More precisely, we let z be a uniformizing parameter on C at ∞ and we fix a flag $\mathcal{G}_\bullet$ of $\mathbb{F}_q[z]/z^k$-submodules

$$(0) \subset \mathcal{G}_1 \subset \cdots \subset \mathcal{G}_{\ell-1} \subset \mathcal{G}_\ell = (\mathbb{F}_q[z]/z^k)^{\oplus r} \qquad (3.1)$$

such that the successive quotients all have dimension d over $\mathbb{F}_q$. We say that an abelian sheaf $\underline{\mathcal{F}}$ of rank r and dimension d over S has *isomorphy type* $\mathcal{G}_\bullet$ *of* Π if for every point $s \in S$ the flag of $\mathcal{O}_{C_s}$-modules

$$(0) \subset \mathcal{F}_1/\mathcal{F}_0 \subset \cdots \subset \mathcal{F}_{\ell-1}/\mathcal{F}_0 \subset \mathcal{F}_\ell/\mathcal{F}_0 = \mathcal{F}_0(k \cdot \infty)/\mathcal{F}_0$$

is isomorphic to the flag $\mathcal{G}_\bullet \otimes_{\mathbb{F}_q} \kappa(s)$.

To fix the isomorphy type of τ let S be a scheme with characteristic morphism $c : S \to C$. We denote by $\mathcal{J} \subset \mathcal{O}_{C_S}$ the ideal sheaf defining the graph of c, by Γ_d the closed subscheme of C_S defined by $\mathcal{J}^d$, and by $\mathcal{O}_d$ the structure sheaf $\mathcal{O}_{C_S}/\mathcal{J}^d$ of Γ_d. We fix integers

$$(e_1 \leq e_2 \leq \cdots \leq e_r) = \underline{e} \qquad (3.2)$$

between 0 and d with $e_1 + \cdots + e_r = d$. We say that an abelian sheaf $\underline{\mathcal{F}}$ of rank r and dimension d over S has *isomorphy type* $\underline{e}$ *of* τ if for every point $s \in S$ the $\mathcal{O}_{C_s}$-module coker τ_{-1} is isomorphic to

$$\bigoplus_{\nu=1}^r \mathcal{O}_{C_s}/\mathcal{J}^{e_\nu}\mathcal{O}_{C_s}.$$

Note that if the characteristic is disjoint from ∞, then the cokernels of all τ_i are isomorphic.

We consider the locally closed subset (see Laumon–Moret-Bailly [32, Section 5]) of Ab-$\mathcal{S}h_H^{r,d}$ consisting of the points over which the universal abelian sheaf has isomorphy types $\mathcal{G}_\bullet$ of Π and $\underline{e}$ of τ. We give this subset the reduced induced structure [32, 4.10] and obtain a substack Ab-$\mathcal{S}h_H^{r,d,\mathcal{G}_\bullet,\underline{e}}$ of Ab-$\mathcal{S}h_H^{r,d}$.

Theorem 3.2. *The substack Ab-$\mathcal{S}h_H^{r,d,\mathcal{G}_\bullet,\underline{e}} \times_C C'$ is smooth over C' of relative dimension $\sum_{\mu>\nu}(e_\mu - e_\nu)$ if nonempty.*

The proof of these theorems follows Laumon–Rapoport–Stuhler [33]. For a given compact open subgroup $H \subset \mathrm{GL}_r(\widehat{A})$ consider a closed subscheme $I \subset C'$ which is supported on the places v for which $H_v \neq \mathrm{GL}_r(A_v)$ and satisfies

$$H \supset H_I := \ker(\mathrm{GL}_r(\widehat{A}) \to \mathrm{GL}_r(\mathcal{O}_I)).$$

Then $Ab\text{-}Sh_{H_I}^{r,d}$ is finite étale over $Ab\text{-}Sh_H^{r,d}$ by Proposition 1.3. In fact it is even a H/H_I-torsor. Hence $Ab\text{-}Sh_H^{r,d}$ is a quotient of $Ab\text{-}Sh_{H_I}^{r,d}$ in the sense of stacks; cf. [32, 4.6.1]. Therefore, it suffices to prove the two theorems for $H = H_I$. We assume this situation from now on.

We will cover $Ab\text{-}Sh_H^{r,d}$ by open substacks corresponding to stable vector bundles with additional level structure. To be precise we proceed as follows. Let $\mathcal{F}$ be a locally free sheaf of rank r on C_S. An I-level structure on $\mathcal{F}$ is an isomorphism $\bar{\eta} : \mathcal{F}|_{I \times S} \xrightarrow{\sim} \mathcal{O}_{I \times S}^r$ of $\mathcal{O}_{I \times S}$-modules.

Definition 3.3. *We say that the pair $(\mathcal{F}, \bar{\eta})$ is stable if for all algebraically closed points $s \in S$ and all locally free $\mathcal{O}_{C_s}$-modules $\mathcal{G}$ properly contained in $\mathcal{F}_s$, we have*

$$\frac{\deg(\mathcal{G}) - \deg(I)}{\mathrm{rk}(\mathcal{G})} < \frac{\deg(\mathcal{F}_s) - \deg(I)}{\mathrm{rk}(\mathcal{F}_s)}$$

(compare Seshadri [39, 4.I, Définition 2]).

We denote by $Ab\text{-}Sh_{H,st}^{r,d}$ the open substack of $Ab\text{-}Sh_H^{r,d}$ of those abelian sheaves for which $(\mathcal{F}_0, \bar{\eta})$ is stable. We will show that $Ab\text{-}Sh_{H,st}^{r,d}$ is representable by a disjoint union of quasi-projective schemes of relative dimension $d(r-1)$ over $C \smallsetminus I$ if $I \neq \emptyset$. For this purpose we need to introduce some additional stacks.

We denote by Vec_I^r the stack classifying locally free sheaves of rank r on C with I-level structure and by $Vec_{I,st}^r$ the substack of such locally free sheaves which are stable. Seshadri [39, 4.III] proves the following fact (see also [33, 4.3]). Note that there is a typing error in the formula for the dimension in [39].

Proposition 3.4. *If $I \neq \emptyset$ the stack $Vec_{I,st}^r$ is representable by a disjoint union of quasi-projective schemes over $\mathbb{F}_q$ which are smooth of dimension $r^2(g - 1 + \deg I)$. Here g is the genus of C.*

Definition 3.5. *Let S be a scheme and let $(\mathcal{F}_i, \Pi_i)$ be a sequence of locally free sheaves on C_S as in the first row of the Definition 1.1 of an abelian sheaf. Suppose that $(\mathcal{F}_i, \Pi_i)$ satisfies the conditions 2 and 3 of Definition 1.1. An I-level structure on $(\mathcal{F}_i, \Pi_i)$ is a collection of I-level structures $\bar{\eta}_i$ on the $\mathcal{F}_i$ that are compatible with the morphisms Π_i. We denote by $Seq_I^{r,d}$ the stack classifying sequences $(\mathcal{F}_i, \Pi_i)$ as above together with I-level structures $\bar{\eta}_i$ and by $Seq_{I,st}^{r,d}$ the open substack of those sequences for which $\mathcal{F}_0$ with its level structure is stable.*

Lemma 3.6. *The natural 1-morphism*

$$Seq_I^{r,d} \to Vec_I^r, \qquad (\mathcal{F}_i, \Pi_i, \bar{\eta}_i) \mapsto (\mathcal{F}_0, \bar{\eta}_0)$$

is representable by a closed subscheme of a flag variety. The strata $Seq_I^{r,d,\mathcal{G}_\bullet}$ with fixed isomorphy type $\mathcal{G}_\bullet$ of Π are smooth over Vec_I^r if nonempty. In particular, $Seq_{I,st}^{r,d}$ and $Seq_{I,st}^{r,d,\mathcal{G}_\bullet}$ are representable by a disjoint union of quasi-projective schemes (respectively, quasi-projective and smooth schemes) over $\mathbb{F}_q$ if $I \neq \emptyset$.

Proof. Let $\mathcal{F}_0$ on C_S be given corresponding to a 1-morphism $S \to \mathcal{V}ec'_I$. Due to the periodicity condition 2, the sequence $(\mathcal{F}_i, \Pi_i)$ corresponds to a flag of length ℓ of $\mathcal{O}_{C_S}$-submodules

$$(0) \subset \mathcal{F}_1/\mathcal{F}_0 \subset \cdots \subset \mathcal{F}_{\ell-1}/\mathcal{F}_0 \subset \mathcal{F}_\ell/\mathcal{F}_0 = \mathcal{F}_0(k \cdot \infty)/\mathcal{F}_0 \qquad (3.3)$$

such that the successive quotients are all locally free $\mathcal{O}_S$-modules of rank d. Hence the first assertion follows.

To prove the statement about $\mathcal{S}eq_I^{r,d,\mathcal{G}_\bullet}$ note that locally on S the sheaf $\mathcal{F}_\ell/\mathcal{F}_0$ is isomorphic to $\mathcal{O}_S[z]/z^k$. By the elementary divisor theorem a point Spec K of $\mathcal{S}eq_I^{r,d}$ belongs to $\mathcal{S}eq_I^{r,d,\mathcal{G}_\bullet}$ if and only if the flag (3.3) is conjugate to $\mathcal{G}_\bullet$ under $\mathrm{GL}_r(K[z]/z^k)$. Therefore, $\mathcal{S}eq_I^{r,d,\mathcal{G}_\bullet}$ is relatively representable over $\mathcal{V}ec_I^r$ by the homogeneous space

$$\mathrm{GL}_r(\mathbb{F}_q[z]/z^k)/ \mathrm{Stab}(\mathcal{G}_\bullet).$$

The group $\mathrm{GL}_r(\mathbb{F}_q[z]/z^k)$ is the Weil restriction $\mathcal{R}_{(\mathbb{F}_q[z]/z^k)/\mathbb{F}_q} \mathrm{GL}_r$ and hence a smooth connected algebraic group over $\mathbb{F}_q$. The stabilizer of $\mathcal{G}_\bullet$ corresponds to a closed algebraic subgroup defined over $\mathbb{F}_q$. Thus the above homogeneous space is a smooth algebraic variety over $\mathbb{F}_q$. From this the lemma follows. $\qquad\square$

Definition 3.7. *We let $\mathcal{H}ecke_I^{r,d}$ be the stack classifying the commutative diagrams with I-level structures*

$$
\begin{array}{ccccccc}
\cdots \longrightarrow & \mathcal{F}_{i-1} & \xrightarrow{\Pi_{i-1}} & \mathcal{F}_i & \xrightarrow{\Pi_i} & \mathcal{F}_{i+1} & \xrightarrow{\Pi_{i+1}} \cdots \\
& \uparrow{t_{i-2}} & & \uparrow{t_{i-1}} & & \uparrow{t_i} & \\
\cdots \longrightarrow & \mathcal{F}'_{i-2} & \xrightarrow{\Pi'_{i-2}} & \mathcal{F}'_{i-1} & \xrightarrow{\Pi'_{i-1}} & \mathcal{F}'_i & \xrightarrow{\Pi'_i} \cdots
\end{array}
$$

such that the sequences $(\mathcal{F}_i, \Pi_i)$ and $(\mathcal{F}'_i, \Pi'_i)$ with their I-level structures belong to $\mathcal{S}eq_I^{r,d}$ and such that the t_i satisfy conditions 1 and 4 of Definition 1.1 and respect the I-level structures. Assigning to such a diagram over S the morphism $S \to C \smallsetminus I$ on whose graph the cokernels of the t_i are supported, defines a 1-morphism of stacks $\mathcal{H}ecke_I^{r,d} \to C \smallsetminus I$.

The above stacks fit into the following 2-cartesian diagram of stacks

$$
\begin{array}{ccc}
\mathcal{A}b\text{-}\mathcal{S}h_H^{r,d} & \longrightarrow & \mathcal{S}eq_I^{r,d} \\
\downarrow & & \downarrow{(\mathrm{id},\sigma_{Seq})} \\
\mathcal{H}ecke_I^{r,d} & \xrightarrow{(1^{\mathrm{st}}\mathrm{row},2^{\mathrm{nd}}\mathrm{row})} & \mathcal{S}eq_I^{r,d} \times \mathcal{S}eq_I^{r,d} \\
\downarrow & & \\
C \smallsetminus I. &
\end{array}
\qquad (3.4)
$$

On the stack $\mathcal{H}ecke_I^{r,d}$ the cokernel of t_{-1} is a quotient of $\mathcal{F}_0$ which is locally free of rank d over the base and supported on the graph of the characteristic morphism c. We analyze this property. Let $\mathcal{T}$ be the stack

$$(C \smallsetminus I) \times \mathcal{S}eq_I^{r,d} .$$

On $C \times \mathcal{T}$ consider the ideal sheaf $\mathcal{J}$ defining the graph $\Gamma(c)$ of the characteristic morphism $c : \mathcal{T} \to C \smallsetminus I \subset C$ (see [32, Section 12]). We denote by Γ_d the closed substack of $C \times \mathcal{T}$ defined by $\mathcal{J}^d$ and by $\mathcal{O}_d$ the sheaf $\mathcal{O}_{C \times \mathcal{T}} / \mathcal{J}^d$. The quotients of $\mathcal{F}_0$ that are supported on $\Gamma(c)$ and are locally free over $\mathcal{T}$ of rank d are classified by Grothendieck's Quot-scheme relative to $\mathcal{T}$ [19, number 221, Théorème 3.1]

$$Quot^d_{\mathcal{F}_0 \otimes \mathcal{O}_d / \Gamma_d / \mathcal{T}} .$$

It is a stack projective over $\mathcal{T}$.

Lemma 3.8. *The* 1-*morphism*

$$\mathcal{H}ecke_I^{r,d} \to Quot^d_{\mathcal{F}_0 \otimes \mathcal{O}_d / \Gamma_d / \mathcal{T}} \times \mathcal{S}eq_I^{r,d}$$

given by the cokernel of t_{-1} and the second row, is representable by a closed immersion. Over $C \smallsetminus (I \cup \infty)$ the 1-*morphism*

$$\mathcal{H}ecke_I^{r,d} \to Quot^d_{\mathcal{F}_0 \otimes \mathcal{O}_d / \Gamma_d / \mathcal{T}}$$

obtained by projection onto the first factor is a 1-*isomorphism. In particular, the* 1-*morphism $\mathcal{H}ecke_I^{r,d} \to \mathcal{S}eq_I^{r,d} \times \mathcal{S}eq_I^{r,d}$ from (3.4) is representable by a quasi-projective morphism.*

Proof. The substack $\mathcal{H}ecke_I^{r,d}$ of $Quot^d_{\mathcal{F}_0 \otimes \mathcal{O}_d / \Gamma_d / \mathcal{T}} \times \mathcal{S}eq_I^{r,d}$ is defined by the following conditions:

1. $\mathcal{F}'_{-1}$ equals the kernel of the morphism from $\mathcal{F}_0$ to the universal quotient;
2. for each $i = -\ell, \dots, -2$, the sheaf $\mathcal{F}'_i$ is contained in the intersection of $\mathcal{F}_{i+1}$ and $\mathcal{F}'_{-1}$, which we view as subsheaves of $\mathcal{F}_0$ via $\Pi_{-1} \circ \cdots \circ \Pi_{i+1}$ and t_{-1};
3. if we let t_i be the inclusion $\mathcal{F}'_i \subset \mathcal{F}_{i+1}$, then coker t_i is annihilated by $\mathcal{J}^d$;
4. t_{-1} is compatible with the I-level structures on $\mathcal{F}'_{-1}$ and $\mathcal{F}_0$.

Namely, by descending induction on i, the short exact sequences of $\mathcal{O}_S$-modules

$$
\begin{array}{ccccccccc}
0 & \longrightarrow & \operatorname{coker} t_{i-1} & \longrightarrow & \operatorname{coker}(\Pi_i \circ t_{i-1}) & \longrightarrow & \operatorname{coker} \Pi_i & \longrightarrow & 0 \\
 & & & & \| & & & & \\
0 & \longrightarrow & \operatorname{coker} \Pi'_{i-1} & \longrightarrow & \operatorname{coker}(t_i \circ \Pi'_{i-1}) & \longrightarrow & \operatorname{coker} t_i & \longrightarrow & 0
\end{array}
$$

imply that coker t_i is a locally free $\mathcal{O}_S$-module of rank d. Now clearly the above conditions are represented by a closed immersion.

Over $C \smallsetminus (I \cup \infty)$ defining $\mathcal{F}'_i$ as the intersection $\mathcal{F}_{i+1} \cap \mathcal{F}'_{-1}$ in conditions 1 and 2 automatically gives an object $(\mathcal{F}'_i, \Pi'_i)$ in $\mathcal{S}eq_I^{r,d}$. This proves that the projection onto the first factor is a 1-isomorphism there. $\square$

Proof of Theorem 3.1. Recall that we have assumed $H = H_I$. Considering the diagram (3.4), we conclude from the previous lemmas that $Ab\text{-}Sh_{H,st}^{r,d}$ is representable by a disjoint union of quasi-projective schemes over $C \setminus I$ if $I \neq \emptyset$.

Now we let $I \subset I' \subset C'$ be two finite closed subschemes with $I' \neq \emptyset$ and we set $H' = H_{I'}$. By restricting I'-level structures to I-level structures we obtain a 1-morphism of stacks

$$r_{I',I} : Ab\text{-}Sh_{H'}^{r,d} \to Ab\text{-}Sh_H^{r,d} .$$

Over $C \setminus I'$ this 1-morphism is a torsor under the finite group

$$G_{I',I} := \ker(GL_r(\mathcal{O}_{I'}) \to GL_r(\mathcal{O}_I)) \cong H/H'$$

due to Proposition 1.3. Since $r_{I',I}^{-1}(Ab\text{-}Sh_{H,st}^{r,d}) \subset Ab\text{-}Sh_{H',st}^{r,d}$, the open substack $Ab\text{-}Sh_{H',st}^{r,d}$ which is stable under $G_{I',I}$, gives as a quotient in the sense of stacks an open substack

$$Ab\text{-}Sh_{H',st}^{r,d} / G_{I',I} \subset Ab\text{-}Sh_H^{r,d}$$

that contains $Ab\text{-}Sh_{H,st}^{r,d} \times_{C \setminus I} C \setminus I'$. It is an algebraic stack in the sense of Deligne–Mumford. If we let I' vary among finite closed subschemes of C' containing I these open substacks cover $Ab\text{-}Sh_H^{r,d}$, since every vector bundle becomes stable for a sufficiently high level structure. This proves Theorem 3.1 except for the assertion on the dimension which follows from Theorem 3.2.   □

Proof of Theorem 3.2. We denote by $Quot^{\underline{e}}$ the reduced, locally closed substack of $Quot_{\mathcal{F}_0 \otimes \mathcal{O}_d / \Gamma_d / \mathcal{T}}^d$ consisting of those points for which the universal quotient is isomorphic to

$$\bigoplus_{\nu=1}^r \mathcal{O}_{C_s} / \mathcal{J}^{e_\nu} \mathcal{O}_{C_s} .$$

We claim that $Quot^{\underline{e}}$ is smooth over $\mathcal{T}$ of relative dimension $\sum_{\mu > \nu}(e_\mu - e_\nu)$. Indeed, locally on $\mathcal{T}$ the sheaf $\mathcal{F}_0 \otimes \mathcal{O}_d$ is isomorphic to $\mathcal{O}_d^r$. Let $\mathcal{H} \subset \mathcal{O}_d^r$ be the kernel of the morphism from $\mathcal{O}_d^r$ to the universal quotient on $Quot^{\underline{e}}$. The condition on the isomorphy type of the universal quotient implies that $\mathcal{H}$ is conjugate to $\oplus_{\nu=1}^r \mathcal{J}^{e_\nu}/\mathcal{J}^d$ under $GL_r(\mathcal{O}_d)$. Therefore, $Quot^{\underline{e}}$ is locally isomorphic to the homogeneous space

$$GL_r(\mathcal{O}_d) / \text{Stab}(\oplus_{\nu=1}^r \mathcal{J}^{e_\nu}/\mathcal{J}^d).$$

As in the proof of Lemma 3.6, this homogeneous space is smooth and the claim follows. The theorem can now be deduced using Lemma 3.8 and applying [33, Lemma 4.2] to diagram (3.4).   □

4 An example

In this section let $C = \mathbb{P}^1_{\mathbb{F}_q}$ and $A = \mathbb{F}_q[t]$. Let $I = V(t) \subset C$ and $H = H_I$ and set $z = \frac{1}{t}$. Then $C \setminus I = \text{Spec } \mathbb{F}_q[z]$. We consider the case where $d = r = 2, k = \ell = 1$ and describe the algebraic stack $Ab\text{-}Sh_H^{2,2}$. It decomposes

$$Ab\text{-}Sh_H^{2,2} = \coprod_{n \in \mathbb{Z}} Ab\text{-}Sh_H^{2,2}(n)$$

into the open and closed substacks on which the vector bundle $\mathcal{F}_0$ has degree n. The shift by 1 from Definition 1.7 yields a 1-isomorphism $Ab\text{-}Sh_H^{2,2}(n) \to Ab\text{-}Sh_H^{2,2}(n + 2)$. So it suffices to describe $Ab\text{-}Sh_H^{2,2}(n)$ for $n = 0, 1$. We want to treat the case $n = 0$ here.

Let $M_I^{2,2}$ be the scheme

$$\operatorname{Spec} \mathbb{F}_q[\zeta][a_{\mu\nu} : 1 \leq \mu, \nu \leq 2]/(a_{11} + a_{22} + 2\zeta, a_{11}a_{22} - a_{12}a_{21} - \zeta^2).$$

We view $(a_{\mu\nu})$ as a 2×2 matrix with trace -2ζ and determinant ζ^2. Mapping z to ζ defines a morphism $c : M_I^{2,2} \to C$. On $S = M_I^{2,2}$, we set for $i \in \mathbb{Z}$

$$\mathcal{F}_i = \mathcal{O}_{C_S}(i \cdot \infty)^{\oplus 2} \quad \text{and} \quad \tau_i = (1 + t(a_{\mu\nu})) \cdot \sigma^* : \sigma^* \mathcal{F}_i \to \mathcal{F}_{i+1},$$

and we let $\Pi_i : \mathcal{F}_i \to \mathcal{F}_{i+1}$ be the morphism induced by the inclusion $\mathcal{O}_{C_S} \subset \mathcal{O}_{C_S}(\infty)$. Due to the trace and determinant condition on the matrix $(a_{\mu\nu})$, the data $(\mathcal{F}_i, \Pi_i, \tau_i)$ is an abelian sheaf of rank 2, dimension 2, and characteristic c over S.

We want to define an I-level structure on $\underline{\mathcal{F}} = (\mathcal{F}_i, \Pi_i, \tau_i)$. Note that $\underline{\mathcal{F}}|_{I \times S}$ is canonically isomorphic to $\mathcal{O}_S^2$ with $\tau|_{I \times S} = \operatorname{Id}_2 \cdot \sigma^*$. Hence the identity morphism on $\mathcal{O}_S^2$ defines an I-level structure $\bar{\eta}$ on $\underline{\mathcal{F}}$.

Proposition 4.1. *The 1-morphism $M_I^{2,2} \to Ab\text{-}Sh_H^{2,2}$ induced by $(\underline{\mathcal{F}}, \bar{\eta})$ identifies $M_I^{2,2}$ with the (representable) open substack of $Ab\text{-}Sh_H^{2,2}$ on which the underlying vector bundle with level structure $(\mathcal{F}_0, \bar{\eta})$ is stable (Definition 3.3) and has degree zero.*

We will see below that $(\mathcal{F}_0, \bar{\eta})$ is stable and of degree zero if and only if $\mathcal{F}_0 \cong \mathcal{O}_{C_S}^2$.

Proof. Let T be a scheme together with a characteristic morphism $c' : T \to \operatorname{Spec} \mathbb{F}_q[z]$ disjoint from I. Denote the image $c'^*(z)$ in $\mathcal{O}_T$ by ζ'. Let $(\mathcal{F}_i', \Pi_i', \tau_i', \bar{\eta}')$ be an abelian sheaf of rank 2, dimension 2, and characteristic c' over T with I-level structure such that $(\mathcal{F}_0', \bar{\eta}')$ is stable and of degree zero. We have to exhibit a uniquely defined morphism $f : T \to M_I^{2,2}$ such that $f^*(\underline{\mathcal{F}}, \bar{\eta}) \cong (\mathcal{F}_i', \Pi_i', \tau_i', \bar{\eta}')$. Since the morphisms Π_i' identify $\mathcal{F}_i'$ with $\mathcal{F}_0'(i \cdot \infty)$ it suffices to concentrate on $\mathcal{F}_0'$ and τ_0'.

We claim that the stability condition implies $\mathcal{F}_0' \cong \mathcal{O}_{C_T}^2$ globally on T. Indeed, let $\pi : C_T \to T$ be the projection onto the second factor. We first show that $\pi_* \mathcal{F}_0'$ is locally free of rank 2 on T and that $\pi^* \pi_* \mathcal{F}_0' \to \mathcal{F}_0'$ is an isomorphism. Let $s \in T$ be an algebraically closed point. The stability implies that every invertible $\mathcal{O}_{C_s}$-module $\mathcal{G} \subset \mathcal{F}_0'|_{C_s}$ has degree at most 0. Hence $\mathcal{F}_0'|_{C_s} \cong \mathcal{O}_{C_s}^2$ and thus $H^1(C_s, \mathcal{F}_0'|_{C_s}) = (0)$. By the theorem on cohomology and base change [25, III.12.11] this implies that $\mathrm{R}^1 \pi_* \mathcal{F}_0'$ vanishes and that $\pi_* \mathcal{F}_0'$ is locally free of rank 2 on T. Moreover, $\pi^* \pi_* \mathcal{F}_0' \to \mathcal{F}_0'$ is an isomorphism in the fiber over s and hence on all of T by Nakayama. Now the level structure $\bar{\eta}'$ induces an isomorphism $\pi_* \mathcal{F}_0' \cong (\pi^* \pi_* \mathcal{F}_0')|_{I \times T} \xrightarrow{\sim} \mathcal{O}_T^2$. From this our claim follows.

As τ_0' maps $\sigma^* \mathcal{F}_0'$ into $\mathcal{F}_1' = \mathcal{F}_0'(\infty)$, it is represented with respect to a basis of $\mathcal{F}_0'$ by a matrix

$$\tau_0' = (U_0 + t U_1) \cdot \sigma^* \quad \text{with} \quad U_0 \in \mathrm{GL}_2(\mathcal{O}_T) \quad \text{and} \quad U_1 \in M_2(\mathcal{O}_T).$$

Identifying $\mathcal{F}_0'|_{I \times T}$ with $\mathcal{O}_T^2$, we can express $\bar{\eta}'$ by a matrix in $\mathrm{GL}_2(\mathcal{O}_T)$. There is a uniquely defined change of basis of $\mathcal{F}_0'$ such that this matrix becomes the identity. Then we also have $U_0 = \mathrm{Id}$. The condition on coker τ_0' now implies that

$$\det(\mathrm{Id} + t U_1) = (1 - \zeta' t)^2.$$

Then the required morphism $f : T \to M_I^{2,2}$ is given by $f^*(a_{\mu\nu}) = U_1$ and $f^*(\zeta) = \zeta'$. $\qquad\square$

Remark 4.2. From this example one sees that $\mathcal{A}b\text{-}\mathcal{S}h_H^{r,d}$ need not be smooth over C. Namely, $M_I^{2,2}$ is not smooth at the points with $a_{12} = a_{21} = 0$, $a_{11} = a_{22} = -\zeta$. The reason for this is that at these points the $\mathcal{O}_{C_s}$-module coker τ_0 is isomorphic to $(\kappa(s)[z]/(z - \zeta))^{\oplus 2}$ whereas at all other points it is isomorphic to $\kappa(s)[z]/(z - \zeta)^2$. Compare with Theorem 3.2.

This example also shows that in general one cannot hope that the stacks $\mathcal{A}b\text{-}\mathcal{S}h_H^{r,d}$ are schemes. Namely, for any level I, there are abelian sheaves of rank 2 and dimension 2 with an I-level structure that have nontrivial automorphisms. Indeed, let $I = V(a) \subset C'$ for an $a \in \mathbb{F}_q[t]$ with $\deg a = n$. Let $c : S \to C \smallsetminus I$ be arbitrary and denote the image of z in $\mathcal{O}_S$ by ζ. Let $f \in \mathcal{O}_S[t]$ have degree $\leq n + 1$. Then the abelian sheaf with

$$\mathcal{F}_i = \mathcal{O}_{C_S}((i + n) \cdot \infty) \oplus \mathcal{O}_{C_S}(i \cdot \infty), \quad \tau_i = \begin{pmatrix} 1 - \zeta t & f \\ 0 & 1 - \zeta t \end{pmatrix} \cdot \sigma^*$$

admits an I-level structure after a finite étale extension of S. It has nontrivial automorphisms compatible with this level structure of the form

$$\begin{pmatrix} 1 & x a \\ 0 & 1 \end{pmatrix}$$

for $x \in \mathbb{F}_q$.

As a consequence $\mathcal{A}b\text{-}\mathcal{S}h_H^{r,d}$ is not quasi-compact in general. Indeed, recall from the proof of Theorem 3.1 the covering of $\mathcal{A}b\text{-}\mathcal{S}h_H^{r,d}$ by open substacks

$$U_I := \mathcal{A}b\text{-}\mathcal{S}h_{H_I, st}^{r,d} / (H / H_I)$$

where $I \subset C'$ runs through all finite subschemes with $H_I \subset H$. Note that $U_I \subset U_{I'}$ if $I \subset I'$. Now if $\mathcal{A}b\text{-}\mathcal{S}h_H^{r,d}$ were quasi-compact this covering would have a finite refinement. So $\mathcal{A}b\text{-}\mathcal{S}h_H^{r,d}$ would equal a single U_I for a large enough I. From this we could deduce that

$$\mathcal{A}b\text{-}\mathcal{S}h_{H_I}^{r,d} = \mathcal{A}b\text{-}\mathcal{S}h_{H_I, st}^{r,d}$$

is in fact a scheme. Since for $r = d = 2$ the latter is not the case, $\mathcal{A}b\text{-}\mathcal{S}h_H^{2,2}$ is not quasi-compact.

5 Isomorphism classes versus isogeny classes

The general yoga that mediates between isomorphism classes of abelian varieties over $\mathbb{Z}_p$-schemes and prime-to-p isogeny classes of such can be transfered to abelian sheaves. This will allow us to define H-level structures for arbitrary compact open subgroups $H \subset \mathrm{GL}_r(\mathbb{A}_f)$. We begin by defining the notion of isogeny for abelian sheaves.

Definition 5.1. *A morphism between abelian sheaves* $(\mathcal{F}_i, \Pi_i, \tau_i)$ *and* $(\mathcal{F}'_i, \Pi'_i, \tau'_i)$ *over S is called an* isogeny *if*

1. *all morphisms $\mathcal{F}_i \to \mathcal{F}'_i$ are injective,*
2. *coker$(\mathcal{F}_i \to \mathcal{F}'_i)$ is supported on $D \times S$ for an effective divisor $D \subset C$, and*
3. *coker$(\mathcal{F}_i \to \mathcal{F}'_i)$ is locally free of finite rank as an $\mathcal{O}_S$-module.*

The isogeny is called finite *(or* prime to ∞*) if for all i the support of coker$(\mathcal{F}_i \to \mathcal{F}'_i)$ is disjoint from ∞. A (finite) quasi-isogeny between $(\mathcal{F}_i, \Pi_i, \tau_i)$ and $(\mathcal{F}'_i, \Pi'_i, \tau'_i)$ is a (finite) isogeny between $(\mathcal{F}_i, \Pi_i, \tau_i)$ and $(\mathcal{F}'_i(D), \Pi'_i, \tau'_i)$ for some effective divisor $D \subset C$ (respectively, $D \subset C \smallsetminus \infty$).*

Note that an isogeny between two abelian sheaves can only exist if they both have the same rank and dimension. The following proposition is evident. It justifies our definition of quasi-isogenies.

Proposition 5.2. *Let $\alpha : \underline{\mathcal{F}} \to \underline{\mathcal{F}}'$ be an isogeny of abelian sheaves over S. Then there exists an effective divisor $D \subset C$ and an isogeny $\alpha^\vee : \underline{\mathcal{F}}' \to \underline{\mathcal{F}}(D)$ with $\alpha \circ \alpha^\vee$ and $\alpha^\vee \circ \alpha$ being the isogenies induced by the inclusion $\mathcal{O}_C \subset \mathcal{O}_C(D)$. If D is chosen minimal, then D and $\alpha^\vee$ are uniquely determined.*

Example 5.3. Let $\underline{\mathcal{F}} = (\mathcal{F}_i, \Pi_i, \tau_i)$ be an abelian sheaf of rank r and dimension d over S. Then the collection of the Π_i defines an isogeny

$$(\Pi_i) : \underline{\mathcal{F}}[1] \to \underline{\mathcal{F}}$$

where [1] denotes the shift by 1 (cf. Definition 1.7). Similarly, let P be a point on C and let $S \in \mathcal{N}ilp_{A_P}$ via the characteristic morphism $c : S \to C$. Then the collection of the τ_i defines an isogeny

$$(\tau_i) : \sigma^* \underline{\mathcal{F}}[1] \to \underline{\mathcal{F}}.$$

For abelian varieties there is a general principle relating isomorphism classes of abelian varieties with level structure over $\mathbb{Z}_p$-schemes to prime-to-p isogeny classes of such. This principle also applies to abelian sheaves. We need to introduce the analogue for $\mathbb{Z}_p$ in our situation.

Notation 5.4. The local ring $\mathcal{O}_{C,\infty}$ is a discrete valuation ring. Let z be a uniformizing parameter. We identify the completion of $\mathcal{O}_{C,\infty}$ with $\mathbb{F}_q[\![z]\!]$ and Q_∞ with $\mathbb{F}_q(\!(z)\!)$. Let ζ be an indeterminant over $\mathbb{F}_q$ and denote by $\mathbb{F}_q[\![\zeta]\!]$ the ring of formal power series in ζ. Fix the characteristic morphism $c : \operatorname{Spec} \mathbb{F}_q[\![\zeta]\!] \to C$ defined by $c^*(z) = \zeta$. Clearly, this morphism identifies $\mathbb{F}_q[\![z]\!]$ with $\mathbb{F}_q[\![\zeta]\!]$. However, since $z - \zeta$ does not necessarily act trivially on coker τ_i we use two different symbols to separate the two roles played by z as a uniformizing parameter at ∞ and as an element of $\mathcal{O}_S$. From now on all base schemes for abelian sheaves will be schemes over $\operatorname{Spec} \mathbb{F}_q[\![\zeta]\!]$. We consider the base change of our stacks

$$\mathcal{A}b\text{-}\mathcal{S}h_H^{r,d} \times_C \operatorname{Spec} \mathbb{F}_q[\![\zeta]\!].$$

For the sake of brevity we denote them again by $\mathcal{A}b\text{-}\mathcal{S}h_H^{r,d}$. This should not cause confusion since from now on we work entirely in the local situation over $\operatorname{Spec} \mathbb{F}_q[\![\zeta]\!]$.

Working with isogeny classes instead of isomorphism classes we have to modify our definition of H-level structures. The new definition will have the additional advantage that it extends to arbitrary compact open subgroups $H \subset \operatorname{GL}_r(\mathbb{A}_f)$ which are not necessarily contained in $\operatorname{GL}_r(\widehat{A})$. Let $\underline{\mathcal{F}}$ be an abelian sheaf of rank r over S. We define the functor

$$(\underline{\mathcal{F}}|_{\mathbb{A}_f})^\tau : \mathcal{S}ch_S \longrightarrow \quad \mathbb{A}_f\text{-modules}$$
$$T/S \longmapsto \mathbb{A}_f \otimes_{\widehat{A}} (\underline{\mathcal{F}}|_{\widehat{A}})^\tau(T).$$

Assume that S is connected. Choosing an algebraically closed base point $\iota : s \to S$ we may view $(\underline{\mathcal{F}}|_{\mathbb{A}_f})^\tau$ as the $\mathbb{A}_f[\pi_1(S, s)]$-module $(\iota^*\underline{\mathcal{F}}|_{\mathbb{A}_f})^\tau(s)$.

We consider the set $\operatorname{Isom}_{\mathbb{A}_f}((\underline{\mathcal{F}}|_{\mathbb{A}_f})^\tau, \mathbb{A}_f^r) := \operatorname{Isom}_{\mathbb{A}_f}((\iota^*\underline{\mathcal{F}}|_{\mathbb{A}_f})^\tau(s), \mathbb{A}_f^r)$ of isomorphisms of $\mathbb{A}_f$-modules. Via its natural action on $\mathbb{A}_f^r$ the group $\operatorname{GL}_r(\mathbb{A}_f)$ acts on this set from the left. Via its action on $(\iota^*\underline{\mathcal{F}}|_{\mathbb{A}_f})^\tau(s)$ the group $\pi_1(S, s)$ acts on it from the right.

Definition 5.5. *Let $H \subset \operatorname{GL}_r(\mathbb{A}_f)$ be a compact open subgroup. A rational H-level structure on $\underline{\mathcal{F}}$ over S is an H-orbit in $\operatorname{Isom}_{\mathbb{A}_f}((\underline{\mathcal{F}}|_{\mathbb{A}_f})^\tau, \mathbb{A}_f^r)$ which is fixed by $\pi_1(S, s)$. (Again the latter condition implies that the notion of level structure is independent of the chosen base point.)*

Every quasi-isogeny $\alpha : \underline{\mathcal{F}} \to \underline{\mathcal{F}}'$ induces an isomorphism

$$(\alpha|_{\mathbb{A}_f})^\tau : (\iota^*\underline{\mathcal{F}}|_{\mathbb{A}_f})^\tau(s) \xrightarrow{\sim} (\iota^*\underline{\mathcal{F}}'|_{\mathbb{A}_f})^\tau(s)$$

and thus carries rational H-level structures on $\underline{\mathcal{F}}$ to rational H-level structures on $\underline{\mathcal{F}}'$.

Theorem 5.6. *If $H \subset \operatorname{GL}_r(\widehat{A})$ is a compact open subgroup, then the stack $\mathcal{A}b\text{-}\mathcal{S}h_H^{r,d}$ is canonically 1-isomorphic to the stack $\mathcal{X}$ whose category of S-valued points has*

as objects   all pairs $(\underline{\mathcal{F}}, \bar{\gamma})$ consisting of an abelian sheaf $\underline{\mathcal{F}}$ of rank r and dimension d and a rational H-level structure $\bar{\gamma}$ on $\underline{\mathcal{F}}$ over S, and

as morphisms   all finite quasi-isogenies that are compatible with the rational H-level structures.

Proof. Let S be a Spec $\mathbb{F}_q[\![\zeta]\!]$-scheme and let $(\underline{\mathcal{F}}, \bar{\eta})$ be an object of $Ab\text{-}Sh_H^{r,d}(S)$; i.e., $\underline{\mathcal{F}}$ is an abelian sheaf and $\bar{\eta}$ is an H-level structure on $\underline{\mathcal{F}}$ over S (in the sense of Definition 1.6). Then $\mathbb{A}_f \otimes_{\widehat{A}} \bar{\eta}$ is a rational H-level structure on $\underline{\mathcal{F}}$. This defines a canonical 1-morphism of stacks $f : Ab\text{-}Sh_H^{r,d} \to \mathcal{X}$.

For it to be a 1-isomorphism we have to show that $f(S) : Ab\text{-}Sh_H^{r,d}(S) \to \mathcal{X}(S)$ is an equivalence of categories for all S. Let $(\underline{\mathcal{F}} = (\mathcal{F}_i, \Pi, i, \tau_i), \bar{\gamma})$ be an object of $\mathcal{X}(S)$. Choose an algebraically closed base point $\iota : s \to S$ and a representative

$$\gamma : (\iota^*\underline{\mathcal{F}}|_{\mathbb{A}_f})^\tau(s) \xrightarrow{\sim} \mathbb{A}_f^r$$

of $\bar{\gamma}$. There is an element $a \in A$ with $\gamma^{-1}(a\widehat{A}^r) \subset (\iota^*\underline{\mathcal{F}}|_{\widehat{A}})^\tau(s)$. Then the sheaf

$$(\iota^*\underline{\mathcal{F}}|_{\widehat{A}})^\tau(s)/\gamma^{-1}(a\widehat{A}^r)$$

on C_s has finite length and support disjoint from ∞. Via the identification of $(\underline{\mathcal{F}}|_{\widehat{A}})^\tau$ with $(\iota^*\underline{\mathcal{F}}|_{\widehat{A}})^\tau(s)$ this sheaf can be viewed as a quotient sheaf of $\mathcal{F}_i$ for all i. Its support is of the form $D \times S$ where $D \subset C'$ is the divisor of zeros of a. Note that $D \times S$ is disjoint from the characteristic of $\underline{\mathcal{F}}$. Therefore, the kernel of this quotient map is an abelian sheaf $(\mathcal{F}_i', \Pi_i', \tau_i')$ of rank r and dimension d over S which is isogenous to $\underline{\mathcal{F}}$. Consider the abelian sheaf $\underline{\mathcal{F}}'(D) := (\mathcal{F}_i'(D), \Pi_i', \tau_i')$ and the induced finite quasi-isogeny α from $\underline{\mathcal{F}}'(D)$ to $\underline{\mathcal{F}}$. By construction $\gamma \circ (\alpha|_{\mathbb{A}_f})^\tau$ induces an isomorphism

$$\bar{\eta} : (\underline{\mathcal{F}}'(D)|_{\widehat{A}})^\tau \xrightarrow{\sim} \widehat{A}^r.$$

This yields an H-level structure $\bar{\eta}$ on $\underline{\mathcal{F}}'(D)$. Clearly, this construction is independent of the choice of γ.

The pair $(\underline{\mathcal{F}}'(D), \mathbb{A}_f \otimes_{\widehat{A}} \bar{\eta})$ is isogenous to $(\underline{\mathcal{F}}, \bar{\gamma})$ by construction. This proves that the functor $f(S)$ is essentially surjective. Analyzing the above construction further shows that it is also fully faithful. Hence f is a 1-isomorphism of stacks.   □

The reader should note that if v is a finite place of C, the analogous statement holds for abelian sheaves over A_v-schemes and prime-to-v isogenies.

The above theorem enables us to define the stacks of abelian sheaves with H-level structure for arbitrary compact open subgroups $H \in \mathrm{GL}_r(\mathbb{A}_f)$.

Definition 5.7. $Ab\text{-}Sh_H^{r,d}$ *is the stack over* Spec $\mathbb{F}_q[\![\zeta]\!]$ *whose category of S-valued points has*

as objects	*all pairs $(\underline{\mathcal{F}}, \bar{\gamma})$ consisting of an abelian sheaf $\underline{\mathcal{F}}$ of rank r and dimension d and a rational H-level structure $\bar{\gamma}$ on $\underline{\mathcal{F}}$ over S, and*
as morphisms	*all finite quasi-isogenies that are compatible with the rational H-level structures.*

For varying H, the stacks $Ab\text{-}Sh_H^{r,d}$ form a projective system of stacks. The transition maps $Ab\text{-}Sh_{H'}^{r,d} \to Ab\text{-}Sh_H^{r,d}$ for $H' \subset H$ are representable by finite étale morphisms of schemes. We define a right action of $\mathrm{GL}_r(\mathbb{A}_f)$ on this projective system by letting $g \in \mathrm{GL}_r(\mathbb{A}_f)$ act through the 1-isomorphisms

$$g : Ab\text{-}Sh_H^{r,d} \xrightarrow{\sim} Ab\text{-}Sh_{gHg^{-1}}^{r,d}$$

which are defined by $(\underline{\mathcal{F}}, \bar{\gamma}) \mapsto (\underline{\mathcal{F}}, \overline{g^{-1}\gamma})$ on S-valued points. Using these 1-isomorphisms it follows from Theorem 3.1 that all the stacks $Ab\text{-}Sh_H^{r,d}$ are algebraic in the sense of Deligne–Mumford and locally of finite type over $\mathrm{Spec}\,\mathbb{F}_q[\![\zeta]\!]$.

Part Two: z-Divisible Groups

Our ultimate goal in this article is to study the uniformization of the stacks $Ab\text{-}Sh_H^{r,d}$ at ∞. Classically, this corresponds to the p-adic uniformization of moduli spaces of abelian varieties. For those uniformization questions the associated p-divisible groups are an indispensable tool. In the same manner we are thus lead to the idea of "z-divisible groups." These groups were studied in detail in Hartl [23]. But they already appeared in special cases in the work of Drinfeld [11], Genestier [18], Laumon [31], Taguchi [41] and Rosen [38]. For the uniformization of $Ab\text{-}Sh_H^{r,d}$ these z-divisible groups are of equal importance as p-divisible groups are for abelian varieties. Therefore, the next few sections are devoted to them. We first review some facts from [23] in Sections 6 and 7.

As z-divisible groups are of most use over schemes on which z is not a unit we will from now on work over $\mathrm{Spf}\,\mathbb{F}_q[\![\zeta]\!]$ (see Notation 5.4). Since we want to relate z-divisible groups to abelian sheaves, we should also consider abelian sheaves over $\mathrm{Spf}\,\mathbb{F}_q[\![\zeta]\!]$-schemes. Hence we right away introduce the base change

$$Ab\text{-}Sh_H^{r,d} \times_C \mathrm{Spf}\,\mathbb{F}_q[\![\zeta]\!].$$

This is no longer an algebraic stack. But it is a *formal algebraic* $\mathrm{Spf}\,\mathbb{F}_q[\![\zeta]\!]$-*stack* (Definition A.5). Formal algebraic stacks are related to algebraic stacks in the same way as formal schemes are related to usual schemes. For the necessary background, we refer the reader to the appendix.

6 Definition of z-divisible groups

We continue to work in the local situation introduced in Notation 5.4. In particular, z is a uniformizing parameter at ∞ and we identify Q_∞ with $\mathbb{F}_q(\!(z)\!)$. Let Δ be the central skew field over Q_∞ of invariant k/ℓ and let $\mathcal{O}_\Delta$ be its ring of integers. We identify $\mathcal{O}_\Delta$ with the $\mathbb{F}_q$-algebra $\mathbb{F}_{q^\ell}[\![z, \Pi]\!]$ of noncommutative power series subject to the relations

$$\Pi^\ell = z^k, \quad z\Pi = \Pi z, \quad z\lambda = \lambda z, \quad \Pi\lambda^q = \lambda\Pi \quad \text{for all } \lambda \in \mathbb{F}_{q^\ell}.$$

Denote by $\mathcal{N}ilp_{\mathbb{F}_q[\![\zeta]\!]}$ the category of schemes over $\mathrm{Spec}\,\mathbb{F}_q[\![\zeta]\!]$ on which ζ is locally nilpotent. From now on in this article the scheme S will be in $\mathcal{N}ilp_{\mathbb{F}_q[\![\zeta]\!]}$.

Definition 6.1. *Let R be a unitary ring. We define an R-module scheme over S to be a flat commutative S-group scheme E together with a unitary ring homomorphism $R \to \mathrm{End}_S E$. It is called* finite of order d *if it is so as an S-group scheme. A morphism of R-module schemes is a morphism of the underlying S-group schemes which is compatible with the R-action.*

For an R-module scheme E over S we define its *co-Lie module* $\mathrm{Lie}^* E$ as the $\mathcal{O}_S$-module of invariant differentials. It is canonically isomorphic to $e^* \Omega_{E/S}$ where $e : S \to E$ is the zero section. We have $\mathrm{Lie}\, E = \mathcal{H}om_S(\mathrm{Lie}^* E, \mathcal{O}_S)$ as $\mathcal{O}_S$-module.

The additive group scheme $\mathbb{G}_{a,S}$ is an example for an $\mathbb{F}_q$-module scheme over S. Likewise every S-group scheme which locally on S is isomorphic to $\mathbb{G}_{a,S}^d$ for some integer $d \geq 0$ is an $\mathbb{F}_q$-module scheme. Such a scheme is called an $\mathbb{F}_q$-*vector group scheme of dimension d* over S. For every $a \in \mathbb{F}_q$ the endomorphism induced on its co-Lie module equals the multiplication with a viewed as an element of $\Gamma(S, \mathcal{O}_S)$.

Definition 6.2. *Let $h, d \geq 1$ be integers. A z-divisible group of height h and dimension d over S is an inductive system of finite $\mathbb{F}_q[\![z]\!]$-module schemes over S,*

$$E = (E_1 \xrightarrow{i_1} E_2 \xrightarrow{i_2} E_3 \xrightarrow{i_3} \cdots),$$

such that for each integer $n \geq 1$,

1. *the $\mathbb{F}_q$-module scheme E_n can be embedded into an $\mathbb{F}_q$-vector group scheme over S;*
2. *the order of E_n is q^{hn};*
3. *the following sequence of $\mathbb{F}_q[\![z]\!]$-module schemes over S is exact:*

$$0 \to E_n \xrightarrow{i_n} E_{n+1} \xrightarrow{z^n} E_{n+1};$$

4. *$(z - \zeta)^d = 0$ on $\mathrm{Lie}^* E_n$;*
5. *$d = \max\{\dim_{\kappa(s)}(\mathrm{Lie}^* E_n \otimes_{\mathcal{O}_S} \kappa(s)) : s \in S, n \geq 1\}$.*

A morphism of z-divisible groups over S is a morphism of inductive systems of $\mathbb{F}_q[\![z]\!]$-module schemes.

We set $\mathrm{Lie}^* E = \varprojlim \mathrm{Lie}^* E_n$. Conditions 1–4 imply that this is a locally free $\mathcal{O}_S$-module. Condition 5 asserts that its rank is the dimension of E.

The reader should observe that we do not require that $z - \zeta$ acts trivially on $\mathrm{Lie}^* E$. This is in conformity with the previous sections. In this respect our notion of z-divisible group is more general than the variants considered in [11, 18, 41, 38] and different from the classical case of p-divisible groups.

The group of morphisms $\mathrm{Hom}_S(E, E')$ between two z-divisible groups E and E' over S is a torsion free $\mathbb{F}_q[\![z]\!]$-module. Let $\underline{\mathrm{Hom}}_S(E, E')$ denote the sheaf of germs of morphisms on S.

Definition 6.3. *The category of z-divisible groups over S up to isogeny has as objects the z-divisible groups over S and as morphisms from E to E' all global sections of* the sheaf $\underline{\mathrm{Hom}}_S(E, E') \otimes_{\mathbb{F}_q[\![z]\!]} \mathbb{F}_q(\!(z)\!)$ on S. An isomorphism in this new category is called a quasi-isogeny. An isogeny between z-divisible groups *is a morphism between z-divisible groups which also is a quasi-isogeny.*

In particular, for every quasi-isogeny α there exists locally on S an integer n such that $z^n \alpha$ is an isogeny. Quasi-isogenies have the following rigidity property.

Proposition 6.4. *Let $\iota : S' \hookrightarrow S$ be a closed subscheme defined by a sheaf of ideals which is locally nilpotent. Let E and E' be two z-divisible groups over S. Then every quasi-isogeny from $\iota^* E$ to $\iota^* E'$ lifts in a unique way to a quasi-isogeny from E to E'.*

Proposition 6.5. *Let $\alpha : E \to E'$ be a quasi-isogeny of z-divisible groups over S. Then the functor on $\mathcal{N}ilp_{\mathbb{F}_q[\![\zeta]\!]}$,*

$$T \mapsto \{\varphi \in \mathrm{Hom}_{\mathbb{F}_q[\![\zeta]\!]}(T, S) : \varphi^* \alpha \text{ is an isogeny}\},$$

is representable by a closed subscheme of S.

Definition 6.6. *A z-divisible $\mathcal{O}_\Delta$-module over S is a z-divisible group E over S with an action $\mathcal{O}_\Delta \to \mathrm{End}_S E$ of $\mathcal{O}_\Delta$, which prolongs the natural action of $\mathbb{F}_q[\![z]\!]$. A morphism of z-divisible $\mathcal{O}_\Delta$-modules which is an isogeny of z-divisible groups is called an* isogeny.

Definition 6.7. *If S belongs to $\mathcal{N}ilp_{\mathrm{Spf}\,\mathbb{F}_{q^\ell}[\![\zeta]\!]}$ a z-divisible $\mathcal{O}_\Delta$-module E of height $r\ell$ and dimension $d\ell$ over S is called* special *if the action of $\mathcal{O}_\Delta$ induced on $\mathrm{Lie}^* E$, makes $\mathrm{Lie}^* E$ into a locally free $\mathbb{F}_{q^\ell} \otimes \mathcal{O}_S$-module of rank d.*

At the end of the next section, we will show that the latter are precisely the z-divisible groups that arise from abelian sheaves.

Proposition 6.8. *For a z-divisible $\mathcal{O}_\Delta$-module E over S, the condition of being special is represented by an open and closed immersion into S.*

Proof. Clearly, $\mathrm{Lie}^* E$ decomposes into a direct sum $\sum_{i=0}^{\ell-1} (\mathrm{Lie}^* E)_i$ of components $(\mathrm{Lie}^* E)_i$ on which $\lambda \in \mathbb{F}_{q^\ell} \subset \mathcal{O}_\Delta$ acts via $\lambda^{q^i} \in \mathcal{O}_S$. Then E is special if and only if all $(\mathrm{Lie}^* E)_i$ are locally free $\mathcal{O}_S$-modules of rank d. The proposition follows. $\qquad\square$

7 Dieudonné modules of z-divisible groups

We continue with the notation from Section 6. Let S be a scheme in $\mathcal{N}ilp_{\mathbb{F}_q[\![\zeta]\!]}$ and consider the completion of C_S along the closed subscheme $\infty \times V(\zeta)$. Its structure sheaf is the sheaf $\mathcal{O}_S[\![z]\!]$ on S of formal power series in z. We denote the sheaf $\mathcal{O}_\Delta \otimes_{\mathbb{F}_q[\![z]\!]} \mathcal{O}_S[\![z]\!]$ on S by $\mathcal{O}_\Delta \widehat{\otimes} \mathcal{O}_S$.

Consider the additive group $\mathbb{G}_{a,S} = \mathrm{Spec}\,\mathcal{O}_S[\xi]$ over S. On $\mathbb{G}_{a,S}$ we have the *Frobenius isogeny* $\mathrm{Frob}_q : \mathbb{G}_{a,S} \to \mathbb{G}_{a,S}$ defined by $\mathrm{Frob}_q^*(\xi) = \xi^q$. Let $E = (E_n, i_n)$ be a z-divisible group over S. We associate to E the sheaf

$$M_E = \varprojlim_n \mathcal{H}om_S(E_n, \mathbb{G}_{a,S})$$

on S. We make M_E into a sheaf of $\mathcal{O}_S[\![z]\!]$-modules by letting z act through the isogeny z on E. The σ_S-linear multiplication with Frob_q on the left defines a morphism

$$F_E : \sigma^* M_E \to M_E.$$

If E is, moreover, a z-divisible $\mathcal{O}_\Delta$-module, then M_E becomes an $\mathcal{O}_\Delta \widehat{\otimes} \mathcal{O}_S$-module through the action of $\mathcal{O}_\Delta$ from the right. The module M_E may be viewed as the analogue of the *contravariant Dieudonné module* associated to a p-divisible group. See [23] for a general discussion of this analogy. We recall the following facts.

Definition 7.1. *A Dieudonné $\mathbb{F}_q[\![z]\!]$-module over S of dimension d and rank r is a sheaf $\widehat{\mathcal{F}}$ of $\mathcal{O}_S[\![z]\!]$-modules on S equipped with an $\mathcal{O}_S[\![z]\!]$-module homomorphism $F : \sigma^* \widehat{\mathcal{F}} \to \widehat{\mathcal{F}}$ such that locally on S in the Zariski topology,*

1. *$\widehat{\mathcal{F}}$ is free of rank r as an $\mathcal{O}_S[\![z]\!]$-module,*
2. *coker F is free of rank d as an $\mathcal{O}_S$-module,*
3. *$(z - \zeta)^d = 0$ on coker F.*

A morphism *between Dieudonné $\mathbb{F}_q[\![z]\!]$-modules is a morphism of sheaves of $\mathcal{O}_S[\![z]\!]$-modules which is compatible with F.*

Note that F is automatically injective. As for p-divisible groups the z-divisible groups are classified by their Dieudonné $\mathbb{F}_q[\![z]\!]$-modules.

Theorem 7.2. *The functor $E \mapsto (M_E, F_E)$ is an antiequivalence between the category of z-divisible groups of height r and dimension d over S and the category of Dieudonné $\mathbb{F}_q[\![z]\!]$-modules of rank r and dimension d over S. There is a canonical isomorphism $\mathrm{Lie}^* E = \mathrm{coker}\, F_E$.*

If one seeks a classification of p-divisible groups up to isogeny, one works with isocrystals instead of crystals. For z-divisible groups we do the same.

Definition 7.3. *A Dieudonné $\mathbb{F}_q(\!(z)\!)$-module over S is a finite locally free $\mathcal{O}_S[\![z]\!][\frac{1}{z}]$-module $\mathcal{V}$ together with an isomorphism $F : \sigma^* \mathcal{V} \xrightarrow{\sim} \mathcal{V}$.*

To every Dieudonné $\mathbb{F}_q[\![z]\!]$-module $\widehat{\mathcal{F}} = (\widehat{\mathcal{F}}, F)$ over S we associate its Dieudonné $\mathbb{F}_q(\!(z)\!)$-module

$$\widehat{\mathcal{F}}\left[\frac{1}{z}\right] := \left(\widehat{\mathcal{F}} \otimes_{\mathcal{O}_S[\![z]\!]} \mathcal{O}_S[\![z]\!]\left[\frac{1}{z}\right], F \otimes \mathrm{id} \right).$$

Definition 7.4. *A morphism $\alpha : \widehat{\mathcal{F}} \to \widehat{\mathcal{F}}'$ of Dieudonné $\mathbb{F}_q[\![z]\!]$-modules is called an* isogeny *if α is injective and coker α is a locally free $\mathcal{O}_S$-module of finite rank. We define a* quasi-isogeny *between Dieudonné $\mathbb{F}_q[\![z]\!]$-modules $\widehat{\mathcal{F}}$ and $\widehat{\mathcal{F}}'$ to be an isomorphism between $\widehat{\mathcal{F}}[\frac{1}{z}]$ and $\widehat{\mathcal{F}}'[\frac{1}{z}]$.*

Proposition 7.5. *The functor $E \mapsto (M_E, F_E)$ maps isogenies to isogenies and quasi-isogenies to quasi-isogenies.*

Now let m/n be a rational number written in lowest terms with $n > 0$. Then we define the Dieudonné $\mathbb{F}_q((z))$-module $\mathcal{V}(m/n)$ over $\operatorname{Spec} \mathbb{F}_q$ as

$$\mathcal{V} = \mathbb{F}_q((z))^n, \qquad F = \begin{pmatrix} 0 & \cdots & & z^m \\ 1 & \ddots & & \vdots \\ & \ddots & \ddots & \\ & & 1 & 0 \end{pmatrix} \cdot \sigma^* : \sigma^* \mathcal{V} \to \mathcal{V}.$$

There is the following analogue of Dieudonné's Theorem [34].

Theorem 7.6. *Let K be an algebraically closed field with $\operatorname{Spec} K \in \mathcal{N}ilp_{\mathbb{F}_q[\![\zeta]\!]}$. Then every Dieudonné $\mathbb{F}_q((z))$-module over $\operatorname{Spec} K$ is isomorphic to a direct sum*

$$\bigoplus_i \mathcal{V}(m_i/n_i) \otimes_{\mathbb{F}_q((z))} K((z))$$

for uniquely determined rational numbers $m_1/n_1 \leq m_2/n_2 \leq \cdots$.

This result allows us to define the *Newton polygon* of a Dieudonné $\mathbb{F}_q[\![z]\!]$-module $\underline{\widehat{\mathcal{F}}} = (\widehat{\mathcal{F}}, F)$ over a field K in $\mathcal{N}ilp_{\mathbb{F}_q[\![\zeta]\!]}$. Namely, over an algebraically closed extension, its Dieudonné $\mathbb{F}_q((z))$-module decomposes as in the theorem. Then the Newton polygon is the polygon which passes through the points

$$(n_1 + \cdots + n_i, m_1 + \cdots + m_i)$$

for all i and is extended linearly between them. It is independent of the chosen algebraically closed extension. We also define the *Hodge polygon* as usual by the elementary divisors of the $K[\![z]\!]$-module $\operatorname{coker} F$. The Hodge polygon lies below the Newton polygon. They both have the same initial point $(0, 0)$ and the same terminal point $(\operatorname{rk} \underline{\widehat{\mathcal{F}}}, \dim \underline{\widehat{\mathcal{F}}})$. We obtain the analogue of the theorem of Grothendieck–Katz [21, 27].

Theorem 7.7. *Let $\underline{\widehat{\mathcal{F}}}$ be a Dieudonné $\mathbb{F}_q[\![z]\!]$-module of rank r over S and let P be the graph of a continuous real-valued function on $[0, r]$ which is linear between successive integers. Then the set of points in S at which the Hodge (respectively, Newton) polygon of $\underline{\widehat{\mathcal{F}}}$ lies above P is Zariski closed.*

Note that the stratification of $\mathcal{A}b\text{-}\mathcal{S}h_H^{r,d}$ considered in Section 3 is related to the stratification according to the Hodge-polygon. This relation comes from Construction 7.13 below.

Definition 7.8. *Let $S \in \mathcal{N}ilp_{\mathbb{F}_q[\![\zeta]\!]}$ be the spectrum of a field. We say that a Dieudonné $\mathbb{F}_q[\![z]\!]$-module over S is isoclinic if its Newton polygon has only a single slope.*

Proposition 7.9. *Let $\underline{\widehat{\mathcal{F}}}$ be an isoclinic Dieudonné $\mathbb{F}_q[\![z]\!]$-module of rank r and dimension d over a perfect field K. Then $\underline{\widehat{\mathcal{F}}}$ is isogenous over K to a Dieudonné $\mathbb{F}_q[\![z]\!]$-module satisfying $\operatorname{im} F^r = z^d \widehat{\mathcal{F}}$.*

Later we will also need the analogue of Katz's Constancy Theorem.

Theorem 7.10. *Let* $\operatorname{Spec} K[\![\pi]\!] \in \mathcal{N}ilp_{\mathbb{F}_q[\![\zeta]\!]}$ *be the spectrum of a power series ring over an algebraically closed field* K *and let* $\widehat{\mathcal{F}}$ *be a Dieudonné* $\mathbb{F}_q[\![z]\!]$-*module over* $\operatorname{Spec} K[\![\pi]\!]$. *Suppose that at the two points of* $\operatorname{Spec} K[\![\pi]\!]$ *the Newton polygons coincide, and that this common Newton polygon has only a single slope. Then* $\widehat{\mathcal{F}}$ *is isogenous to a constant Dieudonné* $\mathbb{F}_q[\![z]\!]$-*module (i.e., one obtained by pullback under the morphism* $\operatorname{Spec} K[\![\pi]\!] \to \operatorname{Spec} K$).

We now want to apply this theory to *special z-divisible $\mathcal{O}_\Delta$-modules*. Since for every z-divisible group E there is a canonical isomorphism $\operatorname{Lie}^* E = \operatorname{coker} F_E$ the property of being special reflects on M_E. We consider the following class of Dieudonné $\mathbb{F}_q[\![z]\!]$-modules. Let S be in $\mathcal{N}ilp_{\mathbb{F}_{q^\ell}[\![\zeta]\!]}$.

Definition 7.11. *A formal abelian sheaf of rank r and dimension d over S is a sheaf $\widehat{\mathcal{F}}$ of $\mathcal{O}_\Delta \widehat{\otimes} \mathcal{O}_S$-modules on S together with a morphism of $\mathcal{O}_\Delta \widehat{\otimes} \mathcal{O}_S$-modules $F : \sigma^* \widehat{\mathcal{F}} \to \widehat{\mathcal{F}}$ such that*

1. *$(\widehat{\mathcal{F}}, F)$ is a Dieudonné $\mathbb{F}_q[\![z]\!]$-module over S of rank $r\ell$,*
2. *$\operatorname{coker} F$ is locally free of rank d as an $\mathbb{F}_{q^\ell} \otimes \mathcal{O}_S$-module.*

In the situation of special z-divisible $\mathcal{O}_\Delta$-modules, Theorem 7.2 takes the following form.

Theorem 7.12. *The functor $E \mapsto M_E$ is an antiequivalence between the category of special z-divisible $\mathcal{O}_\Delta$-modules of height $r\ell$ and dimension $d\ell$ over S and the category of formal abelian sheaves of rank r and dimension d over S.*

Next, we want to describe the relation with abelian sheaves. We will obtain the analogue of the classical functor which assigns to every abelian variety its p-divisible group.

Construction 7.13. Let S be in $\mathcal{N}ilp_{\mathbb{F}_q[\![\zeta]\!]}$ and assume that there is a morphism $\beta : S \to \operatorname{Spec} \mathbb{F}_{q^\ell}[\![\zeta]\!]$. Let $(\mathcal{F}_i, \Pi_i, \tau_i)$ be an abelian sheaf of rank r and dimension d over S. Consider the completions $\widehat{\mathcal{F}}_i = \mathcal{F}_i \otimes_{\mathcal{O}_{C_S}} \mathcal{O}_S[\![z]\!]$ of the sheaves $\mathcal{F}_i$ at ∞. Using the periodicity $\mathcal{F}_\ell \cong \mathcal{F}_0(k \cdot \infty)$ we obtain morphisms

$$\Pi_i : \quad \widehat{\mathcal{F}}_i \to \widehat{\mathcal{F}}_{i+1} \quad \text{for all } i = 0, \ldots, \ell - 2, \quad z^k \Pi_{\ell-1} : \quad \widehat{\mathcal{F}}_{\ell-1} \to \widehat{\mathcal{F}}_0 \quad (7.1)$$

and

$$\tau_i : \sigma^* \widehat{\mathcal{F}}_i \to \widehat{\mathcal{F}}_{i+1} \quad \text{for all } i = 0, \ldots, \ell - 2, \quad z^k \tau_{\ell-1} : \sigma^* \widehat{\mathcal{F}}_{\ell-1} \to \widehat{\mathcal{F}}_0. \quad (7.2)$$

We set $\widehat{\mathcal{F}} = \widehat{\mathcal{F}}_0 \oplus \cdots \oplus \widehat{\mathcal{F}}_{\ell-1}$ and let the endomorphism $\Pi : \widehat{\mathcal{F}} \to \widehat{\mathcal{F}}$ be given by the morphisms (7.1). We let $\lambda \in \mathbb{F}_{q^\ell}$ act on $\widehat{\mathcal{F}}_i$ as the scalar $\beta^* \lambda^{q^i}$ and we let the σ-linear endomorphism $F : \sigma^* \widehat{\mathcal{F}} \to \widehat{\mathcal{F}}$ be given by the morphisms (7.2); i.e., Π, λ, and F are expressed by the block matrices

$$\Pi = \begin{pmatrix} 0 & \cdots & & z^k \Pi_{\ell-1} \\ \Pi_0 & \ddots & & \vdots \\ & \ddots & \ddots & \\ & & \Pi_{\ell-2} & 0 \end{pmatrix}, \qquad \lambda = \begin{pmatrix} \beta^*\lambda\,\mathrm{Id}_\ell & & & \\ & \beta^*\lambda^q\,\mathrm{Id}_\ell & & \\ & & \ddots & \\ & & & \beta^*\lambda^{q^{\ell-1}}\,\mathrm{Id}_\ell \end{pmatrix},$$

and

$$F = \begin{pmatrix} 0 & \cdots & & z^k \tau_{\ell-1} \\ \tau_0 & \ddots & & \vdots \\ & \ddots & \ddots & \\ & & \tau_{\ell-2} & 0 \end{pmatrix}.$$

In this way, $(\widehat{\mathcal{F}}, F)$ is a formal abelian sheaf of rank r and dimension d over S. By Theorem 7.12, we obtain a functor from $\mathcal{A}b\text{-}\mathcal{S}h^{r,d}(S)$ to the category of special z-divisible $\mathcal{O}_\Delta$-modules of height $r\ell$ and dimension $d\ell$ over S.

Moreover, every isogeny of abelian sheaves over S induces an isogeny of the associated formal abelian sheaves and an isogeny of the associated special z-divisible $\mathcal{O}_\Delta$-modules.

8 The Serre–Tate Theorem

Classically the Serre–Tate Theorem relates the deformation theory of an abelian variety in characteristic p to the deformation theory of its p-divisible group. In the case of abelian sheaves, the same principle prevails. We begin with the analogue of the following classical construction. Let A be a fixed abelian variety over a scheme $S \in \mathcal{N}ilp_{\mathbb{Z}_p}$ and let $A[p^\infty]$ be its p-divisible group. Then to every pair $(X, \widehat{\alpha})$ consisting of a p-divisible group X over S and an isogeny $\widehat{\alpha} : A[p^\infty] \to X$ of p-divisible groups there exists a uniquely determined abelian variety $\widehat{\alpha}_* A$ and a p-isogeny $\alpha : A \to \widehat{\alpha}_* A$ which induces $\widehat{\alpha}$ on p-divisible groups. This construction can be applied to abelian sheaves as well.

Proposition 8.1. *Let* $S \in \mathcal{N}ilp_{\mathbb{F}_{q^\ell}[\![\zeta]\!]}$ *and let* $\underline{\mathcal{F}}$ *be an abelian sheaf of rank* r *and dimension* d *over* S. *Let* $\widehat{\underline{\mathcal{F}}}$ *be the associated formal abelian sheaf and let* $\widehat{\alpha} : \widehat{\underline{\mathcal{F}}}' \to \widehat{\underline{\mathcal{F}}}$ *be a quasi-isogeny. Then there exists an abelian sheaf* $\underline{\mathcal{F}}'$ *and a quasi-isogeny* $\alpha : \underline{\mathcal{F}}' \to \underline{\mathcal{F}}$ *over* S *giving rise to* $\widehat{\alpha}$. *If we require that* α *is an isomorphism over* C' *then* $(\underline{\mathcal{F}}', \alpha)$ *is unique up to canonical isomorphism. In this case, we denote* $\underline{\mathcal{F}}'$ *by* $\widehat{\alpha}^* \underline{\mathcal{F}}$.

Proof. We denote by $\beta : S \to \mathrm{Spf}\,\mathbb{F}_{q^\ell}[\![\zeta]\!]$ the structure morphism of S. For each $i = 0, \ldots, \ell-1$ we extract from the sheaf $\widehat{\mathcal{F}}'$ of $\mathcal{O}_S[\![z]\!]$-modules underlying $\widehat{\underline{\mathcal{F}}}'$ the locally free subsheaf $\widehat{\mathcal{F}}_i' \subset \widehat{\mathcal{F}}'$ of rank r on which $\lambda \in \mathbb{F}_{q^\ell}$ acts through the character $\lambda \mapsto \beta^*\lambda^{q^i}$. The quasi-isogeny $\widehat{\alpha}$ gives an inclusion $\widehat{\mathcal{F}}_i' \hookrightarrow \mathcal{F}_i \otimes_{\mathcal{O}_{C_S'}} \mathcal{O}_S[\![z]\!][\frac{1}{z}]$. This permits us to glue $\widehat{\mathcal{F}}_i'$ with the corresponding sheaf $\mathcal{F}_i|_{C_S'}$ to obtain a locally free sheaf $\mathcal{F}_i'$ of rank r on C_S. For arbitrary $i \in \mathbb{Z}$ we take $n \in \mathbb{Z}$ such that $0 \leq i - n\ell < \ell$ and define

$$\mathcal{F}'_i := \mathcal{F}'_{i-n\ell} \otimes \mathcal{O}_{C_S}(nk \cdot \infty).$$

From the morphisms Π' and F' of $\widehat{\underline{\mathcal{F}}}'$ and Π_i and τ_i of $\underline{\mathcal{F}}$, we obtain the morphisms

$$\Pi'_i : \mathcal{F}'_i \to \mathcal{F}'_{i+1} \quad \text{and} \quad \tau'_i : \sigma^* \mathcal{F}'_i \to \mathcal{F}'_{i+1}.$$

The morphism $\Pi'_{\ell-1} : \mathcal{F}'_{\ell-1} \to \mathcal{F}'_\ell$ is induced from $z^{-k}\Pi' : \widehat{\mathcal{F}}'_{\ell-1} \to \widehat{\mathcal{F}}'_\ell(k \cdot \infty)$. The same applies to $\tau'_{\ell-1}$. These morphisms make $\widehat{\alpha}^* \underline{\mathcal{F}} := (\mathcal{F}'_i, \Pi'_i, \tau'_i)$ into an abelian sheaf of rank r and dimension d over S whose associated formal abelian sheaf is $\widehat{\underline{\mathcal{F}}}'$. By construction there is a quasi-isogeny $\alpha : \widehat{\alpha}^* \underline{\mathcal{F}} \to \underline{\mathcal{F}}$ which is an isomorphism over C' and induces the quasi-isogeny $\widehat{\alpha}$ on formal abelian sheaves. $\square$

Proposition 8.2. *Let $S \in \mathcal{N}ilp_{\mathbb{F}_{q^\ell}[\![\zeta]\!]}$ and let $j : \bar{S} \hookrightarrow S$ be a closed subscheme defined by a sheaf of ideals which is locally nilpotent. Let $\underline{\mathcal{F}}$ and $\underline{\mathcal{F}}'$ be two abelian sheaves of rank r and dimension d over S. Then every quasi-isogeny from $j^* \underline{\mathcal{F}}$ to $j^* \underline{\mathcal{F}}'$ lifts in a unique way to a quasi-isogeny from $\underline{\mathcal{F}}$ to $\underline{\mathcal{F}}'$.*

Proof. Let $\bar{\alpha} : j^* \underline{\mathcal{F}} \to j^* \underline{\mathcal{F}}'$ be a quasi-isogeny. It suffices to treat the case where the qth power of the ideal sheaf defining $\bar{S}$ is zero. In this case, the morphisms σ_S and $\sigma_{\bar{S}}$ factor through j,

$$\sigma_S = j \circ \bar{\sigma} : S \to \bar{S} \to S \quad \text{and} \quad \sigma_{\bar{S}} = \bar{\sigma} \circ j : \bar{S} \to S \to \bar{S}.$$

Consider the quasi-isogeny $\bar{\sigma}^* \bar{\alpha}[1] : \bar{\sigma}^* j^* \underline{\mathcal{F}}[1] \to \bar{\sigma}^* j^* \underline{\mathcal{F}}'[1]$, where $[1]$ denotes the shift by 1 (cf. Definition 1.7). We view the morphisms τ_i as an isogeny $(\tau_i) : \sigma^* \underline{\mathcal{F}}[1] \to \underline{\mathcal{F}}$ and obtain a commutative diagram

$$
\begin{array}{ccc}
\bar{\sigma}^* j^* \underline{\mathcal{F}}[1] & \xrightarrow{\bar{\sigma}^* \bar{\alpha}[1]} & \bar{\sigma}^* j^* \underline{\mathcal{F}}'[1] \\
{\scriptstyle (\tau_i)} \downarrow & & \downarrow {\scriptstyle (\tau'_i)} \\
\underline{\mathcal{F}} & \xrightarrow{\alpha} & \underline{\mathcal{F}}',
\end{array}
$$

which defines a quasi-isogeny α. Pulling back this diagram under j, we see that $j^* \alpha = \bar{\alpha}$. Moreover, the diagram shows that α is uniquely determined by $\bar{\alpha}$. This proves the proposition. $\square$

From the proof, we even see the following.

Corollary 8.3. *Keep the situation of the proposition.*

1. *If $\bar{\alpha} : j^* \underline{\mathcal{F}} \to j^* \underline{\mathcal{F}}'$ is an isogeny, then the lift is an isogeny $\alpha : \underline{\mathcal{F}} \to \underline{\mathcal{F}}'(n \cdot \infty)$ for some integer $n \geq 0$.*
2. *If $\bar{\alpha}$ is an isomorphism over C', then the same holds for α.*

Next, we come to the analogue of the Serre–Tate Theorem. Let $S \in \mathcal{N}ilp_{\mathbb{F}_{q^\ell}[\![\zeta]\!]}$ and let $j : S' \hookrightarrow S$ be a closed subscheme defined by a sheaf of ideals which is locally nilpotent. Let $\underline{\mathcal{F}}'$ be an abelian sheaf of rank r and dimension d over S' and let $\widehat{\underline{\mathcal{F}}}'$ be the associated formal abelian sheaf. The *category of lifts of $\underline{\mathcal{F}}'$ to S* has as objects   all pairs $(\underline{\mathcal{F}}, \alpha : j^*\underline{\mathcal{F}} \xrightarrow{\sim} \underline{\mathcal{F}}')$ where $\underline{\mathcal{F}}$ is an abelian sheaf over S and α an isomorphism of abelian sheaves over S',

as morphisms   isomorphisms between the $\underline{\mathcal{F}}$s that are compatible with the αs.

Similarly, we define the *category of lifts of* $\widehat{\underline{\mathcal{F}}}$ to S. By Propositions 8.2 and 6.4, all Hom-sets in these categories contain at most one element.

Theorem 8.4 (analogue of the Serre–Tate Theorem). *The category of lifts of* $\underline{\mathcal{F}}'$ *to* S *and the category of lifts of* $\widehat{\underline{\mathcal{F}}}'$ *to* S *are equivalent.*

Proof. Let $\widehat{}$ be the functor that assigns to a lift of $\underline{\mathcal{F}}'$ the corresponding lift of $\widehat{\underline{\mathcal{F}}}'$. Full faithfulness of $\widehat{}$ follows from Corollary 8.3. It remains to show that $\widehat{}$ is essentially surjective. So let $(\widehat{\underline{\mathcal{F}}}, \widehat{\alpha} : j^*\widehat{\underline{\mathcal{F}}} \xrightarrow{\sim} \widehat{\underline{\mathcal{F}}}')$ be a lift of $\widehat{\underline{\mathcal{F}}}'$ to S.

It suffices to treat the case where the qth power of the ideal sheaf defining S' is zero. In this case, the morphism σ_S factors through j,

$$\sigma_S = j \circ \sigma' : S \to S' \to S.$$

Consider the abelian sheaf $\widetilde{\underline{\mathcal{F}}} := (\sigma'^*\underline{\mathcal{F}}')[1]$ over S; i.e.,

$$(\widetilde{\mathcal{F}}_i, \widetilde{\Pi}_i, \widetilde{\tau}_i) = (\sigma'^*\mathcal{F}'_{i-1}, \sigma'^*\Pi'_{i-1}, \sigma'^*\tau'_{i-1}).$$

The morphisms τ'_i constitute an isogeny $\tau' := (\tau'_i) : j^*\widetilde{\underline{\mathcal{F}}} \to \underline{\mathcal{F}}'$ which is an isomorphism over C'. We let

$$\widehat{\gamma}' = \widehat{\tau}'^{-1} \circ \widehat{\alpha} : j^*\widehat{\underline{\mathcal{F}}} \to \widehat{\underline{\mathcal{F}}}' \to j^*\widehat{\widetilde{\underline{\mathcal{F}}}}$$

be the resulting quasi-isogeny of formal abelian sheaves. By Proposition 6.4 it lifts to a quasi-isogeny $\widehat{\gamma} : \widehat{\underline{\mathcal{F}}} \to \widehat{\widetilde{\underline{\mathcal{F}}}}$. We put $\underline{\mathcal{F}} := \widehat{\gamma}^*\widetilde{\underline{\mathcal{F}}}$ (Prop. 8.1) and we let $\gamma : \underline{\mathcal{F}} \to \widetilde{\underline{\mathcal{F}}}$ be the induced quasi-isogeny of abelian sheaves. Then $(\underline{\mathcal{F}}, \tau' \circ j^*\gamma)$ is the desired lift of $\underline{\mathcal{F}}$. $\qquad\square$

In the remainder of this section, we give an example for an abelian sheaf and we compute the associated formal abelian sheaf. This example will be crucial for the uniformization of $\mathcal{A}b\text{-}\mathcal{S}h_H^{r,d}$ in Part Three.

Example 8.5. On the scheme $\bar{S} = \operatorname{Spec} \mathbb{F}_q$ we set for $i = 0, \ldots, \ell$

$$\mathcal{M}_i = \mathcal{O}_{C_{\bar{S}}}(k \cdot \infty)^{\oplus i} \oplus \mathcal{O}_{C_{\bar{S}}}^{\oplus \ell - i}.$$

We let $\Pi_i : \mathcal{M}_i \to \mathcal{M}_{i+1}$ be the morphism coming from the natural inclusion $\mathcal{O}_{C_{\bar{S}}} \subset \mathcal{O}_{C_{\bar{S}}}(k \cdot \infty)$ in the $(i+1)$st summand. We define the σ-linear morphism

$$\tau_i = \begin{pmatrix} 0 \cdots & & 1 \\ 1 & \ddots & \vdots \\ & \ddots & \vdots \\ & & 1 & 0 \end{pmatrix} \cdot \sigma^* : \sigma^*\mathcal{M}_i \to \mathcal{M}_{i+1}.$$

For arbitrary $i \in \mathbb{Z}$, we take $n \in \mathbb{Z}$ such that $0 \leq i - n\ell < \ell$ and set

$$\mathcal{M}_i := \mathcal{M}_{i-n\ell} \otimes \mathcal{O}_{C_{\bar{S}}}(nk \cdot \infty), \qquad \Pi_i := \Pi_{i-n\ell} \otimes \mathrm{id}, \qquad \tau_i := \tau_{i-n\ell} \otimes \mathrm{id}.$$

Then $\underline{\mathcal{M}} := (\mathcal{M}_i, \Pi_i, \tau_i)$ is an abelian sheaf of rank ℓ and dimension k over $\mathrm{Spec}\,\mathbb{F}_q$. We let $e = \frac{r}{\ell}$ and set $\overline{\mathbb{M}} = \underline{\mathcal{M}}^{\oplus e}$. This is an abelian sheaf of rank r and dimension d. We compute the formal abelian sheaf $\widehat{\overline{\mathbb{M}}}$ associated to $\overline{\mathbb{M}}$.

In order to do this, we have to extend the base scheme $\bar{S}$ to $\bar{S}' = \mathrm{Spec}\,\mathbb{F}_{q^\ell}$. Since we afterwards want to lift $\widehat{\overline{\mathbb{M}}}$ to $S' = \mathrm{Spf}\,\mathbb{F}_{q^\ell}[\![\zeta]\!]$, we describe this lift right away. For $i = 0, \ldots, \ell - 1$, we consider the $\ell \times \ell$-matrices

$$\Pi_i = \begin{pmatrix} 1 & & & & \\ & \ddots & & & \\ & & z^k & & \\ & & & \ddots & \\ & & & & 1 \end{pmatrix} \quad \text{and} \quad T = \begin{pmatrix} 0 & \cdots & & (z-\zeta)^k \\ 1 & \ddots & & \vdots \\ & \ddots & \ddots & \vdots \\ & & 1 & 0 \end{pmatrix}, \qquad (8.1)$$

where the z^k in Π_i sits in the $(i+1)$st row. We let the formal abelian sheaf $\widehat{\mathcal{M}}$ of rank ℓ and dimension k be the $\mathcal{O}_{S'}[\![z]\!]$-module $\widehat{\mathcal{M}} = \mathcal{O}_{S'}[\![z]\!]^{\ell^2}$ together with the morphisms

$$\Pi = \begin{pmatrix} 0 & \cdots & & \Pi_{\ell-1} \\ \Pi_0 & \ddots & & \vdots \\ & \ddots & \ddots & \vdots \\ & & \Pi_{\ell-2} & 0 \end{pmatrix}, \qquad \lambda = \begin{pmatrix} \lambda\,\mathrm{Id}_r & & & \\ & \lambda^q\,\mathrm{Id}_r & & \\ & & \ddots & \\ & & & \lambda^{q^{\ell-1}}\,\mathrm{Id}_r \end{pmatrix} : \widehat{\mathcal{M}} \to \widehat{\mathcal{M}},$$

and

$$F = \begin{pmatrix} 0 & \cdots & & T \\ T & \ddots & & \vdots \\ & \ddots & \ddots & \vdots \\ & & T & 0 \end{pmatrix} \cdot \sigma^* : \sigma^* \widehat{\mathcal{M}} \to \widehat{\mathcal{M}}.$$

We set $\widehat{\mathbb{M}} = \widehat{\mathcal{M}}^{\oplus e}$. If $j : \bar{S}' \hookrightarrow S'$ denotes the inclusion, then $j^* \widehat{\mathbb{M}}$ is the formal abelian sheaf associated to $\overline{\mathbb{M}}$. We denote it by $\widehat{\overline{\mathbb{M}}}$. Via Theorem 7.12 we obtain from $\widehat{\overline{\mathbb{M}}}$ a special z-divisible $\mathcal{O}_\Delta$-module $\overline{\mathbb{E}}$ over $\mathrm{Spec}\,\mathbb{F}_{q^\ell}$ whose Dieudonné $\mathbb{F}_q[\![z]\!]$-module is $\widehat{\overline{\mathbb{M}}}$. One easily checks $F^\ell = z^k \cdot \sigma^{\ell*}$ on $\widehat{\overline{\mathbb{M}}}$. In the terminology of [23] this means that $\overline{\mathbb{E}}$ is *descent*. The Newton-polygon of $\widehat{\overline{\mathbb{M}}}$ is a straight line between the end points $(0,0)$ and $(r\ell, d\ell)$.

Via the Serre–Tate Theorem (Theorem 8.4), we obtain from the lift $\widehat{\mathbb{M}}$ of the formal abelian sheaf of $\overline{\mathbb{M}}$ an abelian sheaf $\mathbb{M}$ over $\mathrm{Spf}\,\mathbb{F}_{q^\ell}$ which lifts $\overline{\mathbb{M}}$.

Let us compute the group of quasi-isogenies of these (formal) abelian sheaves.

Proposition 8.6. *Fix a generator $\lambda \in \mathbb{F}_{q^\ell}$ of the extension $\mathbb{F}_{q^\ell}/\mathbb{F}_q$. Let V be the Vandermonde matrix*

$$V = \begin{pmatrix} 1 & \lambda & \lambda^2 & \dots & \lambda^{\ell-1} \\ 1 & \lambda^q & \lambda^{2q} & \dots & \lambda^{(\ell-1)q} \\ \vdots & \vdots & \vdots & & \vdots \\ 1 & \lambda^{q^{\ell-1}} & \lambda^{2q^{\ell-1}} & \dots & \lambda^{(\ell-1)q^{\ell-1}} \end{pmatrix} \quad and \quad V_e = \begin{pmatrix} V & & \\ & \ddots & \\ & & V \end{pmatrix} \quad (8.2)$$

be the block diagonal matrix of dimension r. Then the group of quasi-isogenies of $\overline{\mathbb{M}}$ over $\mathbb{F}_q^{\mathrm{alg}}$ is the group of Q-valued points of the algebraic group $J = V_e \, \mathrm{GL}_r \, V_e^{-1}$ over Q.

Proof. Let $g \in J(Q)$ and let $D \subset C$ be an effective divisor satisfying $\mathrm{div}(g_{\mu\nu}) \geq -D$ for all entries of g. Then multiplication with g defines morphisms $\iota^*\mathcal{M}_i^{\oplus e} \to \iota^*\mathcal{M}_i^{\oplus e}(D)$ which form a quasi-isogeny of $\overline{\mathbb{M}}$ (defined over $\mathrm{Spec}\,\mathbb{F}_{q^\ell}$). Conversely, one computes that every quasi-isogeny of $\overline{\mathbb{M}}$ arises in this way. □

If $T \in \mathcal{N}ilp_{\mathbb{F}_{q^\ell}[[\zeta]]}$ is a scheme, we let $\beta : T \to \mathrm{Spf}\,\mathbb{F}_{q^\ell}[[\zeta]]$ be its structure morphism and $\bar{\beta} : \bar{T} \to \mathrm{Spec}\,\mathbb{F}_{q^\ell}$ be the reduction modulo ζ.

Proposition 8.7. *Let* $T \in \mathcal{N}ilp_{\mathbb{F}_{q^\ell}[[\zeta]]}$ *be connected. Then there is a canonical isomorphism* $g \mapsto g_T$ *of* $J(Q)$ *to the group* $\mathrm{QIsog}_T(\beta^*\mathbb{M})$ *of quasi-isogenies of* $\beta^*\mathbb{M}$ *over* T.

Proof. The map $g \mapsto g_T$ is defined as

$$J(Q) \longrightarrow \mathrm{QIsog}_{\bar{T}}(\bar{\beta}^*\overline{\mathbb{M}}) \overset{\sim}{\longrightarrow} \mathrm{QIsog}_T(\beta^*\mathbb{M})$$
$$g \longmapsto \bar{\beta}^*g \longmapsto g_T,$$

the last isomorphism coming from Proposition 8.2. Clearly, this defines a monomorphism of groups.

We show that $g \mapsto \bar{\beta}^*g$ is surjective. Let $\alpha : \overline{\mathbb{M}}_{\bar{T}} \to \overline{\mathbb{M}}_{\bar{T}}(D)$ be an isogeny for some effective divisor $D \subset C$. There exists an $a \in A$ whose divisor is $\geq D$ on C'. We view $\alpha|_{C'_{\bar{T}}}$ as a matrix $U \in M_r(\mathcal{O}_{\bar{T}} \otimes A[\frac{1}{a}])$. Then the matrix $V_e^{-1}UV_e$ satisfies

$$^\sigma(V_e^{-1}UV_e) = V_e^{-1}UV_e$$

and hence lies in $\mathrm{GL}_r(Q)$. We conclude that $U \in J(Q)$ and $\bar{\beta}^*U = \alpha$. □

Proposition 8.8. *The group of quasi-isogenies of the formal abelian sheaf* $\widehat{\mathbb{M}}$ *is isomorphic to* $J(Q_\infty)$.

Proof. A straightforward calculation shows that this isomorphism can be described as follows. Let

$$W_i = \begin{pmatrix} z^k \, \mathrm{Id}_i & 0 \\ 0 & \mathrm{Id}_{\ell-i} \end{pmatrix} \quad and \quad W = \begin{pmatrix} g & & \\ & W_1 g W_1^{-1} & \\ & & \ddots \\ & & & W_{\ell-1} g W_{\ell-1}^{-1} \end{pmatrix}.$$

Then an element $\in (Q_\infty)$ is mapped to the quasi-isogeny $W^{\oplus e}$ of $\widehat{\mathcal{M}}$. □

9 Moduli spaces for z-divisible groups

Consider the formal abelian sheaf $\widehat{\mathbb{M}}$ from Example 8.5. We will define a moduli problem for formal abelian sheaves which are quasi-isogenous to $\widehat{\mathbb{M}}$. This is a higher dimensional variant of a moduli problem studied by Drinfeld [11]. At the same time it is a close analogue of a moduli problem for p-divisible groups considered by Rapoport–Zink [36]. Like these two problems our moduli problem too will be solved by a formal scheme over $\mathrm{Spf}\,\mathbb{F}_{q^\ell}[\![\zeta]\!]$. Following [11, 36] we will use this formal moduli scheme in Section 12 to (partly) uniformize the stacks of abelian sheaves.

For a scheme S in $\mathcal{N}ilp_{\mathbb{F}_q[\![\zeta]\!]}$, we denote by $\bar{S}$ the closed subscheme defined by the sheaf of ideals $\zeta\mathcal{O}_S$. We call $\bar{S}$ the special fiber of S. If E is a z-divisible group over S we denote by $E_{\bar{S}} = E \times_S \bar{S}$ the base change to the special fiber. A similar notation will be applied to Dieudonné $\mathbb{F}_q[\![z]\!]$-modules, etc. If $\beta : S \to \mathrm{Spf}\,\mathbb{F}_{q^\ell}[\![\zeta]\!]$ is a morphism of formal schemes we denote by $\bar{\beta} : \bar{S} \to \mathrm{Spec}\,\mathbb{F}_{q^\ell}$ its restriction to special fibers. We define the following moduli problem for formal abelian sheaves.

Definition 9.1. *Let G be the contravariant functor* $\mathcal{N}ilp_{\mathbb{F}_q[\![\zeta]\!]} \longrightarrow Sets$,

$$S \longmapsto \{Isomorphism\ classes\ of\ triples\ (\beta, \widehat{\mathcal{F}}, \widehat{\alpha}),\ where$$

- $\beta : S \to \mathrm{Spf}\,\mathbb{F}_{q^\ell}[\![\zeta]\!]$ *is a morphism of formal schemes,*

- $\widehat{\mathcal{F}}$ *is a formal abelian sheaf of rank r and dimension d over S,*

- $\widehat{\alpha} : \widehat{\mathcal{F}}_{\bar{S}} \to \bar{\beta}^*\widehat{\overline{\mathbb{M}}}$ *is a quasi-isogeny of formal abelian sheaves.*$\}$

Thereby two triples $(\beta, \widehat{\mathcal{F}}, \widehat{\alpha})$ and $(\beta', \widehat{\mathcal{F}}', \widehat{\alpha}')$ are isomorphic if $\beta = \beta'$ and if there is an isomorphism between $\widehat{\mathcal{F}}$ and $\widehat{\mathcal{F}}'$ over S which is compatible with $\widehat{\alpha}$ and $\widehat{\alpha}'$.

Using the equivalence between special z-divisible $\mathcal{O}_\Delta$-modules and formal abelian sheaves from Theorem 7.12 we see that

Proposition 9.2. *G can be described as the functor* $\mathcal{N}ilp_{\mathbb{F}_q[\![\zeta]\!]} \longrightarrow Sets$,

$$S \longmapsto \{Isomorphism\ classes\ of\ triples\ (\beta, E, \rho),\ where$$

- $\beta : S \to \mathrm{Spf}\,\mathbb{F}_{q^\ell}[\![\zeta]\!]$ *is a morphism of formal schemes,*

- E *is a special z-divisible $\mathcal{O}_\Delta$-module of height $r\ell$ and dimension $d\ell$*
  over S,

- $\rho : \bar{\beta}^*\overline{\mathbb{E}} \to E_{\bar{S}}$ *is a quasi-isogeny of z-divisible $\mathcal{O}_\Delta$-modules.*$\}$

Sending $(\beta, \widehat{\mathcal{F}}, \widehat{\alpha})$ to β gives a morphism $G \to \mathrm{Spf}\,\mathbb{F}_{q^\ell}[\![\zeta]\!]$. We define an action of the Galois group $\mathrm{Gal}(\mathbb{F}_{q^\ell}/\mathbb{F}_q)$ on G over $\mathrm{Spf}\,\mathbb{F}_{q^\ell}[\![\zeta]\!]$. Namely, for each $\pi^* \in \mathrm{Gal}(\mathbb{F}_{q^\ell}/\mathbb{F}_q) = \mathrm{Gal}(\mathbb{F}_{q^\ell}[\![\zeta]\!]/\mathbb{F}_q[\![\zeta]\!])$, consider the cartesian square

$$
\begin{array}{ccc}
S^\pi & \xrightarrow{\ \pi_S\ } & S \\
\downarrow{\scriptstyle \pi^*\beta} & & \downarrow{\scriptstyle \beta} \\
\mathrm{Spf}\,\mathbb{F}_{q^\ell}[\![\zeta]\!] & \xrightarrow{\ \pi\ } & \mathrm{Spf}\,\mathbb{F}_{q^\ell}[\![\zeta]\!].
\end{array}
$$

Then we let π^* act by mapping the element $(\beta, \widehat{\underline{\mathcal{F}}}, \widehat{\alpha}) \in G(S)$ to

$$(\pi^*\beta, \pi_S^*\widehat{\underline{\mathcal{F}}}, \pi_S^*\widehat{\alpha}) \in G(S^\pi).$$

Following the arguments given in Rapoport–Zink [36] in the case of p-divisible groups, one can prove the following representability theorem. Recall that an adic formal scheme G over $\mathrm{Spf}\ \mathbb{F}_{q^\ell}[\![\zeta]\!]$ is called *locally formally of finite type* if G_{red} is locally of finite type over $\mathrm{Spec}\ \mathbb{F}_{q^\ell}$.

Theorem 9.3. *The functor that assigns to a scheme $S \in \mathcal{N}ilp_{\mathbb{F}_q[\![\zeta]\!]}$ the set of isomorphism classes of triples (β, E, ρ), where*

- $\beta : S \to \mathrm{Spf}\ \mathbb{F}_{q^\ell}[\![\zeta]\!]$ *is a morphism of formal schemes,*
- E *is a z-divisible group of dimension $d\ell$ and height $r\ell$ over S,*
- $\rho : \bar{\beta}^*\mathbb{E} \to E_{\bar{S}}$ *is a quasi-isogeny of z-divisible groups,*

is representable by a quasi-separated, locally noetherian, adic formal scheme which is locally formally of finite type over $\mathrm{Spf}\ \mathbb{F}_{q^\ell}[\![\zeta]\!]$.

The proof of this theorem is given in [23]. It makes use of Dieudonné $\mathbb{F}_q[\![z]\!]$-modules which replace the crystals of the p-divisible groups in [36].

Corollary 9.4. *The functor G is representable by a quasi-separated, locally noetherian, adic formal scheme which is locally formally of finite type over $\mathrm{Spf}\ \mathbb{F}_{q^\ell}[\![\zeta]\!]$.*

Proof. Let $\widetilde{G}$ be the formal scheme whose existence is stated in Theorem 9.3. Let E be the universal z-divisible group over $\widetilde{G}$. We transport the $\mathcal{O}_\Delta$-action from $\mathbb{E}$ to E via ρ. Then G is the closed formal subscheme of $\widetilde{G}$ on which E is special and $\mathcal{O}_\Delta$ acts through isogenies (Propositions 6.5 and 6.8). $\qquad\square$

9.5. We define an action of the group $J(Q_\infty)$ on G. By Proposition 8.8 there is an isomorphism ε_∞ from the group $J(Q_\infty)$ to the group of quasi-isogenies of $\widehat{\mathbb{M}}$. We let $g \in J(Q_\infty)$ act on G through ε_∞,

$$(\beta, \widehat{\underline{\mathcal{F}}}, \widehat{\alpha}) \mapsto (\beta, \widehat{\underline{\mathcal{F}}}, \bar{\beta}^*\varepsilon_\infty(g) \circ \widehat{\alpha}).$$

This action commutes with the Galois action on G.

Let now $\Gamma \subset J(Q_\infty)$ be a discrete subgroup. We say that Γ is *separated* if it is separated in the profinite topology. This means that for every $g \in \Gamma$ there is a normal subgroup $\Gamma' \subset \Gamma$ of finite index that does not contain g.

Again following the arguments of Rapoport–Zink, we prove the following in [23].

Theorem 9.6. *Let $\Gamma \subset J(Q_\infty)$ be a separated discrete subgroup. Then the quotient $\Gamma \backslash G$ is a locally noetherian, adic formal algebraic $\mathrm{Spf}\ \mathbb{F}_{q^\ell}[\![\zeta]\!]$-stack locally formally of finite type over $\mathrm{Spf}\ \mathbb{F}_{q^\ell}$. Moreover, the 1-morphism $G \to \Gamma \backslash G$ is adic.*

See A.5–A.9 in the appendix for an explanation of this statement.

Part Three: Uniformization

We now turn towards the uniformization of the stacks $Ab\text{-}Sh_H^{r,d}$ at ∞. Therefore, we again consider their base change

$$Ab\text{-}Sh_H^{r,d} \times_C \operatorname{Spf} \mathbb{F}_q[\![\zeta]\!].$$

As remarked in Part Two these are *formal algebraic* $\operatorname{Spf} \mathbb{F}_q[\![\zeta]\!]$-*stacks* (Definition A.5). In analogy with the work of Rapoport–Zink [36] and Drinfeld [11], the space used to uniformize these stacks will be the formal scheme G. We view it as a formal algebraic $\operatorname{Spf} \mathbb{F}_{q^\ell}[\![\zeta]\!]$-stack (Example A.10). The uniformization we obtain will only be partial. To be precise, we find a closed subset Z in $Ab\text{-}Sh_H^{r,d} \times_C \infty$ and we consider the formal completion $Ab\text{-}Sh_H^{r,d}/Z$ of $Ab\text{-}Sh_H^{r,d}$ along Z. It is a formal algebraic $\operatorname{Spf} \mathbb{F}_q[\![\zeta]\!]$-stack (Proposition A.14). We will uniformize $Ab\text{-}Sh_H^{r,d}/Z$. This uniformization therefore takes place in the 2-category of formal algebraic $\operatorname{Spf} \mathbb{F}_q[\![\zeta]\!]$-stacks.

10 Algebraizations

We first give still another interpretation of the moduli space G. Namely, since the formal abelian sheaf $\widehat{\overline{M}}$ comes from the abelian sheaf $\overline{M}$, the universal formal abelian sheaf on G and its quasi-isogeny $\widehat{\alpha}$ to $\widehat{\overline{M}}$ can be algebraized; i.e., they too come from an abelian sheaf, namely, from $\widehat{\alpha}^*\overline{M}$. Now consider a scheme $S \in \mathcal{N}ilp_{\mathbb{F}_q[\![\zeta]\!]}$ and denote by $\bar{S}$ the closed subscheme of S defined by $\zeta = 0$.

Definition 10.1. *Let G' be the contravariant functor $\mathcal{N}ilp_{\mathbb{F}_q[\![\zeta]\!]} \longrightarrow \mathcal{S}ets$.*

$S \longmapsto \big\{$*Isomorphism classes of pairs* $(\underline{\mathcal{F}}, \alpha)$, *where*

- $\underline{\mathcal{F}}$ *is an abelian sheaf of rank r and dimension d over S,*

- $\alpha : \underline{\mathcal{F}}_{\bar{S}} \to \overline{M}_{\bar{S}}$ *is a quasi-isogeny which is an isomorphism over $C'\big\}$.*

Thereby two such pairs $(\underline{\mathcal{F}}, \alpha)$ and $(\underline{\mathcal{F}}', \alpha')$ are isomorphic if there is an isomorphism between $\underline{\mathcal{F}}$ and $\underline{\mathcal{F}}'$ over S which is compatible with α and α'.

Theorem 10.2. *The functors G and $G' \times_{\operatorname{Spf} \mathbb{F}_q[\![\zeta]\!]} \operatorname{Spf} \mathbb{F}_{q^\ell}[\![\zeta]\!]$ are canonically isomorphic as $\operatorname{Gal}(\mathbb{F}_{q^\ell}/\mathbb{F}_q)$-modules (where $\operatorname{Gal}(\mathbb{F}_{q^\ell}/\mathbb{F}_q)$ acts trivially on G').*

Proof. We will exhibit two mutually inverse maps between these two functors. We start by describing the morphism $G' \times_{\operatorname{Spf} \mathbb{F}_q[\![\zeta]\!]} \operatorname{Spf} \mathbb{F}_{q^\ell}[\![\zeta]\!] \to G$.

So let $(\underline{\mathcal{F}}, \alpha) \in G'(S)$ and $\beta : S \to \operatorname{Spf} \mathbb{F}_{q^\ell}[\![\zeta]\!]$. By Construction 7.13, we obtain from $\underline{\mathcal{F}}$ and β a formal abelian sheaf $\widehat{\underline{\mathcal{F}}}$ of rank r and dimension d over S. The quasi-isogeny α induces a quasi-isogeny of formal abelian sheaves $\widehat{\alpha} : \widehat{\underline{\mathcal{F}}}_{\bar{S}} \to \beta^*\widehat{\overline{M}}$. The triple $(\beta, \widehat{\underline{\mathcal{F}}}, \widehat{\alpha})$ defines an S-valued point of the functor G. One easily checks that the morphism just constructed is $\operatorname{Gal}(\mathbb{F}_{q^\ell}/\mathbb{F}_q)$-equivariant.

Conversely, let $(\beta, \widehat{\underline{\mathcal{F}}}, \widehat{\alpha}) \in G(S)$. By Proposition 6.4 there is a unique lift of $\widehat{\widehat{\alpha}}$ to a quasi-isogeny $\widehat{\alpha} : \widehat{\underline{\mathcal{F}}} \to \beta^*\widehat{\mathbb{M}}$. From Proposition 8.1 we obtain an abelian sheaf $\underline{\mathcal{F}} := \widehat{\alpha}^*(\beta^*\mathbb{M})$ whose formal abelian sheaf is $\widehat{\underline{\mathcal{F}}}$, and a quasi-isogeny $\alpha : \underline{\mathcal{F}} \to \beta^*\mathbb{M}$ which is an isomorphism over C' and which induces $\widehat{\alpha}$ on the formal abelian sheaves. We have $(\underline{\mathcal{F}}, \alpha_{\bar{s}}) \in G'(S)$. Thus we have constructed a morphism $G \to G'$. Again one checks that this morphism is $\mathrm{Gal}(\mathbb{F}_{q^\ell}/\mathbb{F}_q)$-invariant. The two morphisms just described are mutually inverse and yield the desired isomorphism between G and $G' \times_{\mathrm{Spf}\,\mathbb{F}_q[\![\zeta]\!]} \mathrm{Spf}\,\mathbb{F}_{q^\ell}[\![\zeta]\!]$. $\qquad\square$

The action of $J(Q_\infty)$ on G from 9.5 induces an action of $J(Q_\infty)$ on G', since it is compatible with the Galois action on G.

Definition 10.3. *The pair* $(\underline{\mathcal{F}}, \alpha) \in G'(S)$ *which is associated to an element* $(\beta, \widehat{\underline{\mathcal{F}}}, \widehat{\alpha}) \in G(S)$ *by Theorem* 10.2 *will be called the* algebraization *of* $(\beta, \widehat{\underline{\mathcal{F}}}, \widehat{\alpha})$.

Example 10.4. We want to explain how a quasi-isogeny $\alpha : \underline{\mathcal{F}}_{\bar{s}} \to \overline{\mathbb{M}}_{\bar{s}}$ which is an isomorphism over C' induces H-level structures on $\underline{\mathcal{F}}$ for compact open subgroups $H \subset \mathrm{GL}_r(\mathbb{A}_f)$.

Consider the abelian sheaf $\overline{\mathbb{M}} = (\mathcal{M}_i^{\oplus e}, \Pi_i^{\oplus e}, \tau_i^{\oplus e})$ from Example 8.5 pulled back to $\bar{S} = \mathrm{Spec}\,\mathbb{F}_{q^\ell}$. The restrictions $\mathcal{M}_i^{\oplus e}|_{C'_{\bar{S}}}$ are all isomorphic via the morphisms Π_i. We denote this restriction by $\overline{\mathbb{M}}|_{C'_{\bar{S}}}$. The same holds for the morphisms τ_i. So we obtain a morphism

$$\tau|_{C'_{\bar{S}}} : \sigma^*\overline{\mathbb{M}}|_{C'_{\bar{S}}} \to \overline{\mathbb{M}}|_{C'_{\bar{S}}}.$$

Via the canonical identification $\overline{\mathbb{M}}|_{C'_{\bar{S}}} = \mathcal{O}^r_{C'_{\bar{S}}}$ this morphism is expressed by the block diagonal matrix

$$\tau|_{C'_{\bar{S}}} = \begin{pmatrix} T & & \\ & \ddots & \\ & & T \end{pmatrix} \cdot \sigma^*, \quad \text{where} \quad T = \begin{pmatrix} 0 & \cdots & & 1 \\ 1 & \ddots & & \vdots \\ & \ddots & \ddots & \\ & & 1 & 0 \end{pmatrix} \in \mathrm{GL}_\ell(\mathbb{F}_q).$$

Therefore, multiplication with the matrix V_e from (8.2) defines an isomorphism

$$\psi : (\overline{\mathbb{M}}|_{C'_{\bar{S}}}, \tau|_{C'_{\bar{S}}}) \xrightarrow{\sim} (\mathcal{O}^r_{C'_{\bar{S}}}, \mathrm{Id}_r \cdot \sigma^*)$$

that commutes with the σ-linear endomorphisms $\tau|_{C'_{\bar{S}}}$ on one and $\mathrm{Id}_r \cdot \sigma^*$ on the other side. This ψ induces an isomorphism

$$\gamma = (\psi|_{\mathbb{A}_f})^\tau : (\overline{\mathbb{M}}|_{\mathbb{A}_f})^\tau \xrightarrow{\sim} \mathbb{A}^r_f$$

which gives rise to an H-level structure on $\overline{\mathbb{M}}$.

Now let $S \in \mathcal{N}ilp_{\mathbb{F}_q[\![\zeta]\!]}$ be an arbitrary scheme equipped with a fixed morphism $\beta : S \to \mathrm{Spf}\,\mathbb{F}_{q^\ell}[\![\zeta]\!]$. Let $\underline{\mathcal{F}}$ be an abelian sheaf over S and let $\alpha : \underline{\mathcal{F}}_{\bar{s}} \to \overline{\mathbb{M}}_{\bar{s}}$ be a quasi-isogeny which is an isomorphism over C'. For an algebraically closed base point $\iota : s \to S$ the composition of α with $\beta^*\psi$ gives rise to an isomorphism

$$\gamma \circ (\alpha|_{\mathbb{A}_f})^\tau := ((\beta^*\psi \circ \alpha)|_{\mathbb{A}_f})^\tau : (\iota^*\underline{\mathcal{F}}|_{\mathbb{A}_f})^\tau(s) \xrightarrow{\sim} \mathbb{A}_f^r$$

which is fixed under the action of $\pi_1(S, s)$ on the source. This induces an H-level structure on $\underline{\mathcal{F}}$.

11 Closedness of the uniformizable locus

In this section we show that an abelian sheaf of rank r and dimension d over an algebraically closed field in $\mathcal{N}ilp_{\mathbb{F}_{q^\ell}[\![\zeta]\!]}$ is isogenous to $\mathbb{M}$ if and only if its formal abelian sheaf is isoclinic. The locus of these points inside $\mathcal{A}b\text{-}\mathcal{S}h_H^{r,d} \times_C \text{Spf } \mathbb{F}_{q^\ell}[\![\zeta]\!]$ will be the one uniformized by G. We prove that this locus is formally closed. A couple of lemmas are needed beforehand.

Lemma 11.1. *Let $\underline{\mathcal{F}}$ be an abelian sheaf of rank r and dimension d over a finite field in $\mathcal{N}ilp_{\mathbb{F}_{q^\ell}[\![\zeta]\!]}$. Assume that the formal abelian sheaf associated to $\underline{\mathcal{F}}$ is isoclinic. Then for some integer $n > 0$ divisible by r there is an isomorphism $(\sigma^n)^*\underline{\mathcal{F}} \cong \underline{\mathcal{F}}$ which maps the isogeny*

$$T := (\tau_i) \circ \sigma^*(\tau_i) \circ \cdots \circ (\sigma^{n-1})^*(\tau_i) : (\sigma^n)^*\underline{\mathcal{F}}[n] \to \underline{\mathcal{F}}$$

to the isogeny $(\Pi_i)^n : \underline{\mathcal{F}}[n] \to \underline{\mathcal{F}}$.

Proof. It suffices to prove the assertion for an abelian sheaf isogenous to $\underline{\mathcal{F}}$. By Proposition 7.9, the formal abelian sheaf associated to $\underline{\mathcal{F}}$ is isogenous to a formal abelian sheaf $(\widehat{\mathcal{F}}, F)$ which satisfies $\text{im } F^r = z^d\widehat{\mathcal{F}}$. Pulling back $\underline{\mathcal{F}}$ along this isogeny (Proposition 8.1), we can assume that $(\widehat{\mathcal{F}}, F)$ is the formal abelian sheaf associated to $\underline{\mathcal{F}}$. Since $\underline{\mathcal{F}}$ is defined over a finite field, we certainly obtain $(\sigma^n)^*\underline{\mathcal{F}} \cong \underline{\mathcal{F}}$ for a suitable $n > 0$. We may even assume that r divides n. The isogeny T is an isomorphism over C'. Now the equation $\text{im } F^r = z^d\widehat{\mathcal{F}}$ implies that the image of T is $\underline{\mathcal{F}}(-\frac{nk}{\ell} \cdot \infty)$. Hence T factors as $T = \alpha \circ (\Pi_i)^n$ for an automorphism α of $\underline{\mathcal{F}}$. Since the automorphism group of $\underline{\mathcal{F}}$ is finite by Proposition 1.5, the lemma follows. $\square$

Definition 11.2. *Let $\underline{\mathcal{F}}$ and $\underline{\mathcal{F}}'$ be abelian sheaves of rank r and dimension d over a field K in $\mathcal{N}ilp_{\mathbb{F}_{q^\ell}[\![\zeta]\!]}$. We define the Q-vector space*

$$\text{Hom}_K^0(\underline{\mathcal{F}}, \underline{\mathcal{F}}') = \varinjlim_D \text{Hom}_K(\underline{\mathcal{F}}, \underline{\mathcal{F}}'(D)),$$

where the limit is taken over all effective divisors $D \subset C$. Furthermore, for formal abelian sheaves $\widehat{\mathcal{F}}$ and $\widehat{\mathcal{F}}'$ over K, we define the Q_∞-vector space

$$\text{Hom}_K^0(\widehat{\mathcal{F}}, \widehat{\mathcal{F}}') = \text{Hom}_K(\widehat{\mathcal{F}}, \widehat{\mathcal{F}}') \otimes_{\mathbb{F}_q[\![z]\!]} Q_\infty.$$

Note that each time the invertible elements in Hom^0 are precisely the quasi-isogenies.

Corollary 11.3. *Let $\mathcal{F}$ and $\mathcal{F}'$ be abelian sheaves of rank r and dimension d over a finite field K in $\mathcal{N}ilp_{\mathbb{F}_{q^\ell}[\![\zeta]\!]}$. Assume that their associated formal abelian sheaves $\widehat{\mathcal{F}}$ and $\widehat{\mathcal{F}}'$ are isoclinic. Then for a suitable finite extension K'/K, we have an isomorphism*

$$\mathrm{Hom}^0_{K'}(\mathcal{F}, \mathcal{F}') \otimes_Q Q_\infty \xrightarrow{\sim} \mathrm{Hom}^0_{K^{\mathrm{alg}}}(\widehat{\mathcal{F}}, \widehat{\mathcal{F}}').$$

Proof. We first assume that $C = \mathbb{P}^1_{\mathbb{F}_q}$. Let $\mathrm{Spec}\, K[z] \subset \mathbb{P}^1_K$ be a neighborhood of $\infty = V(z)$. Restricted to this neighborhood, all the sheaves $\mathcal{F}_i$ and $\mathcal{F}'_i$ are free of rank r. We fix bases for $i = 0, \ldots, \ell - 1$ and consider for the other i the bases induced by the periodicity $\mathcal{F}_i = \mathcal{F}_{i-\ell}(k \cdot \infty)$. Then the morphism

$$\bigoplus_{i=0}^{\ell-1} \tau_i : \bigoplus_{i=0}^{\ell-1} \sigma^* \mathcal{F}_i \longrightarrow \bigoplus_{i=1}^{\ell} \mathcal{F}_i$$

is represented by a matrix $U \in M_{r\ell}(K[z])$. We endow the formal abelian sheaf $(\widehat{\mathcal{F}}, F)$ of $\mathcal{F}$ with the induced basis. With respect to this basis, F is of the form $F = \widehat{U} \cdot \sigma^*$ for the matrix

$$\widehat{U} = \begin{pmatrix} 0 & \cdots & \mathrm{Id}_r \\ \mathrm{Id}_r & \ddots & \vdots \\ & \ddots & \ddots & \vdots \\ & & \mathrm{Id}_r & 0 \end{pmatrix} \cdot U = \sum_{\nu=0}^{N} \widehat{U}_\nu z^\nu \in M_{r\ell}(K[z]).$$

The same holds for $\mathcal{F}'$, where we denote the corresponding matrix by $\widehat{U}'$. Let us for a moment forget the structure of the (formal) abelian sheaves that is given by the Πs. Let $n = rm$ be the integer from Lemma 11.1 and let K' be the compositum of $\mathbb{F}_{q^n}$ and K inside K^{alg}. We claim that there is an isomorphism of Q_∞-vector spaces

$$\{\Phi \in M_{r\ell}(Q \otimes_{\mathbb{F}_q} K') : \Phi\widehat{U} = \widehat{U}'^\sigma \Phi\} \otimes_Q Q_\infty \qquad (11.1)$$
$$\xrightarrow{\sim} \{\Phi \in M_{r\ell}(K^{\mathrm{alg}}((z))) : \Phi\widehat{U} = \widehat{U}'^\sigma \Phi\},$$

where the superscript $^\sigma\Phi$ denotes the application of σ^* to the entries of the matrix Φ. The injectivity is obvious. We have to prove the surjectivity. So let an element of the right-hand side be given. After multiplying it with a power of z, it is represented by a matrix

$$\Phi = \sum_{\mu=0}^{\infty} \Phi_\mu z^\mu \in M_{r\ell}(K^{\mathrm{alg}}[\![z]\!]).$$

We expand the equation $\Phi\widehat{U} = \widehat{U}'^\sigma \Phi$ into powers of z and get, for all μ,

$$\Phi_\mu \widehat{U}_0 - \widehat{U}'^\sigma_0 \Phi_\mu = \sum_{\nu=1}^{N} (\Phi_{\mu-\nu} \widehat{U}_\nu - \widehat{U}'^\sigma_\nu \Phi_{\mu-\nu}). \qquad (11.2)$$

Now by Lemma 11.1, we have

$$\widehat{U}^\sigma \widehat{U} \cdots {}^{\sigma^{rm-1}}\widehat{U} = z^{dm} \cdot \mathrm{Id}_{r\ell} \quad \text{and} \quad \widehat{U}'^\sigma \widehat{U}' \cdots {}^{\sigma^{rm-1}}\widehat{U}' = z^{dm} \cdot \mathrm{Id}_{r\ell}.$$

This implies the equation $z^{dm} \cdot \Phi = z^{dm} \cdot {}^{\sigma^{rm}}\Phi$, whence $\Phi_\mu = {}^{\sigma^{rm}}\Phi_\mu$ for all μ.

We find $\Phi_\mu \in M_{r\ell}(K')$. Now we consider for an integer i the sequence of matrices $(\Phi_i, \ldots, \Phi_{i+N})$. As i varies, these sequences run through a finite set. Therefore, there are infinitely many i giving rise to the same sequence. Let j be the difference of two such i and consider the matrix

$$\Phi - z^j \Phi =: \sum_{\mu=0}^{\infty} \widetilde{\Phi}_\mu z^\mu.$$

In this matrix, we find a sequence with $\widetilde{\Phi}_i = \cdots = \widetilde{\Phi}_{i+N} = 0$. Looking at equation (11.2), we see that we may set all $\widetilde{\Phi}_\mu = 0$ for $\mu > i + N$ to obtain a matrix

$$\widetilde{\Phi} = \sum_{\mu=0}^{i-1} \widetilde{\Phi}_\mu z^\mu \in M_{r\ell}(K'[z])$$

which satisfies $\widetilde{\Phi}\widehat{U} = \widehat{U}'^\sigma \widetilde{\Phi}$ and is congruent to Φ modulo z^j. As j can be chosen arbitrarily large, the surjectivity of (11.1) is established.

Now note that the Πs on the (formal) abelian sheaves induce endomorphisms of the Q_∞-vector spaces in (11.1). So the compatibility with the Πs is a condition that cuts out isomorphic linear subspaces on both sides of (11.1). From this the corollary follows in the case $C = \mathbb{P}^1$. For arbitrary C, consider a finite flat morphism $\pi : C \to \mathbb{P}^1$ mapping ∞_C to ∞. We have just proved the assertion for $\pi_*\underline{\mathcal{F}}$ and $\pi_*\underline{\mathcal{F}}'$. Since again the elements of $\mathcal{O}_C$ induce endomorphisms of the Q_∞-vector spaces in (11.1), we may deduce the assertion for $\underline{\mathcal{F}}$ and $\underline{\mathcal{F}}'$. $\square$

Recall the abelian sheaf $\mathbb{M}$ over $\mathrm{Spf}\,\mathbb{F}_{q^\ell}[\![\zeta]\!]$ and the level structures on $\mathbb{M}$ from Example 10.4.

Proposition 11.4. *Let $S \in \mathcal{N}ilp_{\mathbb{F}_{q^\ell}[\![\zeta]\!]}$ be locally of finite type over $\mathbb{F}_{q^\ell}[\![\zeta]\!]$ and let $H \subset \mathrm{GL}_r(\mathbb{A}_f)$ be a compact open subgroup. Let $\underline{\mathcal{F}}$ be an abelian sheaf of rank r and dimension d with a rational H-level structure over S. Then for a point $s \in S$ the following assertions are equivalent:*

1. *The formal abelian sheaf associated to $\underline{\mathcal{F}}_s$ is isoclinic.*
2. *Over a finite extension of the residue field $\kappa(s)$ of s there is a quasi-isogeny between $\underline{\mathcal{F}}_s$ and $\mathbb{M}_s$ which is compatible with the H-level structures on both sides.*
3. *There is an abelian sheaf $\underline{\mathcal{F}}'$ over a finite field $\mathbb{F} \subset \overline{\mathbb{F}}_{q^\ell}$, a finite quasi-isogeny $\alpha : \overline{\mathbb{M}}_{\mathbb{F}} \to \underline{\mathcal{F}}'$ over $\mathbb{F}$ compatible with the H-level structures, and a quasi-isogeny $\varphi_s : \underline{\mathcal{F}}'_s \to \underline{\mathcal{F}}_s$ which is an isomorphism over C' and which is defined over a finite extension of $\kappa(s)$.*

Proof. Note that the formal abelian sheaf of $\underline{\mathcal{F}}_s$ is isoclinic if and only if it is isogenous over an algebraically closed extension of $\kappa(s)$ to the formal abelian sheaf $\widehat{\mathbb{M}}$ of $\mathbb{M}$. Therefore, 2 implies 1. Clearly, 2 follows from 3.

To prove that 1 implies 3, we proceed by induction on the transcendence degree of the residue field $\kappa(s)$ of s over $\mathbb{F}_q$. Observe that $\kappa(s)$ is finitely generated since S

is locally of finite type over $\mathbb{F}_{q^\ell}[[\zeta]]$. If $\kappa(s) \subset \mathbb{F}_q^{\mathrm{alg}}$ we obtain from Corollary 11.3 a quasi-isogeny $\alpha : \mathbb{M}_{\mathbb{F}} \to \underline{\mathcal{F}}_{\mathbb{F}}$ over a finite field $\mathbb{F} \supset \kappa(s)$. Altering α by a quasi-isogeny of $\mathbb{M}_{\mathbb{F}}$ we can achieve that α is compatible with the H-level structures. Let $\widehat{\alpha}$ be the induced quasi-isogeny on formal abelian sheaves. We set $\underline{\mathcal{F}}' = \widehat{\alpha}^* \underline{\mathcal{F}}_{\mathbb{F}}$. Then α factors as

$$\overline{\mathbb{M}}_{\mathbb{F}} \xrightarrow{\alpha'} \underline{\mathcal{F}}' \xrightarrow{\alpha''} \underline{\mathcal{F}}_{\mathbb{F}},$$

where α' is a finite quasi-isogeny compatible with the H-level structures, and α'' is a quasi-isogeny which is an isomorphism over C'.

If $\kappa(s) \not\subset \mathbb{F}_q^{\mathrm{alg}}$, we choose a point s' of codimension 1 in the closure of s inside S. By Theorem 7.7 the formal abelian sheaf of $\underline{\mathcal{F}}_{s'}$ is also isoclinic. So the induction hypothesis asserts that there is an abelian sheaf $\underline{\mathcal{F}}'$ over a finite field $\mathbb{F} \subset \mathbb{F}_{q^\ell}$, a finite quasi-isogeny $\alpha : \overline{\mathbb{M}}_{\mathbb{F}} \to \underline{\mathcal{F}}'$ over $\mathbb{F}$ compatible with the H-level structures, and a quasi-isogeny $\varphi_K : \underline{\mathcal{F}}'_K \to \underline{\mathcal{F}}_K$ over an algebraic closure K of $\kappa(s')$, which is an isomorphism over C'. A suitable extension of $\mathcal{O}_{\overline{\{s\}},s'}$ is a power series ring in one variable $K[[\pi]]$ over K.

Now consider the abelian sheaf $\underline{\mathcal{F}}_{S'}$ over $S' := \operatorname{Spec} K[[\pi]]$ which is the pullback of $\underline{\mathcal{F}}$ under the morphism $S' \to S$. Let $\widehat{\underline{\mathcal{F}}}_{S'}$ be the formal abelian sheaf associated to $\underline{\mathcal{F}}_{S'}$. The Newton polygon of $\widehat{\underline{\mathcal{F}}}_{S'}$ is constant over S'. So by Theorem 7.10 there is a quasi-isogeny of formal abelian sheaves $\widehat{\varphi} : \widehat{\underline{\mathcal{F}}}'_{S'} \to \widehat{\underline{\mathcal{F}}}_{S'}$. After changing $\widehat{\varphi}$ by a quasi-isogeny of $\widehat{\underline{\mathcal{F}}}'$, we may assume that φ_K gives rise to $\widehat{\varphi}|_K$. Due to the Serre–Tate Theorem (Theorem 8.4) $\widehat{\varphi}$ induces for all n a lift of φ_K to a quasi-isogeny φ_n over $S'_n := \operatorname{Spec} K[[\pi]]/(\pi^{n+1})$. By Corollarly 8.3, φ_n is an isomorphism over C'. Let D be the divisor of the pole of $\widehat{\varphi}$ at ∞. Then $\varphi_n : \underline{\mathcal{F}}'_{S'_n} \to \underline{\mathcal{F}}(D)_{S'_n}$ is a true isogeny. So in the limit we obtain a quasi-isogeny between $\underline{\mathcal{F}}'_{S'}$ and $\underline{\mathcal{F}}_{S'}$ over $\operatorname{Spf} K[[\pi]]$ which is an isomorphism over C'. By Grothendieck's existence theorem it comes from a quasi-isogeny over $\operatorname{Spec} K[[\pi]]$ which gives us a quasi-isogeny $\varphi_s : \underline{\mathcal{F}}'_s \to \underline{\mathcal{F}}_s$ over $K((\pi))$. From the following lemma we obtain the desired quasi-isogeny over a finite extension of $\kappa(s)$. $\qquad\square$

Lemma 11.5. *Let $K \supset \mathbb{F}_q$ be an arbitrary field. Consider two abelian sheaves $\underline{\mathcal{F}}$ and $\underline{\mathcal{F}}'$ over K and a quasi-isogeny $\varphi : \underline{\mathcal{F}}'_L \to \underline{\mathcal{F}}_L$ defined over some extension L of K. Then there is a quasi-isogeny $\varphi' : \underline{\mathcal{F}}' \to \underline{\mathcal{F}}$ defined over a finite extension of K. If φ is an isomorphism over C' we can find a φ' which is also an isomorphism over C'.*

Proof. Let $D \subset C$ be an effective divisor such that $\varphi : \underline{\mathcal{F}}'_L \to \underline{\mathcal{F}}_L(D)$ is an isogeny. In the description of the morphisms $\varphi_i : \mathcal{F}'_i \to \mathcal{F}_i(D)$, there are only finitely many coefficients from L involved due to the periodicity. Let R be the K-subalgebra of L generated by these coefficients. Now consider the locally closed subscheme S of $\operatorname{Spec} R$ defined by the conditions that the φ_i give an isogeny. These are the equations $\Pi_i \circ \varphi_i = \varphi_{i+1} \circ \Pi'_i$ and $\tau_i \circ \sigma^* \varphi_i = \varphi_{i+1} \circ \tau'_i$ and the conditions that φ_i is injective and that $\operatorname{coker} \varphi_i$ is locally free of finite rank and supported on $\widetilde{D} \times S$ for an effective divisor $\widetilde{D} \subset C$. As S is of finite type over K we find a K'-rational point on it for a finite extension K'/K. The data over this point defines a quasi-isogeny $\varphi' : \underline{\mathcal{F}}'_{K'} \to \underline{\mathcal{F}}_{K'}$. If φ is an isomorphism over C', it is clear that we can achieve the same for φ'. $\qquad\square$

Corollary 11.6. Let $S \in \mathcal{N}ilp_{\mathbb{F}_{q^\ell}[\![\zeta]\!]}$ and let $H \subset GL_r(\mathbb{A}_f)$ be a compact open subgroup. Let $\underline{\mathcal{F}}$ be an abelian sheaf of rank r and dimension d with a rational H-level structure over S. Then the set of points in S over which there is a quasi-isogeny between $\underline{\mathcal{F}}$ and $\mathbb{M}$ which is compatible with the H-level structures, is closed.

Proof. By Theorem 7.7, the set of points over which the formal abelian sheaf associated to $\underline{\mathcal{F}}$ is isoclinic is closed in S. If S is locally of finite type over $\mathbb{F}_{q^\ell}[\![\zeta]\!]$, then the corollary follows from Proposition 11.4.

Let S be arbitrary. We only need to treat the case where S is reduced. Then the abelian sheaf $\underline{\mathcal{F}}$ induces a 1-morphism $f : S \rightarrow \mathcal{A}b\text{-}\mathcal{S}h_H^{r,d} \times_C \infty$ of algebraic Spec $\mathbb{F}_{q^\ell}$-stacks. As the question is local on S we can assume that S is quasi-compact. Since $\mathcal{A}b\text{-}\mathcal{S}h_H^{r,d}$ is locally of finite type over C we may further assume that f factors through a presentation $X \rightarrow \mathcal{A}b\text{-}\mathcal{S}h_H^{r,d} \times_C \infty$, where X is a scheme of finite type over $\mathbb{F}_{q^\ell}[\![\zeta]\!]$. Then the closedness of the set on X implies the closedness of the set on S. $\square$

Example 11.7. Consider the universal abelian sheaf over $M_I^{2,2}$ from the example in Section 4. Let $S = M_I^{2,2} \times_C \infty$. Pink has computed that the closed set from Corollary 11.6 is the proper subset

$$\bigcup g \, V(\zeta, a_{11}, a_{22}, a_{21}) \subset S,$$

where the union runs over all $g \in GL_2(\mathbb{F}_q)$ which act on the points $(a_{\mu\nu}) \in S$ by conjugation $(a_{\mu\nu}) \mapsto g(a_{\mu\nu})g^{-1}$; cf. [3, Section 7]. (Note that $\mathcal{A}b\text{-}\mathcal{S}h_{H_I}^{2,2}$ is a $GL_2(\mathbb{F}_q)$-torsor over $\mathcal{A}b\text{-}\mathcal{S}h^{2,2}$.)

Remark 11.8. There is an interesting consequence for the uniformizability of t-motives. Namely, by work of Gardeyn [17], the t-motive associated to an abelian sheaf over a complete extension K of $\mathbb{F}_q(\!(\zeta)\!)$ is uniformizable in the sense of Anderson [1], if and only if firstly the abelian sheaf extends to an abelian sheaf over the valuation ring R of a finite extension of K, and secondly its reduction modulo the maximal ideal of R is isogenous to $\mathbb{M}$.

Let X be an admissible formal scheme in the sense of Raynaud [37, 4] and let $(\underline{\mathcal{F}}, \bar{\eta})$ be an abelian sheaf with H-level structure over X. Then we deduce that the set of points on the associated rigid-analytic space X^{rig} over which the abelian sheaf $\underline{\mathcal{F}}$ is uniformizable, is formally closed. In particular, if X^{rig} is quasi-compact, then the complement of this set is also quasi-compact. For more elaboration on this issue see Böckle–Hartl [3].

In the remainder of this section, we prove a weak result on the uniform existence of the quasi-isogeny between $\underline{\mathcal{F}}$ and $\mathbb{M}$ which above has been studied point-wise. It will suffice for our purposes in this article. Undoubtedly there should be much stronger results in this direction.

Lemma 11.9. Let $\mathbb{F} \supset \mathbb{F}_q$ be a finite field and let $S = \operatorname{Spec} R \in \mathcal{N}ilp_{\mathbb{F}_q[\![\zeta]\!]}$ be a reduced noetherian affine scheme. Consider abelian sheaves $\underline{\mathcal{F}}$ and $\underline{\mathcal{F}}'$ of rank r and dimension d over S and $\mathbb{F}$, respectively. Let $s : \operatorname{Spec} L \to S$ be an algebraically closed point over which a quasi-isogeny $\varphi_s : \underline{\mathcal{F}}'_s \to \underline{\mathcal{F}}_s$ is given. Assume that φ_s is an isomorphism over C'. Then there exists a quasi-compact reduced scheme S' of finite type over S containing a lift of s, such that φ_s extends over all of S' to a quasi-isogeny $\varphi : \underline{\mathcal{F}}'_{S'} \to \underline{\mathcal{F}}_{S'}$ which is an isomorphism over C'.

Proof. We first treat the case where $C = \mathbb{P}^1_{\mathbb{F}_q}$ and $C' = \operatorname{Spec} \mathbb{F}_q[t]$. Then the projective $R[t]$-module $\mathcal{F}|_{C'_S}$ underlying $\underline{\mathcal{F}}$ is a direct summand of a free $R[t]$-module

$$\mathcal{F}|_{C'_S} \oplus \mathcal{F}^{\mathrm{nil}} = \widetilde{\mathcal{F}}$$

of rank $\widetilde{r}$. We extend $\tau_{\mathcal{F}}$ by zero to the endomorphism $\widetilde{\tau} := \tau_{\underline{\mathcal{F}}}|_{C'_S} \oplus 0 : \sigma^* \widetilde{\mathcal{F}} \to \widetilde{\mathcal{F}}$. After choosing bases of $\widetilde{\mathcal{F}}$ and $\underline{\mathcal{F}}'|_{C'_S}$, the coherent sheaves on C'_S with σ-linear endomorphism $(\widetilde{\mathcal{F}}, \widetilde{\tau})$ and $\underline{\mathcal{F}}'|_{C'_S}$ are isomorphic to

$$(R[t]^{\widetilde{r}}, U \cdot \sigma^*) \quad \text{and} \quad (R[t]^r, U' \cdot \sigma^*)$$

for suitable matrices $U = \sum_{\nu} U_{\nu} t^{\nu} \in M_{\widetilde{r}}(R[t])$ and $U' = \sum_{\nu} U'_{\nu} t^{\nu} \in \mathrm{GL}_r(\mathbb{F}[t])$. The quasi-isogeny $\varphi : \underline{\mathcal{F}}'_{S'} \to \underline{\mathcal{F}}_{S'}$ we are looking for, then corresponds to a matrix $\Phi = \sum_{\mu} \Phi_{\mu} t^{\mu} \in M_{\widetilde{r} \times r}(\mathcal{O}_{S'}[t])$ satisfying the condition $\Phi U' = U^{\sigma} \Phi$. We expand this condition according to powers of t to get

$$\Phi_{\mu} U'_0 - U_0{}^{\sigma} \Phi_{\mu} = \sum_{\nu=1}^{\mu} (U_{\nu}{}^{\sigma} \Phi_{\mu-\nu} - \Phi_{\mu-\nu} U'_{\nu}). \tag{11.3}$$

The quasi-isogeny φ_s over $\operatorname{Spec} L$ corresponds to a matrix for some integer N

$$\overline{\Phi} = \sum_{\mu=0}^{N} \overline{\Phi}_{\mu} t^{\mu} \in M_{\widetilde{r} \times r}(L[t])$$

satisfying (11.3). In order to extend $\overline{\Phi}$ to a neighborhood of s, we simply adjoin the entries of indeterminant matrices $\Phi_0, \ldots, \Phi_N$ to the ring R and divide out the relations (11.3) for $\mu = 0, \ldots, N$. Thus we obtain an R-algebra R'' and an étale morphism $S'' = \operatorname{Spec} R'' \to S$. The point s lifts to a point $s'' : \operatorname{Spec} L \to S''$ by mapping Φ_{μ} to $\overline{\Phi}_{\mu}$.

We let $\widetilde{S} \subset S''$ be the closed subset on which the right-hand side of (11.3) and the matrices Φ_{μ} are zero for all $\mu > N$. Clearly, s'' lies in $\widetilde{S}$. Over $\widetilde{S}$ the matrix Φ defines a morphism $f : \underline{\mathcal{F}}'|_{C'_{\widetilde{S}}} \to (\widetilde{\mathcal{F}}, \widetilde{\tau})_{\widetilde{S}}$. Since $\tau_{\mathcal{F}'}|_{C'}$ is an isomorphism and $\widetilde{\tau}|_{\mathcal{F}^{\mathrm{nil}}} = 0$ we see that f factors through a morphism $\varphi : \underline{\mathcal{F}}'|_{C'_{\widetilde{S}}} \to \underline{\mathcal{F}}|_{C'_{\widetilde{S}}}$. The coherent sheaves $\ker \varphi$ and $\operatorname{coker} \varphi$ are supported on closed subschemes of $C'_{\widetilde{S}}$. Observe that $\widetilde{S}$ is noetherian. So the projection of these closed subschemes to $\widetilde{S}$ are constructible subsets of $\widetilde{S}$ by Chevalley's theorem [20, Corollaire IV.1.8.5]. Their complement S' is also constructible. Hence S' is a finite union of locally close reduced subschemes of $\widetilde{S}$.

We replace S' with the *disjoint* union of these subschemes. Then S' is a quasi-compact reduced scheme of finite type over S. Now φ defines a quasi-isogeny $\underline{\mathcal{F}}'_{S'} \to \underline{\mathcal{F}}_{S'}$ which is an isomorphism over $C'_{S'}$. Moreover, $s'' \in S'$ and φ specializes to φ_s at s''.

If C is arbitrary, we consider a finite flat morphism $\pi : C \to \mathbb{P}^1_{\mathbb{F}_q}$ mapping ∞_C to ∞. Then we have just proved the existence of a quasi-isogeny $\pi_*\underline{\mathcal{F}}_{S'} \to \pi_*\underline{\mathcal{F}}'_{S'}$ over S'. In order for this to give a quasi-isogeny $\underline{\mathcal{F}}_{S'} \to \underline{\mathcal{F}}'_{S'}$ it must commute with the elements of $\mathcal{O}_C$. Clearly, this condition cuts out a reduced closed subscheme of S' which contains s''. We replace S' by this subset. This concludes the proof of the lemma.   □

Proposition 11.10. *Let $S \in \mathcal{N}ilp_{\mathbb{F}_{q^\ell}[\![\zeta]\!]}$ be a quasi-compact and reduced scheme and let $H \subset \mathrm{GL}_r(\mathbb{A}_f)$ be a compact open subgroup. Consider an abelian sheaf $(\underline{\mathcal{F}}, \bar{\gamma})$ of rank r and dimension d with rational H-level structure over S. Assume that for every point $s \in S$ the formal abelian sheaf associated to $\underline{\mathcal{F}}_s$ is isoclinic. Then there exists a surjective morphism of $\mathbb{F}_{q^\ell}$-schemes $S' \to S$ with S' quasi-compact and reduced, and a quasi-isogeny over S' between $\underline{\mathcal{F}}_{S'}$ and $\mathbb{M}_{S'}$ compatible with the H-level structures.*

Proof. Since the data $(\underline{\mathcal{F}}, \bar{\gamma})$ involves only finitely many coefficients from $\mathcal{O}_S$ we may assume that S is of finite type over $\mathbb{F}_{q^\ell}$. In particular, S is noetherian. (We could also use the fact that $\mathcal{A}b\text{-}\mathcal{S}h_H^{r,d}$ is locally of finite type over C.)

Let $s \in S$ be a point. From Proposition 11.4 we obtain an abelian sheaf $\underline{\mathcal{F}}'$ over a finite field $\mathbb{F}$, a finite quasi-isogeny $\alpha : \overline{\mathbb{M}}_{\mathbb{F}} \to \underline{\mathcal{F}}'$ over $\mathbb{F}$ compatible with the H-level structures, and a quasi-isogeny $\varphi_s : \underline{\mathcal{F}}'_s \to \underline{\mathcal{F}}_s$ which is an isomorphism over C'. The quasi-isogeny φ_s is defined over an algebraically closed extension of $\kappa(s)$. So by the previous lemma there is a morphism $S'_s \to S$ of finite type from a quasi-compact reduced scheme S'_s, such that s lifts to a point of S'_s, and φ_s extends to a quasi-isogeny φ over all of S'_s which is an isomorphism over C'. Clearly, the quasi-isogeny $\varphi \circ \alpha : \overline{\mathbb{M}}_{S'_s} \to \underline{\mathcal{F}}_{S'_s}$ over S'_s is compatible with the H-level structures. By Chevalley's theorem, the image of the morphism $S'_s \to S$ is a constructible subset S_s of S containing s. Since S is of finite type over $\mathbb{F}_{q^\ell}$ the S_s form a countable covering of S by constructible subsets. By [20, Corollaire 0.9.2.4] finitely many of the S_s suffice to cover S. We let S' be the finite disjoint union of the corresponding S'_s.   □

12 The Uniformization Theorem

Let $H \subset \mathrm{GL}_r(\mathbb{A}_f)$ be a compact open subgroup. We will define 1-morphisms of formal algebraic $\mathrm{Spf}\,\mathbb{F}_{q^\ell}[\![\zeta]\!]$-stacks from G to the stacks $\mathcal{A}b\text{-}\mathcal{S}h_H^{r,d} \times_C \mathrm{Spf}\,\mathbb{F}_{q^\ell}[\![\zeta]\!]$ of abelian sheaves with rational H-level structure (Definition 5.7). For this purpose recall the H-level structures on $\overline{\mathbb{M}}$ constructed in Example 10.4.

12.1. Let S be in $\mathcal{N}ilp_{\mathbb{F}_{q^\ell}[\![\zeta]\!]}$ and denote by $\bar{S}$ its special fiber. The morphism $G \to G'$ from Theorem 10.2 associates to an element $(\beta, \widehat{\underline{\mathcal{F}}}, \widehat{\alpha}) \in G(S)$ an abelian sheaf $\underline{\mathcal{F}}$ and a quasi-isogeny $\alpha : \underline{\mathcal{F}}_{\bar{S}} \to \bar{\beta}^*\overline{\mathbb{M}}$ which is an isomorphism over C'. For an algebraically closed base point $\iota : s \to S$ the quasi-isogeny α has lead in Example 10.4 to an isomorphism

$$\gamma \circ (\alpha|_{\mathbb{A}_f})^\tau : (\iota^* \underline{\mathcal{F}}|_{\mathbb{A}_f})^\tau (s) \xrightarrow{\;\sim\;} \mathbb{A}_f^r$$

which is fixed under the action of $\pi_1(S, s)$ on the source. Then we define a 1-morphism Θ of formal algebraic $\operatorname{Spf} \mathbb{F}_{q^\ell}[\![\zeta]\!]$-stacks by the following map on S-valued points:

$$\Theta : G \times \operatorname{GL}_r(\mathbb{A}_f)/H \longrightarrow Ab\text{-}\mathcal{S}h_H^{r,d} \times_C \operatorname{Spf} \mathbb{F}_{q^\ell}[\![\zeta]\!], \qquad (12.1)$$

$$(\beta, \widehat{\underline{\mathcal{F}}}, \widehat{\alpha}) \times hH \mapsto (\underline{\mathcal{F}}, h^{-1}\gamma(\alpha|_{\mathbb{A}_f})^\tau) \times \beta.$$

It is equivariant with respect to the right $\operatorname{GL}_r(\mathbb{A}_f)$-action on the projective systems on both sides of (12.1).

12.2. There is an action of $J(Q)$ on the source, which we describe next. Recall that we have defined in 9.5 an action of $J(Q_\infty)$ on G through the isomorphism ε_∞ from $J(Q_\infty)$ to the group of quasi-isogenies of $\widehat{\mathbb{M}}$. We let $J(Q)$ act on G via the inclusion $J(Q) \subset J(Q_\infty)$.

On the other hand, we have a morphism

$$\varepsilon^\infty : J(Q) \hookrightarrow \operatorname{GL}_r(\mathbb{A}_f)$$

which is defined by the commutative diagram

$$
\begin{array}{ccc}
(\overline{\mathbb{M}}|_{\mathbb{A}_f})^\tau & \xrightarrow{\;(g|_{\mathbb{A}_f})^\tau\;} & (\overline{\mathbb{M}}|_{\mathbb{A}_f})^\tau \\
\gamma \downarrow & & \downarrow \gamma \\
\mathbb{A}_f^r & \xrightarrow{\;\varepsilon^\infty(g)\;} & \mathbb{A}_f^r
\end{array}
$$

for $g \in J(Q)$. Note that ε^∞ identifies $J(Q)$ with the diagonal embedding of $\operatorname{GL}_r(Q)$ into $\operatorname{GL}_r(\mathbb{A}_f)$.

We define a left action of the group $J(Q)$ on $G \times \operatorname{GL}_r(\mathbb{A}_f)/H$,

$$(\beta, \widehat{\underline{\mathcal{F}}}, \widehat{\alpha}) \times hH \longmapsto (\beta, \widehat{\underline{\mathcal{F}}}, \bar{\beta}^* \varepsilon_\infty(g) \circ \widehat{\alpha}) \times \varepsilon^\infty(g)hH.$$

Proposition 12.3. *We abbreviate $Y = G \times \operatorname{GL}_r(\mathbb{A}_f)/H$. The action of $J(Q)$ induces a 1-isomorphism of formal algebraic $\operatorname{Spf} \mathbb{F}_{q^\ell}[\![\zeta]\!]$-stacks*

$$Y \times J(Q) \xrightarrow{\;\sim\;} Y \times_{Ab\text{-}\mathcal{S}h_H^{r,d} \times \operatorname{Spf} \mathbb{F}_{q^\ell}[\![\zeta]\!]} Y.$$

Proof. We have to show that the functor between the categories of S-valued points on both sides is an equivalence. Note that in the stack Y the only morphisms are the identities. From this full faithfulness follows. For essential surjectivity, we have to show that two S-valued points of Y lie in the same orbit if and only if they are mapped to the same point by Θ. So let $(\beta, \widehat{\underline{\mathcal{F}}}, \widehat{\alpha}, hH)$ and $(\beta, \widehat{\underline{\mathcal{F}}}', \widehat{\alpha}', h'H)$ be in $Y(S)$ and let $(\underline{\mathcal{F}}, \alpha, hH)$ and $(\underline{\mathcal{F}}', \alpha, h'H)$ be their algebraizations.

First we assume that

$$(\beta, \widehat{\underline{\mathcal{F}}}', \widehat{\alpha}', h'H) = (\beta, \widehat{\underline{\mathcal{F}}}, \bar{\beta}^* \varepsilon_\infty(g) \circ \widehat{\alpha}, \varepsilon^\infty(g) hH)$$

in $Y(S)$ for a $g \in J(Q)$. Consider the following diagram of quasi-isogenies over $\bar{S}$:

$$
\begin{array}{ccc}
\mathcal{F}_{\bar{S}} & \xrightarrow{\ \alpha\ } & \overline{\mathbb{M}}_{\bar{S}} \\
\bar{\varphi} \downarrow & & \downarrow g_{\bar{S}} \\
\mathcal{F}'_{\bar{S}} & \xrightarrow{\ \alpha'\ } & \overline{\mathbb{M}}_{\bar{S}},
\end{array}
\tag{12.2}
$$

where $g_{\bar{S}}$ is the quasi-isogeny obtained from g by Proposition 8.7. The quasi-isogeny $\bar{\varphi}$ defined by this diagram is finite by construction. By Proposition 8.2 it lifts uniquely to a quasi-isogeny $\varphi : \underline{\mathcal{F}} \to \underline{\mathcal{F}}'$. We claim that φ is finite. Namely, the induced quasi-isogeny $\widehat{\varphi}$ on formal abelian sheaves satisfies $\widehat{\varphi}_{\bar{S}} = \mathrm{id}_{\widehat{\underline{\mathcal{F}}}_{\bar{S}}}$. Hence we find $\widehat{\varphi} = \mathrm{id}_{\widehat{\underline{\mathcal{F}}}}$ by the uniqueness of the lift. This shows that φ is a finite quasi-isogeny.

Now $\bar{\varphi}$ induces on $\underline{\mathcal{F}}$ the H-level structure

$$h^{-1} \varepsilon^\infty(g)^{-1} \gamma (\alpha'|_{\mathbb{A}_f})^\tau (\bar{\varphi}|_{\mathbb{A}_f})^\tau = h^{-1} \gamma (\alpha|_{\mathbb{A}_f})^\tau.$$

Therefore, the two points have the same image under Θ.

Conversely, let $\varphi : \underline{\mathcal{F}} \to \underline{\mathcal{F}}'$ be a finite quasi-isogeny inducing an equality of H-level structures on $\underline{\mathcal{F}}$,

$$h'^{-1} \gamma (\alpha'|_{\mathbb{A}_f})^\tau (\varphi|_{\mathbb{A}_f})^\tau = h^{-1} \gamma (\alpha|_{\mathbb{A}_f})^\tau.$$

This time diagram (12.2) defines a quasi-isogeny $g_{\bar{S}}$ from $\overline{\mathbb{M}}_{\bar{S}}$ to itself. By Proposition 8.7 it comes from an element $g \in J(Q)$. Then we have

$$(\beta, \widehat{\underline{\mathcal{F}}}', \widehat{\alpha}', h'H) = (\beta, \widehat{\underline{\mathcal{F}}}, \bar{\beta}^* \varepsilon_\infty(g) \circ \widehat{\alpha}, \varepsilon^\infty(g) hH). \qquad \square$$

Hence the map Θ factors through a 1-morphism of formal algebraic Spf $\mathbb{F}_{q^\ell}[\![\zeta]\!]$-stacks

$$J(Q) \backslash G \times \mathrm{GL}_r(\mathbb{A}_f)/H \longrightarrow Ab\text{-}Sh_H^{r,d} \times_C \mathrm{Spf}\, \mathbb{F}_{q^\ell}[\![\zeta]\!].$$

Note that the quotient $J(Q) \backslash G \times \mathrm{GL}_r(\mathbb{A}_f)/H$ is a formal algebraic Spf $\mathbb{F}_{q^\ell}[\![\zeta]\!]$-stack due to Theorem 9.6. Indeed, the subgroup $J(Q) \hookrightarrow J(Q_\infty) \times \mathrm{GL}_r(\mathbb{A}_f)$ is discrete. Hence

$$J(Q) \backslash G \times \mathrm{GL}_r(\mathbb{A}_f)/H \cong \coprod_\Gamma \Gamma \backslash G,$$

where Γ runs through a countable set of subgroups of $J(Q_\infty)$ of the form

$$(J(Q_\infty) \times gHg^{-1}) \cap J(Q) \subset J(Q_\infty).$$

These are separated discrete subgroups.

Proposition 12.4. *The* 1-*morphism* Θ *defines a* 1-*morphism of formal algebraic* Spf $\mathbb{F}_q[\![\zeta]\!]$-*stacks*

$$\Theta' : G' \times \mathrm{GL}_r(\mathbb{A}_f)/H \longrightarrow \mathcal{A}b\text{-}\mathcal{S}h_H^{r,d} \times_C \mathrm{Spf}\,\mathbb{F}_q[\![\zeta]\!],$$

which is invariant with respect to the action of $J(Q)$ *on the source defined in Section* 10.

Proof. We have to show that Θ and the action of $J(Q)$ commute with the Galois-descent data on the source and the target of the 1-morphism Θ. For the $J(Q)$-action this was already observed in 9.5. For Θ we must check that

$$(\underline{\mathcal{F}}, h^{-1}\gamma(\alpha|_{\mathbb{A}_f})^\tau) = (\underline{\mathcal{F}}, h^{-1\sigma}\gamma(\alpha|_{\mathbb{A}_f})^\tau)$$

in $\mathcal{A}b\text{-}\mathcal{S}h_H^{r,d}$, where $^\sigma\gamma$ is obtained from γ by the action of the generator Frob_q of $\mathrm{Gal}(\mathbb{F}_{q^\ell}/\mathbb{F}_q)$. Now $^\sigma\gamma^{-1} \circ \gamma$ comes in fact from an automorphism of $\overline{\mathbb{M}}$ (cf. Example 10.4). This induces an automorphism of $\underline{\mathcal{F}}$ which carries the H-level structure $h^{-1}\gamma(\alpha|_{\mathbb{A}_f})^\tau$ to $h^{-1\sigma}\gamma(\alpha|_{\mathbb{A}_f})^\tau$. $\quad\square$

Let Z be the set of points $s : \mathrm{Spec}\,L \to \mathcal{A}b\text{-}\mathcal{S}h_H^{r,d} \times_C \infty$ such that the universal abelian sheaf over s is isogenous to $\overline{\mathbb{M}}$ over an algebraic closure of L. Consider the preimage $Z' \subset \mathcal{A}b\text{-}\mathcal{S}h_H^{r,d} \times_{\mathbb{F}_q} \mathbb{F}_{q^\ell}$ of Z under the base change morphism coming from $\mathbb{F}_q \subset \mathbb{F}_{q^\ell}$.

Lemma 12.5. *The set* Z' *can also be described as the set of points* s *over which the associated formal abelian sheaf is isoclinic. In particular,* Z *and* Z' *are closed subsets.*

Proof. The formal abelian sheaf $\widehat{\overline{\mathbb{M}}}$ of $\overline{\mathbb{M}}$ is isoclinic. Therefore, Z' is contained in the set of the lemma. Conversely, let s belong to this latter set. Since $\mathcal{A}b\text{-}\mathcal{S}h_H^{r,d}$ is locally of finite type over C we can assume that s comes from a point on a local presentation $X \to \mathcal{A}b\text{-}\mathcal{S}h_H^{r,d} \times_C \infty$ where X is a scheme of finite type over $\mathbb{F}_q$. From Proposition 11.4 we obtain a quasi-isogeny $\alpha : \underline{\mathcal{F}}_s \to \mathbb{M}_s$ over an algebraic closure of L. Hence s belongs to Z'.

Now Theorem 7.7 implies that the subset Z' is closed. Namely, Z' is the complement of the open substack on which the associated Newton polygon lies strictly below the Newton polygon of $\widehat{\overline{\mathbb{M}}}$. Therefore, also the image Z of Z' is closed. $\quad\square$

We denote by $\mathcal{A}b\text{-}\mathcal{S}h_H^{r,d}/Z$ the formal completion of $\mathcal{A}b\text{-}\mathcal{S}h_H^{r,d}$ along Z (Definition A.12). It is a formal algebraic Spf $\mathbb{F}_q[\![\zeta]\!]$-stack. By its definition the 1-morphism Θ' factors through $\mathcal{A}b\text{-}\mathcal{S}h_H^{r,d}/Z$. Indeed, if a point $s \in \mathcal{A}b\text{-}\mathcal{S}h_H^{r,d} \times_C \infty$ lies in the image of Θ' the formal abelian sheaf associated to s is isogenous to $\widehat{\overline{\mathbb{M}}}$ by definition and hence isoclinic. We can now formulate our Uniformization Theorem.

Uniformization Theorem 12.6. *There are* $\mathrm{GL}_r(\mathbb{A}_f)$-*equivariant* 1-*isomorphisms of formal algebraic* Spf $\mathbb{F}_q[\![\zeta]\!]$-*stacks*

$$\bar{\Theta} : \ J(Q)\backslash G \times \mathrm{GL}_r(\mathbb{A}_f)/H \xrightarrow{\sim} Ab\text{-}\mathcal{S}h_H^{r,d}/Z \times_{\mathbb{F}_q} \mathbb{F}_{q^\ell},$$

$$\bar{\Theta}' : \ J(Q)\backslash G' \times \mathrm{GL}_r(\mathbb{A}_f)/H \xrightarrow{\sim} Ab\text{-}\mathcal{S}h_H^{r,d}/Z.$$

Example 12.7. In the case of elliptic sheaves,

$$Ab\text{-}\mathcal{S}h_H^{r,1}/Z \times_{\mathbb{F}_q} \mathbb{F}_{q^\ell} = Ab\text{-}\mathcal{S}h_H^{r,1} \times_C \mathrm{Spf}\,\mathbb{F}_{q^\ell}[\![\zeta]\!].$$

This follows from the fact that there is only one polygon between the points $(0, 0)$ and $(r, 1)$ with nonnegative slopes and integral break points, namely, the straight line. So all formal abelian sheaves are isoclinic and Z is all of $Ab\text{-}\mathcal{S}h_H^{r,d} \times_C \infty$. In this case, the open and closed subscheme of G on which the universal quasi-isogeny has height zero is the formal scheme $\Omega^{(r)}$ used by Drinfeld [11]. A detailed account on this can be found in Genestier [18]. Therefore, we have an isomorphism of formal schemes $G \cong \mathbb{Z} \times \Omega^{(r)}$. This decomposition of G is compatible with the decomposition of $Ab\text{-}\mathcal{S}h_{H_I}^{r,1} \cong \mathbb{Z} \times \mathcal{D}r\text{-}\mathcal{M}od_I^r$ from Example 1.8. So Drinfeld's uniformization theorem, which announces an isomorphism of formal schemes

$$\mathrm{GL}_r(Q)\backslash\Omega^{(r)} \times \mathrm{GL}_r(\mathbb{A}_f)/H_I \xrightarrow{\sim} \mathcal{D}r\text{-}\mathcal{M}od_I^r \times_C \mathrm{Spf}\,\mathbb{F}_{q^\ell}[\![\zeta]\!],$$

is equivalent to ours.

Example 12.8. Consider the algebraic stack $Ab\text{-}\mathcal{S}h_H^{2,2}$ from Section 4. We describe $Z(0) := Z \cap Ab\text{-}\mathcal{S}h_H^{2,2}(0)$. In Example 11.7, we have remarked that

$$Z \cap M_I^{2,2} \times_C \infty = \bigcup g\, \mathrm{V}(\zeta, a_{11}, a_{22}, a_{21}).$$

Now we claim that

$$Z(0) = (Z \cap M_I^{2,2} \times_C \infty) \cup (Ab\text{-}\mathcal{S}h_H^{2,2}(0) \smallsetminus M_I^{2,2}) \times_C \infty.$$

Indeed, let $s : \mathrm{Spec}\,L \to (Ab\text{-}\mathcal{S}h_H^{2,2}(0) \smallsetminus M_I^{2,2}) \times_C \infty$ be a point. We must show that the abelian sheaf $\underline{\mathcal{F}}$ over L is isogenous to $\overline{\mathbb{M}}$. By what was said in Section 4 we have $\mathcal{F}_0 = \mathcal{O}_{\mathbb{P}_L^1}(m \cdot \infty) \oplus \mathcal{O}_{\mathbb{P}_L^1}(-m \cdot \infty)$ for an integer $m \geq 1$. With respect to this basis, τ_0 is described by a matrix

$$\tau_0 = \begin{pmatrix} a_0 + a_1 t & b_0 + \cdots + b_{2m+1}t^{2m+1} \\ 0 & d_0 + d_1 t \end{pmatrix} \cdot \sigma^*.$$

Due to the presence of the I-level structure, we may assume $a_0 = d_0 = 1$ and $b_0 = 0$. Since coker τ_0 is supported at ∞ we must have $a_1 = d_1 = 0$. Now let $u_i \in L^{\mathrm{alg}}$ be solutions of the equations $u_i^q - u_i + b_i = 0$ for $i = 1, \ldots, 2m+1$. Then the isomorphism

$$\overline{\mathbb{M}}|_{C'} \to \underline{\mathcal{F}}|_{C'}, \quad \begin{pmatrix} x \\ y \end{pmatrix} \mapsto \begin{pmatrix} 1 & u_0 t + \cdots + u_{2m+1}t^{2m+1} \\ 0 & 1 \end{pmatrix} \cdot \begin{pmatrix} x \\ y \end{pmatrix}$$

extends to a quasi-isogeny $\overline{\mathbb{M}} \to \underline{\mathcal{F}}$ over L^{alg}. Hence s belongs to $Z(0)$.

Remark 12.9. In [36] Rapoport–Zink study the uniformization of Shimura varieties of EL- and PEL-type. One of their theorems yields the uniformization of the formal completion of a Shimura variety along the most supersingular isogeny class [36, 6.30]. The Newton polygon in this isogeny class is maximal. In this sense our Uniformization Theorem is closely analogous to theirs. Beyond this, Rapoport–Zink also obtain uniformization theorems of other isogeny classes. There is no doubt that these theorems too have counterparts for abelian sheaves.

The remainder of this article is devoted to the proof of the Uniformization Theorem.

13 Proof of the Uniformization Theorem

13.1. By Proposition 12.4, it suffices to prove the assertion for the 1-morphism $\bar{\Theta}$. We fix the following notation. On the formal scheme G we let $\mathcal{J}$ be the largest ideal of definition of G; cf. [20, I_{new}, 10.5.4]. For an integer $n \geq 0$ we denote by G_n the scheme $(G, \mathcal{O}_G/\mathcal{J}^{n+1})$. We set

$$Y := G \times \mathrm{GL}_r(\mathbb{A}_f)/H,$$
$$Y_n := G_n \times \mathrm{GL}_r(\mathbb{A}_f)/H,$$
$$\mathcal{Y} := J(Q)\backslash G \times \mathrm{GL}_r(\mathbb{A}_f)/H,$$
$$\mathcal{X} := \mathcal{A}b\text{-}\mathcal{S}h_H^{r,d}/Z \times_{\mathrm{Spf}\,\mathbb{F}_q[\![\zeta]\!]} \mathrm{Spf}\,\mathbb{F}_{q^\ell}[\![\zeta]\!].$$

Let $S \in \mathcal{N}ilp_{\mathbb{F}_q[\![\zeta]\!]}$ and let $(\underline{\mathcal{F}}, \bar{\gamma}, \beta) \in \mathcal{X}(S)$ which we consider as a 1-morphism $S \to \mathcal{X}$. We have to show that the stack

$$\mathcal{Y} \times_{\mathcal{X}} S$$

is a scheme mapping isomorphically to S. The proof relies on several intermediate lemmas.

Lemma 13.2. *The 1-morphism $\bar{\Theta} : \mathcal{Y} \to \mathcal{X}$ is a 1-monomorphism of formal algebraic stacks, i.e., for every $S \in \mathcal{N}ilp_{\mathbb{F}_{q^\ell}[\![\zeta]\!]}$, the functor $\bar{\Theta}(S) : \mathcal{Y}(S) \to \mathcal{X}(S)$ is fully faithful.*

Proof. We can view the formal algebraic Spf $\mathbb{F}_{q^\ell}[\![\zeta]\!]$-stacks $\mathcal{X}$ and $\mathcal{Y}$ as Spec $\mathbb{F}_{q^\ell}[\![\zeta]\!]$-stacks $\widetilde{\mathcal{X}}$ and $\widetilde{\mathcal{Y}}$ by setting for an Spec $\mathbb{F}_{q^\ell}[\![\zeta]\!]$-scheme S,

$$\widetilde{\mathcal{X}}(S) = \begin{cases} \mathcal{X}(S) & \text{if } S \in \mathcal{N}ilp_{\mathbb{F}_{q^\ell}[\![\zeta]\!]}, \\ \emptyset & \text{if } S \notin \mathcal{N}ilp_{\mathbb{F}_{q^\ell}[\![\zeta]\!]}. \end{cases}$$

Then the assertion follows from Proposition 12.3, Lemma A.13 and [32, Proposition 3.8]. □

Lemma 13.3. *The 1-morphism of algebraic Spec $\mathbb{F}_{q^\ell}$-stacks $\bar{\Theta}_{\mathrm{red}} : \mathcal{Y}_{\mathrm{red}} \to \mathcal{X}_{\mathrm{red}}$ (see A.6 in the appendix) is representable by a morphism of schemes.*

Proof. This follows from the fact that every 1-monomorphism of algebraic stacks is representable by a morphism of schemes; cf. [32, Théorème A.2 and Corollaire 8.1.3].□

Lemma 13.4. $\bar{\Theta}_{\mathrm{red}} : \mathcal{Y}_{\mathrm{red}} \to \mathcal{X}_{\mathrm{red}}$ *is surjective.*

Proof. Note that $\mathcal{X}_{\mathrm{red}}$ is the closed substack $Z \subset \mathcal{A}b\text{-}\mathcal{S}h_H^{r,d} \times_C \infty$ with its induced reduced structure. Let $s \in \mathcal{X}_{\mathrm{red}}$ be a point. By definition of Z there is a quasi-isogeny $\alpha : \underline{\mathcal{F}}_s \to \mathbb{M}_s$. We can multiply it with a quasi-isogeny of $\mathbb{M}$ and thus assume that α is compatible with the H-level structures. The induced quasi-isogeny $\underline{\mathcal{F}}_s \to \widehat{\alpha}^* \mathbb{M}_s$ is finite and also compatible with the H-level structures. Therefore, s lies in the image of $\bar{\Theta}_{\mathrm{red}}$.   □

Lemma 13.5. *Let* $S \in \mathcal{N}ilp_{\mathbb{F}_{q^\ell}[\![\zeta]\!]}$ *be quasi-compact and reduced, and consider a 1-morphism* $S \to \mathcal{X}_{\mathrm{red}}$. *Then there exists a surjective morphism of* $\mathbb{F}_{q^\ell}$-*schemes* $S' \to S$ *with* S' *quasi-compact and reduced, and a 2-commutative diagram*

$$
\begin{array}{ccc}
S' & \longrightarrow & Y_{\mathrm{red}} \\
\downarrow & & \downarrow \\
S & \longrightarrow & \mathcal{X}_{\mathrm{red}}.
\end{array}
$$

Proof. This is just a reformulation of Proposition 11.10.   □

Lemma 13.6. $\bar{\Theta}_{\mathrm{red}} : \mathcal{Y}_{\mathrm{red}} \to \mathcal{X}_{\mathrm{red}}$ *is quasi-compact.*

Proof. Let $S \in \mathcal{N}ilp_{\mathbb{F}_{q^\ell}[\![\zeta]\!]}$ be quasi-compact and reduced, and let $S \to \mathcal{X}_{\mathrm{red}}$ be a 1-morphism. Let $S' \to S$ be the surjective morphism from Lemma 13.5. It gives rise to a surjective morphism of schemes over $\mathbb{F}_{q^\ell}$,

$$
\mathcal{Y}_{\mathrm{red}} \times_{\mathcal{X}_{\mathrm{red}}} S \twoheadleftarrow \mathcal{Y}_{\mathrm{red}} \times_{\mathcal{X}_{\mathrm{red}}} S' \twoheadleftarrow Y_{\mathrm{red}} \times_{\mathcal{X}_{\mathrm{red}}} S'.
$$

Since $Y_{\mathrm{red}} \times_{\mathcal{X}_{\mathrm{red}}} S' \cong S' \times J(Q)$ by Proposition 12.3, we obtain an epimorphism $S' \twoheadrightarrow \mathcal{Y}_{\mathrm{red}} \times_{\mathcal{X}_{\mathrm{red}}} S$. Now the lemma follows from the quasi-compactness of S'.   □

Lemma 13.7. $\bar{\Theta}_{\mathrm{red}} : \mathcal{Y}_{\mathrm{red}} \to \mathcal{X}_{\mathrm{red}}$ *is proper.*

Proof. Since $\mathcal{Y}_{\mathrm{red}}$ and $\mathcal{X}_{\mathrm{red}}$ are locally of finite type and $\bar{\Theta}_{\mathrm{red}}$ is quasi-compact, we see that $\bar{\Theta}_{\mathrm{red}}$ is of finite type. Being a 1-monomorphism it is also separated. So it remains to prove that it is universally closed. For this we use the valuative criterion [32, Théorème 7.3]. Let R be a valuation ring with $\operatorname{Spec} R \in \mathcal{N}ilp_{\mathbb{F}_{q^\ell}[\![\zeta]\!]}$ and let K be its field of fractions. We have to show that for every 2-commutative diagram

$$
\begin{array}{ccc}
\operatorname{Spec} K & \longrightarrow & \mathcal{Y}_{\mathrm{red}} \\
\downarrow & & \downarrow \\
\operatorname{Spec} R & \longrightarrow & \mathcal{X}_{\mathrm{red}},
\end{array}
$$

there exists a finite extension R'/R of valuation rings and a 1-morphism of $\operatorname{Spec} \mathbb{F}_{q^\ell}$-stacks $\operatorname{Spec} R' \to \mathcal{Y}_{\mathrm{red}}$ which 2-commutes with the above diagram. However, the existence of this data follows from Lemma 13.5.   □

Lemma 13.8. *The 1-morphism* $\bar{\Theta} : \mathcal{Y} \to \mathcal{X}$ *is adic (Definition A.9).*

Proof. We will show that $\bar{\Theta}_{\text{red}}$ is a 1-isomorphism. Since both $\mathcal{X}$ and $\mathcal{Y}$ are adic (see A.7 in the appendix), this suffices.

Let $P : X \to \mathcal{X}_{\text{red}}$ be a presentation. Then P is an epimorphism since $\mathcal{X}_{\text{red}}$ is an algebraic $\text{Spec}\,\mathbb{F}_{q^\ell}$-stack. Thus it suffices to show that $\mathcal{Y}_{\text{red}} \times_{\mathcal{X}_{\text{red}}} X \to X$ is a 1-isomorphism. From the previous lemmas, we know that it is a proper monomorphism of schemes, hence a closed immersion. Since it is also surjective and X is reduced it is an isomorphism as desired.   □

Lemma 13.9. *The 1-morphism* $\bar{\Theta} : \mathcal{Y} \to \mathcal{X}$ *is étale.*

Proof. Since quasi-isogenies of z-divisible groups lift to infinitesimal neighborhoods we first see that $Y \to \mathcal{X}$ is étale.

Now let $\mathcal{J}$ be an ideal of definition of $\mathcal{X}$ and let $\mathcal{Z}$ be the closed substack defined by $\mathcal{J}$. Since $\bar{\Theta}$ and also the presentation $Y \to \mathcal{Y}$ are adic (Theorem 9.6), we obtain 1-morphisms

$$Y \times_{\mathcal{X}} \mathcal{Z} \to \mathcal{Y} \times_{\mathcal{X}} \mathcal{Z} \to \mathcal{Z}$$

of algebraic $\text{Spec}\,\mathbb{F}_{q^\ell}[\![\zeta]\!]$-stacks. Since these 1-morphisms are representable by morphisms of schemes, we can apply [20] to see that $\mathcal{Y} \times_{\mathcal{X}} \mathcal{Z} \to \mathcal{Z}$ is étale. This proves the lemma.   □

We can now finish the proof of the Uniformization Theorem.

Proof of Uniformization Theorem 12.6. Keep the notation of the proof of Lemma 13.9. We have to show that

$$\mathcal{Y} \times_{\mathcal{X}} \mathcal{Z} \to \mathcal{Z}$$

is a 1-isomorphism of algebraic $\text{Spec}\,\mathbb{F}_q[\![\zeta]\!]$-stacks. Let S be a $\text{Spec}\,\mathbb{F}_q[\![\zeta]\!]$-scheme and let $S \to \mathcal{Z}$ be a 1-morphism. Then from the previous lemmas we conclude that

$$\mathcal{Y} \times_{\mathcal{X}} S \to S$$

is an étale monomorphism of schemes, hence an open immersion. Being also surjective, it is indeed an isomorphism. This completes the proof of the Uniformization Theorem.   □

Appendix: Background on formal algebraic stacks

For a general introduction to the theory of algebraic stacks we refer to Laumon–Moret-Bailly [32] or Deligne–Mumford [9]. In this appendix we propose the notion of formal algebraic stacks which generalizes the notion of algebraic stacks in the same way as formal schemes are a generalization of usual schemes. In fact much of the theory of algebraic stacks can be developed also for formal algebraic stacks. See Hartl [24] for details.

For a scheme S let $\mathcal{S}ch_S$ be the category of S-schemes equipped with the étale topology. An *algebraic stack* $\mathcal{X}$ over S is defined as a category fibered in groupoids over $\mathcal{S}ch_S$ satisfying further conditions [32, Definition 4.1]. We transfer this concept to the formal category.

For the rest of this appendix we let S be a formal scheme. We denote by $\mathcal{N}ilp_S$ the category of schemes over S on which an ideal of definition of S is locally nilpotent. We remind the reader that every scheme may be considered as a formal scheme having (0) as an ideal of definition. In this sense every $U \in \mathcal{N}ilp_S$ is itself a formal scheme. We equip $\mathcal{N}ilp_S$ with the étale topology. We make the following definitions (compare [31, 28]).

Definition A.1. A formal S-space *is a sheaf of sets on the site* $\mathcal{N}ilp_S$.

Definition A.2. A (quasi-separated) formal algebraic S-space *is a formal S-space X such that*

1. *the diagonal morphism $X \to X \times_S X$ is relatively representable by a quasi-compact morphism of formal schemes, and*
2. *there is a formal scheme X' over S and a morphism of formal S-spaces $X' \to X$ which is representable (automatic because of 1) by an étale surjective morphism of formal schemes.*

Definition A.3. A formal S-stack $\mathcal{X}$ *is a category $\mathcal{X}$ fibered in groupoids over $\mathcal{N}ilp_S$ such that*

1. *for every $U \in \mathcal{N}ilp_S$ and every $x, y \in \mathcal{X}(U)$, the presheaf*

$$\mathcal{I}som(x, y) : \mathcal{N}ilp_U \to \mathcal{S}ets,$$
$$(V \to U) \mapsto \mathrm{Hom}_{\mathcal{X}(V)}(x_V, y_V)$$

   is, in fact, a sheaf on $\mathcal{N}ilp_U$;
2. *for every covering $U_i \to U$ in $\mathcal{N}ilp_S$, all descent data for this covering are effective.*

Definition A.4. *A formal S-stack is called* representable *if it is 1-isomorphic to a formal algebraic space.*

   A 1-morphism $\mathcal{X} \to \mathcal{Y}$ of formal S-stacks is called representable *if for every $U \in \mathcal{N}ilp_S$ and every $y \in \mathcal{Y}(U)$ viewed as a 1-morphism $U \to \mathcal{Y}$ of formal S-stacks the fiber product $\mathcal{X} \times_{\mathcal{Y}} U$ (in the 2-category of formal S-stacks) is representable.*

Definition A.5. A (quasi-separated) formal algebraic S-stack *is a formal S-stack $\mathcal{X}$ such that*

1. *the diagonal 1-morphism of formal S-stacks*

$$\mathcal{X} \to \mathcal{X} \times_S \mathcal{X}$$

   is representable, separated, and quasi-compact;
2. *there exists a formal algebraic S-space X and a 1-morphism of formal S-stacks*

$$P : X \to \mathcal{X}$$

which is representable (automatic because of 1*) by a smooth and surjective morphism of formal algebraic S-spaces.*

The 1*-morphism P is called a* presentation *of* $\mathcal{X}$. *We say that* $\mathcal{X}$ *is of* DM-type *if the presentation P in* 2 *can be chosen étale.*

A.6. Let $\mathcal{X}$ be a formal algebraic S-stack and let $P : X \to \mathcal{X}$ be a presentation. We define the underlying reduced stack $\mathcal{X}_{\mathrm{red}}$ as follows: For every $U \in \mathcal{N}ilp_S$ we let $\mathcal{X}_{\mathrm{red}}(U)$ be the full subcategory of $\mathcal{X}(U)$ whose objects are the $x \in \mathcal{X}(U)$ such that there is a covering $U' \to U$ in $\mathcal{N}ilp_S$, an element $x' \in X_{\mathrm{red}}(U')$, and an isomorphism in $\mathcal{X}(U')$ between $x_{U'}$ and $P(x')$. Then $\mathcal{X}_{\mathrm{red}}$ is an algebraic S_{red}-stack. If, moreover, $\mathcal{X}$ is of DM-type, then $\mathcal{X}_{\mathrm{red}}$ is an algebraic S_{red}-stack in the sense of Deligne–Mumford. In this way we obtain from every 1-morphism $f : \mathcal{Y} \to \mathcal{X}$ of formal algebraic S-stacks a 1-morphism $f_{\mathrm{red}} : \mathcal{Y}_{\mathrm{red}} \to \mathcal{X}_{\mathrm{red}}$ of algebraic S_{red}-stacks. We say that f is *locally formally of finite type* if f_{red} is locally of finite type.

A.7. For an algebraic stack $\mathcal{X}$ one can define its *structure sheaf* $\mathcal{O}_{\mathcal{X}}$, which is a sheaf on the lisse-étale site of $\mathcal{X}$; cf. [32, Section 12]. Then one has the usual bijection between closed substacks of $\mathcal{X}$ and quasi-coherent sheaves of ideals of $\mathcal{O}_{\mathcal{X}}$.

The same can be done for formal algebraic S-stacks $\mathcal{X}$. In this setting we say that a sheaf of ideals $\mathcal{J}$ of $\mathcal{O}_{\mathcal{X}}$ is an *ideal of definition of* $\mathcal{X}$ if for some (any) presentation $P : X \to \mathcal{X}$ of $\mathcal{X}$ the ideal sheaf $P^*\mathcal{J}$ is an ideal of definition of X. The formal algebraic S-stack is called $\mathcal{J}$-*adic* if $\mathcal{J}^n$ is an ideal of definition for every n. It is called *adic* if it is $\mathcal{J}$-adic for some $\mathcal{J}$. If $\mathcal{X}$ (i.e., X) is locally noetherian then there exists a unique *largest ideal of definition* $\mathcal{K}$ of $\mathcal{X}$, namely, the one defining the closed substack $\mathcal{X}_{\mathrm{red}}$ of $\mathcal{X}$. Note that if $\mathcal{X}$ is $\mathcal{J}$-adic for some $\mathcal{J}$ then it is also $\mathcal{K}$-adic. One easily verifies the following proposition which generalizes A.6.

Proposition A.8. *Let* $\mathcal{I}$ *be an ideal of definition of S and let* $\mathcal{J}$ *be an ideal of definition of a formal algebraic S-stack* $\mathcal{X}$ *with* $\mathcal{I} \cdot \mathcal{O}_{\mathcal{X}} \subset \mathcal{J}$. *Then the closed substack of* $\mathcal{X}$ *which is defined by the ideal* $\mathcal{J}$ *is an algebraic* $(S, \mathcal{O}_S/\mathcal{I})$*-stack.*

Definition A.9. *A* 1*-morphism* $f : \mathcal{Y} \to \mathcal{X}$ *of locally noetherian formal algebraic S-stacks is called* adic *if for some (any) ideal of definition* $\mathcal{J}$ *of* $\mathcal{X}$ *the ideal* $f^*\mathcal{J}$ *is an ideal of definition of* $\mathcal{Y}$.

We discuss some examples.

Example A.10. Every formal scheme G over S can be viewed as a sheaf of sets on the category $\mathcal{N}ilp_S$. Moreover, every sheaf on $\mathcal{N}ilp_S$ is a formal S-stack. Therefore, we can view every quasi-separated formal scheme G over S as a formal algebraic S-stack of DM-type.

Example A.11. Let S^{alg} be a scheme and let S_0 be a closed subscheme. (We allow the case $S_0 = S^{\mathrm{alg}}$.) We let the formal scheme S be the formal completion of S^{alg} along S_0. If $\mathcal{X}$ is an (algebraic) S^{alg}-stack (in the sense of Deligne–Mumford) then $\mathcal{X} \times_{S^{\mathrm{alg}}} S$ is an adic formal (algebraic) S-stack (of DM-type).

We generalize this example as follows.

Definition A.12. *Let S^{alg} be a scheme and let S_0 be a closed subscheme. We let the formal scheme S be the formal completion of S^{alg} along S_0. Let $\mathcal{X}$ be an algebraic S^{alg}-stack and let $\mathcal{Z} \subset \mathcal{X}$ be a closed substack, contained in $\mathcal{X} \times_{S^{alg}} S_0$. We view the objects of $\mathcal{X}$ as 1-morphisms $U \to \mathcal{X}$ for varying $U \in Sch_{S^{alg}}$. We define the formal completion $\widehat{\mathcal{X}}_{\mathcal{Z}}$ of $\mathcal{X}$ along $\mathcal{Z}$ as the full subcategory of $\mathcal{X}$ consisting of those objects $U \to \mathcal{X}$ such that $U_{red} \to \mathcal{X}$ factors through $\mathcal{Z}$, i.e., such that there exists a 2-commutative diagram of 1-morphisms*

$$
\begin{array}{ccc}
U & \longrightarrow & \mathcal{X} \\
\uparrow & & \uparrow \\
U_{red} & \longrightarrow & \mathcal{Z}.
\end{array}
$$

Note that if $\mathcal{X}$ is an S^{alg}-scheme this definition coincides with the usual definition of formal completion along a closed subscheme.

One verifies directly that $\widehat{\mathcal{X}}_{\mathcal{Z}}$ is an S^{alg}-stack. The embedding $\widehat{\mathcal{X}}_{\mathcal{Z}} \to \mathcal{X}$ is a 1-morphism of S^{alg}-stacks which is a 1-monomorphism, i.e., the functors $\widehat{\mathcal{X}}_{\mathcal{Z}}(U) \to \mathcal{X}(U)$ are fully faithful. This implies that the diagonal 1-morphism

$$
\widehat{\mathcal{X}}_{\mathcal{Z}} \to \widehat{\mathcal{X}}_{\mathcal{Z}} \times_{\mathcal{X}} \widehat{\mathcal{X}}_{\mathcal{Z}}
$$

is a 1-isomorphism of S^{alg}-stacks [32, 2.3]. As a consequence we obtain

Lemma A.13. *Let $\mathcal{Y} \to \widehat{\mathcal{X}}_{\mathcal{Z}}$ and $\mathcal{Y}' \to \widehat{\mathcal{X}}_{\mathcal{Z}}$ be two 1-morphisms of S^{alg}-stacks. Then there is a 1-isomorphism of S^{alg}-stacks*

$$
\mathcal{Y} \times_{\widehat{\mathcal{X}}_{\mathcal{Z}}} \mathcal{Y}' \xrightarrow{\sim} \mathcal{Y} \times_{\mathcal{X}} \mathcal{Y}'.
$$

The fibration $\widehat{\mathcal{X}}_{\mathcal{Z}} \to Sch_{S^{alg}}$ factors through $\mathcal{N}ilp_S \hookrightarrow Sch_{S^{alg}}$. Note the fact that for an étale covering $U_i \to U$ in $Sch_{S^{alg}}$ we have $U \in \mathcal{N}ilp_S$ if and only if $U_i \in \mathcal{N}ilp_S$ for all i. This implies that $\widehat{\mathcal{X}}_{\mathcal{Z}}$ is a formal S-stack. We show that $\widehat{\mathcal{X}}_{\mathcal{Z}}$ is even a formal algebraic S-stack. Namely, condition 1 of Definition A.5 can be read off from the following 2-cartesian diagram of S^{alg}-stacks

$$
\begin{array}{ccccccc}
\widehat{\mathcal{X}}_{\mathcal{Z}} & \xrightarrow{\sim} & \widehat{\mathcal{X}}_{\mathcal{Z}} \times_{\mathcal{X}} \widehat{\mathcal{X}}_{\mathcal{Z}} & \longrightarrow & \widehat{\mathcal{X}}_{\mathcal{Z}} \times_{S^{alg}} \widehat{\mathcal{X}}_{\mathcal{Z}} & \xleftarrow{\sim} & \widehat{\mathcal{X}}_{\mathcal{Z}} \times_S \widehat{\mathcal{X}}_{\mathcal{Z}} \\
& & \downarrow & & \downarrow & & \\
& & \mathcal{X} & \longrightarrow & \mathcal{X} \times_{S^{alg}} \mathcal{X}. & &
\end{array}
$$

Condition 2 results from the fact that every presentation $X \to \mathcal{X}$ of $\mathcal{X}$ induces a presentation $\widehat{X}_Z \to \widehat{\mathcal{X}}_{\mathcal{Z}}$ of $\widehat{\mathcal{X}}_{\mathcal{Z}}$ by the formal completion $\widehat{X}_Z$ of X along the closed subscheme $Z = X \times_{\mathcal{X}} \mathcal{Z}$. Thus we have proved

Proposition A.14. *Keep the notation of Definition A.12. Then the formal completion $\widehat{\mathcal{X}}_{\mathcal{Z}}$ of $\mathcal{X}$ along $\mathcal{Z}$ is a formal algebraic S-stack. If $\mathcal{X}$ is an algebraic S^{alg}-stack in the sense of Deligne–Mumford then $\widehat{\mathcal{X}}_{\mathcal{Z}}$ is of DM-type. If $\mathcal{J}$ is the ideal sheaf on $\mathcal{X}$ defining the closed substack $\mathcal{Z}$ then $\widehat{\mathcal{X}}_{\mathcal{Z}}$ is $\mathcal{J} \cdot \mathcal{O}_{\widehat{\mathcal{X}}_{\mathcal{Z}}}$-adic.*

A.15. Let S^{alg} be a scheme and let $\mathcal{X}$ be an algebraic S^{alg}-stack. There is a notion of *points* of $\mathcal{X}$. Namely, a point of $\mathcal{X}$ is given by a 1-morphism Spec $K \to \mathcal{X}$ for an S^{alg}-field K. The set $|\mathcal{X}|$ of points of $\mathcal{X}$ forms a topological space; cf. [32, Section 5]. Let $Z \subset |\mathcal{X}|$ be a closed subset, i.e., there is an open substack $\mathcal{U} \subset \mathcal{X}$ such that $Z = |\mathcal{X}| \smallsetminus |\mathcal{U}|$. We can equip Z in a unique way with a structure of reduced closed substacks [32, 4.10]. By Definition A.12, we can consider the formal completion of $\mathcal{X}$ along Z.

Example A.16 (quotients). Let $U \in \mathcal{N}ilp_S$ and let G be a formal U-group space (i.e., a group object in the category of formal U-spaces). A (left) G-*torsor* is a formal U-space P with an action of G (from the left) such that there is a covering $U' \to U$ in $\mathcal{N}ilp_S$ for which $P \times_U U'$ is $G \times_U U'$-isomorphic to $G \times_U U'$ which acts on itself by left translation.

Let X be a formal S-space, Y an X-space (i.e., a formal S-space equipped with a morphism $Y \to X$) and G an X-group space which acts on Y from the left. We define the quotient stack $G\backslash Y$ as the following category fibered in groupoids over $\mathcal{N}ilp_S$: For every $U \in \mathcal{N}ilp_S$ the category $(G\backslash Y)(U)$ consists of all triples (x, P, α) where $x \in X(U)$, P is a $G \times_{X,x} U$-torsor and $\alpha : P \to Y_{X,x}U$ is a $G \times_{X,x} U$-equivariant morphism of formal U-spaces. One easily verifies that the quotient $G\backslash Y$ is a formal S-stack.

In particular, if X is an adic formal algebraic S-space and G a finite étale S-group scheme, then the quotient $G\backslash X$ is even an adic formal algebraic S-stack of DM-type. In this case, the canonical projection $X \to G\backslash X$ is an étale presentation of $G\backslash X$.

References

[1] G. Anderson, *t*-motives, *Duke Math. J.*, **53** (1986), 457–502.

[2] A. Blum and U. Stuhler, Drinfeld modules and elliptic sheaves, in S. Kumar, G. Laumon, U. Stuhler, and M. S. Narasimhan, eds., *Vector Bundles on Curves: New Directions*, Lecture Notes in Mathematics 1649, Springer-Verlag, Berlin, New York, 1991, 110–188.

[3] G. Böckle and U. Hartl, Uniformizable families of *t*-motives, preprint, 2004; arXiv: math.NT/0411262.

[4] S. Bosch and W. Lütkebohmert: Formal and rigid geometry I: Rigid spaces, *Math. Ann.*, **295** (1993), 291–317.

[5] J.-F. Boutot and H. Carayol, Uniformisation *p*-adique des courbes de Shimura: Les théorèmes de Čerednik et de Drinfel'd, in *Courbes modulaires et courbes de Shimura (Orsay, 1987/1988)*, Astérisque 196–197, Société Mathématique de France, Paris, 1991, 45–158.

[6] V. Čerednik, Uniformization of algebraic curves by discrete arithmetic subgroups of $PGL_2(k_w)$ with compact quotients, *Math. USSR. Sb.*, **29** (1976), 55–78.

[7] G. Cornell, J. Silverman, and G. Stevens, eds., *Modular Forms and Fermat's Last Theorem*, Springer-Verlag, New York, 1997.

[8] P. Deligne and D. Husemöller, Survey of Drinfeld modules, in *Current Trends in Arithmetical Algebraic Geometry (Arcata, California, 1985)*, Contemporary Mathematics 67, American Mathematical Society, Providence, RI, 1987, 25–91.

[9] P. Deligne and D. Mumford, The irreducibility of the space of curves of given genus, *Publ. Math. IHES*, **36** (1969), 75–110.

[10] V. G. Drinfeld, Elliptic modules, *Math. USSR-Sb.*, **23** (1976), 561–592.

[11] V. G. Drinfeld, Coverings of p-adic symmetric domains, *Functional Anal. Appl.*, **10** (1976), 107–115.

[12] V. G. Drinfeld, Commutative subrings of certain noncommutative rings, *Functional Anal. Appl.*, **11** (1977), 9–12.

[13] V. G. Drinfeld, A proof of Langlands' global conjecture for GL(2) over a function field, *Functional Anal. Appl.*, **11**-3 (1977), 223–225.

[14] V. G. Drinfeld, Moduli variety of F-sheaves, *Functional Anal. Appl.*, **21**-2 (1987), 107–122.

[15] D. Eisenbud, *Commutative Algebra with a View Toward Algebraic Geometry*, Graduate Texts in Mathematics 150, Springer-Verlag, Berlin, New York, 1995.

[16] G. Faltings, Endlichkeitssätze für abelsche Varietäten über Zahlkörpern, *Invent. Math.*, **73** (1983), 349–366.

[17] F. Gardeyn, New criteria for uniformization of t-motives, preprint, 2001.

[18] A. Genestier, *Espaces symétriques de Drinfeld*, Astérisque 234, Société Mathématique de France, Paris, 1996.

[19] A. Grothendieck, Fondements de la géométrie algébrique, in *Extraits du Séminaire Bourbaki 1957–1962*, Secrétariat Mathématique, Paris, 1962.

[20] A. Grothendieck, *Élements de Géométrie Algébrique*, Publications Mathématiques IHES 4, 8, 11, 17, 20, 24, 28, 32, Institut des Hautes Études Scientifiques, Bures-sur-Yvette, France, 1960–1967; see also Grundlehren 166, Springer-Verlag, Berlin, Heidelberg, 1971.

[21] A. Grothendieck, *Groupes de Barsotti-Tate et cristaux de Dieudonné*, Séminaire de Mathématiques Supérieures 45, Les Presses de l'Université de Montréal, Montreal, 1974.

[22] M. Harris and R. Taylor, *The Geometry and Cohomology of Some Simple Shimura Varieties*, Annals of Mathematics Studies 151, Princeton University Press, Princeton, NJ, 2001.

[23] U. Hartl, Local Shtuka and divisible local Anderson modules, in preparation.

[24] U. Hartl, Formal algebraic stacks, in preparation.

[25] R. Hartshorne, *Algebraic Geometry*, Graduate Texts in Mathematics 52, Springer-Verlag, Berlin, New York, Heidelberg, 1977.

[26] Th. Hausberger, *Uniformisations des Variétés de Laumon-Rapoport-Stuhler et application à la correspondance de Langlands locale*, Ph.D. thesis, Université Louis Pasteur (Strasbourg I), Strasbourg, 2001.

[27] N. Katz, Slope filtration of F-crystals, in *Journées de Géométrie Algébrique de Rennes* (*Rennes, 1978*, Vol. I, Astérisque 63, Société Mathématique de France, Paris, 1979, 113–163.

[28] D. Knutson, *Algebraic Spaces*, Lecture Notes in Mathematics 203, Springer-Verlag, Berlin, New York, Heidelberg, 1971.

[29] L. Lafforgue, *Chtoucas de Drinfeld et conjecture de Ramanujan-Petersson*, Astérisque 243, Société Mathématique de France, Paris, 1997.

[30] L. Lafforgue, Chtoucas de Drinfeld et correspondance de Langlands, *Invent. Math.*, **147** (2002), 1–241.

[31] G. Laumon, *Cohomology of Drinfeld Modular Varieties I*, Cambridge Studies in Advanced Mathematics 41, Cambridge University Press, Cambridge, UK, 1996.

[32] G. Laumon and L. Moret-Bailly, *Champs algébriques*, Ergebnisse der Mathematik und ihrer Grenzgebiete 39, Springer-Verlag, Berlin, New York, 2000.

[33] G. Laumon, M. Rapoport, and U. Stuhler, $\mathcal{D}$-elliptic sheaves and the Langlands correspondence, *Invent. Math.*, **113** (1993), 217–338.

[34] Y. I. Manin, The theory of commutative formal groups over fields of finite characteristic, *Russian Math. Surveys*, **18**-6 (1963), 1–83.

[35] I. Y. Potemine, Drinfeld-Anderson motives and multicomponent KP hierarchy, in *Recent Progress in Algebra: An International Conference on Recent Progress in Algebra, August 11-15, 1997, Kaist, Taejon, South Korea*, Contemporary Mathematics 224, American Mathematical Society, Providence, RI, 1999.

[36] M. Rapoport and T. Zink, *Period Spaces for p-Divisible Groups*, Annals of Mathematics Studies 141, Princeton University Press, Princeton, NJ, 1996.

[37] M. Raynaud, Géométrie analytique rigide d'apres Tate, Kiehl, ..., *Bul. Soc. Math. France Mém.*, **39/40** (1974), 319–327.

[38] M. Rosen, Formal Drinfeld modules, *J. Number Theory*, **103** (2003), 234–256.

[39] C. S. Seshadri, *Fibrés vectoriels sur les courbes algébriques*, Astérisque 96, Société Mathématique de France, Paris, 1982.

[40] U. Stuhler, *P*-adic homogeneous spaces and moduli problems, *Math. Z.*, **192** (1986), 491–540.

[41] Y. Taguchi, Semi-simplicity of the Galois representations attached to drinfeld modules over fields of "infinite characteristics", *J. Number Theory*, **44**-3 (1993), 292–314.

Faltings' Delta-Invariant of a Hyperelliptic Riemann Surface

Robin de Jong

Mathematisch Instituut
Universiteit Leiden
Niels Bohrweg 1
2333 CA Leiden
The Netherlands
rdejong@math.leidenuniv.nl

1 Introduction

The analogy between number fields and function fields extends to varieties defined over such fields. Indeed, if one adds in on the arithmetic side contributions that correspond to the complex embeddings of a number field, it becomes possible to construct intersection theories with properties analogous to those that we have on the function field side. The major ingredients of arithmetic intersection theory are the product formula for number fields and complex differential geometry to deal with the local intersections "at infinity."

In this paper we consider the original intersection theory for arithmetic surfaces developed by Arakelov [1] and Faltings [7]. Already in this theory, one encounters interesting complex differential geometric invariants, and it is worthwhile to study these invariants in more detail. We focus on the delta-invariant, which was defined in [7], and give an explicit formula for it in the case of a hyperelliptic Riemann surface of arbitrary genus. We note that [7] treats the case of elliptic curves, and that the case of Riemann surfaces of genus 2 has been considered before in [3].

In order to state our result, let us recall some notation and earlier results. Let X be a compact and connected Riemann surface of genus $g > 0$. Let G be the Arakelov-Green function of X and let μ be the fundamental $(1, 1)$-form on X as defined in [1, 7]. Let $S(X)$ be the invariant defined by

$$\log S(X) := - \int_X \log \|\vartheta\|(gP - Q) \cdot \mu(P).$$

Here $\|\vartheta\|$ is the function on $\mathrm{Pic}_{g-1}(X)$ defined as on [7, p. 401] and Q can be any point on X. The integral is well defined and is independent of the choice of the point Q. In our paper [10], we gave an explicit formula for the Arakelov–Green function of X.

Theorem 1.1. *Let $\mathcal{W}$ be the classical divisor of Weierstrass points on X. For P, Q points on X, with P not a Weierstrass point, we have*

$$G(P, Q)^g = S(X)^{1/g^2} \cdot \frac{\|\vartheta\|(gP - Q)}{\prod_{W \in \mathcal{W}} \|\vartheta\|(gP - W)^{1/g^3}}.$$

Here the product runs over the Weierstrass points of X, counted with their weights. The formula is also valid if P is a Weierstrass point, provided that we take the leading coefficients of a power series expansion about P in both numerator and denominator.

In the same paper, we also gave an explicit formula for the delta-invariant of X. The delta-invariant is a fundamental invariant of X, expressing the proportionality between two natural metrics on the determinant of the Hodge bundle. For P on X, not a Weierstrass point, and z a local coordinate about P, we put

$$\|F_z\|(P) := \lim_{Q \to P} \frac{\|\vartheta\|(gP - Q)}{|z(P) - z(Q)|^g}.$$

Further, we let $W_z(\omega)(P)$ be the Wronskian at P in z of an orthonormal basis $\{\omega_1, \dots, \omega_g\}$ of the differentials $H^0(X, \Omega_X^1)$ provided with the standard hermitian inner product $(\omega, \eta) \mapsto \frac{i}{2} \int_X \omega \wedge \bar{\eta}$. We define an invariant $T(X)$ of X by

$$T(X) := \|F_z\|(P)^{-(g+1)} \cdot \prod_{W \in \mathcal{W}} \|\vartheta\|(gP - W)^{(g-1)/g^3} \cdot |W_z(\omega)(P)|^2,$$

where again the product runs over the Weierstrass points of X, counted with their weights. It can be checked that this depends on neither the choice of P nor the choice of local coordinate z about P. A more intrinsic definition is possible (see [10]), but the above formula will be convenient for us. We remark that $T(X)$ does not involve an integral over X, contrary to the invariant $S(X)$.

Theorem 1.2. *For Faltings' delta-invariant $\delta(X)$ of X, the formula*

$$\exp(\delta(X)/4) = S(X)^{-(g-1)/g^2} \cdot T(X)$$

holds.

In this paper, we make the invariant $T(X)$ explicit in the case that X is a hyperelliptic Riemann surface of genus $g \geq 2$. We relate it to a nonzero invariant $\|\varphi_g\|(X)$ of X, the Petersson norm of the modular discriminant associated to X, which we introduce in Section 2. As we will see, for hyperelliptic Riemann surfaces, this is a very natural invariant to consider. Unfortunately, it is not so clear how to extend its definition to the general Riemann surface of genus g.

Definition 1.3. *We denote by G' the modified Arakelov–Green function*

$$G'(P, Q) := S(X)^{-1/g^3} \cdot G(P, Q)$$

on $X \times X$.

We prove the following theorem dealing with G' and $T(X)$. Recall that the Weierstrass points of X are just the ramification points of a hyperelliptic map $X \to \mathbb{P}^1$.

Theorem 1.4. *Let W be a Weierstrass point of X. Let $n = \binom{2g}{g+1}$. Consider the product $\prod_{W' \neq W} G'(W, W')$ running over all Weierstrass points W' different from W, ignoring their weights. Then $\prod_{W' \neq W} G'(W, W')$ is independent of the choice of W and the formula*

$$\prod_{W' \neq W} G'(W, W')^{(g-1)^2} = 2^{(g-1)^2} \cdot \pi^{2g+2} \cdot T(X)^{\frac{g+1}{g}} \cdot \|\varphi_g\|(X)^{\frac{1}{2n}}$$

holds.

The next theorem will be derived in a forthcoming article [11]. The result looks similar to the formula in Theorem 1.4, but the proof is very different.

Theorem 1.5. *Let $m = \binom{2g+2}{g}$. Then we have*

$$\prod_{(W, W')} G'(W, W')^{n(g-1)} = \pi^{-2g(g+2)m} \cdot T(X)^{-(g+2)m} \cdot \|\varphi_g\|(X)^{-\frac{3}{2}(g+1)},$$

the product running over all ordered pairs of distinct Weierstrass points of X.

Combining the above two theorems yields a simple closed formula for the invariant $T(X)$ in terms of $\|\varphi_g\|(X)$.

Theorem 1.6. *Consider the modified discriminant $\|\Delta_g\|(X) := 2^{-(4g+4)n} \cdot \|\varphi_g\|(X)$. Then the formula*

$$T(X) = (2\pi)^{-2g} \cdot \|\Delta_g\|(X)^{-\frac{3g-1}{8ng}}$$

holds.

Combining this with Theorem 1.2, we obtain the following corollary.

Corollary 1.7. *For Faltings' delta-invariant $\delta(X)$ of X, the formula*

$$\exp(\delta(X)/4) = (2\pi)^{-2g} \cdot S(X)^{-(g-1)/g^2} \cdot \|\Delta_g\|(X)^{-\frac{3g-1}{8ng}}$$

holds.

The significance of this result is that it makes the efficient calculation of the delta-invariant possible for hyperelliptic Riemann surfaces. We have given a demonstration of this in our paper [10].

We remark that in the case $g = 2$, an explicit formula for the delta-invariant has been given already by Bost [3]. Apart from the Petersson norm of the modular discriminant, his formula involves an invariant $\|H\|(X)$. This invariant has properties similar to our $S(X)$.

The idea of the proof of Theorem 1.4 is quite straightforward: we start with the definition of the invariant $T(X)$ and the formula for G in Theorem 1.1 and observe what happens if we let P approach the Weierstrass point W on X. Thus we have to perform a local study around W of the function $\prod_{W'} \|\vartheta\|(gP - W')$ and of the functions $\|F_z\|(P)$ and $W_z(\omega)(P)$ for a suitable local coordinate z. In Section 3, we find a suitable local coordinate on an embedding of X into its jacobian. In Section 6, we collect the local information that we need in order to complete the proof in Section 7. Some preliminary work on this local information is carried out in Sections 4 and 5. These two sections form the technical heart of the paper.

2 The modular discriminant

In this section, we introduce the modular discriminant φ_g and its Petersson norm $\|\varphi_g\|$. The modular discriminant generalises the usual discriminant function Δ for elliptic curves.

Let $g \geq 2$ be an integer and let $\mathcal{H}_g$ be the Siegel upper half-space of symmetric complex $g \times g$-matrices with positive definite imaginary part. For $z \in \mathbb{C}^g$ (viewed as a column vector), a matrix $\tau \in \mathcal{H}_g$ and $\eta', \eta'' \in \frac{1}{2}\mathbb{Z}^g$, we have the theta function with characteristic $\eta = \left[\begin{smallmatrix}\eta'\\\eta''\end{smallmatrix}\right]$ given by

$$\vartheta[\eta](z; \tau) := \sum_{n \in \mathbb{Z}^g} \exp(\pi i^t(n + \eta')\tau(n + \eta') + 2\pi i^t(n + \eta')(z + \eta'')).$$

For any subset S of $\{1, 2, \ldots, 2g + 1\}$, we define a theta characteristic η_S as in [14, Chapter IIIa]: let

$$\eta_{2k-1} = \begin{bmatrix} {}^t(0, \ldots, 0, \frac{1}{2}, 0, \ldots, 0) \\ {}^t(\frac{1}{2}, \ldots, \frac{1}{2}, 0, 0, \ldots, 0) \end{bmatrix}, \quad 1 \leq k \leq g + 1,$$

$$\eta_{2k} = \begin{bmatrix} {}^t(0, \ldots, 0, \frac{1}{2}, 0, \ldots, 0) \\ {}^t(\frac{1}{2}, \ldots, \frac{1}{2}, \frac{1}{2}, 0, \ldots, 0) \end{bmatrix}, \quad 1 \leq k \leq g,$$

where the nonzero entry in the top row occurs in the kth position. Then we put $\eta_S := \sum_{k \in S} \eta_k$ where the sum is taken modulo 1.

Definition 2.1 (cf. [13, Section 3]). *Let $\mathcal{T}$ be the collection of subsets of $\{1, 2, \ldots, 2g + 1\}$ of cardinality $g + 1$. Write $U = \{1, 3, \ldots, 2g + 1\}$ and let $\circ$ denote the symmetric difference. The modular discriminant φ_g is defined to be the function*

$$\varphi_g(\tau) := \prod_{T \in \mathcal{T}} \vartheta[\eta_{T \circ U}](0; \tau)^8$$

on $\mathcal{H}_g$. The function φ_g is a modular form on $\Gamma_g(2) := \{\gamma \in \mathrm{Sp}(2g, \mathbb{Z}) | \gamma \equiv I_{2g} \bmod 2\}$ of weight $4r$, where $r = \binom{2g+1}{g+1}$.

Consider an equation $y^2 = f(x)$, where $f \in \mathbb{C}[X]$ is a monic and separable polynomial of degree $2g + 1$. Write $f(x) = \prod_{k=1}^{2g+1}(x - a_k)$ and denote by $D = \prod_{k<l}(a_k - a_l)^2$ the discriminant of f. Let X be the hyperelliptic Riemann surface of genus g defined by $y^2 = f(x)$. Then X carries a basis of holomorphic differentials $\mu_k := x^{k-1}dx/2y$, where $k = 1, \ldots, g$. Further, in [14, Chapter IIIa, Section 5], it is shown how, given an ordering of the roots of f, one can construct a canonical symplectic basis of the homology of X. Throughout this paper, we will always work with such a canonical basis of homology, i.e., a certain ordering of the roots of a hyperelliptic equation will always be taken for granted.

Let $(\mu|\mu')$ be the period matrix of the differentials μ_k with respect to a chosen canonical basis of homology, and let $\tau = \mu^{-1}\mu'$.

Proposition 2.2. *We have the formula*

$$D^n = \pi^{4gr}(\det \mu)^{-4r}\varphi_g(\tau)$$

relating the discriminant D of the polynomial f to the value $\varphi_g(\tau)$ of the modular discriminant.

Proof. See [13, Proposition 3.2]. □

Definition 2.3. *Let X be a hyperelliptic Riemann surface of genus $g \geq 2$ and let τ be a period matrix for X formed on a canonical symplectic basis, given by an ordering of the roots of an equation $y^2 = f(x)$ for X. Then we write $\|\varphi_g\|(\tau)$ for the Petersson norm $(\det \operatorname{Im} \tau)^{2r} \cdot |\varphi_g(\tau)|$ of $\varphi_g(\tau)$. This does not depend on the choice of τ and hence it defines an invariant $\|\varphi_g\|(X)$ of X.*

It follows from Proposition 2.2 that $\|\varphi_g\|(X)$ is nonzero.

3 A local coordinate

For our local computations on our hyperelliptic Riemann surface we need a convenient local coordinate. We find one by embedding the Riemann surface into its jacobian and by taking one of the euclidean coordinates.

Let X be a hyperelliptic Riemann surface of genus $g \geq 2$, let $y^2 = f(x)$ with f monic of degree $2g + 1$ be an equation for X, let μ_k be the differential given by $\mu_k = x^{k-1}dx/2y$ for $k = 1, \ldots, g$, and let $(\mu|\mu')$ be their period matrix formed on a canonical basis of homology. Let L be the lattice in $\mathbb{C}^g$ generated by the columns of $(\mu|\mu')$. We have an embedding $\iota : X \hookrightarrow \mathbb{C}^g/L$ given by integration $P \mapsto \int_\infty^P(\mu_1, \ldots, \mu_g)$. We want to express the coordinates $z_1, \ldots, z_g$, restricted to $\iota(X)$, in terms of a local coordinate about $0 = \iota(\infty)$. This is established by the following lemma. In general, we denote by $O(w_1, \ldots, w_s; d)$ a Laurent series in the variables $w_1, \ldots, w_s$ all of whose terms have total degree at least d. We owe the argument to [12].

Lemma 3.1. *The coordinate z_g is a local coordinate about 0 on $\iota(X)$, and we have*

$$z_k = \frac{1}{2(g-k)+1} z_g^{2(g-k)+1} + O(z_g; 2(g-k)+2)$$

on $\iota(X)$ for $k = 1, \ldots, g$.

Proof. We can choose a local coordinate t about ∞ on X such that $x = t^{-2}$ and $y = -t^{-(2g+1)} + O(t; -2g)$. For $P \in X$ in a small enough neighborhood of ∞ on X and for a suitable integration path on X, we then have

$$z_k(P) = \int_{\infty}^{P} \frac{x^{k-1}dx}{2y} = \int_0^{t(P)} \frac{t^{-2(k-1)} \cdot (-2t^{-3}dt)}{-2t^{-(2g+1)} + O(t; -2g)}$$

$$= \int_0^{t(P)} (t^{2(g-k)} + O(t; 2(g-k)+1))dt$$

$$= \frac{1}{2(g-k)+1} t(P)^{2(g-k)+1} + O(t(P); 2(g-k)+2).$$

By taking $k = g$, we find $z_g = t + O(t; 2)$, and for $k = 1, \ldots, g-1$, then

$$z_k = \frac{1}{2(g-k)+1} z_g^{2(g-k)+1} + O(z_g; 2(g-k)+2),$$

which is what we wanted.   □

4 Schur polynomials

In this section, we assemble some facts on Schur polynomials. We will need these facts at various places in the next sections. Fix a positive integer g. Consider the ring of symmetric polynomials with integer coefficients in the variables $x_1, \ldots, x_g$. The elementary symmetric functions e_r are defined by means of the generating function $E(t) = \sum_{r \geq 0} e_r t^r = \prod_{k=1}^{g} (1 + x_k t)$.

Definition 4.1. *Let d be a positive integer and let $\pi = \{\pi_1, \ldots, \pi_h\}$ with $\pi_1 \geq \cdots \geq \pi_h$ be a partition of d. The Schur polynomial associated to π is the polynomial*

$$S_\pi := \det(e_{\pi'_k - k + l})_{1 \leq k, l \leq h},$$

where h is the length of the partition π, and where π' is the conjugate partition of π given by $\pi'_k = \#\{l : \pi_l \geq k\}$, i.e., the partition obtained by switching the associated Young diagram around its diagonal. The polynomial S_π is symmetric and has total degree d. We denote by S_g the Schur polynomial in g variables associated to the partition $\pi = \{g, g-1, \ldots, 2, 1\}$. Thus the formula

$$S_g = \det(e_{g-2k+l+1})_{1 \leq k, l \leq g}$$

holds, and the polynomial S_g has total degree $g(g+1)/2$.

We denote by p_r the elementary Newton functions (power sums) given by the generating function $P(t) = \sum_{r \geq 1} p_r t^{r-1} = \sum_{k \geq 1} x_k / (1 - x_k t)$. The following proposition is then a special case of [5, Theorem 4.1].

Proposition 4.2. *The Schur polynomial S_g can be expressed as a polynomial in the g functions $p_1, p_3, \ldots, p_{2g-1}$ only. This polynomial is unique.*

Definition 4.3. *We define s_g to be the unique polynomial in g variables given by Proposition 4.2.*

The next proposition is a special case of [5, Theorem 6.2].

Proposition 4.4. *Let $s(x_1, \ldots, x_g) \in \mathbb{C}[x_1, \ldots, x_g]$ be a polynomial in g variables such that for any set of g complex numbers $w_1, \ldots, w_g$, the polynomial $s(z_1 - w, z_2 - w^3, \ldots, z_g - w^{2g-1})$ in w either has exactly g roots $w_1, \ldots, w_g$, or vanishes identically, if we give z the value*

$$z = (p_1(w_1, \ldots, w_g), p_3(w_1, \ldots, w_g), \ldots, p_{2g-1}(w_1, \ldots, w_g)).$$

Then s is equal to the polynomial s_g up to a constant factor.

Definition 4.5. *We define σ_g to be the polynomial in g variables given by the equation*

$$\sigma_g(z_1, \ldots, z_g) = s_g(z_g, 3z_{g-1}, \ldots, (2g-1)z_1).$$

The following proposition is then the result of a simple calculation.

Proposition 4.6. *Up to a sign, the homogeneous part of least total degree of σ_g is equal to the Hankel determinant*

$$H(z) = \det \begin{pmatrix} z_1 & z_2 & \cdots & z_{(g+1)/2} \\ z_2 & z_3 & \cdots & z_{(g+3)/2} \\ \vdots & \vdots & \ddots & \vdots \\ z_{(g+1)/2} & z_{(g+3)/2} & \cdots & z_g \end{pmatrix}$$

if g is odd, or

$$H(z) = \det \begin{pmatrix} z_1 & z_2 & \cdots & z_{g/2} \\ z_2 & z_3 & \cdots & z_{(g+2)/2} \\ \vdots & \vdots & \ddots & \vdots \\ z_{g/2} & z_{(g+2)/2} & \cdots & z_{g-1} \end{pmatrix}$$

if g is even.

We conclude with some more general facts. These can all be found, for example, in [8, Appendix A].

Proposition 4.7. *Let $\pi = \{\pi_1, \ldots, \pi_h\}$ with $\pi_1 \geq \cdots \geq \pi_h$ be a partition. Then the formula*

$$S_\pi(1, \ldots, 1) = \prod_{k<l} \frac{\pi_k - \pi_l + l - k}{l - k}$$

holds. In particular, $S_g(1, \ldots, 1) = 2^{g(g-1)/2}$.

Definition 4.8. *Let $\mathbf{i} = (i_1, \ldots, i_d)$ be a d-tuple of nonnegative integers. The $\mathbf{i}$th generalized Newton function $p^{(\mathbf{i})}$ is defined to be the polynomial*

$$p^{(\mathbf{i})} := p_1^{i_1} \cdot p_2^{i_2} \cdots \cdots p_d^{i_d},$$

where the p_r are the elementary Newton functions.

Proposition 4.9. *The set of generalized Newton functions $p^{(\mathbf{i})}$, where $\mathbf{i}$ runs through the d-tuples $\mathbf{i} = (i_1, \ldots, i_d)$ of nonnegative integers with $\sum \alpha i_\alpha = d$, forms a basis of the $\mathbb{Q}$-vector space of symmetric polynomials with rational coefficients of total degree d.*

Proposition 4.10. *For a partition π of d and a d-tuple $\mathbf{i} = (i_1, \ldots, i_d)$, denote by $\omega_\pi(\mathbf{i})$ the coefficient of the monomial $x_1^{\pi_1} \cdots \cdots x_d^{\pi_d}$ in $p^{(\mathbf{i})}$. Then the polynomial S_π can be expanded on the basis $\{p^{(\mathbf{i})}\}$ of generalized Newton functions of total degree d as $S_\pi = \sum_{\mathbf{i}} \frac{1}{z(\mathbf{i})} \cdot \omega_\pi(\mathbf{i}) \cdot p^{(\mathbf{i})}$. Here $z(\mathbf{i}) = i_1! 1^{i_1} \cdot i_2! 2^{i_2} \cdots \cdots i_d! d^{i_d}$.*

5 The sigma function

We consider again hyperelliptic Riemann surfaces of genus $g \geq 2$, defined by equations $y^2 = f(x)$ with f monic and separable of degree $2g + 1$. We write $f(x) = x^{2g+1} + \lambda_1 x^{2g} + \cdots + \lambda_{2g} x + \lambda_{2g+1}$ and denote by λ the vector of coefficients $(\lambda_1, \ldots, \lambda_{2g+1})$. In this section, we study the sigma function $\sigma(z; \lambda)$ with argument $z \in \mathbb{C}^g$ and parameter λ. This is a modified theta function, studied extensively in the 19th Century. Klein observed that the sigma function serves very well to study the function theory of hyperelliptic Riemann surfaces. For us it will be a convenient technical tool for obtaining the local expansions that we need. We will give the definition of the sigma function, as well as its power series expansion in z, λ. For more details we refer to the *Enzyklopädie der mathematischen Wissenschaften*, Band II, Teil 2, Kapitel 7.XII. A modern reference is [4], where one also finds applications of the sigma function in the theory of the Korteweg–de Vries differential equation.

As before, let μ_k be the holomorphic differential given by $\mu_k = x^{k-1} dx/2y$ for $k = 1, \ldots, g$, and let $(\mu|\mu')$ be their period matrix formed on a canonical basis of homology. Let L be the lattice in $\mathbb{C}^g$ generated by the columns of $(\mu|\mu')$. By the theorem of Abel–Jacobi, we have a bijective map $\mathrm{Pic}_{g-1}(X) \xrightarrow{\sim} \mathbb{C}^g/L$ given by $\sum_k m_k P_k \longmapsto \sum_k m_k \int_\infty^{P_k} (\mu_1, \ldots, \mu_g)$. Denote by Θ the image of the theta divisor of classes of effective divisors of degree $g - 1$, and let $q : \mathbb{C}^g \to \mathbb{C}^g/L$ be the projection map. Let $\tau = \mu^{-1}\mu'$. By a fundamental theorem of Riemann, there exists a unique theta characteristic δ such that $\vartheta[\delta](z; \tau)$ vanishes to order 1 precisely along $q^{-1}(\Theta)$.

Definition 5.1. *Let ν be the matrix of A-periods of the differentials of the second kind* $\nu_k := \frac{1}{4y} \sum_{l=k}^{2g-k}(l+1-k)\lambda_{l+k+1}x^k dx$ *for $k = 1, \ldots, g$. These differentials have a second-order pole at ∞ and no other poles. The sigma function is then the function*

$$\sigma(z; \lambda) := \exp\left(-\frac{1}{2}z\nu\mu^{-1}{}^t z\right) \cdot \vartheta[\delta](\mu^{-1}z; \tau).$$

Using some of the facts on Schur polynomials from the previous section, we can give the power series expansion of $\sigma(z; \lambda)$. The result is probably well known to specialists, although we couldn't find an explicit reference in the literature. For the formulation and the proof, we were inspired by [12], as well as by a private communication with the author. For the special case $g = 2$, a somewhat stronger version of the result has been obtained by Grant; see [9, Theorem 2.11].

Proposition 5.2. *The power series expansion of $\sigma(z; \lambda)$ about $z = 0$ is of the form*

$$\sigma(z; \lambda) = \gamma \cdot \sigma_g(z) + O(\lambda),$$

where σ_g is the polynomial given by Definition 4.5 and where the symbol $O(\lambda)$ denotes a power series in z, λ in which each term contains a λ_k raised to a positive integral power. The constant γ satisfies the formula

$$\gamma^{8n} = \pi^{4g(r-n)}(\det \mu)^{-4(r-n)}\varphi_g(\tau).$$

If we assign the variable z_k a weight $2(g - k) + 1$, and the variable λ_k a weight $-2k$, then the power series expansion in z, λ of $\sigma(z; \lambda)$ is homogeneous of weight $g(g + 1)/2$.

Proof. First of all, the homogeneity of the power series expansion in z, λ with respect to the assigned weights follows from an explicit formula for $\sigma(z; \lambda)$ given in [6]. This homogeneity is also mentioned there; cf. the concluding remarks after Corollary 1. Write $\sigma(z; \lambda) = \sigma_0(z) + O(\lambda)$ where $O(\lambda)$ denotes a power series in z, λ in which each term contains a λ_k raised to a positive integral power. Because of the homogeneity, the series $\sigma_0(z)$ is necessarily a polynomial in the variables $z_1, \ldots, z_g$. By the Riemann vanishing theorem, there is a dense open subset $U \subset \mathbb{C}^{2g+1}$ such that for any $\lambda \in U$, the function $\sigma(z; \lambda)$ satisfies the following property: for any set of g points $P_1, \ldots, P_g$ on the hyperelliptic Riemann surface $X = X_\lambda$ corresponding to λ, the function $\sigma(z - \int_\infty^P (\mu_1, \ldots, \mu_g); \lambda)$ in P on X either has exactly g roots $P_1, \ldots, P_g$, or vanishes identically, when we give the argument z the value $z = \sum_k \int_\infty^{P_k}(\mu_1, \ldots, \mu_g)$. In the limit $\lambda \to 0$, we find, then, as in the proof of Lemma 3.1, that for any set of g complex numbers $w_1, \ldots, w_g$, the polynomial

$$\sigma_0\left(\frac{1}{2g-1}(z_g - w^{2g-1}), \frac{1}{2g-3}(z_{g-1} - w^{2g-3}), \ldots, \frac{1}{3}(z_2 - w^3), z_1 - w\right)$$

in w either has exactly g roots $w_1, \ldots, w_g$, or vanishes identically, if z takes the value $z = (p_1(w_1, \ldots, w_g), p_3(w_1, \ldots, w_g), \ldots, p_{2g-1}(w_1, \ldots, w_g))$. By Proposition 4.4, the polynomial σ_0 must be equal to the polynomial σ_g up to a constant factor γ. As to this constant γ, we find in [2, Section IX] a calculation of a constant γ' such that $\sigma(z; \lambda) = \gamma' \cdot H(z) + O(z; \lfloor (g + 3)/2 \rfloor)$, where $H(z)$ is the Hankel determinant from Proposition 4.6 and where now we consider the power series expansion only with respect to the variables $z_1, \ldots, z_g$ and with respect to their usual weight $\deg(z_k) = 1$. By Proposition 4.6, this γ' is equal to our constant γ, up to a sign. We just quote the result of Baker's computation:

$$\gamma^4 = \vartheta(0; \tau)^4 \cdot \prod_{\substack{k<l \\ k,l \in U}} (a_k - a_l)^2 / (\ell_1 \ell_3 \cdots \ell_{2g+1}),$$

where

$$\ell_r := -i \cdot \prod_{\substack{k \in U \\ k \neq r}} (a_k - a_r) / \prod_{k \notin U} (a_k - a_r).$$

Thomae's formula (cf. [14, Chapter IIIa, Section 8]) says that

$$\vartheta(0; \tau)^8 = (\det \mu)^4 \pi^{-4g} \prod_{\substack{k<l \\ k,l \in U}} (a_k - a_l)^2 \prod_{\substack{k<l \\ k,l \notin U}} (a_k - a_l)^2.$$

Combining, we obtain , $\gamma^8 = D \cdot \pi^{-4g} \cdot (\det \mu)^4$. The formula for γ that we gave then follows from Proposition 2.2.   □

Example 5.3. By way of illustration, we have computed σ_g for small g:

g	σ_g
1	z_1
2	$-z_1 + \frac{1}{3} z_2^3$
3	$z_1 z_3 - z_2^2 - \frac{1}{3} z_2 z_3^3 + \frac{1}{45} z_3^6$
4	$z_1 z_3 - z_2^2 - z_3^2 z_4 + z_2 z_3 z_4^2 - \frac{1}{3} z_1 z_4^3 + \frac{1}{15} z_2 z_4^5 - \frac{1}{105} z_3 z_4^7 + \frac{1}{4725} z_4^{10}$

Remark 5.4. As can be seen from Proposition 4.6, the homogeneous part of least total degree (with respect to the usual weight $\deg(z_k) = 1$) of $\sigma_g(z)$ has degree $\lfloor (g+1)/2 \rfloor$. Hence, by a fundamental theorem of Riemann, the theta-characteristic δ gives rise to a linear system of dimension $\lfloor (g - 1)/2 \rfloor$ on X.

6 Leading coefficients

In this section, we calculate the leading coefficients of the power series expansions in z_g of the holomorphic functions $\vartheta[\delta](g\mu^{-1}z; \tau)|_{\iota(X)}$ and $W_{z_g}(\mu)$, the Wronskian in z_g of the basis $\{\mu_1, \ldots, \mu_g\}$.

Proposition 6.1. *In the power series expansions of $\sigma(gz; \lambda)|_{\iota(X)}$, and hence also of $\vartheta[\delta](g\mu^{-1}z; \tau)|_{\iota(X)}$, the leading coefficient is equal to $\gamma \cdot 2^{g(g-1)/2}$.*

Proof. By Lemma 3.1 and Proposition 5.2, we know that the power series expansion of $\sigma(gz; \lambda)|_{\iota(X)}$ has the form

$$\sigma(gz; \lambda)|_{\iota(X)} = \gamma \cdot \sigma_g \left(\frac{g}{2g-1} z_g^{2g-1}, \frac{g}{2g-3} z_g^{2g-3}, \ldots, \frac{g}{3} z_g^3, gz_g \right)$$
$$+ O(z_g; g(g+1)/2 + 1).$$

Hence we need to calculate $\sigma_g(\frac{g}{2g-1}, \frac{g}{2g-3}, \ldots, \frac{g}{3}, g)$. By Definition 4.5, this is equal to $s_g(g, g, \ldots, g)$. But we have $s_g(g, g, \ldots, g) = S_g(1, 1, \ldots, 1)$ by Proposition 4.2 and Definition 4.3, and by Proposition 4.7 we have $S_g(1, \ldots, 1) = 2^{g(g-1)/2}$. The proposition follows. $\qquad \square$

Proposition 6.2. *The leading coefficient of the power series expansion of the Wronskian $W_{z_g}(\mu)$ is equal to $\pm 2^{g(g-1)/2}$.*

Proof. Expanding the Wronskian yields

$$W_{z_g}(\mu) = \det \left(\frac{1}{(k-1)!} \frac{d^k z_l}{dz_g^l} \right)_{1 \leq k,l \leq g}$$

$$= \begin{pmatrix} z_g^{2g-2} & z_g^{2g-4} & \cdots & z_g^2 & 1 \\ (2g-2)z_g^{2g-3} & (2g-4)z_g^{2g-5} & \cdots & 2z_g & 0 \\ \vdots & \vdots & \ddots & \vdots & \vdots \\ \binom{2g-2}{g-1}z_g^g & \binom{2g-4}{g-1}z_g^{g-2} & \cdots & 0 & 0 \end{pmatrix} + O(z_g; \frac{g(g-1)}{2} + 1).$$

Let A be the matrix of binomial coefficients $A := ((\binom{2g-2k}{g-l}))_{1 \leq k,l \leq g-1}$. From the expansion it follows that the required leading coefficient is equal to $\det A$. We will compute this number. First of all, note that

$$\det A = \frac{(2g-2)!(2g-4)! \cdots 2!}{(g-1)!(g-2)! \cdots 1!} \det \left(\frac{1}{(g-2k+l)!} \right)_{1 \leq k,l \leq g-1},$$

where we define $1/n! := 0$ for $n < 0$. Now let $d = g(g-1)/2$ and consider the ring of symmetric polynomials with integer coefficients in $g-1$ variables. It is well known that for the elementary symmetric functions e_r, we have an expansion

$$e_r = \frac{1}{r!} \det \begin{pmatrix} p_1 & 1 & 0 & \cdots & 0 \\ p_2 & p_1 & 2 & \cdots & 0 \\ \cdots & \cdots & \cdots & \cdots & \cdots \\ p_{r-1} & p_{r-2} & p_{r-3} & \cdots & r-1 \\ p_r & p_{r-1} & p_{r-2} & \cdots & p_1 \end{pmatrix},$$

with p_r the elementary Newton functions. From Definition 4.1 and this expansion, it follows that $\det(1/(g-2k+l)!)$ is the coefficient of p_1^d in the expansion of S_{g-1} with respect to the basis of generalized Newton functions. By Proposition 4.10, this coefficient is equal to $\omega_{g-1}(d)/d!$, where $\omega_{g-1}(d)$ is the coefficient of $x_1^{g-1}x_2^{g-2}\cdots x_{g-1}^2 x_g$ in p_1^d. It immediately follows that $\det(1/(g-2k+l)!) = 1/(g-1)!(g-2)!\cdots 1!$. Combining, one finds $\det A = 2^{g(g-1)/2}$. $\square$

7 Proof of Theorem 1.4

Now we are ready to prove Theorem 1.4. Let X be a hyperelliptic Riemann surface of genus $g \geq 2$, and let W be one of its Weierstrass points.

Proof of Theorem 1.4. Fix a hyperelliptic equation $y^2 = f(x)$ for X with f monic and separable of degree $2g+1$ that puts W at infinity. Choose a canonical basis of the homology of X, and form the period matrix $(\mu|\mu')$ of the differentials $x^{k-1}dx/2y$ for $k = 1, \ldots, g$ on this basis. Let L be the lattice in $\mathbb{C}^g$ generated by the columns of $(\mu|\mu')$, and embed X into $\mathbb{C}^g/L$ with base point W as in Section 3. We have the standard euclidean coordinates $z_1, \ldots, z_g$ on $\mathbb{C}^g/L$ and according to Lemma 3.1 we have that z_g is a local coordinate about W on X. The weight w of W is given by $w = g(g-1)/2$. Consider then the following quantities:

$$A(W') := \lim_{Q\to W} \frac{\|\vartheta\|(gQ-W')}{|z_g|^g} \quad \text{for Weierstrass points } W' \neq W;$$

$$A(W) := \lim_{Q\to W} \frac{\|\vartheta\|(gQ-W)}{|z_g|^{w+g}} = \lim_{Q\to W} \frac{\|F_{z_g}\|(Q)}{|z_g|^w};$$

$$B(W) := \lim_{Q\to W} \frac{|W_{z_g}(\omega)(Q)|}{|z_g|^w},$$

where $W_{z_g}(\omega)$ is the Wronskian in z_g of an orthonormal basis $\{\omega_1, \ldots, \omega_g\}$ of $H^0(X, \Omega_X^1)$. We have by Theorem 1.1 that

$$G'(W, W')^g = \frac{A(W')}{\prod_{W''} A(W'')^{w/g^3}} \quad \text{for Weierstrass points } W' \neq W;$$

hence

$$\prod_{W'\neq W} G'(W, W')^g = \frac{1}{A(W)} \cdot \left(\prod_{W'} A(W')\right)^{\frac{g+1}{2g^2}}.$$

Further, we have by the definition of $T(X)$, letting P approach W,

$$T(X) = A(W)^{-(g+1)} \cdot \left(\prod_{W'} A(W')\right)^{\frac{w(g-1)}{g^3}} \cdot B(W)^2.$$

Eliminating the factor $\prod_{W'} A(W')$ yields

$$\prod_{W' \neq W} G'(W, W')^{(g-1)^2} = A(W)^4 \cdot B(W)^{-\frac{2g+2}{g}} \cdot T(X)^{\frac{g+1}{g}}.$$

Now we use the results obtained in Section 6. Let $\tau = \mu^{-1}\mu'$. A simple calculation gives that $A(W)$ is $(\det \operatorname{Im} \tau)^{1/4}$ times the absolute value of the leading coefficient of the power series expansion of $\vartheta[\delta](g\mu^{-1}z; \tau)|_{\iota(X)}$ in z_g. Hence by Propositions 5.2 and 6.1, we have

$$A(W) = 2^{g(g-1)/2} \cdot \pi^{g\frac{r-n}{2n}} \cdot (\det \operatorname{Im} \tau)^{1/4} \cdot |\det \mu|^{-\frac{r-n}{2n}} \cdot |\varphi_g(\tau)|^{\frac{1}{8n}}.$$

Next let $\| \cdot \|$ be the metric on $\wedge^g H^0(X, \Omega_X^1)$ derived from the hermitian inner product $(\omega, \eta) \mapsto \frac{i}{2} \int_X \omega \wedge \overline{\eta}$ on $H^0(X, \Omega_X^1)$. Riemann's second bilinear relations tell us that $\|\mu_1 \wedge \cdots \wedge \mu_g\|^2 = (\det \operatorname{Im} \tau) \cdot |\det \mu|^2$. This gives that $|W_{z_g}(\omega)| = |W_{z_g}(\mu)| \cdot (\det \operatorname{Im} \tau)^{-1/2} \cdot |\det \mu|^{-1}$. From Proposition 6.2, we derive then

$$B(W) = 2^{g(g-1)/2} \cdot (\det \operatorname{Im} \tau)^{-1/2} \cdot |\det \mu|^{-1}.$$

Plugging in our results for $A(W)$ and $B(W)$ finally gives the theorem. $\square$

Remark 7.1. The fact that the product from Theorem 1.4 is independent of the choice of the Weierstrass point W follows a fortiori from the computations in the above proof. It would be interesting to have an *a priori* reason for this independence.

Remark 7.2. We have not been able to find in general a formula for $G'(W, W')$ with W, W' just two Weierstrass points. In the case $g = 2$, it can be shown that

$$G'(W, W')^2 = 2^{1/4} \cdot \|\varphi_2\|(X)^{-3/64} \cdot \prod_{W'' \neq W, W'} \|\vartheta\|(W - W' + W'').$$

This formula should be compared with the explicit formula for $G(W, W')$ given in [3], Proposition 4. We guess that in general $G'(W, W')^g$ is equal to

$$A(X) \cdot \prod_{\substack{\{W_1, \ldots, W_{g-1}\}, \\ W, W' \notin \{W_1, \ldots, W_{g-1}\}}} \|\vartheta\|(W - W' + W_1 + \cdots + W_{g-1}),$$

with $A(X)$ some invariant of X. Such a result is consistent with Theorems 1.4 and 1.5.

Acknowledgments. The author wishes to thank Gerard van der Geer for his encouragement and helpful remarks. He also kindly thanks Professor Yoshihiro Ônishi for his remarks pertaining to Proposition 5.2.

References

[1] S. Y. Arakelov, An intersection theory for divisors on an arithmetic surface, *Izv. Akad. Nauk.*, **38** (1974), 1179–1192; cf. *Math. USSR Izv.*, **8** (1974), 1167–1180.

[2] H. F. Baker, On the hyperelliptic sigma functions, *Amer. J. Math.*, **20** (1898), 301–384.

[3] J.-B. Bost, Fonctions de Green-Arakelov, fonctions thêta et courbes de genre 2, *C. R. Acad. Sci. Paris Ser.* I, **305** (1987), 643–646.

[4] V. M. Buchstaber, V. Z. Enolskii, and D. V. Leykin, Kleinian functions, hyperelliptic jacobians and applications, *Rev. Math. Math. Phys.*, **10** (1997), 1–125.

[5] V. M. Buchstaber, V. Z. Enolskii, and D. V. Leykin, Rational analogs of abelian functions, *Functional Anal. Appl.*, **33**-2 (1999), 83–94.

[6] V. M. Buchstaber, D. V. Leykin, and V. Z. Enolskii, σ-functions of (n, s)-curves, *Russian Math. Surveys*, **54** (1999), 628–629.

[7] G. Faltings, Calculus on arithmetic surfaces, *Ann. Math.*, **119** (1984), 387–424.

[8] W. Fulton and J. Harris, *Representation Theory: A First Course*, Graduate Texts in Mathematics 129, Springer-Verlag, New York, 1991.

[9] D. Grant, A generalization of Jacobi's derivative formula to dimension two, *J. Reine Angew. Math.*, **392** (1988), 125–136.

[10] R. de Jong, Arakelov invariants of Riemann surfaces, preprint, submitted.

[11] R. de Jong, Explicit Mumford Isomorphism for Hyperelliptic Curves, Prépublication M/04/51, Institut des Hautes Études Scientifiques, Bures-sur-Yvette, France, 2004; submitted.

[12] Y. Ônishi, Determinant expressions for hyperelliptic functions (with an appendix by Shigeki Matsutani), preprint.

[13] P. Lockhart, On the discriminant of a hyperelliptic curve, *Trans. Amer. Math. Soc.*, **342**-2 (1994), 729–752.

[14] D. Mumford, *Tata Lectures on Theta* I, II, Progress in Mathematics 28, 43, Birkhäuser Boston, Cambridge, MA, 1984.

A Hirzebruch Proportionality Principle in Arakelov Geometry

Kai Köhler

Mathematisches Institut
Heinrich-Heine-Universität Düsseldorf
Universitätsstraße 1
Gebäude 25.22
D-40225 Düsseldorf
Germany
koehler@math.uni-duesseldorf.de

Summary. We describe a tautological subring in the arithmetic Chow ring of bases of abelian schemes. Among the results are an Arakelov version of the Hirzebruch proportionality principle and a formula for a critical power of $\widehat{c}_1$ of the Hodge bundle.

Subject Classifications: 14G40, 58J52, 20G05, 20G10, 14M17

1 Introduction

The purpose of this note is to exploit some implications of a fixed-point formula in Arakelov geometry when applied to the action of the (-1) involution on abelian schemes of relative dimension d. It is shown that the fixed-point formula's statement in this case is equivalent to giving the values of arithmetic Pontrjagin classes of the Hodge bundle $\overline{E} := (R^1\pi_*\mathcal{O}, \|\cdot\|_{L^2})^*$, where these Pontrjagin classes are defined as polynomials in the arithmetic Chern classes defined by Gillet and Soulé. The resulting formula (see Theorem 3.4) is

$$\widehat{p}_k(\overline{E}) = (-1)^k \left(\frac{2\zeta'(1-2k)}{\zeta(1-2k)} + \sum_{j=1}^{2k-1} \frac{1}{j} - \frac{2\log 2}{1-4^{-k}} \right) (2k-1)! a(\mathrm{ch}(E)^{[2k-1]}) \quad (1)$$

with the canonical map a defined on classes of differential forms. When combined with the statement of the Gillet–Soulé's nonequivariant arithmetic Grothendieck–Riemann–Roch formula [GS8, Fal], one obtains a formula for the class $\widehat{c}_1^{1+d(d-1)/2}$ of the d-dimensional Hodge bundle in terms of topological classes and a certain special differential form γ (Theorem 5.1), which represents an Arakelov Euler class. Morally, this should be regarded as a formula for the height of complete cycles of codimension d in the moduli space (but the nonexistence of such cycles for $d \geq 3$ has been shown by Keel and Sadun [KS]). Still, it might serve as a model for the noncomplete case. Finally, we derive an Arakelov version of the Hirzebruch proportionality principle (not to be confused with its extension by Mumford [M]), namely a ring homomorphism from the Arakelov Chow ring $\mathrm{CH}^*(\overline{L}_{d-1})$ of Lagrangian Grassmannians to the arithmetic Chow ring of bases of abelian schemes $\widehat{\mathrm{CH}}^*(B)$ (Theorem 5.5).

Theorem 1.1. *Let S denote the tautological bundle on L_{d-1}. There is a ring homomorphism*

$$h\colon \mathrm{CH}^*(\overline{L}_{d-1})_{\mathbb{Q}} \to \widehat{\mathrm{CH}}^*(B)_{\mathbb{Q}}/(a(\gamma))$$

with

$$h(\widehat{c}(\overline{S})) = \widehat{c}(\overline{E}) \left(1 + a \left(\sum_{k=1}^{d-1} (\frac{\zeta'(1-2k)}{\zeta(1-2k)} - \frac{\log 2}{1-4^{-k}})(2k-1)!\,\mathrm{ch}^{[2k-1]}(E) \right) \right)$$

and

$$h(a(c(\overline{S}))) = a(c(\overline{E})).$$

In the last section, we investigate the Fourier expansion of the Arakelov Euler class γ of the Hodge bundle on the moduli space of principally polarized abelian varieties.

A fixed-point formula for maps from arithmetic varieties to Spec D has been proven by Roessler and the author in [KR1], where D is a regular arithmetic ring. In [KR2, Appendix], we described a conjectural generalization to flat equivariantly projective maps between arithmetic varieties over D. The missing ingredient to the proof of this conjecture was the equivariant version of Bismut's formula for the behavior of analytic torsion forms under the composition of immersions and fibrations [B4], i.e., a merge of [B3] and [B4]. This formula has meanwhile been shown by Bismut and Ma [BM].

There is a gap in our proof of this result (Conjecture 3.2), insofar as we only give a sketch. While our sketch is quite exhaustive and provides a rather complete guideline to an extension of a previous proof in [KR1] to the one required here, a fully written-up version of the proof would still be basically a copy of [KR1] and thus be quite lengthy. This is not the subject of this article.

We work only with regular schemes as bases; extending these results to moduli stacks and their compactifications remains an open problem, as Arakelov geometry for such situations has not yet fully been developed. A corresponding Arakelov intersection ring has been established in [BKK] by Burgos, Kramer and Kühn, but the associated K-theory of vector bundles does not exist yet; see [MR] for associated conjectures. In particular, one could search an analogue of the Hirzebruch-Mumford proportionality principle in Arakelov geometry. Van der Geer investigated the classical Chow ring of the moduli stack of abelian varieties and its compactifications [G] with a different method. The approach there to determine the tautological subring uses the nonequivariant Grothendieck–Riemann–Roch theorem applied to line bundles associated to theta divisors. Thus it might be possible to avoid the use of the fixed-point formula in our situation by mimicking this method, possibly by extending the methods of Yoshikawa [Y]; but computing the occurring objects related to the theta divisor is presumably not easy.

Results extending some parts of an early preprint form of this article [K2] in the spirit of Mumford's extension of the proportionality principle have been conjectured in [MR]. That article also exploits the case in which more special automorphisms exist than the (-1) automorphism. Their conjectures and results are mainly generalizing Corollary 4.1.

2 Torsion forms

Let $\pi : E^{1,0} \to B$ denote a d-dimensional holomorphic vector bundle over a complex manifold. Let Λ be a lattice subbundle of the underlying real vector bundle $E^{1,0}_{\mathbb{R}}$ of rank $2d$. Thus the quotient bundle $M := E^{1,0}/\Lambda \to B$ is a holomorphic fibration by tori Z. Let

$$\Lambda^* := \{\mu \in (E^{1,0}_{\mathbb{R}})^* \mid \mu(\lambda) \in 2\pi\mathbb{Z} \text{ for all } \lambda \in \Lambda\}$$

denote the dual lattice bundle. Assume that $E^{1,0}$ is equipped with an Hermitian metric such that the volume of the fibers is constant. Any polarization induces such a metric.

Let N_V be the number operator acting on $\Gamma(Z, \Lambda^q T^{*0,1}Z)$ by multiplication with q. Let Tr_s denote the supertrace with respect to the $\mathbb{Z}/2\mathbb{Z}$-grading on $\Lambda T^*B \otimes \mathrm{End}(\Lambda T^{*0,1}Z)$. Let ϕ denote the map acting on $\Lambda^{2p}T^*B$ as multiplication by $(2\pi i)^{-p}$. We write $\widetilde{\mathfrak{A}}(B)$ for $\widetilde{\mathfrak{A}}(B) := \bigoplus_{p\geq 0}(\mathfrak{A}^{p,p}(B)/(\mathrm{Im}\,\partial + \mathrm{Im}\,\overline{\partial}))$, where $\mathfrak{A}^{p,p}(B)$ denotes the C^∞ differential forms of type (p, p) on B. We shall denote a vector bundle F together with an Hermitian metric h by $\overline{F}$. Then $\mathrm{ch}_g(\overline{F})$ shall denote the Chern–Weil representative of the equivariant Chern character associated to the restriction of (F, h) to the fixed-point subvariety. Recall (see, e.g., [B3]) also that $\mathrm{Td}_g(\overline{F})$ is the differential form

$$\mathrm{Td}_g(\overline{F}) := \frac{c_{\mathrm{top}}(\overline{F^g})}{\sum_{k\geq 0}(-1)^k \mathrm{ch}_g(\Lambda^k\overline{F})}.$$

In [K1, Section 3], a superconnection A_t acting on the infinite-dimensional vector bundle $\Gamma(Z, \Lambda T^{*0,1}Z)$ over B has been introduced, depending on $t \in \mathbb{R}^+$. For a fiberwise acting holomorphic isometry g the limit

$$\lim_{t\to\infty} \phi\,\mathrm{Tr}_s\, g^* N_H e^{-A_t^2} =: \omega_\infty$$

exists and is given by the respective trace restricted to the cohomology of the fibers. The equivariant analytic torsion form $T_g(\pi, \overline{\mathcal{O}_M}) \in \widetilde{\mathfrak{A}}(B)$ was defined there as the derivative at zero of the zeta function with values in differential forms on B given by

$$-\frac{1}{\Gamma(s)} \int_0^\infty (\phi\,\mathrm{Tr}_s\, g^* N_H e^{-A_t^2} - \omega_\infty)t^{s-1}dt$$

for $\mathrm{Re}\, s > d$.

Theorem 2.1. *Let an isometry g act fiberwise with isolated fixed points on the fibration by tori $\pi: M \to B$. Then the equivariant torsion form $T_g(\pi, \mathcal{O}_M)$ vanishes.*

Proof. Let $f_\mu: M \to \mathbb{C}$ denote the function $e^{i\mu}$ for $\mu \in \Lambda^*$. As is shown in [K1, Section 5], the operator A_t^2 acts diagonally with respect to the Hilbert space decomposition

$$\Gamma(Z, \Lambda T^{*0,1}Z) = \bigoplus_{\mu \in \Lambda^*} \Lambda E^{*0,1} \otimes \{f_\mu\}.$$

As in [KR4, Lemma 4.1] the induced action by g maps a function f_μ to a multiple of itself if and only if $\mu = 0$ because g acts fixed point free on $E^{1,0}$ outside the zero section. In that case, f_μ represent an element in the cohomology. Thus the zeta function defining the torsion vanishes.   □

Remark. As in [KR4, Lemma 4.1], the same proof shows the vanishing of the equivariant torsion form $T_g(\pi, \overline{\mathcal{L}})$ for coefficients in a g-equivariant line bundle $\overline{\mathcal{L}}$ with vanishing first Chern class.

We shall also need the following result of [K1] for the nonequivariant torsion form $T(\pi, \mathcal{O}_M) := T_{\mathrm{id}}(\pi, \mathcal{O}_M)$: Assume for simplicity that π is Kähler. Consider for $\mathrm{Re}\, s < 0$ the zeta function with values in $(d-1, d-1)$-forms on B,

$$Z(s) := \frac{\Gamma(2d - s - 1)\,\mathrm{vol}(M)}{\Gamma(s)(d-1)!} \sum_{\lambda \in \Lambda \setminus \{0\}} \left(\frac{\overline{\partial}\partial}{4\pi i}\|\lambda^{1,0}\|^2\right)^{\wedge(d-1)} (\|\lambda^{1,0}\|^2)^{s+1-2d},$$

where $\lambda^{1,0}$ denotes a lattice section in $E^{1,0}$. (In [K1] the volume is equal to 1.) Then the limit $\gamma := \lim_{s \to 0^-} Z'(0)$ exists and it transgresses the Chern–Weil form $c_d(\overline{E^{0,1}})$ representing the Euler class $c_d(E^{0,1})$,

$$\frac{\overline{\partial}\partial}{2\pi i}\gamma = c_d(\overline{E^{0,1}}).$$

In [K1, Theorem 4.1], the torsion form is shown to equal

$$T(\pi, \overline{\mathcal{O}_M}) = \frac{\gamma}{\mathrm{Td}(\overline{E^{0,1}})}$$

in $\widetilde{\mathfrak{A}}(B)$. The differential form γ was intensively studied in [K1].

3 Abelian schemes and the fixed-point formula

We shall use the Arakelov geometric concepts and notation of [SABK] and [KR1]. In this article, we shall only give a brief introduction to Arakelov geometry, and we refer to [SABK] for details. Let D be a regular arithmetic ring, i.e., a regular, excellent, Noetherian integral ring, together with a finite set S of ring monomorphisms of $D \to \mathbb{C}$, invariant under complex conjugation. We shall denote by $G := \mu_n$ the diagonalizable group scheme over D associated to $\mathbb{Z}/n\mathbb{Z}$. We choose once and for all a primitive nth root of unity $\zeta_n \in \mathbb{C}$. Let $f : Y \to \operatorname{Spec} D$ be an equivariant arithmetic variety, i.e., a regular integral scheme, endowed with a μ_n-projective action over $\operatorname{Spec} D$. The groups of nth roots of unity acts on the d-dimensional manifold $Y(\mathbb{C})$ by holomorphic automorphisms and we shall write g for the automorphism corresponding to ζ_n.

We write f^{μ_n} for the map $Y_{\mu_n} \to \operatorname{Spec} D$ induced by f on the fixed-point subvariety. Complex conjugation induces an antiholomorphic automorphism of $Y(\mathbb{C})$ and $Y_{\mu_n}(\mathbb{C})$, both of which we denote by F_∞. The space $\widetilde{\mathfrak{A}}(Y)$ is the sum over p of the subspaces of $\widetilde{\mathfrak{A}}^{p,p}(Y(\mathbb{C}))$ of classes of differential (p,p)-forms ω such that $F_\infty^* \omega = (-1)^p \omega$. Let $D^{p,p}(Y(\mathbb{C}))$ denote similarly the F_∞-equivariant currents as duals of differential forms of type $(d-p, d-p)$. It contains in particular the Dirac currents $\delta_{Z(\mathbb{C})}$ of p-codimensional subvarieties Z of Y.

Gillet–Soulé's arithmetic Chow ring $\widehat{\operatorname{CH}}^*(Y)$ is the quotient of the $\mathbb{Z}$-module generated by pairs (Z, g_Z) with Z an arithmetic subvariety of codimension p, $g_Z \in D^{p-1,p-1}(Y(\mathbb{C}))$ with $\frac{\partial\bar{\partial}}{2\pi i} g_Z + \delta_{Z(\mathbb{C})}$ being a smooth differential form by the submodule generated by the pairs $(\operatorname{div} f, -\log \|f\|^2)$ for rational functions f on Y. Let $\operatorname{CH}^*(Y)$ denote the classical Chow ring. Then there is an exact sequence, in any degree p,

$$\operatorname{CH}^{p,p-1}(Y) \xrightarrow{\rho} \widetilde{\mathfrak{A}}^{p-1,p-1}(Y) \xrightarrow{a} \widehat{\operatorname{CH}}^p(Y) \xrightarrow{\zeta} \operatorname{CH}^p(Y) \longrightarrow 0. \qquad (2)$$

For Hermitian vector bundles $\overline{E}$ on Y, Gillet and Soulé defined arithmetic Chern classes $\widehat{c}_p(\overline{E}) \in \widehat{\operatorname{CH}}^p(Y)_{\mathbb{Q}}$.

By "product of Chern classes," we shall understand in this article any product of at least two equal or nonequal Chern classes of degree greater than 0 of a given vector bundle.

Lemma 3.1. *Let*

$$\widehat{\phi} = \sum_{j=0}^{\infty} a_j \widehat{c}_j + \text{products of Chern classes}$$

denote an arithmetic characteristic class with $a_j \in \mathbb{Q}$ and $a_j \neq 0$ for $j > 0$. Assume that for a vector bundle $\overline{F}$ on an arithmetic variety Y, we have $\widehat{\phi}(\overline{F}) = m + a(\beta)$, where β is a differential form on $Y(\mathbb{C})$ with $\partial\bar{\partial}\beta = 0$ and $m \in \widehat{\operatorname{CH}}^0(Y)_{\mathbb{Q}}$. Then

$$\sum_{j=0}^{\infty} a_j \widehat{c}_j(\overline{F}) = m + a(\beta).$$

Proof. We use induction. For the term in $\widehat{\operatorname{CH}}^0(Y)_{\mathbb{Q}}$, the formula is clear. Assume now for $k \in \mathbb{N}_0$ that

$$\sum_{j=0}^{k} a_j \widehat{c}_j(\overline{F}) = m + \sum_{j=0}^{k} a(\beta)^{[j]}.$$

Then $\widehat{c}_j(\overline{F}) \in a(\ker \partial\overline{\partial})$ for $1 \leq j \leq k$, thus products of these $\widehat{c}_j$s vanish by [SABK, Remark III.2.3.1]. Thus the term of degree $k + 1$ of $\widehat{\phi}(\overline{F})$ equals $a_{k+1}\widehat{c}_{k+1}(\overline{F})$.   □

We define *arithmetic Pontrjagin classes* $\widehat{p}_j \in \widehat{\mathrm{CH}}^{2j}$ of arithmetic vector bundles by the relation

$$\sum_{j=0}^{\infty}(-z^2)^j\widehat{p}_j := (\sum_{j=0}^{\infty}z^j\widehat{c}_j)(\sum_{j=0}^{\infty}(-z)^j\widehat{c}_j).$$

Thus,

$$\widehat{p}_j(\overline{F}) = (-1)^j\widehat{c}_{2j}(\overline{F} \oplus \overline{F}^*) = \widehat{c}_j^2(\overline{F}) + 2\sum_{l=1}^{j}(-1)^l\widehat{c}_{j+l}(\overline{F})\widehat{c}_{j-l}(\overline{F})$$

for an arithmetic vector bundle $\overline{F}$ (compare [MiS, Section 15]). Similarly to the construction of Chern classes via the elementary symmetric polynomials, the Pontrjagin classes can be constructed using the elementary symmetric polynomials in the squares of the variables. Thus many formulae for Chern classes have an easily deduced analogue for Pontrjagin classes. In particular, Lemma 3.1 holds with Chern classes replaced by Pontrjagin classes.

Now let Y, B be μ_n-equivariant arithmetic varieties over some fixed arithmetic ring D and let $\pi : Y \to B$ be a map over D, which is flat, μ_n-projective, and smooth over the complex numbers. Fix a $\mu_n(\mathbb{C})$-invariant Kähler metric on $Y(\mathbb{C})$. We recall [KR1, Definition 4.1] extending the definition of Gillet–Soulé's arithmetic K_0-theory to the equivariant setting: Let $\widetilde{\mathrm{ch}}_g(\overline{\mathcal{E}})$ be an equivariant Bott–Chern secondary class as introduced in [KR1, Theorem 3.4]. The arithmetic equivariant Grothendieck group $\widehat{K}^{\mu_n}(Y)$ of Y is the sum of the abelian group $\widetilde{\mathfrak{A}}(Y_{\mu_n})$ and the free abelian group generated by the equivariant isometry classes of Hermitian vector bundles, together with the following relations: For every short exact sequence $\overline{\mathcal{E}}: 0 \to E' \to E \to E'' \to 0$ and any equivariant metrics on E, E', and E'', we have the relation $\widetilde{\mathrm{ch}}_g(\overline{\mathcal{E}}) = \overline{E}' - \overline{E} + \overline{E}''$ in $\widehat{K}^{\mu_n}(Y)$. We remark that $\widehat{K}^{\mu_n}(Y)$ has a natural ring structure. We denote the canonical map $\widetilde{\mathfrak{A}}(Y_{\mu_n}) \to \widehat{K}^{\mu_n}(Y)$ by a; the canonical trivial Hermitian line bundle $\overline{\mathcal{O}}$ shall often be denoted by 1.

If $\overline{E}$ is a π-acyclic (meaning that $R^k\pi_*E = 0$ if $k > 0$) μ_n-equivariant Hermitian bundle on Y, let $\pi_*\overline{E}$ be the direct image sheaf (which is locally free), endowed with its natural equivariant structure and L_2-metric. Consider the rule which associates the element $\pi_*\overline{E} - T_g(\pi, \overline{E})$ of $\widehat{K}_0^{\mu_n}(B)$ to every π-acyclic equivariant Hermitian bundle $\overline{E}$ and the element

$$\int_{Y(\mathbb{C})_g/B(\mathbb{C})_g} \mathrm{Td}_g(\overline{T\pi})\eta \in \widetilde{\mathfrak{A}}(B_{\mu_n})$$

to every $\eta \in \widetilde{\mathfrak{A}}(Y_{\mu_n})$. This rule induces a group homomorphism $\pi_! : \widehat{K}_0^{\mu_n}(Y) \to \widehat{K}_0^{\mu_n}(B)$ [KR2, Proposition 3.1].

Let $\mathcal{R}$ be a ring as appearing in the statement of [KR1, Theorem 4.4] (in the cases considered in this paper, we can choose $\mathcal{R} = D[1/2]$) and let $R(\mu_n)$ be the

Grothendieck group of finitely generated projective μ_n-comodules. Let $\lambda_{-1}(E)$ denote the alternating sum $\sum_k (-1)^k \Lambda^k E$ of a vector bundle E. Consider the zeta function $L(\alpha, s) = \sum_{k=1}^{\infty} k^{-s} \alpha^k$ for $\mathrm{Re}\, s > 1$, $|\alpha| = 1$. It has a meromorphic continuation to $s \in \mathbb{C}$, which shall be denoted by L, too. Then $L(-1, s) = (2^{1-s} - 1)\zeta(s)$ and the function

$$\widetilde{R}(\alpha, x) := \sum_{k=0}^{\infty} \left(\frac{\partial L}{\partial s}(\alpha, -k) + L(\alpha, -k) \sum_{j=1}^{k} \frac{1}{2j} \right) \frac{x^k}{k!}$$

defines the Bismut equivariant R-class of an equivariant holomorphic hermitian vector bundle $\overline{E}$ with $\overline{E}_{|X_g} = \sum_\zeta \overline{E}_\zeta$ as

$$R_g(\overline{E}) := \sum_{\zeta \in S^1} \left(\mathrm{Tr}\, \widetilde{R}\left(\zeta, -\frac{\Omega^{\overline{E}_\zeta}}{2\pi i} \right) - \mathrm{Tr}\, \widetilde{R}\left(1/\zeta, \frac{\Omega^{\overline{E}_\zeta}}{2\pi i} \right) \right).$$

The following result was stated as a conjecture in [KR2, Conjecture 3.2].

Conjecture 3.2. Set

$$\mathrm{td}(\pi) := \frac{\lambda_{-1}(\pi^* \overline{N}^*_{B/B_{\mu_n}})}{\lambda_{-1}(\overline{N}^*_{Y/Y_{\mu_n}})}(1 - a(R_g(N_{Y/Y_{\mu_n}})) + a(R_g(\pi^* N_{B/B_{\mu_n}}))).$$

Then the following diagram commutes:

$$
\begin{array}{ccc}
\widehat{K}_0^{\mu_n}(Y) & \xrightarrow{\mathrm{td}(\pi)\rho'} & \widehat{K}_0^{\mu_n}(Y_{\mu_n}) \otimes_{R(\mu_n)} \mathcal{R} \\
\downarrow{\scriptstyle \pi_!} & & \downarrow{\scriptstyle \pi_!^{\mu_n}} \\
\widehat{K}_0^{\mu_n}(B) & \xrightarrow{\rho'} & \widehat{K}_0^{\mu_n}(B_{\mu_n}) \otimes_{R(\mu_n)} \mathcal{R},
\end{array}
$$

where ρ' denotes the restriction to the fixed-point subscheme.

As this result is not the main aim of this paper, we only outline the proof; details shall appear elsewhere.

Sketch of the proof. As explained in [KR2, Conjecture 3.2] the proof of the main statement of [KR1] was already written with this general result in mind and it holds without any major change for this situation, when using the generalization of Bismut's equivariant immersion formula for the holomorphic torsion [KR1, Theorem 3.11] to torsion forms. The latter has now been established by Bismut and Ma [BM]. The proof in [KR1] holds when using [BM] instead of [KR1, Theorem 3.11] and [KR2, Proposition 3.1] instead of [KR1, Proposition 4.3].

Also one has to replace in Sections 5, 6.1, and 6.2 the integrals over Y_g, X_g, etc. by integrals over Y_g/B_g, X_g/B_g, while replacing the maps occurring there by corresponding relative versions. As direct images can occur as non-locally-free coherent sheaves, one has to consider at some steps suitable resolutions of vector bundles such that the higher direct images of the vector bundles in this resolution are locally free as, e.g., on [Fal, p. 74]. □

Let $f : B \rightarrow \operatorname{Spec} D$ denote a quasi-projective arithmetic variety and let $\pi : Y \rightarrow B$ denote a principally polarized abelian scheme of relative dimension d. For simplicity, we assume that the volume of the fibers over $\mathbb{C}$ is scaled to equal 1; it would be 2^d for the metric induced from the polarization. We shall explain the effect of rescaling the metric later (after Theorem 5.1). Set $\overline{E} := (R^1 \pi_* \mathcal{O}, \| \cdot \|_{L^2})^*$. This bundle $E = \mathbf{Lie}(Y/B)^*$ is the Hodge bundle. Then by [BBM, Proposition 2.5.2], the full direct image of $\mathcal{O}$ under π is given by $R^\bullet \pi_* \mathcal{O} = \Lambda^\bullet E^*$ and the relative tangent bundle is given by $T\pi = \pi^* E^*$. By similarly representing the cohomology of the fibers Y/B by translation-invariant differential forms, one shows that these isomorphisms induce isometries if and only if the volume of the fibers equals 1 (e.g., as in [K1, Lemma 3.0]); thus

$$\overline{R^\bullet \pi_* \mathcal{O}} = \Lambda^\bullet \overline{E}^* \tag{3}$$

and

$$\overline{T\pi} = \pi^* \overline{E}^*. \tag{4}$$

See also [FC, Theorem VI.1.1], where these properties are extended to toroidal compactifications. For an action of $G = \mu_n$ on Y Conjecture 3.2 combined with the arithmetic Grothendieck–Riemann–Roch theorem in all degrees for π^G states (analogous to [KR1, Section 7.4]).

Theorem 3.3.

$$\widehat{\mathrm{ch}}_G(\overline{R^\bullet \pi_* \mathcal{O}}) - a(T_g(\pi_{\mathbb{C}}, \overline{\mathcal{O}})) = \pi_*^G(\widehat{\mathrm{Td}}_G(\overline{T\pi})(1 - a(R_g(T\pi_{\mathbb{C}})))).$$

As in [KR1], $G = \mu_n$ is used as the index for equivariant arithmetic classes, while the chosen associated automorphism g over the points at infinity is used for objects defined there. We shall mainly consider the case where π^G is actually a smooth covering, Riemannian over $\mathbb{C}$; thus the statement of the arithmetic Grothendieck–Riemann–Roch is, in fact, very simple in this case. We obtain the equation

$$\widehat{\mathrm{ch}}_G(\Lambda^\bullet \overline{E}^*) - a(T_g(\pi_{\mathbb{C}}, \overline{\mathcal{O}})) = \pi_*^G(\widehat{\mathrm{Td}}_G(\pi^* \overline{E}^*)(1 - a(R_g(\pi^* E_{\mathbb{C}}^*)))).$$

Using the equation

$$\widehat{\mathrm{ch}}_G(\Lambda^\bullet \overline{E}^*) = \frac{\widehat{c}_{\mathrm{top}}(\overline{E}^G)}{\widehat{\mathrm{Td}}_G(\overline{E})}$$

this simplifies to

$$\frac{\widehat{c}_{\mathrm{top}}(\overline{E}^G)}{\widehat{\mathrm{Td}}_G(\overline{E})} - a(T_g(\pi_{\mathbb{C}}, \overline{\mathcal{O}})) = \widehat{\mathrm{Td}}_G(\overline{E}^*)(1 - a(R_g(E_{\mathbb{C}}^*)))\pi_*^G \pi^* 1,$$

or, using that $a(\ker \overline{\partial}\partial)$ is an ideal of square zero,

$$\widehat{c}_{\mathrm{top}}(\overline{E}^G)(1 + a(R_g(E_{\mathbb{C}}^*))) - a(T_g(\pi_{\mathbb{C}}, \overline{\mathcal{O}}) \mathrm{Td}_g(\overline{E}_{\mathbb{C}})) = \widehat{\mathrm{Td}}_G(\overline{E})\, \widehat{\mathrm{Td}}_G(\overline{E}^*)\pi_*^G \pi^* 1. \tag{5}$$

Remarks. 1. If G acts fiberwise with isolated fixed points (over $\mathbb{C}$), by Theorem 2.1 the left-hand side of equation (5) is an element of $\widehat{\mathrm{CH}}^0(B)_{\mathbb{Q}(\zeta_n)} + a(\ker \partial\overline{\partial})$. Set for an equivariant bundle F in analogy to the classical $\widehat{A}$-class

$$\widehat{A}_g(F) := \mathrm{Td}_g(F) \exp\left(-\frac{c_1(F) + \mathrm{ch}_g(F)^{[0]}}{2}\right), \tag{6}$$

and let $\widehat{\overline{A}}_G$ denote the corresponding arithmetic class (an unfortunate clash of notations); in particular $\widehat{A}_g(F^*) = (-1)^{\mathrm{rk}(F/F^G)}\widehat{A}_g(F)$. For isolated fixed points, by comparing the components in degree 0 in equation (5), one obtains

$$\pi_*^G \pi^* 1 = (-1)^d (\widehat{A}_g(E)^{[0]})^{-2},$$

and thus by Theorem 2.1,

$$1 + a(R_g(E_{\mathbb{C}}^*)) = \left(\frac{\widehat{\overline{A}}_G(\overline{E})}{\widehat{A}_g(E)^{[0]}}\right)^2. \tag{7}$$

(compare [KR4, Proposition 5.1]). Both sides can be regarded as products over the occurring eigenvalues of g of characteristic classes of the corresponding bundles E_ζ. One can wonder whether the equality holds for the single factors, similar to [KR4]. Related work is announced by Maillot and Roessler in [MR].

2. If $G(\mathbb{C})$ does not act with isolated fixed points, then the right-hand side vanishes, $c_{\mathrm{top}}(E^G)$ vanishes and we find

$$\widehat{c}_{\mathrm{top}}(\overline{E}^G) = a(T_g(\pi_{\mathbb{C}}, \overline{\mathcal{O}}) \mathrm{Td}_g(\overline{E}_{\mathbb{C}})). \tag{8}$$

As was mentioned in [K1, equation (7.8)], one finds, in particular,

$$\widehat{c}_d(\overline{E}) = a(\gamma). \tag{9}$$

For this statement, we need Gillet–Soulé's arithmetic Grothendieck–Riemann–Roch [GS8] in all degrees, while the above statements use this theorem only in degree 0. The full result was stated in [S, Section 4]; a proof of an analogue statement is given in [R2, Section 8]. Another proof was sketched in [Fal] using a possibly different direct image. If one wants to avoid the use of this strong result, one can at least show the existence of some $(d-1, d-1)$ differential form γ' with $\widehat{c}_d(\overline{E}) = a(\gamma')$ the following way: The analogue proof of equation (9) in the classical algebraic Chow ring $\mathrm{CH}^*(B)$ using the classical Riemann–Roch–Grothendieck Theorem shows the vanishing of $c_d(E)$. Thus by the exact sequence

$$\widetilde{\mathfrak{A}}^{d-1,d-1}(B) \xrightarrow{a} \widehat{\mathrm{CH}}^d(B) \xrightarrow{\zeta} \mathrm{CH}^d(B) \longrightarrow 0$$

we see that (9) holds with some form γ'.

Now we restrict ourself to the action of the automorphism (-1). We need to assume that this automorphism corresponds to a μ_2-action. This condition can always be satisfied by changing the base $\mathrm{Spec}\, D$ to $\mathrm{Spec}\, D[\frac{1}{2}]$ (see [KR1, Introduction] or [KR4, Section 2]).

Theorem 3.4. *Let* $\pi : Y \rightarrow B$ *denote a principally polarized abelian scheme of relative dimension d over an arithmetic variety B. Set* $\overline{E} := (R^1\pi_*\mathcal{O}, \|\cdot\|_{L^2})^*$. *Then the Pontrjagin classes of* $\overline{E}$ *are given by*

$$\widehat{p}_k(\overline{E}) = (-1)^k \left(\frac{2\zeta'(1-2k)}{\zeta(1-2k)} + \sum_{j=1}^{2k-1} \frac{1}{j} - \frac{2\log 2}{1-4^{-k}} \right) (2k-1)! a(\mathrm{ch}(E)^{[2k-1]}).$$

$$(10)$$

The log 2-term actually vanishes in the arithmetic Chow ring over Spec $D[1/2]$.

Remark. The occurrence of R-class-like terms in Theorem 3.4 makes it very unlikely that there is an easy proof of this result which does not use arithmetic Riemann–Roch Theorems. This is in sharp contrast to the classical case over $\mathbb{C}$, where the analogues formulae are a trivial topological result: The underlying real vector bundle of $E_{\mathbb{C}}$ is flat, as the period lattice determines a flat structure. Thus the topological Pontrjagin classes $p_j(E_{\mathbb{C}})$ vanish.

Proof. Let $Q(z)$ denote the power series in z given by the Taylor expansion of

$$4(1+e^{-z})^{-1}(1+e^z)^{-1} = \frac{1}{\cosh^2 \frac{z}{2}}$$

at $z = 0$. Let $\widehat{Q}$ denote the associated multiplicative arithmetic characteristic class. Thus by definition for $G = \mu_2$,

$$4^d \widehat{\mathrm{Td}}_G(\overline{E}) \widehat{\mathrm{Td}}_G(\overline{E}^*) = \widehat{Q}(\overline{E})$$

and $\widehat{Q}$ can be represented by Pontrjagin classes, as the power series Q is even. Now we can apply Lemma 3.1 for Pontrjagin classes to equation (5) of equation (7). By a formula by Cauchy [Hi3, Section 1, equation (10)], the summand of $\widehat{Q}$ consisting only of single Pontrjagin classes is given by taking the Taylor series in z at $z = 0$ of

$$Q(\sqrt{-z})\frac{d}{dz}\frac{z}{Q(\sqrt{-z})} = \frac{\frac{d}{dz}(z\cosh^2\frac{\sqrt{-z}}{2})}{\cosh^2\frac{\sqrt{-z}}{2}} = 1 + \frac{\sqrt{-z}}{2}\tanh\frac{\sqrt{-z}}{2} \quad (11)$$

and replacing every power z^j by $\widehat{p}_j$. The bundle $\overline{E}^G$ is trivial, hence $\widehat{c}_{\mathrm{top}}(\overline{E}^G) = 1$. Thus by equation (5) with $\pi_*^G \pi^* 1 = 4^d$, we obtain

$$\sum_{k=1}^{\infty} \frac{(4^k-1)(-1)^{k+1}}{(2k-1)!} \zeta(1-2k)\widehat{p}_k(\overline{E}) = -a(R_g(E_{\mathbb{C}})).$$

The function $\widetilde{R}(\alpha, x)$ by which the Bismut equivariant R-class is constructed satisfies for $\alpha = -1$ the relation

$$\widetilde{R}(-1, x) - \widetilde{R}(-1, -x) = \sum_{k=1}^{\infty} \left[(4^k - 1) \left(2\zeta'(1-2k) + \zeta(1-2k) \sum_{j=1}^{2k-1} \frac{1}{j} \right) \right.$$

$$\left. - 2\log 2 \cdot 4^k \zeta(1-2k)\right] \cdot \frac{x^{2k-1}}{(2k-1)!}. \qquad (12)$$

Thus we finally obtain the desired result.   □

The first Pontrjagin classes are given by

$$\widehat{p}_1 = -2\widehat{c}_2 + \widehat{c}_1^2, \qquad \widehat{p}_2 = 2\widehat{c}_4 - 2\widehat{c}_3\widehat{c}_1 + \widehat{c}_2^2, \qquad \widehat{p}_3 = -2\widehat{c}_6 + 2\widehat{c}_5\widehat{c}_1 - 2\widehat{c}_4\widehat{c}_2 + \widehat{c}_3^2.$$

In general, $\widehat{p}_k = (-1)^k 2\widehat{c}_{2k} + $ products of Chern classes. Thus knowing the Pontrjagin classes allows us to express the Chern classes of even degree by the Chern classes of odd degree.

Corollary 3.5. *The Chern–Weil form representing the total Pontrjagin class vanishes (except in degree 0):*

$$c(\overline{E} \oplus \overline{E}^*) = 1, \quad i.e., \quad \det(1 + (\Omega^E)^{\wedge 2}) = 1$$

for the curvature Ω^E of the Hodge bundle. The Pontrjagin classes in the algebraic Chow ring $\mathrm{CH}(B)$ vanish:

$$c(E \oplus E^*) = 1.$$

Proof. These facts follow from applying the forget-functors $\omega \colon \widehat{\mathrm{CH}}(B) \to \mathfrak{A}(B(\mathbb{C}))$ and $\zeta \colon \widehat{\mathrm{CH}}(B) \to \mathrm{CH}(B)$.   □

The first fact can also be deduced by "linear algebra," e.g., using the Mathai–Quillen calculus, but it is not that easy. The second statement was obtained in [G, Theorem 2.5] using the nonequivariant Grothendieck–Riemann–Roch theorem and the geometry of theta divisors.

4 A K-theoretical proof

The Pontrjagin classes form one set of generators of the algebra of even classes; another important set of generators is given by $(2k)!$ times the Chern character in even degrees $2k$. We give the value of these classes below. Let U denote the additive characteristic class associated to the power series

$$U(x) := \sum_{k=1}^{\infty} \left(\frac{\zeta'(1-2k)}{\zeta(1-2k)} + \sum_{j=1}^{2k-1} \frac{1}{2j} - \frac{\log 2}{1 - 4^{-k}} \right) \frac{x^{2k-1}}{(2k-1)!}$$

and let d again denote the relative dimension of the abelian scheme.

Corollary 4.1. *The part of $\widehat{\mathrm{ch}}(\overline{E})$ in $\widehat{\mathrm{CH}}^{\mathrm{even}}(B)_{\mathbb{Q}}$ is given by the formula*

$$\widehat{\mathrm{ch}}(\overline{E})^{[\mathrm{even}]} = d - a(U(E)).$$

Proof. The part of $\widehat{\mathrm{ch}}(\overline{E})$ of even degree equals

$$\widehat{\mathrm{ch}}(\overline{E})^{[\mathrm{even}]} = \frac{1}{2}\,\widehat{\mathrm{ch}}(\overline{E} \oplus \overline{E}^*),$$

thus it can be expressed by Pontrjagin classes. More precisely, by Newton's formulae [Hi3, Section 10.1],

$$(2k)!\,\widehat{\mathrm{ch}}^{[2k]} - \widehat{p}_1 \cdot (2k-2)!\,\widehat{\mathrm{ch}}^{[2k-2]} + \cdots + (-1)^{k-1}\widehat{p}_{k-1}2!\,\widehat{\mathrm{ch}}^{[2]} = (-1)^{k+1}k\widehat{p}_k$$

for $k \in \mathbb{N}$. As products of the arithmetic Pontrjagin classes vanish in $\widehat{\mathrm{CH}}(Y)_{\mathbb{Q}}$ by Lemma 3.4, we thus observe that the part of $\widehat{\mathrm{ch}}(\overline{E})$ in $\widehat{\mathrm{CH}}^{\mathrm{even}}(Y)_{\mathbb{Q}}$ is given by

$$\widehat{\mathrm{ch}}(\overline{E})^{[\mathrm{even}]} = d + \sum_{k>0} \frac{(-1)^{k+1}\widehat{p}_k(\overline{E})}{2(2k-1)!}.$$

Thus the result follows from Lemma 3.4. $\square$

As Harry Tamvakis pointed out to the author, a similar argument is used in [T, Section 2] and its predecessors.

Now we show how to deduce Corollary 4.1 (and thus the equivalent Theorem 3.4) using only Conjecture 3.2 without combining it with the arithmetic Grothendieck–Riemann–Roch Theorem as in Theorem 3.3. Of course the structure of the proof shall not be too different as the Grothendieck–Riemann–Roch Theorem was very simple in this case; but the following proof is quite instructive as it provides a different point of view on the resulting characteristic classes. We shall use the λ-ring structure on $\widehat{K}$ constructed in [R1].

Conjecture 3.2 applied to the abelian scheme $\pi : Y \to B$ provides the formula

$$\pi_!\overline{\mathcal{O}} = \pi_!^{\mu_2} \frac{1 - a(R_g(N_{Y/Y_{\mu_n}}))}{\lambda_{-1}(\overline{N}^*_{Y/Y_{\mu_n}})}.$$

In our situation, $\overline{N_{Y/Y_{\mu_n}}} = \overline{T\pi}$. Combining this with the fundamental equations (3), (4) and Theorem 2.1 yields

$$\lambda_{-1}\overline{E}^* = \pi_!^{\mu_2}\pi^* \frac{1 - a(R_g(E^*))}{\lambda_{-1}\overline{E}},$$

and using the projection formula, we find

$$\lambda_{-1}\overline{E \oplus E^*} = 4^d(1 - a(R_g(E^*))).$$

Let $\overline{E}'$ denote the vector bundle E equipped with the trivial μ_2-action. Now one can deduce from this that $\overline{E' \oplus E'^*}$ itself has the form $2d + a(\eta)$ with a $\bar{\partial}\partial$-closed form η: Apply the Chern character to both sides. Then use equation (11) and Lemma 3.1 to deduce by induction that all Chern classes of $\overline{E' \oplus E'^*}$ are in $a(\ker \bar{\partial}\partial)$. Thus using the fact that the arithmetic Chern character is an isomorphism up to torsion

[GS3, Theorem 7.3.4] $\overline{E' \oplus E'^*} = 2d + a(\eta)$ with $a(\eta)$ having even degrees, and $\overline{E \oplus E^*} = (2d + a(\eta)) \otimes (-1)$ in $\widehat{K}^{\mu_2}(B)_{\mathbb{Q}}$. One could use the γ-filtration instead to deduce this result; it would be interesting to find a proof which does not use any filtration.

For a $\beta \in \widetilde{\mathfrak{A}}^{p,p}(B)$, the action of the λ-operators can be determined as follows: The action of the kth Adams operator is given by $\psi^k a(\beta) = k^{p+1} a(\beta)$ [GS3, p. 235]. Then with $\psi_t := \sum_{k>0} t^k \psi^k$, $\lambda_t := \sum_{k \geq 0} t^k \lambda^k$ the Adams operators are related to the λ-operators via

$$\psi_t(x) = -t \frac{d}{dt} \log \lambda_{-t}(x)$$

for $x \in \widehat{K}^{\mu_n}(B)$. As $\psi_t(a(\beta)) = \mathrm{Li}_{-1-p}(t) a(\beta)$ with the polylogarithm Li, we find, for $\beta \in \ker \bar{\partial}\partial$,

$$\lambda_t(a(\beta)) = 1 - \mathrm{Li}_{-p}(-t) a(\beta)$$

or $\lambda^k a(\beta) = -(-1)^k k^p a(\beta)$ ($\mathrm{Li}_{-p}(\frac{t}{t-1})$ is actually a polynomial in t; in this context this can be regarded as a relation coming from the γ-filtration). In particular, $\lambda_{-1} a(\beta) = 1 - \zeta(-p) a(\beta)$, and $\lambda_{-1}(a(\beta) \otimes (-1)) = \lambda_1 a(\beta) \otimes 1 = (1 + (1 - 2^{p+1})\zeta(-p) a(\beta)) \otimes 1$ in $\widehat{K}^{\mu_2} \otimes_{R_{\mu_2}} \mathbb{C}$.

By comparing

$$\lambda_{-1}(a(\eta) \otimes (-1)) = a \left(\sum_{k>0} \zeta(1-2k)(1-4^k)\eta^{[2k-1]} \right) \otimes 1 = a(R_{-1}(E^*)) \otimes 1,$$

we finally derive $a(\eta) = a(-2U(E))$ and thus

$$\overline{E' \oplus E'^*} = 2d - 2a(U(E)).$$

In other words, the Hermitian vector bundle $\overline{E' \oplus E'^*}$ equals the $2d$-dimensional trivial bundle plus the class of differential forms given by $U(E)$ in $\widehat{K}^{\mu_2} \otimes_{R_{\mu_2}} \mathbb{C}$. From this Corollary 4.1 follows.

5 A Hirzebruch proportionality principle and other applications

The following formula can be used to express the height of complete subvarieties of codimension d of the moduli space of abelian varieties as an integral over differential forms.

Theorem 5.1. *There is a real number $r_d \in \mathbb{R}$ and a Chern–Weil form $\phi(\overline{E})$ on $B_{\mathbb{C}}$ of degree $(d-1)(d-2)/2$ such that*

$$\widehat{c}_1^{1+d(d-1)/2}(\overline{E}) = a(r_d \cdot c_1^{d(d-1)/2}(E) + \phi(\overline{E})\gamma).$$

The form $\phi(\overline{E})$ is actually a polynomial with integral coefficients in the Chern forms of $\overline{E}$. See Corollary 5.6 for a formula for r_d.

Proof. Consider the graded ring R_d given by $\mathbb{Q}[u_1, \ldots, u_d]$ divided by the relations

$$\left(1 + \sum_{j=1}^{d-1} u_j\right)\left(1 + \sum_{j=1}^{d-1}(-1)^j u_j\right) = 1 \quad \text{and} \quad u_d = 0, \tag{13}$$

where u_j shall have degree j $(1 \leq j \leq d)$. This ring is finite dimensional as a vector space over $\mathbb{Q}$ with basis

$$u_{j_1} \cdots u_{j_m}, \quad 1 \leq j_1 < \cdots < j_m < d, \quad 1 \leq m < d.$$

In particular, any element of R_d has degree $\leq \frac{d(d-1)}{2}$. As the relation (13) is verified for $u_j = \widehat{c}_j(\overline{E})$ up to multiples of the Pontrjagin classes and $\widehat{c}_d(\overline{E})$, any polynomial in the $\widehat{c}_j(\overline{E})$s can be expressed in terms of the $\widehat{p}_j(\overline{E})$s and $\widehat{c}_d(\overline{E})$ if the corresponding polynomial in the u_js vanishes in R_d.

Thus we can express $\widehat{c}_1^{1+d(d-1)/2}(\overline{E})$ as the image under a of a topological characteristic class of degree $d(d-1)/2$ plus γ times a Chern–Weil form of degree $(d-1)(d-2)/2$. As any element of degree $d(d-1)/2$ in R_d is proportional to $u_1^{d(d-1)/2}$, the theorem follows. $\square$

Any other arithmetic characteristic class of $\overline{E}$ vanishing in R_d can be expressed in a similar way.

Example 5.2. We shall compute $\widehat{c}_1^{1+d(d-1)/2}(\overline{E})$ explicitly for small d. Define topological cohomology classes r_j by $\widehat{p}_j(\overline{E}) = a(r_j)$ via Theorem 3.4. For $d = 1$, clearly

$$\widehat{c}_1(\overline{E}) = a(\gamma).$$

In the case $d = 2$, we find by the formula for $\widehat{p}_1$,

$$\widehat{c}_1^2(\overline{E}) = a(r_1 + 2\gamma) = a\left[\left(-1 + \frac{8}{3}\log 2 + 24\zeta'(-1)\right)c_1(E) + 2\gamma\right].$$

Combining the formulae for the first two Pontrjagin classes, we get

$$\widehat{p}_2 = 2\widehat{c}_4 - 2\widehat{c}_3\widehat{c}_1 + \frac{1}{4}\widehat{c}_1^4 - \frac{1}{2}\widehat{c}_1^2\widehat{p}_1 + \frac{1}{4}\widehat{p}_1^2.$$

Thus for $d = 3$ we find, using $c_3(E) = 0$ and $c_1^2(E) = 2c_2(E)$,

$$\widehat{c}_1^4(\overline{E}) = a(2c_1^2(E)r_1 + 4r_2 + 8c_1(E)\gamma)$$
$$= a\left[\left(-\frac{17}{3} + \frac{48}{5}\log 2 + 48\zeta'(-1) - 480\zeta'(-3)\right)c_1^3(E) + 8c_1(\overline{E})\gamma\right].$$

For $d = 4$ one obtains

$$\widehat{c}_1^7(\overline{E}) = a[64c_2(E)c_3(E)r_1 - (8c_1(E)c_2(E) + 32c_3(E))r_2 + 64c_1(E)r_3$$
$$+ 16(7c_1(\overline{E})c_2(\overline{E}) - 4c_3(\overline{E}))\gamma].$$

As in this case $\mathrm{ch}(E)^{[1]} = c_1(E)$, $3!\,\mathrm{ch}(E)^{[3]} = -c_1^3(E)/2 + 3c_3(E)$, and $5!\,\mathrm{ch}(E)^{[5]} = c_1^5(E)/16$, we find

$$\widehat{c}_1^7(\overline{E}) = a\left[\left(-\frac{1063}{60} + \frac{1520}{63}\log 2 + 96\zeta'(-1) - 600\zeta'(-3) + 2016\zeta'(-5)\right)c_1^5(E)\right.$$

$$\left. + 16(7c_1(\overline{E})c_2(\overline{E}) - 4c_3(\overline{E}))\gamma\right].$$

For $d = 5$ one gets

$$\widehat{c}_1^{11}(\overline{E}) = a\left[2816\gamma c_2(3c_1c_3 - 8c_4) + c_1^{10}\left(-\frac{104611}{2520} + \frac{113632}{2295}\log(2)\right.\right.$$

$$\left.\left. - 3280\zeta'(-7) + 2352\zeta'(-5) - 760\zeta'(-3) + 176\zeta'(-1)\right)\right],$$

and for $d = 6$

$$\widehat{c}_1^{16}(\overline{E}) = a\left[425984\gamma(11c_1c_2c_3c_4 - 91c_2c_3c_5) + 40c_1c_4c_5)\right.$$

$$+ c_1^{15}\left(-\frac{3684242}{45045} + \frac{3321026752}{37303695}\log(2) + \frac{36096}{13}\zeta'(-9)\right.$$

$$- \frac{526080}{143}\zeta'(-7) + \frac{395136}{143}\zeta'(-5) - \frac{136320}{143}\zeta'(-3)$$

$$\left.\left. + \frac{3264}{11}\zeta'(-1)\right)\right].$$

Remark. We shall shortly describe the effect of rescaling the metric for the characteristic classes described above. By the multiplicativity of the Chern character and using $\widehat{ch}(\mathcal{O}, \alpha| \cdot |^2) = 1 - a(\log \alpha)$, $\widehat{ch}(\overline{E})$ changes by

$$\log \alpha \cdot a(\mathrm{ch}(E))$$

when multiplying the metric on E^* by a constant $\alpha \in \mathbb{R}^+$ (or with a function $\alpha \in C^\infty(B(\mathbb{C}), \mathbb{R}^+)$). Thus we observe that in our case $\widehat{ch}(\overline{E})^{[\mathrm{odd}]}$ is invariant under rescaling on E^*, and we get an additional term

$$\log \alpha \cdot a(\mathrm{ch}(E)^{[\mathrm{odd}]})$$

on the right-hand side in Corollary 4.1, when the volume of the fibers equals α^d instead of 1. Thus the right-hand side of Theorem 3.4 gets an additional term

$$\frac{(-1)^{k+1}\log \alpha}{2(2k-1)!}a(\mathrm{ch}(E)^{[2k-1]}).$$

Similarly,

$$\widehat{c}_d(\overline{E}) = a(\gamma) + \log \alpha \cdot a(c_{d-1}(E))$$

for the rescaled metric. In Theorem 5.1, we obtain an additional

$$\log \alpha \cdot a\left(\frac{d(d-1)+2}{2} \cdot c_1^{d(d-1)/2}(E)\right)$$

on the right-hand side, and this shows

$$\phi(E)c_{d-1}(E) = \frac{d(d-1)+2}{2} c_1^{d(d-1)/2}(E). \tag{14}$$

Alternatively, one can show the same formulae by investigating directly the Bott–Chern secondary class of $R\pi_*\mathcal{O}$ for the metric change.

Assume that the base space Spec D equals Spec $\mathcal{O}_K[\frac{1}{2}]$ for a number field K. We consider the pushforward map

$$\widehat{\deg}\colon \widehat{\mathrm{CH}}(B) \longrightarrow \widehat{\mathrm{CH}}^1\left(\mathrm{Spec}\left(\mathcal{O}_K\left[\frac{1}{2}\right]\right)\right)$$

$$\longrightarrow \widehat{\mathrm{CH}}^1\left(\mathrm{Spec}\left(\mathbb{Z}\left[\frac{1}{2}\right]\right)\right) \cong \mathbb{R}/(\mathbb{Q}\log 2),$$

where the last identification contains the traditional factor $\frac{1}{2}$.

As Keel and Sadun [KS] have shown by proving a conjecture by Oort, the moduli space of principally polarized complex abelian varieties does not have any projective subvarieties of codimension d, if $d \geq 3$. Thus the following two corollaries have a nonempty content only for $d = 2$. Still it is likely that they serve as models for similar results for nonprojective subvarieties in an extended Arakelov geometry in the spirit of [BKK]. For that reason, we state them together with the short proof.

Using the definition

$$h(B) := \frac{1}{[K:\mathbb{Q}]} \widehat{\deg}\,\widehat{c}_1^{1+\dim B_{\mathbb{C}}}(\overline{E}_{|B})$$

of the *global height* (thus defined modulo rational multiples of $\log 2$ in this case) of a projective arithmetic variety, we find the following.

Corollary 5.3. *If* $\dim B_{\mathbb{C}} = \frac{d(d-1)}{2}$ *and B is projective, then the (global) height of B with respect to* $\det \overline{E}$ *is given by*

$$h(B) = \frac{r_d}{2} \cdot \deg B + \frac{1}{2}\int_{B_{\mathbb{C}}} \phi(\overline{E})\gamma,$$

with deg *denoting the algebraic degree.*

Let $\alpha(E, \Lambda, \omega^E) \in \bigwedge^* T^*B$ be a differential form associated to bundles of principally polarized abelian varieties (E, Λ, ω^E) (with Hodge bundle E, lattice Λ and polarization form ω^E) in a functorial way: If $f\colon B'' \to B$ is a holomorphic map and $(f^*E, f^*\Lambda, f^*\omega^E)$ the induced bundle over B'', then $\alpha(f^*E, f^*\Lambda, f^*\omega^E) = f^*\alpha(E, \Lambda, \omega^E)$; in other words, α shall be a modular form. Choose an open cover (U_i) of B such that the bundle trivializes over U_i. To define the Hecke operator $T(p)$ for p prime, associated to the group $\mathrm{Sp}(n, \mathbb{Z})$, consider on U_i the set $\mathcal{L}(p)$ of all maximal sublattices $\Lambda' \subset \Lambda_{|U_i}$ such that ω^E takes values in $p\mathbb{Z}$ on Λ'. The sums

$$T(p)\alpha(E, \Lambda, \omega^E)_{|U_i} := \sum_{\Lambda' \in \mathcal{L}(p)} \alpha\left(E, \Lambda', \frac{\omega^E}{p}\right)$$

patch together to a globally defined differential form on B. Note that the set $\mathcal{L}(p)$ may be identified with the set of all maximal isotropic subspaces (Lagrangians) $\Lambda'/p\Lambda$ of the symplectic vector space $(\Lambda/p\Lambda, \omega^E)$ over $\mathbb{F}_p$.

Let B' be a disjoint union of abelian schemes with one connected component for each $\Lambda' \in \mathcal{L}(p)$ such that the Hodge bundle over each connected component over Spec $\mathbb{C}$ is isomorphic to the Hodge bundle $E(\mathbb{C})$ over $B(\mathbb{C})$, but the period lattice and polarization form are given by Λ' and ω^E/p.

Corollary 5.4. *For B as in Corollary* 5.3, *set* $h'(B) := \frac{h(B)}{(\dim B_{\mathbb{C}}+1)\deg B}$. *The height of B and B' are related by*

$$h'(B') = h'(B) + \frac{p^d - 1}{p^d + 1} \cdot \frac{\log p}{2}.$$

Proof. For this proof we need that γ is indeed the form determined by the arithmetic Riemann–Roch Theorem in all degrees (compare equation (9)). The action of Hecke operators on γ was investigated in [K1, Section 7]. In particular, it was shown that

$$T(p)\gamma = \prod_{j=1}^{d}(p^j + 1)\left(\gamma + \frac{p^d - 1}{p^d + 1}\log p \cdot c_{d-1}(\overline{E})\right).$$

The action of Hecke operators commutes with multiplication by a characteristic class, as the latter are independent of the period lattice in E. Thus by Corollary 5.3 the height of B' is given by

$$h(B')$$
$$= \prod_{j=1}^{d}(p^j + 1)\left(\frac{r_d}{2}\cdot \deg B_{\mathbb{C}} + \frac{1}{2}\int_{B_{\mathbb{C}}}\phi(\overline{E})\gamma + \frac{p^d - 1}{p^d + 1}\frac{\log p}{2}\int_{B_{\mathbb{C}}}\phi(\overline{E})c_{d-1}(E)\right).$$

Combining this with equation (14) gives the result.   □

Similarly, one obtains a formula for the action of any other Hecke operator using the explicit description of its action on γ in [K1, equation (7.4)].

The choice of B' is modeled after the action of the Hecke operator $T(p)$ on the intersection cohomology on moduli of abelian varieties, as described in [FC, Chapter VII.3], where B should be regarded as a subvariety of the moduli space and B' as representing its image under $T(p)$ in the intersection cohomology. This action is only defined over Spec $\mathbb{Z}[1/p]$ though. As $\widehat{CH}^1(\mathrm{Spec}\,\mathbb{Z}[1/p]) = \mathbb{R}/(\mathbb{Q}\cdot\log p)$, the additional term in the above formula would disappear for this base.

Now we are going to formulate an Arakelov version of Hirzebruch's proportionality principle. In [Hi2, p. 773] it is stated as follows: Let G/K be a noncompact irreducible Hermitian symmetric space with compact dual G'/K and let $\Gamma \subset G$ be a cocompact subgroup such that $\Gamma\backslash G/K$ is a smooth manifold. Then there is a ring monomorphism

$$h\colon H^*(G'/K, \mathbb{Q}) \to H^*(\Gamma\backslash G/K, \mathbb{Q})$$

such that $h(c(TG'/K)) = c(TG/K)$ (and similar for other bundles F', F corresponding to K-representation V', V dual to each other). This implies in particular that Chern numbers on G'/K and $\Gamma\backslash G/K$ are proportional [Hi1, p. 345]. Now in our case think for the moment about B as the moduli space of principally polarized abelian varieties of dimension d. Its projective dual is the Lagrangian Grassmannian L_d over Spec $\mathbb{Z}$ parametrizing maximal isotropic subspaces in symplectic vector spaces of dimension $2d$ over any field, $L_d(\mathbb{C}) = \mathrm{Sp}(d)/\mathrm{U}(d)$. But as the moduli space is a noncompact quotient, the proportionality principle must be altered slightly by considering Chow rings modulo certain ideals corresponding to boundary components in a suitable compactification. For that reason, we consider the Arakelov Chow group $\mathrm{CH}^*(\overline{L}_{d-1})$ with respect to the canonical Kähler metric on L_{d-1}, which is the quotient of $\mathrm{CH}^*(\overline{L}_d)$ modulo the ideal $(\widehat{c}_d(\overline{S}), a(c_d(\overline{S})))$ with $\overline{S}$ being the tautological bundle on L_d, and we map it to $\widehat{\mathrm{CH}}^*(B)/(a(\gamma))$. Here L_{d-1} shall be equipped with the canonical symmetric metric. For the Hermitian symmetric space L_{d-1}, the Arakelov Chow ring is a subring of the arithmetic Chow ring $\widehat{\mathrm{CH}}(L_{d-1})$ [GS2, 5.1.5] such that the quotient abelian group depends only on $L_{d-1}(\mathbb{C})$. Instead of dealing with the moduli space, we continue to work with a general regular base B.

The Arakelov Chow ring $\mathrm{CH}^*(\overline{L}_{d-1})$ has been investigated by Tamvakis in [T]. Consider the graded commutative ring

$$\mathbb{Z}[\widehat{u}_1, \dots, \widehat{u}_{d-1}] \oplus \mathbb{R}[u_1, \dots, u_{d-1}]$$

where the ring structure is such that $\mathbb{R}[u_1, \dots, u_{d-1}]$ is an ideal of square zero. Let $\widehat{R}_d$ denote the quotient of this ring by the relations

$$\left(1 + \sum_{j=1}^{d-1} u_j\right)\left(1 + \sum_{j=1}^{d-1} (-1)^j u_j\right) = 1$$

and

$$\left(1 + \sum_{k=1}^{d-1} \widehat{u}_k\right)\left(1 + \sum_{k=1}^{d-1} (-1)^k \widehat{u}_k\right)$$
$$= 1 - \sum_{k=1}^{d-1}\left(\sum_{j=1}^{2k-1} \frac{1}{j}\right)(2k-1)!\,\mathrm{ch}^{[2k-1]}(u_1, \dots, u_{d-1}), \qquad (15)$$

where $\mathrm{ch}(u_1, \dots, u_{d-1})$ denotes the Chern character polynomial in the Chern classes, taken of $u_1, \dots, u_{d-1}$. Then by [T, Theorem 1], there is a ring isomorphism $\Phi\colon \widehat{R}_d \to \mathrm{CH}^*(\overline{L}_{d-1})$ with $\Phi(\widehat{u}_k) = \widehat{c}_k(\overline{S}^*)$ and $\Phi(u_k) = a(c_k(\overline{S}^*))$. The Chern character term in (15), which strictly speaking should be written as $(0, \mathrm{ch}^{[2k-1]}(u_1, \dots, u_{d-1}))$, is thus mapped to

$$a(\mathrm{ch}^{[2k-1]}(c_1(\overline{S}^*), \dots, c_{d-1}(\overline{S}^*))).$$

Theorem 5.5. *There is a ring homomorphism*

$$h \colon \mathrm{CH}^*(\overline{L}_{d-1})_{\mathbb{Q}} \longrightarrow \widehat{\mathrm{CH}}^*(B)_{\mathbb{Q}}/(a(\gamma))$$

with

$$h(\widehat{c}(\overline{S})) = \widehat{c}(\overline{E})\left(1 + a\left(\sum_{k=1}^{d-1}(\frac{\zeta'(1-2k)}{\zeta(1-2k)} - \frac{\log 2}{1-4^{-k}})(2k-1)!\,\mathrm{ch}^{[2k-1]}(E)\right)\right)$$

and

$$h(a(c(\overline{S}))) = a(c(\overline{E})).$$

Note that S^* and E are ample. One could as well map $a(c(\overline{S^*}))$ to $a(c(\overline{E}))$, but the correction factor for the arithmetic characteristic classes would have additional harmonic number terms.

Remark. For $d \leq 6$ one can, in fact, construct such a ring homomorphism which preserves degrees. Still this seems to be a very unnatural thing to do. This is thus in remarkable contrast to the classical Hirzebruch proportionality principle.

Proof. When writing relation (15) as

$$\widehat{c}(\overline{S})\widehat{c}(\overline{S^*}) = 1 + a(\epsilon_1)$$

and the relation in Theorem 3.4 as

$$\widehat{c}(\overline{E})\widehat{c}(\overline{E^*}) = 1 + a(\epsilon_2),$$

we see that a ring homomorphism h is given by

$$h(\widehat{c}_k(\overline{S})) = \sqrt{\frac{1+h(a(\epsilon_1))}{1+a(\epsilon_2)}}\,\widehat{c}_k(\overline{E}) = \left(1 + \frac{1}{2}h(a(\epsilon_1)) - \frac{1}{2}a(\epsilon_2)\right)\widehat{c}_k(\overline{E})$$

(where h on $\mathrm{im}(a)$ is defined as in the theorem). Here the factor $1+\frac{1}{2}h(a(\epsilon_1))-\frac{1}{2}a(\epsilon_2)$ has even degree, and thus

$$h(\widehat{c}_k(\overline{S^*})) = \sqrt{\frac{1+h(a(\epsilon_1))}{1+a(\epsilon_2)}}\,\widehat{c}_k(\overline{E^*})$$

which provides the compatibility with the cited relations. □

Remarks. 1. Note that this proof does not make any use of the remarkable fact that $h(a(\epsilon_1^{[k]}))$ and $a(\epsilon_2^{[k]})$ are proportional forms for any degree k.

2. It would be favorable to have a more direct proof of Theorem 5.5, which does not use the description of the tautological subring. The R-class-like terms suggest that one has to use an arithmetic Riemann–Roch Theorem somewhere in the proof; one could wonder whether one could obtain the description of $\mathrm{CH}^*(\overline{L}_{d-1})$ by a method similar to Section 3. Also, one might wonder whether the statement holds for other symmetric spaces. Our construction relies on the existence of a universal proper bundle with a fiberwise acting nontrivial automorphism; thus it shall not extend easily to other cases.

In particular, Tamvakis' height formula [T, Theorem 3] provides a combinatorial formula for the real number r_d occurring in Theorem 5.1. Replace each term $\mathcal{H}_{2k-1}$ occurring in [T, Theorem 3] by

$$-\frac{2\zeta'(1-2k)}{\zeta(1-2k)} - \sum_{j=1}^{2k-1}\frac{1}{j} + \frac{2\log 2}{1-4^{-k}}$$

and divide the resulting value by half of the degree of L_{d-1}. Using Hirzebruch's formula

$$\deg L_{d-1} = \frac{(d(d-1)/2)!}{\prod_{k=1}^{d-1}(2k-1)!!}$$

for the degree of L_{d-1} (see [Hi1, p. 364]) and the $\mathbb{Z}_+$-valued function $g^{[a,b]_{d-1}}$ from [T] counting involved combinatorial diagrams, we obtain the following.

Corollary 5.6. *The real number r_d occurring in Theorem 5.1 is given by*

$$r_d = \frac{2^{1+(d-1)(d-2)/2}\prod_{k=1}^{d-1}(2k-1)!!}{(d(d-1)/2)!}$$
$$\cdot \sum_{k=0}^{d-2}\left(-\frac{2\zeta'(-2k-1)}{\zeta(-2k-1)} - \sum_{j=1}^{2k+1}\frac{1}{j} + \frac{2\log 2}{1-4^{-k-1}}\right)$$
$$\cdot \sum_{b=0}^{\min\{k,d-2-k\}}(-1)^b 2^{-\delta_{b,k}} g^{[k-b,b]_{d-1}},$$

where $\delta_{b,k}$ is Kronecker's δ.

One might wonder whether there is a "topological" formula for the height of locally symmetric spaces similar to [KK, Theorem 8.1]. Comparing the fixed-point height formula [KK, Lemma 8.3] with the Schubert calculus expression [T, Theorem 3] for the height of Lagrangian Grassmannians, one finds

$$\sum_{\epsilon_1,\ldots,\epsilon_{d-1}\in\{\pm 1\}}\frac{1}{\prod_{i\leq j}(\epsilon_i i + \epsilon_j j)}$$
$$\sum_{\ell=1}^{\frac{d(d-1)}{2}}\sum_{i\leq j}\frac{\left(\sum\epsilon_\nu\nu\right)^{\frac{d(d-1)}{2}} - \left(\sum\epsilon_\nu\nu\right)^{\frac{d(d-1)}{2}-\ell+1}\left(\sum\epsilon_\nu\nu - (2-\delta_{ij}(\epsilon_i i + \epsilon_j j))\right)^\ell}{2\ell(\epsilon_i i + \epsilon_j j)}$$
$$= \sum_{k=0}^{d-2}\left(\sum_{j=1}^{2k-1}\frac{1}{j}\right)\sum_{b=0}^{\min\{k,d-2-k\}}(-1)^b 2^{-\delta_{b,k}} g^{[k-b,b]_{d-1}}.$$

In [G, Theorem 2.5] van der Geer shows that R_d embeds into the (classical) Chow ring $\mathrm{CH}^*(\mathcal{M}_d)_\mathbb{Q}$ of the moduli stack $\mathcal{M}_d$ of principally polarized abelian varieties. Using this result, one finds the following.

Lemma 5.7. *Let B be a regular finite covering of the moduli space $\mathcal{M}_d$ of principally polarized abelian varieties of dimension d. Then for any nonvanishing polynomial expression $p(u_1, \ldots, u_{d-1})$ in R_d,*

$$h(p(\widehat{c}_1(\overline{S}), \ldots, \widehat{c}_{d-1}(\overline{S}))) \notin \text{im } a.$$

In particular, h is nontrivial in all degrees. Furthermore, h is injective iff $a(c_1(E)^{d(d-1)/2}) \neq 0$ in $\widehat{\text{CH}}^{d(d-1)/2+1}(B)_\mathbb{Q}/(a(\gamma))$.

The need for a regular covering in our context is an unfortunate consequence of the Arakelov geometry of stacks not yet being fully constructed. Eventually this problem might get remedied. Until then, one can resort to base changes to ensure the existence of regular covers as, e.g., the moduli space of p.p. abelian varieties with level-*n* structure for $n \geq 3$ over Spec $\mathbb{Z}[1/n, e^{2\pi i/n}]$ [FC, Chapter IV.6.2c].

Proof. Consider the canonical map $\zeta : \widehat{\text{CH}}^*(B)_\mathbb{Q}/(a(\gamma)) \to \text{CH}^*(B)_\mathbb{Q}$. Then

$$\zeta(h(p(\widehat{c}_1(\overline{S}), \ldots, \widehat{c}_{d-1}(\overline{S})))) = p(c_1(E), \ldots, c_{d-1}(E)),$$

and the latter is nonvanishing according to [G, Theorem 1.5]. This proves the first assertion.

If $a(c_1(E)^{d(d-1)/2}) \neq 0$ in $\widehat{\text{CH}}^{d(d-1)/2+1}(B)_\mathbb{Q}/(a(\gamma))$, then by the same induction argument as in the proof of [G, Theorem 2.5] R_d embeds in $a(\ker \bar{\partial}\partial)$. Finally, by [T, Theorem 2] any element z of $\widehat{R}_d$ can be written in a unique way as a linear combination of

$$\widehat{u}_{j_1} \cdots \widehat{u}_{j_m} \quad \text{and} \quad u_{j_1} \cdots u_{j_m} \quad \text{with} \quad 1 \leq j_1 < \cdots < j_m < d, \quad 1 \leq m < d.$$

Thus if $z \notin \text{im } a$, then $h(z) \neq 0$ follows by van der Geer's result, and if $z \in \text{im } a \setminus \{0\}$, then $h(z) \neq 0$ follows by embedding $R_d \otimes \mathbb{R}$. □

Using the exact sequence (2), the condition in the Lemma is that the cohomology class $c_1(E)^{d(d-1)/2}$ should not be in the image of the Beilinson regulator.

Finally, by comparing Theorem 5.1 with Kühn's result [Kü, Theorem 6.1] (see also Bost [Bo]), we conjecture that the analogue of Theorem 5.5 holds in a yet to be developed Arakelov intersection theory with logarithmic singularities, extending the methods of [Kü, BKK], as described in [MR]. In other words, there should be a ring homomorphism to the Chow ring of the moduli space of abelian varieties

$$h : \text{CH}^*(\overline{L}_d)_\mathbb{Q} \to \widehat{\text{CH}}^*(\mathcal{M}_d)_\mathbb{Q}$$

extending the one in Theorem 5.5, and γ should provide the Green current corresponding to $\hat{c}_d(\overline{E})$. This would imply the following.

Conjecture 5.8. For an Arakelov intersection theory with logarithmic singularities, extending the methods of [Kü], the height of a moduli space $\mathcal{M}_d$ over Spec $\mathbb{Z}$ of principally polarized abelian varieties of relative dimension *d* is given by

$$h(\mathcal{M}_d) = \frac{r_{d+1}}{2} \deg(\mathcal{M}_d).$$

The factor $1/2$ is caused by the degree map in Arakelov geometry.

6 The Fourier expansion of the Arakelov Euler class of the Hodge bundle

In this section, we shall further investigate the differential form γ which played a prominent role in the preceding results. We adapt most notations from [K1]. In particular, we use as the base space the Siegel upper half-space

$$\mathfrak{H}_n := \{Z = X + iY \in \mathrm{End}(\mathbb{C}^d) \mid {}^t Z = Z, Y > 0\},$$

which is the universal covering of the moduli space of principally polarized abelian varieties. Due to an unavoidable clash of notations, we are forced here to use the letters Z and Y again. Choose the trivial $\mathbb{C}^d$-bundle over $\mathfrak{H}_n$ as the holomorphic vector bundle E and define the lattice Λ over a point $Z \in \mathfrak{H}_d$ as

$$\Lambda_{|Z} := (Z, \mathrm{id})\mathbb{Z}^{2n},$$

where (Z, id) denotes a $\mathbb{C}^{d \times 2d}$-matrix. The polarization defines a Kähler form on E; the associated metric is given by

$$\|Zr + s\|_{|Z}^2 = {}^t(Zr + s)Y^{-1}\overline{(Zr + s)} \quad \text{for } r, s \in \mathbb{Z}^n.$$

(One might scale the metric by a constant factor $1/2$ to satisfy the condition $\mathrm{vol}(Z) = 1$. The torsion form is invariant under this scaling.) The crucial ingredient in the construction of γ in [K1] was a series $\bar{\beta}_t$ depending on real parameters $t, b \in \mathbb{R}$, such that the Epstein zeta function $Z(s)$ with $\gamma = Z'(0)$ can be constructed as the Mellin transform of the b-linear term of $\bar{\beta}_t$. More precisely,

$$Z(s) := -\frac{1}{\Gamma(s)} \int_0^\infty t^{s-1} \left(\frac{d}{db}_{|b=0} \bar{\beta}_t + c_{n-1}(\bar{E}) \right) dt,$$

which also leads to other expressions for γ in terms of $\bar{\beta}$. We derive the Fourier expansion for γ by applying the Poisson summation formula to a lattice of half the maximal rank in the Epstein zeta function defining the torsion form. This leaves us with two infinite series which converge at $s = 0$, and another Epstein zeta function for a lattice of half the previous rank. By iterating this procedure $\frac{\log 2d}{\log 2}$ times, one can actually gain a convergent series expression for γ; compare [E, Section 8], where a similar procedure with $2d$ steps is described.

Set $C := \frac{1}{\pi} Y^{-1}(1 - \frac{1}{2\pi i b} \mathrm{Re}\,\Omega^E)$ and $D := \frac{1}{\pi} Y^{-1} \frac{-i}{2\pi i b} \mathrm{Im}\,\Omega^E$. Thus ${}^t C = C$, ${}^t D = -D$. Then by [K1, equation (6.0)],

$$\bar{\beta}_t = \left(\frac{-b}{\pi t}\right)^d \sum_{\lambda \in \Lambda} \exp\left(-\frac{1}{t}\left\langle \lambda^{1,0}, \left(1 + \frac{i}{2\pi b}\Omega^E\right)\lambda^{0,1}\right\rangle\right)$$

$$= \left(\frac{-b}{\pi t}\right)^d \sum_{r,u \in \mathbb{Z}^d} \exp\left(-\frac{\pi}{t}{}^t(Zr + u)(C + D)(\bar{Z}r + u)\right).$$

Now let B be a symmetric integral $d \times d$-matrix. The space $\mathfrak{b}$ of such matrices embeds into $\mathrm{Sp}(d, \mathbb{Z})$ via

$$B \mapsto \begin{pmatrix} \mathrm{id} & B \\ 0 & \mathrm{id} \end{pmatrix}.$$

The induced action of $B \in \mathfrak{b}$ on $\mathfrak{H}$ is given by $Z \mapsto Z + B$. As $\bar{\beta}_t$ is $\mathrm{Sp}(d, \mathbb{Z})$-invariant, it thus has a Fourier decomposition on the torus $\mathfrak{H}/\mathfrak{b}$. Notice that the space $\mathfrak{c}$ of frequencies does not equal $\mathfrak{b}$ but is the space

$$\mathfrak{c} = \left\{ \frac{1}{2}(B +^t B) \mid B \in \mathfrak{gl}(d, \mathbb{Z}) \right\}$$

of symmetric matrices integral along the diagonal and half-integral off the diagonal.

Using the Poisson summation formula applied to $u \in \mathbb{Z}^d$, we find, for $\bar{\beta}_{t|Z}$ at $Z = X + iY$,

$$\bar{\beta}_t = \left(\frac{-b}{\pi t} \right)^d \sum_{r,u \in \mathbb{Z}^d} \exp\left(-\frac{\pi}{t}{}^t(Zr + u)(C + D)(\bar{Z}r + u) \right)$$

$$= \left(\frac{-b}{\pi t} \right)^d \sum_{r,u \in \mathbb{Z}^d} \exp\left(-\frac{\pi}{t}{}^t(Xr + u)C(Xr + u) \right.$$

$$\left. -\frac{\pi}{t}{}^t r Y C Y r - \frac{2\pi i}{t}{}^t r Y D(Xr + u) \right)$$

$$= \left(\frac{-b}{\pi \sqrt{t}} \right)^d \sum_{r,\hat{u} \in \mathbb{Z}^d} \frac{1}{\sqrt{\det C}} \exp\left(-\pi t\, {}^t \hat{u} C^{-1} \hat{u} - 2\pi i\, {}^t \hat{u} X r \right.$$

$$\left. -\frac{\pi}{t}{}^t r Y (C - DC^{-1}D) Y r - 2\pi\, {}^t r Y DC^{-1}\hat{u} \right).$$

For any symmetric $A \in \mathbb{R}^{d \times d}$ and $M = \frac{1}{2}(r \cdot {}^t u + u \cdot {}^t r)$, we have $\langle M, A \rangle = \mathrm{Tr}\, M^t A = {}^t r A u$. Thus the Fourier coefficient of $e^{-2\pi i \langle M, X \rangle}$ for $M \in \mathfrak{c}$ equals

$$\sum \left(\frac{-b}{\pi \sqrt{t}} \right)^d \frac{1}{\sqrt{\det C}}$$

$$\cdot \exp\left(-\pi t\, {}^t u C^{-1} u - \frac{\pi}{t}{}^t r Y (C - DC^{-1}D) Y r - 2\pi\, {}^t r Y DC^{-1} u \right).$$

In particular, the occurring frequency matrices M in the Fourier decomposition are among the matrices in $\mathfrak{c}$ which have at most two nonzero eigenvalues. Note that

$$C - DC^{-1}D = C(\mathrm{Id} - C^{-1}DC^{-1}D) = C(\mathrm{Id} - C^{-1}D)(\mathrm{Id} + C^{-1}D)$$

$$= (C - D)C^{-1}(C + D) = {}^t(C + D)C^{-1}(C + D), \qquad (16)$$

and, in particular, for $a \in \mathbb{R}^d$,

$${}^t a (C - DC^{-1}D)^{-1} a = {}^t a (C + D)^{-1}(C \pm D){}^t(C + D)^{-1} a = {}^t a (C \mp D)^{-1} a$$

(this value does not depend on the choice of $\pm$), or

$$2(C - DC^{-1}D)^{-1} = (C + D)^{-1} + (C - D)^{-1}.$$

6.1 The coefficients of the nonconstant terms

Proposition 6.1. *Two vectors* $r, u \in \mathbb{R}^d \setminus \{0\}$ *are uniquely determined by the matrix*

$$M := \frac{1}{2}(r \cdot^t u + u \cdot^t r)$$

up to order and multiplication by a constant.

Proof. Assume first that u and r are not colinear. The two nonvanishing eigenvalues of M are given by

$$\lambda_{1,2} = \frac{1}{2}(\langle r, u \rangle \pm \|r\|\|u\|)$$

with corresponding eigenvectors $v_{1,2} = c_{1,2}(\|r\|u \pm \|u\|r)$ with $c_{1,2} \in \mathbb{R} \setminus \{0\}$ arbitrary. In fact,

$$Mv_{1,2} = \frac{c_{1,2}}{2}(\|r\|r\langle u, u\rangle \pm \|u\|r\langle r, u\rangle + \|r\|u\langle r, u\rangle \pm \|u\|u\langle r, r\rangle) = \lambda_{1,2}v_{1,2}.$$

Now $\|v_{1,2}\|^2 = \pm 4c_{1,2}^2\|r\|\|s\|\lambda_{1,2}$ and thus

$$\frac{v_{1,2}}{\|v_{1,2}\|}\sqrt{|\lambda_{1,2}|} = \pm\frac{1}{2}\left(\sqrt{\frac{\|r\|}{\|u\|}}u \pm \sqrt{\frac{\|u\|}{\|r\|}}r\right).$$

Without loss of generality, we may assume the sign to be positive; we then have

$$\frac{v_1}{\|v_1\|}\sqrt{|\lambda_1|} + \frac{v_2}{\|v_2\|}\sqrt{|\lambda_2|} = \sqrt{\frac{\|r\|}{\|u\|}}u$$

and

$$\frac{v_1}{\|v_1\|}\sqrt{|\lambda_1|} - \frac{v_2}{\|v_2\|}\sqrt{|\lambda_2|} = \sqrt{\frac{\|u\|}{\|r\|}}r.$$

Thus all possible sets $\{u, r\}$ of solutions are given in terms of M by

$$\left\{\left\{c\left(\frac{v_1}{\|v_1\|}\sqrt{|\lambda_1|} + \frac{v_2}{\|v_2\|}\sqrt{|\lambda_2|}\right), \frac{1}{c}\left(\frac{v_1}{\|v_1\|}\sqrt{|\lambda_1|} - \frac{v_2}{\|v_2\|}\sqrt{|\lambda_2|}\right)\right\} \;\middle|\; c \in \mathbb{R}, \; c \neq 0\right\}.$$

In the case r, u colinear, the eigenvalue λ_2 vanishes and the proof remains the same with this simplification. □

Remarks. 1. Note that $\lambda_1 > 0$ and $\lambda_2 \leq 0$.

2. There is a simpler formula for r and u up to two possibilities in every coordinate: Necessarily one diagonal element of M is nonzero, say, M_{11}. By solving the system of quadratic equations $2M_{1j} = r_1u_j + r_ju_1$, one finds up to the scaling constant

$$r_j = M_{1j} \pm \sqrt{M_{1j}^2 - M_{11}M_{jj}}.$$

Alas determining the $\pm$-choice in every coordinate is not easy.

3. In our case, the condition $r, s \in \mathbb{Z}^d$ implies that for every $M \in \mathfrak{c}$ there are primitive vectors $r_0, u_0 \in \mathbb{Z}^d$ and $c \in \mathbb{Z}^+$ such that all possible sets $\{r, u\}$ are given by $\{\{kr_0, c/k \cdot u_0\} | k \in \mathbb{Z}, k|c\}$.

Using the Taylor expansion of $(1 - x)^{-1}$ at $x = 0$, we find for the term in the exponential function in $\bar{\beta}_{t,M}$ with $r = kr_0$, $u = cu_0/k$,

$$- \pi t\,{}^t u C^{-1} u - \frac{\pi}{t}\,{}^t r Y (C - DC^{-1}D) Yr - 2\pi\,{}^t r Y DC^{-1} u$$

$$= -\frac{\pi^2 c^2}{k^2}\,{}^t u_0 Y u_0 - \frac{k^2}{t}\,{}^t r_0 Y r_0 + \frac{t}{k^2} \sum_{l \geq 1} \frac{\omega_l}{b^l} + \frac{k^2}{t} \sum_{l \geq 1} \frac{\omega_l'}{b^l} + \sum_{l \geq 1} \frac{\omega_l''}{b^l}, \quad (17)$$

where $\omega_l, \omega_l', \omega_l''$ are differential forms of degree (l, l), depending on M but not on k. Thus $\bar{\beta}_{t,M}$ has the form

$$\bar{\beta}_{t,M} = \sum_{k \in \mathbb{Z}, k|c} \left(\frac{-b}{\pi \sqrt{t}}\right)^d \frac{1}{\sqrt{\det C}(1 + \delta_{r_0 = u_0})} \left(\sum_{l \in \mathbb{Z}} \left(\frac{t}{k^2}\right)^l \alpha_l(b)\right)$$

$$\cdot \left(\exp\left(-\frac{\pi^2 c^2}{k^2}\,{}^t u_0 Y u_0 - \frac{k^2}{t}\,{}^t r_0 Y r_0\right) + \exp\left(-\frac{\pi^2 c^2}{k^2}\,{}^t r_0 Y r_0 - \frac{k^2}{t}\,{}^t u_0 Y u_0\right)\right)$$

with $\alpha_l(b)$ being a differential form of degree greater than or equal to $|l|$, with coefficients in polynomials in $1/b$. In particular, the sum over l is finite. Now for $a, b \in \mathbb{R}^+$, $\alpha \in \mathbb{R}$ the Bessel K-functions provide the formula

$$\frac{1}{\Gamma(s)} \int_0^\infty e^{-at - b/t} t^{s-1-\alpha}\, dt = \frac{2}{\Gamma(s)} \sqrt{\frac{a}{b}}^{\alpha-s} K(\alpha - s, 2\sqrt{ab})$$

and thus

$$\frac{\partial}{\partial s}_{|s=0} \left(\frac{1}{\Gamma(s)} \int_0^\infty e^{-at - b/t} t^{s-1-\alpha}\, dt\right) = 2\sqrt{\frac{a}{b}}^{\alpha} K(\alpha, 2\sqrt{ab}).$$

We define

$$\|M, Y\| := \sqrt{{}^t r Y r \cdot {}^t u Y u};$$

by Proposition 6.1, we know that this value does not depend on the choice of r and u. More easily, one can verify this using $\|M, Y\|^2 + \langle M, Y \rangle^2 = 2 \operatorname{Tr} MYMY$. Also we set

$$\rho(r_0, u_0) := \sqrt{\frac{{}^t r_0 Y r_0}{{}^t u_0 Y u_0}}.$$

Hence we find for the derivative at $s = 0$ of the Mellin transform of $\bar{\beta}_{t,M}$,

$$\frac{\partial}{\partial s}_{|s=0} \left(\frac{1}{\Gamma(s)} \int_0^\infty \bar{\beta}_{t,M} t^{s-1} dt \right)$$

$$= \sum_{k\in\mathbb{Z},\,k|c} \sum_{l\in\mathbb{Z}} \alpha_l(b)|k|^{-2l} \left(\frac{-b}{\pi}\right)^d \frac{1}{\sqrt{\det C}(1+\delta_{r_0=u_0})}$$

$$\cdot 2 \left(\sqrt{\frac{\pi^2 c^2 \cdot {}^t u_0 Y u_0}{k^4 \cdot {}^t r_0 Y r_0}}^{\,d/2-l} + \sqrt{\frac{\pi^2 c^2 \cdot {}^t r_0 Y r_0}{k^4 \cdot {}^t u_0 Y u_0}}^{\,d/2-l} \right)$$

$$\cdot K\left(d/2 - l,\, 2\sqrt{\pi^2 \|M,Y\|^2}\right)$$

$$= \sum_{k\in\mathbb{Z},\,k|c} \sum_{l\in\mathbb{Z}} \alpha_l(b) \frac{2c^{d/2-l}(-b)^d}{\pi^{l+d/2}|k|^d \sqrt{\det C}(1+\delta_{r_0=u_0})}$$

$$\cdot \left(\rho(r_0,u_0)^{l-d/2} + \rho(r_0,u_0)^{d/2-l} \right) K(d/2-l, 2\pi\|M,Y\|)$$

$$= \sum_{l\in\mathbb{Z}} \alpha_l(b) \frac{2(\pi c)^{-d/2-l}(-b)^d \sigma_d(c)}{\sqrt{\det C}(1+\delta_{r_0=u_0})}$$

$$\cdot \left(\rho(r_0,u_0)^{l-d/2} + \rho(r_0,u_0)^{d/2-l} \right) K(d/2-l, 2\pi\|M,Y\|)$$

with $\sigma_m(c) := \sum_{k\in\mathbb{Z}^+,\,k|c} k^m$ being the divisor function. For $c = \prod_{p \text{ prime}} p^{v_p}$, one finds

$$\sigma_m(c) = c^m \prod_{p \text{ prime}} \frac{1 - p^{-m(v_p+1)}}{1 - p^{-m}}$$

and thus $\sigma_m(c) \in \,]c^m, \zeta(m)c^m[$. The form γ is given by the linear term in b in the above equation, for which $|l| \leq d - 1$. Set

$$\eta(r_0, u_0)$$

$$:= e^{-2\pi\|M,Y\|} \rho(r_0,u_0)^{-d/2} \exp\Big(-c\rho(r_0,u_0) \cdot {}^t u_0(C^{-1} - \pi Y)u_0$$

$$- 2\pi c^t r_0 Y DC^{-1} u_0 - \pi^2 c\rho(r_0,u_0)^{-1} \cdot {}^t r_0 \Big(Y(C - DC^{-1}D)Y - \frac{1}{\pi}Y \Big)r_0 \Big)$$

$$= \rho(r_0,u_0)^{-d/2} \exp\Big(-c\rho(r_0,u_0) \cdot {}^t u_0 C^{-1} u_0 - 2\pi c^t r_0 Y DC^{-1} u_0$$

$$- \pi^2 c\rho(r_0,u_0)^{-1} \cdot {}^t r_0 Y(C - DC^{-1}D)Y r_0 \Big).$$

The Bessel K-functions have for $|x| \to \infty$ the asymptotics

$$K(v, x) = \sqrt{\frac{\pi}{2x}} e^{-x} \left(1 + O\left(\frac{1}{x}\right) \right)$$

and thus we find for $\|M,Y\| \to \infty$ by setting $t := \frac{k^2}{c\pi}\rho(r_0, u_0)$ in the defining equation for the α_l,

$$\frac{\partial}{\partial s}_{|s=0}\left(\frac{1}{\Gamma(s)}\int_0^\infty \bar{\beta}_{t,M}t^{s-1}dt\right)$$

$$= \frac{(\pi c)^{-d/2}(-b)^d \sigma_d(c)}{\sqrt{\|M,Y\|}\det C(1+\delta_{r_0=u_0})}(\eta(r_0,u_0)+\eta(u_0,r_0))\left(1+O\left(\frac{1}{\|M,Y\|}\right)\right).$$

(18)

For d odd the Bessel K-functions have special values, and one thus finds explicit expressions for the Fourier coefficients similar to (18). Using the *polylogarithm* defined for $|q| < 1, l \in \mathbb{R}$ by

$$\mathrm{Li}_l(q) := \sum_{k=1}^\infty \frac{q^k}{k^l},$$

(19)

we have the equality for $m \in \mathbb{Z}^+$

$$\sum_{c=1}^\infty \frac{\sigma_m(c)}{c^l}q^c = \sum_{n=1}^\infty n^{m-l}\mathrm{Li}_l(q^n).$$

Thus we obtain with

$$q(r_0,u_0) := \exp(-\rho(r_0,u_0)\cdot {}^tu_0 C^{-1}u_0 - 2\pi {}^tr_0 Y DC^{-1}u_0$$
$$- \pi^2 \rho(r_0,u_0)^{-1}\cdot {}^tr_0 Y(C - DC^{-1}D)Yr_0 - 2\pi i {}^tu_0 Xr_0)$$

the following.

Lemma 6.2. *When summing the part of the Fourier expansion corresponding to frequency matrices which have the same pair of primitive vectors r_0, u_0, we obtain, with $M_0 := \frac{1}{2}(r_0 \cdot {}^t u_0 + u_0 \cdot {}^t r_0)$,*

$$\sum_{c\in\mathbb{Z}^+} e^{-2\pi i\langle cM_0, X\rangle}\frac{\partial}{\partial s}_{|s=0}\left(\frac{1}{\Gamma(s)}\int_0^\infty \bar{\beta}_{t,cM_0}t^{s-1}dt\right)$$

$$= \frac{\pi^{-d/2}(-b)^d\rho(r_0,u_0)^{-d/2}}{\sqrt{\|M_0,Y\|}\det C(1+\delta_{r_0=u_0})}$$

$$\cdot \sum_{n=1}^\infty n^{\frac{d-1}{2}}\left(\mathrm{Li}_{\frac{d+1}{2}}(q(r_0,u_0)^n) + O\left(\frac{1}{\|M_0,Y\|}\right)\mathrm{Li}_{\frac{d+3}{2}}(q(r_0,u_0)^n)\right)$$

$$+ \text{ this same term with } r_0, u_0 \text{ exchanged}$$

$$= \frac{\pi^{-d/2}(-b)^d}{\sqrt{\|M_0,Y\|}\det C(1+\delta_{r_0=u_0})}\sum_{n=1}^\infty n^{\frac{d-1}{2}}\left(\rho(r_0,u_0)^{-d/2}\mathrm{Li}_{\frac{d+1}{2}}(q(r_0,u_0)^n)\right.$$

$$+ \rho(r_0,u_0)^{d/2}\mathrm{Li}_{\frac{d+1}{2}}(q(u_0,r_0)^n)\Big)\cdot\left(1+O\left(\frac{1}{\|M_0,Y\|}\right)\right).$$

Here polylogarithms of forms have to be interpreted via the power series in equation (19).

6.2 The coefficient of the constant term

For $M = 0$, we find by applying again the Poisson summation formula to both sums

$$\bar{\beta}_{t,0} = \sum_{r \in \mathbb{Z}^d} \left(\frac{-b}{\pi \sqrt{t}} \right)^d \frac{1}{\sqrt{\det C}} \exp\left(-\frac{\pi}{t} {}^t r Y (C - DC^{-1}D) Y r \right) \tag{20}$$

$$+ \sum_{u \in \mathbb{Z}^d} \left(\frac{-b}{\pi \sqrt{t}} \right)^d \frac{1}{\sqrt{\det C}} \exp(-\pi t\, {}^t u C^{-1} u) - \left(\frac{-b}{\pi \sqrt{t}} \right)^d \frac{1}{\sqrt{\det C}}$$

$$= \sum_{\hat{r} \in \mathbb{Z}^d} \left(\frac{-b}{\pi} \right)^d \frac{\exp(-\pi t\, {}^t \hat{r} Y^{-1}(C - DC^{-1}D)^{-1} Y^{-1} \hat{r})}{\sqrt{\det(C(C - DC^{-1}D))\,\det Y}}$$

$$+ \sum_{\hat{u} \in \mathbb{Z}^d} \left(\frac{-b}{\pi t} \right)^d \exp\left(-\frac{\pi}{t}\, {}^t \hat{u} C \hat{u} \right) - \left(\frac{-b}{\pi \sqrt{t}} \right)^d \frac{1}{\sqrt{\det C}}. \tag{21}$$

Using (16), we find

$$\det(Y^2 C(C - DC^{-1}D)) = \det(YC + YD)^2$$

and (by Corollary 3.5)

$$\frac{1}{\det(\pi Y(C + D))} = \det\left(1 + \frac{1}{2\pi i b} \Omega^E \right) = \sum_{j=0}^{d} (-b)^{-j} c_j(\overline{E}),$$

and thus (21) simplifies to

$$\bar{\beta}_{t,0} = \theta_1(t) + \theta_2(t) - \left(\frac{-b}{\pi \sqrt{t}} \right)^d \frac{1}{\sqrt{\det C}}, \tag{22}$$

where

$$\theta_1(t) := \sum_{\hat{r} \in \mathbb{Z}^d} (-b)^d \det\left(1 + \frac{1}{2\pi i b} \Omega^E \right) \exp(-\pi t \cdot {}^t \hat{r} Y^{-1}(C \pm D)^{-1} Y^{-1} \hat{r}),$$

$$\theta_2(t) := \sum_{\hat{u} \in \mathbb{Z}^d} \left(\frac{-b}{\pi t} \right)^d \exp\left(-\frac{\pi}{t} \cdot {}^t \hat{u} C \hat{u} \right).$$

Note that the term $-(\frac{-b}{\pi \sqrt{t}})^d \frac{1}{\sqrt{\det C}}$ vanishes under Mellin transformation [K1, Remark on p. 12]. The b-linear term of the second summand $\theta_2(t)$ in (22) is

$$\theta_2(t)^{[b]} = \frac{1}{(d-1)!} \sum_{u \in \mathbb{Z}^d} \left(\frac{-1}{\pi t} \right)^d \exp\left(-\frac{1}{t}\, {}^t u Y^{-1} u \right) \left(\frac{1}{2\pi i t}\, {}^t u Y^{-1} \Omega^E u \right)^{\wedge(d-1)}$$

with Mellin transform

$Z_2(s)^{[b]}$

$$:= -\frac{\Gamma(2d-1-s)}{\Gamma(s)(d-1)!} \sum_{u \in \mathbb{Z}^d} \left(\frac{-1}{\pi}\right)^d ({}^t u Y^{-1} u)^{1-2d+s} \left(\frac{1}{2\pi i} {}^t u Y^{-1} \Omega^E u\right)^{\wedge(d-1)},$$

and thus the corresponding summand of γ equals

$$Z_2'(0) = \frac{(2d-2)!}{(d-1)!\pi^d} \sum_{u \in \mathbb{Z}^d \setminus \{0\}} ({}^t u Y^{-1} u)^{1-2d} \left(\frac{-1}{2\pi i} {}^t u Y^{-1} \Omega^E u\right)^{\wedge(d-1)}.$$

This term is homogeneous in Y of degree $2-d$; thus it behaves like $|Y|^{2-d}$ for $|Y| \to \infty$ or $|Y| \to 0$.

By proceeding as in (17), we observe that the first summand $\theta_1(t)$ in (22) has the form

$$\theta_1(t) = \sum_{r \in \mathbb{Z}^d} (-b)^d \det\left(1 + \frac{1}{2\pi i b} \Omega^E\right) \exp(-\pi t\, {}^t r Y^{-1}(C - DC^{-1}D)^{-1} Y^{-1} r)$$

$$= \sum_{r \in \mathbb{Z}^d} (-b)^d \det\left(1 + \frac{1}{2\pi i b} \Omega^E\right) \exp(-\pi^2 t\, {}^t r Y^{-1} r)$$

$$\cdot \exp(-\pi t\, {}^t r Y^{-1}((C - DC^{-1}D)^{-1} - \pi Y) Y^{-1} r)$$

$$= \sum_{r \in \mathbb{Z}^d} (-b)^d \det\left(1 + \frac{1}{2\pi i b} \Omega^E\right) \exp(-\pi^2 t\, {}^t r Y^{-1} r)$$

$$\cdot \left(1 + \sum_{k=1}^{d} \sum_{\ell=1}^{k} t^\ell (-b)^{-k} \omega_{k,\ell}\right)$$

with $\omega_{k,\ell}$ being a (k, k)-form, homogeneous in Y of degree $-\ell - 2k$ and homogeneous in r of degree 2ℓ. The coefficient of b in θ_1 is given by

$$\theta_1(t)^{[b]} = \theta_{11}(t) + \theta_{12}(t),$$

where

$$\theta_{11}(t) := -\sum_{r \in \mathbb{Z}^d} \exp(-\pi^2 t \cdot {}^t r Y^{-1} r) c_{d-1}(\overline{E}),$$

$$\theta_{12}(t) := -\sum_{r \in \mathbb{Z}^d} \exp(-\pi^2 t \cdot {}^t r Y^{-1} r) \sum_{k=1}^{d-1} \sum_{\ell=1}^{k} t^\ell \omega_{k,\ell} c_{d-1-k}(\overline{E}).$$

The Mellin transform of this term thus equals

$$Z_{11}(s) c_{d-1}(\overline{E}) + \frac{1}{\Gamma(s)} Z_{12}(s) := \sum_{r \in \mathbb{Z}^d \setminus \{0\}} (\pi^2 \cdot {}^t r Y^{-1} r)^{-s} c_{d-1}(\overline{E})$$

$$+ \sum_{r \in \mathbb{Z}^d \setminus \{0\}} \sum_{k=1}^{d-1} \sum_{\ell=1}^{k} \frac{\Gamma(s+\ell)}{\Gamma(s)} (\pi^2 \cdot {}^t r Y^{-1} r)^{-s-\ell} \omega_{k,\ell} c_{d-1-k}(\overline{E}),$$

which is homogeneous in Y of degree $2 - 2d + s$. In particular, the Mellin transform of θ_1 converges in (22) for $\operatorname{Re} s > d/2$ when subtracting the $\hat{r} = 0$ summand (and similarly in (20) for $\operatorname{Re} s < 0$ when subtracting the $r = 0$ summand). Notice that $\theta_{22}(t) \to 0$ for $t \to \infty$ and thus $\frac{1}{\Gamma(s)} Z_{22}(s) \to 0$ for $s \to 0$. Hence $\frac{\partial}{\partial s}|_{s=0} \frac{1}{\Gamma(s)} Z_{12}(s) = Z_{12}(0)$. Furthermore, $Z_{11}(0) = -1$. Clearly, for $\alpha \in \mathbb{R}^+$,

$$Z_{11}(s)_{|\alpha Y} = \alpha^s Z_{11}(s)_{|Y},$$

and thus

$$Z'_{11}(0)_{|\alpha Y} = -\log \alpha + Z'_{11}(0)_{|Y}.$$

Concluding, we find the following.

Theorem 6.3. *The differential form γ representing the torsion form verifies, for $|Y| \to \infty$,*

$$\gamma = Z'_{11}(0) c_{d-1}(\overline{E}) + Z_{12}(0) + Z'_2(0) + O(e^{-c|Y|}), \tag{23}$$

where Z_{11} is a classical real-valued Epstein zeta function; Z_{12} is a sum of Epstein zeta functions with polynomials in the numerator; and $Z'_2(0)$ is given by a convergent series. The first term in (23) behaves like $-\log |Y| \cdot |Y|^{2-2d} c_1 + |Y|^{2-2d} c_2$, the second term is homogeneous in Y of degree $2 - 2d$, and the third term is homogeneous in Y of degree $2 - d$.

Acknowledgments. I thank A. Johan de Jong, Damian Roessler, Christophe Soulé, Harry Tamvakis, Emmanuel Ullmo, Torsten Wedhorn, and the referee for helpful discussions and comments. Also I thank the Deutsche Forschungsgemeinschaft which supported me with a Heisenberg fellowship during the preparation of parts of this article.

References

[B3]   J.-M. Bismut, Equivariant immersions and Quillen metrics, *J. Differential Geom.*, **41** (1995), 53–157.

[B4]   J.-M. Bismut, *Holomorphic Families of Immersions and Higher Analytic Torsion Forms*, Astérisque 244, Société Mathématique de France, Paris, 1997.

[BM]   J.-M. Bismut and X. Ma, Holomorphic immersions and equivariant torsion forms, *J. Reine Angew. Math.*, **575** (2004), 189–235.

[BBM]   P. Berthelot, L. Breen, and W. Messing, *Théorie de Dieudonné Cristalline* II, Lecture Notes in Mathematics 930, Springer, Berlin, New York, Heidelberg, 1982.

[Bo]   J.-B. Bost, *Intersection theory on arithmetic surfaces and L_1^2 metrics*, letter dated March 6, 1998.

[BKK]   J. I. Burgos, J. Kramer, and U. Kühn, Cohomological arithmetic Chow groups, e-print, 2004; *J. Inst. Math. Jussieu*, to appear; available online from http://www.arxiv.org/abs/math.AG/0404122.

[E]   P. Epstein, Zur Theorie allgemeiner Zetafunktionen, *Math. Ann.*, **56** (1903), 615–644.

[Fal]   G. Faltings, *Lectures on the Arithmetic Riemann-Roch Theorem*, Princeton University Press, Princeton, NJ, 1992.

[FC]   G. Faltings and C.-L. Chai, *Degeneration of Abelian Varieties* (with an appendix by David Mumford), Ergebnisse der Mathematik und ihrer Grenzgebiete 22, Springer-Verlag, Berlin, 1990.

[GS2]   H. Gillet and C. Soulé, Arithmetic intersection theory, *Publ. Math. IHES*, **72** (1990), 94–174.

[GS3]   H. Gillet and C. Soulé, Characteristic classes for algebraic vector bundles with hermitian metrics I, II, *Ann. Math.*, **131** (1990), 163–203, 205–238.

[GS8]   H. Gillet and C. Soulé, An arithmetic Riemann-Roch theorem, *Invent. Math.*, **110** (1992), 473–543.

[G]   G. van der Geer, Cycles on the moduli space of abelian varieties, in C. Faber and E. Looijenga, eds., *Moduli of Curves and Abelian Varieties: The Dutch Intercity Seminar on Moduli*, Aspects of Mathematics E33, Vieweg, Braunschweig, Germany, 1999, 65–89.

[Hi1]   F. Hirzebruch, *Collected Papers*, Vol. I, Springer-Verlag, Berlin, 1987.

[Hi2]   F. Hirzebruch, *Collected Papers*, Vol. II, Springer-Verlag, Berlin, 1987.

[Hi3]   F. Hirzebruch, *Topological Methods in Algebraic Geometry*, Springer-Verlag, Berlin, 1978.

[KK]   Ch. Kaiser and K. Köhler, A fixed point formula of Lefschetz type in Arakelov geometry III: Representations of Chevalley schemes and heights of flag varieties, *Invent. Math.*, to appear.

[KS]   S. Keel and L. Sadun, Oort's conjecture for $A_g \otimes \mathbb{C}$, *J. Amer. Math. Soc.*, **16** (2003), 887–900.

[K1]   K. Köhler, Torus fibrations and moduli spaces, in A. Reznikov and M. Schappacher, eds., *Regulators in Analysis, Geometry and Number Theory*, Progress in Mathematics 171, Birkhäuser Boston, Cambridge, MA, 2000, 166–195.

[K2]   K. Köhler,: *A Hirzebruch Proportionality Principle in Arakelov Geometry*, Prépublication 284, Institut de Mathématiques de Jussieu, Paris, 2001.

[KR1]   K. Köhler and D. Roessler, A fixed point formula of Lefschetz type in Arakelov geometry I: Statement and proof, *Invent. Math.*, **145** (2001), 333–396.

[KR2]   K. Köhler and D. Roessler, A fixed point formula of Lefschetz type in Arakelov geometry II: A residue formula, *Ann. Inst. Fourier*, **52** (2002), 1–23.

[KR4]   K. Köhler and D. Roessler, A fixed point formula of Lefschetz type in Arakelov geometry IV: Abelian varieties, *J. Reine Angew. Math.*, **556** (2003), 127–148.

[Kü]   U. Kühn, Generalized arithmetic intersection numbers, *J. Reine Angew. Math.*, **534** (2001), 209–236.

[MR]   V. Maillot and D. Roessler, Conjectures sur les dérivées logarithmiques des fonctions *L* d'Artin aux entiers négatifs, *Math. Res. Lett.*, **9** (2002), 715–724.

[MiS]   J. W. Milnor and J. D. Stasheff, *Characteristic Classes*, Annals of Mathematics Studies 76, Princeton University Press, Princeton, NJ, 1974.

[M]   D. Mumford, Hirzebruch proportionality theorem in the non compact case, *Invent. Math.*, **42** (1977), 239–272.

[R1]   D. Roessler, Lambda-structure on Grothendieck groups of Hermitian vector bundles, *Israel J. Math.*, **122** (2001), 279–304.

[R2]   D. Roessler, An Adams-Riemann-Roch theorem in Arakelov geometry, *Duke Math. J.*, **96** (1999), 61–126.

[S]   C. Soulé, Hermitian vector bundles on arithmetic varieties, in *Algebraic Geometry: Santa Cruz* 1995, Part 1, Proceedings of Symposia in Pure Mathematics 62, American Mathematical Society, Providence, RI, 1997, 383–419.

[SABK] C. Soulé, D. Abramovich, J. F. Burnol, and J. Kramer, *Lectures on Arakelov Geometry*, Cambridge Studies in Mathematics 33, Cambridge University Press, Cambridge, UK, 1992.

[T]   H. Tamvakis, Arakelov theory of the Lagrangian Grassmannian, *J. Reine Angew. Math.*, **516** (1999), 207–223.

[Y]   K. Yoshikawa, Discriminant of theta divisors and Quillen metrics, *J. Differential Geom.*, **52** (1999), 73–115.

On the Height Conjecture for Algebraic Points on Curves Defined over Number Fields

Ulf Kühn

Institut für Mathematik
Humboldt Universität zu Berlin
Unter den Linden 6
D-10099 Berlin
Germany
kuehn@math.hu-berlin.de

Abstract. We study the basic height conjecture for points on curves defined over number fields and show: On any algebraic curve defined over a number field the set of algebraic points contains an unrestricted subset of infinite cardinality such that for all of its points their canonical height is bounded in terms of a small power of their root discriminant. In addition, if we assume GRH, then the upper bound is, as it is conjectured, linear in the logarithm of the root discriminant.

1 Introduction

Let X be a smooth projective curve defined over a number field. Then we have the Arakelov height function with respect to the metrized canonical bundle

$$\mathrm{ht}_{\overline{\omega}} \colon X(\overline{\mathbb{Q}}) \longrightarrow \mathbb{R},$$

whose definition will be given in the main text below, and the logarithmic root discriminant

$$\mathrm{disc} \colon X(\overline{\mathbb{Q}}) \longrightarrow \mathbb{R}.$$

For the latter map, we associate to a point $P \in X(\overline{\mathbb{Q}})$ the number field $k(P)$ and we set $\mathrm{disc}(P) = \log(\Delta_{k(P)})$. Here $\Delta_K = |D_{K/\mathbb{Q}}|^{1/[K:\mathbb{Q}]}$ denotes the root discriminant of a number field K. The above two maps are conjecturally related as follows.

Conjecture 1.1. Let X be a smooth projective curve defined over a number field. Let $\varepsilon > 0$. Then there exists a constant $C(X, \varepsilon)$ such that for P varying over all algebraic points of X, we have

$$\mathrm{ht}_{\overline{\omega}}(P) \leq (1 + \varepsilon)\,\mathrm{disc}(P) + C(X, \varepsilon).$$

This conjectural height inequality is a special case of Vojta's conjectures [La] and also referred to as effective Mordell theorem [MB]. We remark that this conjecture is equivalent to a uniform *abc*-conjecture for all number fields [Fr]. For a long list describing the relations of the *abc*-conjecture to other conjectures in arithmetic geometry and analytic number theory, we refer to [Go] and [Ni].

A subset $V \subseteq X(\overline{\mathbb{Q}})$ is called unrestricted if for all $d, r > 0$ the cardinality of the set $V_{d,r} = \{P \in V \mid [k(P) : \mathbb{Q}] \geq d, \operatorname{disc}(P) \geq r\}$ is infinite. The purpose of this note is to show the following theorem.

Theorem 1.2. *Let X be a smooth projective curve of genus $g \geq 2$ defined over a number field. Let $\varepsilon, \delta > 0$. Then there exists an unrestricted subset $V \subseteq X(\overline{\mathbb{Q}})$ and a constant $C(X, \varepsilon, \delta, V)$ such that for all $P \in V$, we have*

$$\operatorname{ht}_{\overline{\omega}}(P) \leq \varepsilon \exp(\delta \operatorname{disc}(P)) + C(X, \varepsilon, \delta, V). \tag{1.1}$$

If, in addition, the Dirichlet series $L(\chi_D, s)$ for the characters $(\frac{D}{\cdot})$, where D is a negative prime number, have no zeros in a ball of radius $1/4$ around 0, then for all $P \in V$, we have

$$\operatorname{ht}_{\overline{\omega}}(P) \leq \varepsilon \operatorname{disc}(P) + C(X, \varepsilon, V). \tag{1.2}$$

We should remark that our results only hold for an infinite subset of $X(\overline{\mathbb{Q}})$ and the method of proof seems not to be general enough to cover all algebraic points simultaneously.

2 Heights

The height of an algebraic point P on a smooth projective curve defined over a number field K can be defined by means of Arakelov theory as follows.

Let $\pi : \mathcal{X} \to \operatorname{Spec} \mathcal{O}_K$ be a regular model for X over the ring of integers $\mathcal{O}_K$ of K, i.e., $\mathcal{X}$ is a projective, regular scheme, flat over $\operatorname{Spec} \mathcal{O}_K$. In this note, a hermitian line bundle $\overline{\mathcal{L}} = (\mathcal{L}, \| \cdot \|)$ on $\mathcal{X}$ is a line bundle on $\mathcal{X}$ together with a continuous hermitian metric on the induced complex line bundle $\mathcal{L}_\infty$ over the complex manifold $\mathcal{X}_\infty = \prod_{\sigma : K \to \mathbb{C}} \mathcal{X}_\sigma(\mathbb{C})$. A particular hermitian line bundle is the canonical bundle equipped with the Arakelov metric. We denote this distinguished hermitian line bundle by $\overline{\omega}$; see, e.g., [La].

In what follows, we also allow that the metric associated with $\overline{\mathcal{L}}$ has logarithmic singularities at a finite set $\mathcal{S}$ of algebraic points on $\mathcal{X}(\overline{\mathbb{Q}})$ of the following type: near a singular point P, any section l of $\mathcal{L}$ has an expansion in a local coordinate t of the form

$$\|l\|(t) = |t|^{\operatorname{ord}_P(l)} \phi(t) (-\log |t|)^\alpha,$$

where $\phi(t)$ is a continuous nonvanishing function and $\alpha \in \mathbb{R}$. If $\alpha > 0$ for all singular points P, then the metric is called a positive logarithmically singular metric.

Let P be an algebraic point on X and $\overline{\mathcal{L}}$ be a hermitian line bundle. Possibly after replacing K by a finite extension, we may assume that the algebraic point P, the points in $\mathcal{S}$, and X are all defined over K. Since the arithmetic surface $\mathcal{X}$ is proper, we have $\mathcal{X}(K) = \mathcal{X}(\mathcal{O}_K)$. Therefore, the Zariski closure $\mathcal{P}$ of P in $\mathcal{X}$ determines a section $s_P : \operatorname{Spec} \mathcal{O}_K \to \mathcal{X}$. With the above notation, we define the height of a point $P \in X(K) \setminus \mathcal{S}$ with respect to $\overline{\mathcal{L}}$ by

$$\mathrm{ht}_{\overline{\mathcal{L}}}(P) = \frac{1}{[K:\mathbb{Q}]} \left(\log \#(s_P^* \mathcal{L}/(s_P^* l)) - \sum_{\sigma:K \to \mathbb{C}} \log \|l\|(P^\sigma) \right),$$

where l is a regular section of $\mathcal{L}$ which is nonzero at P. Observe that the height does not depend on the choices of l or K. If we denote by p a local equation for $\mathcal{P}$, then we have an equality

$$\log \#(s_P^* \mathcal{L}/(s_P^* l)) = \sum_{x \in \mathcal{X}} \log \#(\mathcal{O}_{\mathcal{X},x}/(p,l)).$$

The above quantity is also denoted by $(\mathcal{P}, \mathrm{div}(l))_{\mathrm{fin}}$ and there are only finitely many $x \in \mathcal{X}$ that give nonzero contribution to $(\mathcal{P}, \mathrm{div}(l))_{\mathrm{fin}}$.

We will need the following basic facts on heights.

Proposition 2.1. *Let $\overline{\mathcal{L}}$ and $\overline{\mathcal{M}}$ be hermitian line bundles on $\mathcal{X}$. Assume $\deg(\mathcal{L}) = \deg(\mathcal{M}) > 0$. If the metric on $\mathcal{L}$ is continuous and the metric on $\mathcal{M}$ is positive logarithmically singular, then for all $\varepsilon > 0$ we can find a constant $C(\varepsilon, \mathcal{X}, \overline{\mathcal{L}}, \overline{\mathcal{M}})$ such that*

$$\mathrm{ht}_{\overline{\mathcal{L}}}(P) \le (1 + \varepsilon) \, \mathrm{ht}_{\overline{\mathcal{M}}}(P) + C(\varepsilon, \mathcal{X}, \overline{\mathcal{L}}, \overline{\mathcal{M}}).$$

Proof. It is well known (see, e.g., [Si, Proposition 3.6]) that in the case where both metrics are continuous, we can find a constant $C(\varepsilon, \mathcal{X}, \overline{\mathcal{L}}, \overline{\mathcal{M}})$ such that for all $\varepsilon > 0$,

$$\mathrm{ht}_{\overline{\mathcal{L}}}(P) \le (1 + \varepsilon) \, \mathrm{ht}_{\overline{\mathcal{M}}}(P) + C(\varepsilon, \mathcal{X}, \overline{\mathcal{L}}, \overline{\mathcal{M}}). \tag{2.1}$$

For simplicity of the argument, we assume that the metric $\| \cdot \|$ on $\mathcal{M}$ has only $Q \in X(\overline{\mathbb{Q}})$ as singular point. Let 1_Q be the canonical section of $\mathcal{O}(Q)$. Then we can find continuous hermitian metrics $\| \cdot \|'$ on $\mathcal{M}$ and $\| \cdot \|$ on $\mathcal{O}(Q)$ such that for all $P \in X(\mathbb{C}) \setminus \{Q\}$ and all sections m of $\mathcal{M}$,

$$\|m\|(P) = \|m\|'(P) \cdot (-\log \|1_Q\|(P))^\alpha.$$

Let $\mathcal{Q}$ be the Zariski closure of Q. Then since $\alpha > 0$, we obtain

$$\begin{aligned}
\mathrm{ht}_{\overline{\mathcal{M}}}(P) &= \mathrm{ht}_{\overline{\mathcal{M}}'}(P) - \alpha \log(-\log \|1_Q\|(P)) \\
&\ge \mathrm{ht}_{\overline{\mathcal{M}}'}(P) - \alpha \log(-\log \|1_Q\|(P) + (\mathcal{P}, \mathcal{Q})_{\mathrm{fin}}) \\
&= \mathrm{ht}_{\overline{\mathcal{M}}'}(P) - \alpha \log \mathrm{ht}_{\overline{\mathcal{O}(Q)}}(P) \\
&\ge (1 - \varepsilon') \, \mathrm{ht}_{\overline{\mathcal{L}}}(P) - \alpha\varepsilon' \frac{1 - \varepsilon'}{\deg(\mathcal{L})} \, \mathrm{ht}_{\overline{\mathcal{L}}}(P) - C'(\mathcal{X}, \varepsilon', \overline{\mathcal{L}}, \overline{\mathcal{M}}).
\end{aligned}$$

For the last inequality, we used (2.1) twice. If we take ε such that $1/(1 + \varepsilon) = 1 - \varepsilon'(1 + \alpha(1 - \varepsilon')/\deg(\mathcal{L}))$, we obtain the claim.   □

Proposition 2.2. *Let $f : \mathcal{Y} \to \mathcal{X}$ be a proper morphism of arithmetic surfaces. Then we have*

$$\mathrm{ht}_{f^* \overline{\mathcal{L}}}(P) = \mathrm{ht}_{\overline{\mathcal{L}}}(f(P))$$

for any logarithmically singular hermitian line bundle $\overline{\mathcal{L}}$ on $\mathcal{X}$ and P not in the singular locus of the logarithmically singular metric on $\mathcal{L}$.

Proof. See, e.g., [BGS, formula (3.2.1)].   □

3 Arithmetic properties of Heegner points

Due to the modular description, the points on the modular curve $X(1)$ are well understood. Recall that $X(1)(\mathbb{C}) = \Gamma(1) \setminus \mathbb{H} \cup \{\infty\}$ and that $X(1)$ is isomorphic to $\mathbb{P}^1$. The regular model of $X(1)$ will be denoted by $\mathcal{X}(1)$. This arithmetic surface is canonically isomorphic to $\mathbb{P}^1_{\mathbb{Z}}$. On $\mathcal{X}(1)$ we have the line bundle of modular forms $\mathcal{M}_{12}$. The natural metric on this line bundle is the Petersson metric, where we use the normalization as given in [Kü, Definition 4.8]. This metric gives rise to the positive logarithmically singular hermitian line bundle $\overline{\mathcal{M}}_{12}$ (see, e.g., [Kü, Propositions 4.9 and 4.12]). For any point $P \in X(1)(K) \setminus \{\infty\}$ we have a well-defined height with respect to $\overline{\mathcal{M}}_{12}$. It is called the *modular height*.

Let D be a negative fundamental discriminant and $K = \mathbb{Q}(\sqrt{D})$. We briefly recall some properties of Heegner divisors. Every ideal class $[\mathfrak{a}]$ of K defines a unique point $P_\mathfrak{a}$ on $\Gamma(1) \setminus \mathbb{H}$ by associating with a fractional ideal $\mathfrak{a} = \mathbb{Z}a + \mathbb{Z}b$ with oriented (i.e., $\mathrm{Im}(b\bar{a}) > 0$) $\mathbb{Z}$-basis a, b the point $\rho_\mathfrak{a} = b/a \in \mathbb{H}$. We call $P_\mathfrak{a}$ the Heegner point to $\mathfrak{a}$ and sometimes write $[\rho_\mathfrak{a}]$ instead of $P_\mathfrak{a}$.

The Heegner divisor $H(D)$ on $\Gamma(1) \setminus \mathbb{H}$ consists of the sum of the $P_\mathfrak{a}$, where $\mathfrak{a}$ runs through all ideal classes of K, counted with multiplicity $2/w$, where w is the number of units in K. The cardinality of $H(D)$ is equal to the class number h of K; its degree is $2h(D)/w$.

Proposition 3.1. *Let $f: X \to X(1)$ be a morphism of algebraic curves that is defined over the field over which X is defined. Let $P \in X(\overline{\mathbb{Q}})$ be a point such that $f(P)$ is contained in a Heegner divisor $H(D)$ with prime discriminant D. Then we have*

$$\mathrm{disc}(P) \geq \frac{1}{2} \log |D| - \frac{55}{2}. \tag{3.1}$$

Proof. The composition formula for the discriminant implies that for all morphisms $f: X \to X(1)$ and points $P \in X(\overline{\mathbb{Q}})$, we have the inequality

$$\mathrm{disc}(P) \geq \mathrm{disc}(f(P)).$$

Thus it suffices to bound the discriminant of a Heegner point $P_\mathfrak{a} = f(P)$. We consider the following diagram of field extensions

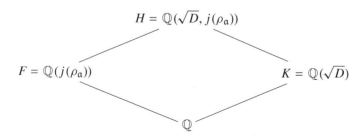

By the theory of complex multiplication, we have $h(D) = [H : K]$ and $D_{H|\mathbb{Q}} = D^{h(D)}$. From [Gr, Lemma 12.1.2], we deduce $\mathrm{Nm}_{F|\mathbb{Q}}(D_{H|F}) = D$. The composition formula $D_{H|\mathbb{Q}} = D^2_{F|\mathbb{Q}} \cdot \mathrm{Nm}_{F|\mathbb{Q}}(D_{H|F})$ gives rise to the equality

$$\text{disc}(P_\mathfrak{a}) = \frac{1}{h(D)} \log |D_{F|\mathbb{Q}}| = \left(\frac{1}{2} - \frac{1}{2h(D)}\right) \log |D|.$$

The class number of an imaginary quadratic number field with prime discriminant satisfies $h(D) > 1/55 \log |D|$ (see, e.g., [Oe]). Thus we have

$$\text{disc}(P_\mathfrak{a}) = \left(\frac{1}{2} - \frac{1}{2h(D)}\right) \log |D| \geq \frac{1}{2} \log |D| - \frac{55}{2}.$$

This proves the proposition.   □

Proposition 3.2. *If $P_\mathfrak{a} \in H(D)$ is a Heegner point then its modular height is given by*

$$\text{ht}_{\overline{\mathcal{M}}_{12}}(P_\mathfrak{a}) = -6\left(\frac{L'(\chi_D, 0)}{L(\chi_D, 0)} + \frac{1}{2} \log |D|\right), \tag{3.2}$$

where $L(\chi_D, s)$ is the Dirichlet L-function for the character $(\frac{D}{\cdot})$.

Proof. Recall that $\Delta(\tau) = q^{24} \prod_{n=1}^{\infty}(1 - q)^n$, where $q = e^{2\pi i \tau}$ with $\tau \in \mathbb{H}$, is a section of $\mathcal{M}_{12}$, whose divisor equals the unique cusp ∞ of $\mathcal{X}(1)$. Its Petersson norm is given by the formula

$$\|\Delta(\tau)\|_{\text{Pet}} = |\Delta(\tau)|(4\pi \, \text{Im}(\tau))^6.$$

Therefore, the modular height of a Heegner point is given by

$$\text{ht}_{\overline{\mathcal{M}}_{12}}(P_\mathfrak{a}) = \frac{1}{[F : \mathbb{Q}]}\left((P_\mathfrak{a}, \infty)_{\text{fin}} - \sum_{\rho_\mathfrak{a} \in H(D)} \log \|\Delta(\rho_\mathfrak{a})\|_{Pet}\right),$$

where $F := \mathbb{Q}(j(\rho_\mathfrak{a}))$ and where for each embedding $\sigma : F \to \overline{\mathbb{Q}}$ the point $\rho_\mathfrak{a} \in \mathbb{H}$ is a lift of $P_\mathfrak{a}^\sigma(\mathbb{C}) \in \Gamma(1) \backslash \mathbb{H}$. We now recall the well-known Kronecker limit formula. If

$$\mathcal{E}(\tau, s) = \frac{1}{2} \sum_{\gamma \in \Gamma_\infty \backslash \Gamma_1} (\text{Im}(\gamma \tau))^s$$

is the real analytic Eisenstein series for $\Gamma(1)$, then the logarithm of the Petersson norm of the Delta function is given by

$$\log(\|\Delta(\tau)\|_{\text{Pet}}^2) = -4\pi \lim_{s \to 1}\left(\mathcal{E}(\tau, s) - \frac{\Gamma(1/2)\Gamma(s - 1/2)\zeta(2s - 1)}{\Gamma(s)\zeta(2s)}\right) + 12 \log(4\pi).$$

We also point to the identity

$$\sum_{\rho_\mathfrak{a} \in H(D)} \mathcal{E}(\rho_\mathfrak{a}, s) = \frac{w}{2}\left|\frac{D}{4}\right|^{s/2} \frac{\zeta_K(s)}{\zeta(2s)},$$

where $\zeta_K(s) = \zeta(s)L(\chi_D, s)$ denotes the Dedekind zeta function of K (see [GZ, p. 210]). In [BK, p. 1726], we have derived from this the formulas

$$\sum_{\rho_\mathfrak{a} \in H(D)} -\log(|\Delta(\rho_\mathfrak{a})|^2 (4\pi \operatorname{Im} \rho_\mathfrak{a})^{12})$$

$$= 4\pi \lim_{s \to 1} \left[\sum_{\rho_\mathfrak{a} \in H(D)} \mathcal{E}(\rho_\mathfrak{a}, s) - h \frac{\Gamma(1/2)\Gamma(s - 1/2)\zeta(2s - 1)}{\Gamma(s)\zeta(2s)} \right] + 12h(D)\log(4\pi)$$

$$= -12h(D) \left(\frac{L'(\chi_D, 0)}{L(\chi_D, 0)} + \frac{1}{2} \log |D| \right).$$

Since $j(\rho_\mathfrak{a})$ is an algebraic integer, we have $(P_\mathfrak{a}, \infty)_{\text{fin}} = 0$. This gives our claim.   □

Remark 3.3. Recall that $\mathcal{X}(1) \cong \mathbb{P}^1_\mathbb{Z}$, with $\mathcal{M}_{12} \cong \mathcal{O}(1)$, and that the line bundle $\mathcal{O}(1)$ equipped with a particular metric gives rise to the naive height $\operatorname{ht}_{\mathbb{P}^1}$. For a Heegner point $P_\mathfrak{a} \in X(1)(K)$, this height is given by

$$\operatorname{ht}_{\mathbb{P}^1}(P_\mathfrak{a}) = \frac{1}{[F : \mathbb{Q}]} \left((P_\mathfrak{a}, \infty)_{\text{fin}} - \sum_{\rho_\mathfrak{a}} \log \max(1, j(\rho_\mathfrak{a})) \right)$$

$$= 6 \left(\frac{L'(\chi_D, 1)}{L(\chi_D, 1)} + \frac{1}{2} \log |D| \right) \left(1 + O \left(\frac{\log \log |D|}{\log |D|} \right) \right)^{-1}.$$

Indeed, since $j(\rho_\mathfrak{a})$ is an algebraic integer, we have $(P_\mathfrak{a}, \infty)_{\text{fin}} = 0$. Now the claim follows immediately from [GS] by combining their equation (7) with their Theorem 3.

Proposition 3.4. *Let $P_\mathfrak{a} \in H(D)$ be a Heegner point with prime discriminant.*

(i) *For all $\delta > 0$, there exists a constant $S(\delta)$ such that*

$$\operatorname{ht}_{\overline{\mathcal{M}}_{12}}(P_\mathfrak{a}) \le S(\delta) \cdot \exp(\delta \operatorname{disc}(P_\mathfrak{a})). \tag{3.3}$$

(ii) *If the Dirichlet L-series $L(\chi_D, s)$ has no zero in the ball of radius $1/4$ around 0, then there exist constants a and b such that the modular height of a Heegner point of discriminant D satisfies*

$$\operatorname{ht}_{\overline{\mathcal{M}}_{12}}(P_\mathfrak{a}) \le a \operatorname{disc}(P_\mathfrak{a}) + b. \tag{3.4}$$

(iii) *Assuming the generalized Riemann hypothesis (GRH) for the Dirichlet L-series $L(\chi_D, s)$ in question, we have*

$$\operatorname{ht}_{\overline{\mathcal{M}}_{12}}(P_\mathfrak{a}) = 6 \operatorname{disc}(P_\mathfrak{a}) + o(\operatorname{disc}(P_\mathfrak{a})). \tag{3.5}$$

Proof. (i)–(ii) Let $E_{\mathcal{O}_K}$ be an elliptic curve with complex multiplication by $\mathcal{O}_K$; then the Faltings height of $E_{\mathcal{O}_K}$ equals 12 times the modular height of its modular point $P_{\mathcal{O}_K}$; see, e.g., [Co, pp. 362 and 365]. By means of the inequality (3.1), we derive that (i) is a reformulation of the corresponding formula in [Co, remark on p. 365] and claim (ii) is a reformulation of [Co, Theorem 6 (ii)].

(iii) Using the functional equation for $L(\chi_D, s)$ we can express the right-hand side of (3.2) as a special value at $s = 1$. Namely, we have

$$-\left(\frac{L'(\chi_D, 0)}{L(\chi_D, 0)} + \frac{1}{2}\log|D|\right) = \left(\frac{L'(\chi_D, 1)}{L(\chi_D, 1)} + \frac{1}{2}\log|D| - \log(2\pi e^\gamma)\right),$$

where γ is the Euler constant. Assuming the GRH, we have

$$\frac{L'(\chi_D, 1)}{L(\chi_D, 1)} = O(\log\log|D|),$$

where the implied constant is uniform in D (see, e.g., [GS, Section 3.1]). This gives

$$\mathrm{ht}_{\overline{\mathcal{M}}_{12}}(P_\mathfrak{a}) = 6\left(\frac{1}{2}\log|D| + O(\log\log|D|)\right).$$

Since $O(\log\log|D|)$ is also of order $o(\log|D|)$, we derive by means of (3.1) the claim.   □

4 Main result

Definition 4.1. *Let X be a curve defined over a number field, and let f be a nonconstant function in the function field of X. We consider f as a morphism $f : X \to \mathbb{P}^1$ and identify $\mathbb{P}^1$ with the modular curve $X(1)$. Then we define*

$$\mathcal{V}(X, f) = \{P \in X(\overline{\mathbb{Q}}) \mid f(P) \text{ is a Heegner point with prime discriminant}\}.$$

Proposition 4.2. *The subset $\mathcal{V}(X, f) \subseteq X(\overline{\mathbb{Q}})$ is unrestricted.*

Proof. The set of Heegner points with prime discriminant on $X(1)$ is, as we have seen already in the proof of Proposition 3.1, unrestricted. The composition formula for the discriminant implies that for all morphisms $f : X \to X(1)$ and points $P \in X(\overline{\mathbb{Q}})$, we have the inequality

$$\mathrm{disc}(f(P)) \le \mathrm{disc}(P).$$

Therefore, the set $\mathcal{V}(X, f)$ is also unrestricted.   □

Theorem 4.3. *Let X be a curve of genus $g \ge 2$ defined over a number field. Let f be a nonconstant function in the function field of X, and let $\varepsilon, \delta > 0$.*

(i) *There exist constants $S(\delta)$ and $C(X, \varepsilon, \mathcal{V}(X, f))$ such that all $P \in \mathcal{V}(X, f)$ satisfy*

$$\mathrm{ht}_{\overline{\omega}}(P) \le (1 + \varepsilon)\frac{S(\delta)(2g - 2)}{\deg(f)}\exp(\delta\,\mathrm{disc}(P)) + C(X, \varepsilon, \mathcal{V}(X, f)). \quad (4.1)$$

(ii) *Assume that* $\mathrm{ht}_{\overline{\mathcal{M}}_{12}}(P_\mathfrak{a}) \leq a\,\mathrm{disc}(P_\mathfrak{a}) + b$ *for all Heegner points* $P_\mathfrak{a}$ *with prime discriminant* D. *Then for all* $P \in \mathcal{V}(X, f)$, *we have*

$$\mathrm{ht}_{\overline{\omega}}(P) \leq (1 + \varepsilon)\frac{a(2g - 2)}{\deg(f)}\,\mathrm{disc}(P) + C(X, \varepsilon, \mathcal{V}(X, f)). \qquad (4.2)$$

Proof. Let $f\colon \mathcal{X} \to \mathcal{X}(1)$ be an extension of the morphism $f\colon X \to X(1)$ given by f. The degrees of the line bundles $\omega^{\otimes \deg(f)}$ and $(f^*\mathcal{M}_{12})^{\otimes(2g-2)}$ are equal and positive. We endow $\mathcal{M}_{12}$ with the Petersson metric and by pullback we obtain the positive logarithmically singular line bundle $f^*\overline{\mathcal{M}}_{12}$ on $\mathcal{X}$. Then by Propositions 2.1 and 2.2, we get, for all $P \in X(\overline{\mathbb{Q}}) \setminus \{f^{-1}(\infty)\}$,

$$\mathrm{ht}_{\overline{\omega}}(P) \leq (1 + \varepsilon')\frac{2g - 2}{\deg(f)}\,\mathrm{ht}_{\overline{\mathcal{M}}_{12}}(f(P)) + C'(X, \varepsilon', \mathcal{V}(X, f)).$$

Here we wrote $C'(X, \varepsilon', \mathcal{V}(X, f))$ instead of $C'(\varepsilon', \mathcal{X}, \overline{\omega}, f^*\overline{\mathcal{M}}_{12})$. If $P \in \mathcal{V}(X, f) \subseteq X(\overline{\mathbb{Q}})$, then $f(P)$ is a Heegner point with prime discriminant. Thus (4.1) follows immediately from (3.3). Finally, (4.2) is an easy consequence of the assumed bound for the modular height of $f(P)$.   □

Remark 4.4.

(i) In Theorem 4.3 we can choose f with arbitrary large degree. If we let $\deg(f) \geq (1 + \varepsilon) \cdot S(\delta) \cdot (2g - 2)/\varepsilon$, we derive formula (1.1) of Theorem 1.2. If we let $\deg(f) \geq (1 + \varepsilon) \cdot a \cdot (2g - 2)/\varepsilon$, we obtain formula (1.2).

(ii) We note that because of [Fr] the exponential height inequality (1.1) should somehow be related to the exponential *abc*-inequality [SY, Su]. We remark also that (1.2) could be seen as a converse to a theorem of Granville and Stark [GS] saying that the *abc*-conjecture implies that there are no Siegel zeros.

References

[BGS]   J.-B. Bost, H. Gillet, and C. Soulé, Heights of projective varieties and positive Green forms, *J. Amer. Math. Soc.*, **7**-4 (1994), 903–1027.

[BK]   J. H. Bruinier and U. Kühn, Integrals of automorphic Green's functions associated to Heegner divisors, *Internat. Math. Res. Not.*, **31** (2003), 1687–1729.

[Co]   P. Colmez, Sur la hauteur de Faltings des variétés abéliennes á multiplication complexe, *Compositio Math.*, **111**-3 (1998), 359–368.

[Fr]   M. Van Frankenhuysen, The *ABC* conjecture implies Vojta's height inequality for curves, *J. Number Theory*, **95**-2 (2002), 289–302.

[Go]   D. Goldfeld, Modular forms, elliptic curves and the *ABC*-conjecture, in *A Panorama of Number Theory, or the View from Baker's Garden* (Zürich, 1999), Cambridge University Press, Cambridge, UK, 2002, 128–147.

[Gr]   B. H. Gross, *Arithmetic on Elliptic Curves with Complex Multiplication*, Lecture Notes in Mathematics 776, Springer-Verlag, Berlin, 1980.

[GS]   A. Granville and H. M. Stark, *abc* implies no "Siegel zeros" for *L*-functions of characters with negative discriminant, *Invent. Math.*, **139**-3 (2000), 509–523.

[GZ]   B. H. Gross and D. B. Zagier, On singular moduli, *J. Reine Angew. Math.*, **355** (1985), 191–220.

[Kü]   U. Kühn, Generalized arithmetic intersection numbers, *J. Reine Angew. Math.*, **534** (2001), 209–236.

[La]   S. Lang, *Introduction to Arakelov Theory*, Springer-Verlag, New York, 1988.

[MB]   L. Moret-Bailly, Hauteurs et classes de Chern sur les surfaces arithmétiques, in *Séminaire sur les Pinceaux de Courbes Elliptiques*, Astérisque 183, Société Mathématique de France, Paris, 1990, 37–58.

[Ni]   A. Nitaji, *The abc Conjecture Home Page*, http://www.math.unicaen.fr/˜nitaj/abc.html.

[Oe]   J. Oesterlé, Nombres de classes des corps quadratiques imaginaires, in *Séminaire Bourbaki*, Vol. 1983/84, Astérisque 121–122, Société Mathématique de France, Paris, 1985, 309–323.

[Si]   J. H. Silverman, The theory of height functions, in G. Cornell and J. H. Silverman, eds., *Arithmetic Geometry*, Springer-Verlag, New York, 1986.

[SY]   C. L. Stewart and K. Yu, On the *abc* conjecture II, *Duke Math. J.*, **108**-1 (2001), 169–181.

[Su]   A. Surroca, Sur l'effectivité du théorème de Siegel et la conjecture *abc*, preprint, 2005; math.NT/0501285.

A Note on Absolute Derivations and Zeta Functions

Jeffrey C. Lagarias

Department of Mathematics
University of Michigan
Ann Arbor, MI 48109–1043
USA
lagarias@umich.edu

1 Introduction

In studying the analogies between zeta functions of number fields and function fields over finite fields, several authors have noted that certain properties of number fields seem well described by viewing them as geometric objects over the "field with one element." Analogies in these directions have been formalized recently in Manin [Ma95], Soulé [So99, So03], Kurokawa, Ochiai, and Wakayama [KOW03], and Deitmar [De04]. There is also earlier work, noting analogies, such as Kurokawa [Ku92], which can be traced in the references in the papers above.

In this direction, Kurokawa, Ochiai, and Wakayama [KOW03] recently introduced a notion of absolute derivation over the rational number field $\mathbb{Q}$. Based on this, they proposed a measure of "quantum noncommutativity" of pairs of primes over the rational field, given as follows. For real variables $x, y > 1$, define

$$F(x, y) = \sum_{k=1}^{\infty} x^{k-1} \frac{y^{-x^k}}{(1 - y^{-x^k})^2}. \tag{1}$$

Now define, for $x, y > 1$,

$$QNC(x, y) := \frac{1}{12xy} (x(y - 1)F(x, y) - y(x - 1)F(y, x)). \tag{2}$$

The "quantum noncommutativity" of two primes p and q is defined to be $QNC(p, q)$. It is easy to see that $QNC(x, y) = -QNC(y, x)$, whence $QNC(x, x) = 0$, and one has $QNC(2, 3) = 0.00220482\ldots$, for example. They then raised questions [KOW03, p. 580] whether there is a connection between the quantum noncommutativity measure and zeta functions. They defined the infinite skew-symmetric matrix $\mathbf{R} = [\mathbf{R}_{ij}]$ whose (i, j)th entry

$$\mathbf{R}_{ij} := QNC(p_i, p_j),$$

where p_i denotes the ith prime listed in increasing order, so that $p_1 = 2$, $p_2 = 3$, $p_3 = 5$, etc.

In question (A), they asked whether it could be true (in some suitable sense) that

$$\det \left(\mathbf{I} - \mathbf{R} \left(s - \frac{1}{2} \right) \right) = c\xi(s), \tag{3}$$

in which $\xi(s) = \frac{1}{2}s(s-1)\pi^{-\frac{s}{2}}\Gamma(\frac{s}{2})\zeta(s)$, is the Riemann ξ-function and c is a nonzero constant. (They proposed $c = 2$.) They also asked a more general question (B) for (suitable) automorphic or Galois representations ρ, which would involve a skew-symmetric matrix $\mathbf{R}(\rho)$ with (i, j)th entry

$$\mathbf{R}_{ij}(\rho) := \frac{\rho(p) + \rho(q)^*}{2} \mathbf{R}_{ij},$$

involving a weighted version of elements $QNC(p_i, p_j)$, and asks whether it could be true that

$$\det \left(\mathbf{I} - \mathbf{R}(\rho) \left(s - \frac{1}{2} \right) \right) = c s^{m(\rho)} (s-1)^{m(\rho)} \hat{L}(s, \rho), \tag{4}$$

where $\hat{L}(s, \rho)$ is the completed L-function attached to the representation ρ, and $m(\rho)$ is the multiplicity of the trivial representation in ρ.

The object of this note is to give a negative answer to question (A); the same method should apply to give a negative answer to question (B). We also offer some (inconclusive) remarks concerning whether the notion of "QNC" can be modified to allow a positive answer to these questions.

In order to make questions (A) and (B) well defined, it is necessary to formulate a definition of infinite determinant in (3). We take as a basic requirement of a definition of such an infinite determinant that any zero s of a determinant (3) must necessarily have $z = \frac{1}{s - \frac{1}{2}}$ belonging to the spectrum of $\mathbf{R}$, i.e., that for this value the resolvent $(zI - \mathbf{R})^{-1}$ is not a bounded operator on the full domain of $\mathbf{R}$, assumed to be a Banach space.

The basic requirement implies that if $\mathbf{R}$ acts as a bounded operator on a Hilbert space in (3), then a positive answer to question (A) would necessarily imply the Riemann hypothesis for $\zeta(s)$, and to question (B) would imply the Riemann hypothesis for $L(s, \pi)$. This follows since $\mathbf{R}$ would then be skew-adjoint, hence have pure imaginary spectrum, whence the determinant (assumed defined) could only vanish when $s - \frac{1}{2}$ is pure imaginary. One can weaken question (A) so that it no longer implies the Riemann hypothesis, by requiring that the left side $\det(\mathbf{I} - \mathbf{R}(s - \frac{1}{2}))$ of (3) detect all the zeta zeros that are on the critical line $\Re(s) = \frac{1}{2}$, and not required to detect zeros off the line, and similarly for question (B). The result below gives a negative answer to question (A) in this weaker formulation as well.

We treat the operator $\mathbf{R}$ as acting on the Hilbert space l_2 of column vectors, and observe it defines a bounded operator. It follows that it is skew-adjoint and so has spectrum confined to the imaginary axis. However, we show that its spectrum cannot detect all the zeta zeros that lie on the critical line, whether or not the Riemann hypothesis holds. Note that $\rho = \frac{1}{2} + i\gamma$ is a zeta zero, the corresponding point of the spectrum of $\mathbf{R}$ is $\lambda = -\frac{i}{\gamma}$.

The main point is that the quantum noncommutativity function is so rapidly decreasing as p, q increase that

$$\sum_{j=1}^{\infty} \sum_{k=1}^{\infty} |\mathbf{R}_{jk}| < \infty. \tag{5}$$

We show this in Section 2, and we deduce that the matrix $\mathbf{R}$ defines a trace class operator on l_2. The weaker condition

$$\sum_{j=1}^{\infty} \sum_{k=1}^{\infty} |\mathbf{R}_{jk}|^2 < \infty \tag{6}$$

already implies that $\mathbf{R}$ is a compact operator (in fact, a Hilbert–Schmidt operator); see Akhiezer and Glazman [AG93, Section 28]. (In Akhiezer and Glazman, the term "completely continuous operator" = "compact operator.") A compact operator necessarily has a pure discrete spectrum with all nonzero eigenvalues of finite multiplicity, with only limit point zero [RS80, Theorem VI.15]. Since we now know $\mathbf{R}$ is skew-adjoint, its eigenvalues, which necessarily occur in complex conjugate pure imaginary pairs, and can be ordered by decreasing absolute value, $\{\pm i\lambda_j : j = 1, 2, \ldots\}$, with $\lambda_1 \geq \lambda_2 \geq \cdots > 0$. A trace class operator $\mathbf{A}$ is a compact operator with the property that its singular values μ_j (eigenvalues of the positive self-adjoint operator $(\mathbf{A}^*\mathbf{A})^{\frac{1}{2}}$) satisfy

$$\sum_{j=1}^{\infty} \mu_j < \infty. \tag{7}$$

For skew-adjoint operators, $\mu_j = |\lambda_j|$, giving the trace-class condition

$$\sum_{j=1}^{\infty} |\lambda_j| < \infty. \tag{8}$$

For trace class operators $\mathbf{A}$, there is an essentially unique definition of $\det(I + \mathbf{A})$ that satisfies the basic requirement, see B. Simon [Si77], who reviews three equivalent definitions of this determinant (see also [Si79, Chapter 3]). He bases his treatment on the formula

$$\det(I - w\mathbf{A}) := \sum_{k=0}^{\infty} \mathrm{Tr}(\wedge^k(w\mathbf{A})) = \sum_{k=0}^{\infty} \mathrm{Tr}(\wedge^k \mathbf{A}) w^k,$$

which is also presented in Reed and Simon [RS78, Section XIII.17, p. 323]. This determinant is an entire function in the variable w, given by a convergent infinite product

$$\det(I - w\mathbf{A}) = \prod_j (1 - w\lambda_j(\mathbf{A})),$$

in which the eigenvalues $\lambda_j(\mathbf{A})$ of $\mathbf{A}$ are counted with their algebraic multiplicity. This determinant satisfies the basic requirement (see Reed and Simon [RS78, Theorems XIII.105(c) and XIII.106]). The truth of (3) for the trace class operator $\mathbf{R}$, taking $w = s - \frac{1}{2}$, would imply that if $s = \frac{1}{2} + i\gamma_j$ is a zeta zero on the critical line, then the two values $\lambda_j = \pm\frac{i}{\gamma_j}$ belong to the spectrum of $\mathbf{R}$. It is well known [TH86, Chapter X] that a positive proportion of zeta zeros lie on the critical line $\Re(s) = \frac{1}{2}$, and the asymptotics of these zeros easily give

$$\sum_{\{\gamma:\zeta(\frac{1}{2}+i\gamma)=0\}} \frac{1}{|\gamma|} = +\infty. \tag{9}$$

This contradicts (8).

In Section 3 we discuss the problem of whether the notion of "QNC" can be modified to give a positive answer to question (A).

2 Trace class operator

Our object is to show the following.

Theorem 1. *The operator $\mathbf{R}$ acting on the column vector space l_2 defines a trace class operator.*

Proof. A bounded operator $\mathbf{A}$ is trace class if $|\mathbf{A}| = (\mathbf{A}^*\mathbf{A})^{\frac{1}{2}}$ is trace class, i.e., the positive operator $|\mathbf{A}|$ has pure discrete spectrum and the sum of its eigenvalues converges; cf. Reed and Simon [RS80, Section VI.6]. A necessary and sufficient condition for an operator $\mathbf{A}$ to be trace class is that for every orthonormal basis $\{\phi_n : 1 \le n < \infty\}$ of l_2, one has

$$\sum_{n=1}^{\infty} |\langle \mathbf{A}\phi_n, \phi_n \rangle| < \infty; \tag{10}$$

see Reed and Simon [RS80, Chapter VI, Example 26, p. 218].

Taking $\mathbf{A} = \mathbf{R}$, since it is skew-symmetric, we have $\mathbf{R}^*\mathbf{R} = -\mathbf{R}^2$. It follows that if $|\mathbf{R}|$ is trace class, then it has pure discrete spectrum and the singular values of $\mathbf{R}$ are just the absolute values of the eigenvalues of $\mathbf{R}$.

We first prove (5). We have

$$|QNC(p,q)| \le \frac{1}{12}(F(p,q) + F(q,p)).$$

Now we have $p, q \ge 2$ so $(1 - p^{-q^k})^2 \ge \frac{9}{16}$, whence

$$F(p,q) \le \frac{16}{9}\sum_{k=1}^{\infty} p^{k-1}q^{-p^k} \le 2q^{-p} + 2q^{-p}\left(\sum_{k=2}^{\infty} p^{k-1}q^{p-p^k}\right) \le 6q^{-p}.$$

In the last step above we used (for $k, p, q \geq 2$)

$$p^{k-1}q^{p-p^k} \leq p^{k-1}2^{-p^{k-1}} \leq 2^{k-1}2^{-2^{k-1}} \leq 2^{-k+2}.$$

(Note that $x2^{-x}$ is decreasing for $x \geq 2 > \frac{1}{\log 2}$.) This yields

$$|QNC(p,q)| \leq \frac{1}{2}(p^{-q} + q^{-p}),$$

from which we obtain

$$\sum_{j=1}^{\infty}\sum_{k=1}^{\infty}|\mathbf{R}_{jk}| \leq \sum_{m=2}^{\infty}\left(\sum_{n=m}^{\infty}m^{-n}\right) < \infty,$$

as asserted.

We use (5) to verify criterion (10). Let $\{e_k : 1 \leq k < \infty\}$ be the standard orthonormal basis of column vectors of l_2, so that $\mathbf{R}(e_k) = \sum_{j=1}^{\infty}\mathbf{R}_{jk}e_j$. Now let $\phi_n = \sum_{k=1}^{\infty}c_{nk}e_k$ be an orthonormal basis of l_2, so that $[c_{nk}]$ is a unitary matrix. Then we have $||\phi_n||^2 = \sum_{k=1}^{\infty}|c_{nk}|^2 = 1$, and unitarity also implies

$$\sum_{n=1}^{\infty}|c_{nk}|^2 = 1. \tag{11}$$

Now we compute

$$\sum_{n=1}^{\infty}|\langle\mathbf{R}\phi_n, \phi_n\rangle| = \sum_{n=1}^{\infty}\left|\left\langle\sum_{j=1}^{\infty}\sum_{k=1}^{\infty}c_{nk}\mathbf{R}_{jk}e_j, \sum_{j=1}^{\infty}c_{nj}e_j\right\rangle\right|$$

$$\leq \sum_{n=1}^{\infty}\sum_{j=1}^{\infty}\sum_{k=1}^{\infty}|c_{nk}\mathbf{R}_{jk}\overline{c_{nj}}|$$

$$\leq \sum_{j=1}^{\infty}\sum_{k=1}^{\infty}|\mathbf{R}_{jk}|\left(\sum_{n=1}^{\infty}|c_{nj}||c_{nk}|\right)$$

$$\leq \sum_{j=1}^{\infty}\sum_{k=1}^{\infty}|\mathbf{R}_{jk}|\left(\sum_{n=1}^{\infty}\frac{1}{2}(|c_{nj}|^2 + |c_{nk}|^2)\right)$$

$$\leq \sum_{j=1}^{\infty}\sum_{k=1}^{\infty}|\mathbf{R}_{jk}| < \infty$$

as required.   □

3 Concluding remarks

It is an interesting question whether the concept of "QNC" has a natural modification to correct the difficulty observed here, and possibly to give a positive answer to question (A). We have no proposal how to do this, but make the following remarks.

The argument made above rests on the following fact: A necessary condition on a skew-symmetric compact operator $\mathbf{R}$ acting on a Hilbert space to have a determinant (3) satisfying the basic requirement that detects the zeta function zeros is that it be a Hilbert–Schmidt operator not of trace class. In order to define an infinite determinant on the full class of Hilbert–Schmidt operators, an extended definition of infinite determinant is required. There are notions of regularized determinant $\det(I + w\mathbf{A})$ that apply to Hilbert–Schmidt operators $\mathbf{A}$ and satisfy the basic requirement. One such, denoted $\det_2(I + w\mathbf{A})$, in discussed in Simon [Si77] and Simon [Si79, Chapter 3]. See Pietsch [Pi87, Chapters 4 and 7] for further work on such questions.

The results of Kurokawa, Ochiai, and Wakayama [KOW03] were motivated in part by the function field case for the absolute function field $K = \mathbb{F}_q(T)$, as noted at the beginning of their paper. We note that one might reconsider the function field analogy, varying the base function field. For the (absolute) function field case $\mathbb{F}_q(T)$ the corresponding matrix (and operator) $\mathbf{R} \equiv 0$, but if one allowed other function fields K of genus 1 or higher, then the function field analogue of the quantity (9) also diverges. This holds because the function field zeta zeros $\frac{1}{2} + i\gamma$ have γ falling in a finite number of arithmetic progressions $(\bmod \frac{2\pi}{\log p})$, so that

$$\sum_\gamma \frac{1}{|\gamma|} = +\infty.$$

Thus the difficulty above manifests itself already in the function field case. It therefore might be useful to look for formulas for quantum noncommutativity for prime ideals in a function field K of genus at least 1, intending to construct an analogous matrix $\mathbf{R}_K$. The operator corresponding to $\mathbf{R}_K$ on l_2 would necessarily be Hilbert–Schmidt, but not of trace class, if it were to have eigenvalues $\pm\frac{i}{\gamma}$, where $\frac{1}{2} + i\gamma$ runs over the function field zeta zeros of K, counted with multiplicity. Perhaps such study could clarify the notion of "QNC."

Finally, we note that if to the sum defining the function $F(x, y)$ in (1) the term $k = 0$ were added, the definition of $QNC(p, q)$ would be modified to add the extra terms

$$\frac{1}{12pq}\left(\frac{1}{q-1} - \frac{1}{p-1}\right).$$

The resulting modified operator $\tilde{\mathbf{R}}$ then has

$$\sum_{i,j} |\tilde{\mathbf{R}}_{ij}| = +\infty,$$

and is a Hilbert–Schmidt operator on l_2 not of trace class.

Acknowledgment. The author thanks the reviewer for helpful comments.

References

[AG93]   N. Akhiezer and I. M. Glazman, *Theory of Linear Operators on Hilbert Space*, reprinted ed., Dover, New York, 1993.

[De04]   A. Deitmar, Schemes over $\mathbb{F}_1$, in G. van der Geer, B. Moonen, and R. Schoof, eds., *Number Fields and Function Fields: Two Parallel Worlds*, Birkhäuser Boston, Cambridge, MA, 2005, 87–100 (this volume).

[Ku92]   N. Kurokawa, Multiple zeta functions: An example, in *Zeta Functions and Geometry (Tokyo* 1990), Advanced Studies in Pure Mathematics 21, Kinokuniya, Tokyo, 1992, 219–226.

[KOW03]   B. Kurokawa, H. Ochiai, and A. Wakayama, Absolute derivations and zeta functions, *Documenta Math.*, **Extra Vol.** (Kazuya Kato's Fiftieth Birthday) (2003), 565–584.

[Ma95]   Y. Manin, *Lectures on Zeta Functions and Motives (According to Deninger and Kurokawa)*, Astérisque 228, Société Mathématique de France, Paris, 1995, 121–163.

[Pi87]   A. Pietsch, *Eigenvalues and s-Numbers*, Cambridge University Press, Cambridge, UK, 1987.

[RS80]   M. Reed and B. Simon, *Methods of Modern Mathematical Mhysics* I: *Functional Analysis*, revised and enlarged ed., Academic Press, Orlando, FL, 1980.

[RS78]   M. Reed and B. Simon, *Methods of Modern Mathematical Mhysics* IV: *Analysis of Operators*, Academic Press, Orlando, FL, 1978.

[Si77]   B. Simon, Notes on infinite determinants of Hilbert space operators, *Adv. Math.*, **24** (1977), 244–273.

[Si79]   B. Simon, *Trace Ideals and Their Applications*, Cambridge University Press, Cambridge, UK, 1979.

[So99]   C. Soulé, *On the Field with One Element*, Arbeitstagung, Max-Planck-Institut für Mathematik, Bonn, 1999; Prépublication M/55, Institut des Hautes Études Scientifiques, Bures-sur-Yvette, France, 1999.

[So03]   C. Soulé, Les variétés sur la corps à un élément, *Moscow Math. J.*, **4** (2004), 217–244, 312.

[TH86]   E. C. Titchmarsh, *The Theory of the Riemann Zeta Function*, 2nd ed. (revised by D. R. Heath-Brown), Oxford University Press, Oxford, UK, 1986.

On the Order of Certain Characteristic Classes of the Hodge Bundle of Semi-Abelian Schemes

Vincent Maillot and Damian Roessler

Institut de Mathématiques de Jussieu
Université Paris 7 Denis Diderot
C.N.R.S.
Case Postale 7012
2 Place Jussieu
F-75251 Paris Cedex 05
France
vmaillot@math.jussieu.fr, dcr@math.jussieu.fr

Summary. We give a new proof of the fact that the even terms (of a multiple of) the Chern character of the Hodge bundles of semi-abelian schemes are torsion classes in Chow theory and we give explicit bounds for almost all the prime powers appearing in their order. These bounds appear in the numerators of modified Bernoulli numbers. We also obtain similar results in an equivariant situation.

1 Introduction

Let $g \geq 1$ (respectively, $n \geq 4$) be an integer (respectively, an even integer) and let $\mathbf{A}_{g,n}$ be the fine moduli scheme of principally polarized abelian varieties over $\mathbb{C}$ with an n-level structure (see [CF, Chapter I]). By $\overline{\mathbf{A}}_{g,n}$ we denote a toroidal compactification of Faltings–Chai type (see [CF, Chapter IV]). Let $G \to \overline{\mathbf{A}}_{g,n}$ be the universal semi-abelian scheme over $\overline{\mathbf{A}}_{g,n}$. We set $\mathbb{E} := e^* \Omega_{G/\overline{\mathbf{A}}_{g,n}}$, where $e : \overline{\mathbf{A}}_{g,n} \to G$ is the zero-section. For any integer $k \geq 0$, we shall write $\mathrm{ch}_0^k(V)$ for the additive characteristic class on vector bundles V, such that $\mathrm{ch}_0^k(V) := c_1(V)^k$ when V is a line bundle. Furthermore, for any integer $l \geq 2$, we shall write B_l' for the numerator of the rational number $(2^l - 1)B_l$, where B_l is the lth Bernoulli number. Recall that the Bernoulli numbers are defined by the formula

$$\frac{t}{\exp(t) - 1} = \sum_{j \geq 0} B_j \frac{t^j}{j!}.$$

Theorem 1. *Let $b : \widetilde{\mathbf{A}}_{g,n} \to \overline{\mathbf{A}}_{g,n}$ be any desingularization and let $l \geq 2$ be an even integer. Then we have the following:*

(1) *The characteristic class $\mathrm{ch}_0^l(b^*\mathbb{E}) \in \mathrm{CH}^l(\widetilde{\mathbf{A}}_{g,n})$ is a torsion class.*

(2) *Let $t \geqslant 1$ be the smallest natural number such that $t \cdot \mathrm{ch}_0^l(b^*\mathbb{E}) = 0$, and let p be a prime number such that $p > l$. If $q \geqslant 0$ is the largest integer such that $p^q \,|\, t$, then $p^q \,|\, B_l'$.*

Here are some numerical examples. Case $l = 2$: the class $\mathrm{ch}_0^2(b^*\mathbb{E})$ is a torsion class of order a power of 2, since $(2^2 - 1)B_2 = 3/6 = 1/2$. Case $l = 12$: there is an integer $r \geqslant 0$ such that $691 \cdot 2310^r \cdot \mathrm{ch}_0^{12}(b^*\mathbb{E}) = 691 \cdot (2 \cdot 3 \cdot 5 \cdot 7 \cdot 11)^r \cdot \mathrm{ch}_0^{12}(b^*\mathbb{E}) = 0$, since $(2^{12} - 1)B_{12} = -2073/2 = -3 \cdot 691/2$.

Recall that $\mathrm{CH}(\widetilde{\mathbf{A}}_{g,n})$ refers to the Chow intersection ring of $\widetilde{\mathbf{A}}_{g,n}$ (see [F]). The ring $\mathrm{CH}(\widetilde{\mathbf{A}}_{g,n})$ carries a natural ring grading and $\mathrm{CH}^l(\widetilde{\mathbf{A}}_{g,n})$ refers to its lth graded term. Characteristic classes with values in a cohomology theory factor via the cycle class map through their counterparts with values in the Chow ring. The Chow ring is thus a universal target for characteristic classes.

If one replaces $\overline{\mathbf{A}}_{g,n}$ by $\mathbf{A}_{g,n}$ in Theorem 1 (so that b becomes an isomorphism), then the statement that the characteristic class $\mathrm{ch}_0^l(b^*\mathbb{E})$ is a torsion class was proven by van der Geer in [VDG]. Prompted by his work, Esnault and Viehweg then proved (1) in [EV1]. The original contribution of Theorem 1 thus consists of the information (2) given about the order of the torsion.

For z belonging to the unit circle S^1, we define the Lerch ζ-function $\zeta_L(z, s) := \sum_{k \geqslant 1} z^k/k^s$ for $s \in \mathbb{C}$ such that $\Re(s) > 1$, and using analytic continuation, we extend it to a meromorphic function of s over $\mathbb{C}$.

In the next theorem, n is an integer $\geqslant 1$ and D is any Dedekind ring containing $\mathcal{O}_{\mathbb{Q}(\mu_n)}$ as a subring. Recall that $\mathcal{O}_{\mathbb{Q}(\mu_n)}$ is the ring of integers of the subfield $\mathbb{Q}(\mu_n)$ of $\mathbb{C}$ generated by the nth roots of unity. Let C be a smooth quasi-projective scheme over $\mathrm{Spec}\, D[\frac{1}{n}]$. Let furthermore $\mathcal{C} \to C$ be a polarized abelian scheme and let ι be an automorphism of finite order n of $\mathcal{C}$ over C. Suppose that the fixed-point scheme $\mathcal{C}_\iota$ of ι is finite and flat over C. Let $\mathcal{H} := H_{\mathrm{dR}}^1(\mathcal{C}/C)$. The automorphism ι induces an automorphism of finite order of $\mathcal{H}$, which we also denote by ι. For each $u \in \mu_n(D)$, let $\mathcal{H}_u := \mathrm{Ker}(\iota - u \cdot \mathrm{Id})$.

Theorem 2. *Let $l \geqslant 1$ be an integer. The meromorphic function $\zeta_L(u, z)$ is regular at $z = 1 - l$ and the complex number $\zeta_L(u, 1 - l)$ lies in $\mathcal{O}_{\mathbb{Q}(\mu_n)}[\frac{1}{n \cdot l!}]$. The equality*

$$\sum_{u \in \mu_n(D)} \zeta_L(u, 1 - l) \,\mathrm{ch}_0^l(\mathcal{H}_u) = 0$$

holds in $\mathrm{CH}^l(C) \otimes \mathcal{O}_{\mathbb{Q}(\mu_n)}[\frac{1}{n \cdot l!}]$.

Theorem 2 is compatible with Theorem 1 in the following sense. Let $C = \mathbf{A}_{g,n}$ and let $\mathcal{C}$ be the restriction of G to $\mathbf{A}_{g,n}$. Let ι be the automorphism of order 2 of $\mathcal{C}$ over C given by taking the inverse in the group scheme. Then the equality statement in Theorem 2 is equivalent to Theorem 1 with $\overline{\mathbf{A}}_{g,n}$ replaced by $\mathbf{A}_{g,n}$.

Theorem 2 overlaps with Stickelberger's theorem; this is explained at the end of Section 4.2.

We shall now describe our methods of proof. Theorems 1 and 2 are both proved by applying a relative coherent Lefschetz fixed-point formula (see Section 2.3) to certain vector bundles and certain fibrations. A formula involving the extended Hodge bundle $b^*\mathbb{E}$ (respectively, the first Gauss–Manin bundle H^1_{dR}) is then obtained and Theorem 1 (respectively, Theorem 2) is deduced from this formula, using some linear algebra and some facts relating the exponential function and the Lerch zeta-function.

In the paper by Esnault and Viehweg quoted above [EV1], the Grothendieck–Riemann–Roch theorem is applied to a quotient of a compactification of the group scheme G to prove that $\mathrm{ch}^l_0(b^*\mathbb{E})$ is a torsion class. This method is conceptually close to ours but its seems difficult to obtain fine information about denominators using it, because it involves the Chow group of the compactification, where the denominators of the Chern character can become large, when the dimension of the compactification is large. By contrast, the relative Lefschetz formula only involves the Chow group of the fixed-point set, which has the same dimension as the base. Another advantage of the fixed-point formula is that it involves fewer denominators at the outset whereas the Grothendieck–Riemann–Roch theorem would probably have to be replaced by the Adams–Riemann–Roch theorem to make control of denominators possible.

The authors were led to Theorems 1 and 2 and the methods of proof presented here by a conjecture on characteristic classes of Hodge bundles in the context of Arakelov theory. For this we refer to Section 4.2.

The structure of the article is as follows. In the second section, we describe some results from the book of Chai and Faltings on the toroidal compactification of the universal semi-abelian family, as presented in the article [EV1]; we use these results to relate the sheaf of relative differentials with logarithmic singularities to the normal bundle of the fixed-point set of -1 (see Proposition 2). We then proceed to describe the relative fixed-point formula which will be our main tool in the proof. In the third section, we first prove Theorem 2 by applying the fixed-point formula to the relative de Rham complex of the relevant abelian scheme; second we prove Theorem 1 by applying the fixed-point formula to the relative logarithmic de Rham complex. In the fourth section, we shall discuss some consequences of the above theorems, as well as some conjectures to which they lead. A salient consequence of Theorem 2 is Corollary 1, which concerns abelian schemes with complex multiplications but possibly no automorphisms of finite order other than -1.

2 Preliminaries

2.1 Differentials with logarithmic singularities

In this subsection, we shall review the definition of a sheaf of differentials with logarithmic singularities along a divisor with normal crossings, as well as its basic properties. Our basic reference is [EV2, Chapter 2].

Let Z be a quasi-projective nonsingular variety over $\mathbb{C}$ and let D be a normal crossings divisor in Z. Let d be the dimension of Z. We set $U := Z \backslash D$ and denote by $j \colon U \hookrightarrow Z$ be the inclusion map. We shall write $\Omega^*_Z(\log D)$ for the complex of sheaves of differential forms with logarithmic singularities along D. The complex $\Omega^*_Z(\log D)$ is a subcomplex of the complex $j_*\Omega^*_U$ and it has the following defining property: if $V \subseteq Z$ is an open set, $p \geqslant 0$ is an integer and $\omega \in j_*\Omega^p_U(V) =$

$\Omega_Z^p(U \cap V)$, then $\omega \in \Omega_Z^p(\log D)(V)$ iff ω and $d\omega$ have simple poles along $D \cap V$. To say that ω has a simple pole along $D \cap V$ in the latter situation means the following: in any affine open subscheme $W \subseteq V$ such that the ideal of $D \cap W$ in $\mathcal{O}_Z(W)$ is principal, the section $e \cdot \omega|_{U \cap W}$ lies in the image of the restriction map $\Omega_Z^p(W) \to \Omega_Z^p(U \cap W)$ for any generator $e \in \mathcal{O}_Z(W)$ of the ideal of $D \cap W$ (this condition does not depend on the choice of e). The definition of $\Omega_Z^*(\log D)$ implies that for each $p \geqslant 0$ the sheaf $\Omega_Z^p(\log D)$ has the structure of $\mathcal{O}_Z$-module, which is compatible with the injection $\Omega_Z^p(\log D) \hookrightarrow j_* \Omega_U^p$; furthermore, $\Omega_Z^p(\log D)$ is then locally free for this $\mathcal{O}_Z$-module structure.

Abusing language, we shall write $\mathcal{O}_{D_0}$ for $\oplus_{i=1}^n \mathcal{O}_{C_i}$, where the C_i run over the irreducible components of D. Let $P \in Z$ and let $\mathbf{m}$ be the maximal ideal of the local ring $\mathcal{O}_P$. We write s for the number of irreducible components of D which contain P. We may suppose without restriction of generality that the components $C_1, \ldots, C_s$ contain P. We denote by $f_1, \ldots, f_s \in \mathbf{m}$ generators of the ideals of the components $C_1, \ldots, C_s$. Since by definition $f_1, \ldots, f_s$ form a regular sequence in $\mathcal{O}_P$ and since $\mathcal{O}_P$ is a regular ring, we may find (see [M, Part 14, Theorem 14.2]) elements $f_{s+1}, \ldots, f_d \in \mathbf{m}$, such that the elements $f_1, \ldots, f_d$ form a regular system of parameters in $\mathcal{O}_P$. There is then a neighborhood V of P, such that the elements $d f_1, \ldots, d f_d$ form a basis of $\Omega_Z(V)$ as a $\mathcal{O}_Z(V)$-module. It is shown in [EV2, Chapter 2, 2.2 (c), p. 11] that in this situation the elements $d f_1/f_1, \ldots, d f_s/f_s, d f_{s+1}, \ldots, d f_d$ form a basis of $\Omega_Z(\log D)(V)$ as an $\mathcal{O}_Z(V)$-module.

Furthermore, there is a canonical exact sequence

$$0 \longrightarrow \Omega_Z^1 \longrightarrow \Omega_Z^1(\log D) \overset{r}{\longrightarrow} \mathcal{O}_{D_0} \longrightarrow 0$$

where the morphism r has the following description. We shall use the terminology of the last paragraph. The homomorphism $\Omega_Z(\log D)(V) \to \mathcal{O}_{D_0}(V) = \oplus_{i=1}^s \mathcal{O}_{C_i}(V)$ sends $\alpha \cdot d f_i/f_i$ (respectively, $\alpha \cdot d f_i$), where $\alpha \in \mathcal{O}_Z(V)$ and $1 \leqslant i \leqslant s$ (respectively, $d \geqslant i > s$), to the image of α in $\mathcal{O}_{C_i}(V)$ (respectively, on 0).

Now let Z' be a nonsingular quasi-projective variety over $\mathbb{C}$ and $g: Z' \to Z$ be a morphism over $\mathbb{C}$. Let $D' := (g^*(D))_{\mathrm{red}}$ and suppose that D' is a divisor with normal crossings. We write U' for its complement and let $j': U' \hookrightarrow Z'$ be the inclusion morphism. Notice that by adjunction, there is a natural morphism of coherent sheaves $j^* \Omega_Z(\log D) \to j^* \Omega_Z = \Omega_U$; this induces a morphism $g|_U^*(j^*\Omega_Z(\log D)) \to g|_U^*(j^*\Omega_Z)$ and since $j \circ g|_U = g \circ j'$, we obtain a morphism $g^*\Omega_Z(\log D) \to j'_* j'^* g^* \Omega_Z$ by adjunction. Composing with the natural morphism $g^*\Omega_Z \to \Omega_{Z'}$, we finally obtain a morphism $g^*\Omega_Z(\log D) \to j'_* j'^* \Omega_{Z'} = j'_* \Omega_{U'}$.

Lemma 1. *The image of the morphism of coherent sheaves* $g^*\Omega_Z(\log D) \to j'_* j'^* \Omega_{Z'} = j'_* \Omega_{U'}$ *just given lies inside* $\Omega_{Z'}(\log D')$.

We shall prove Lemma 1 together with Lemma 2, which we first describe. Consider first the following diagram:

$$
\begin{array}{ccccccc}
 & g^*\Omega_Z & \longrightarrow & g^*\Omega_Z(\log D) & \longrightarrow & g^*\mathcal{O}_{D_0} & \longrightarrow 0 \\
 & & & \downarrow & & \downarrow & \\
0 \longrightarrow & \Omega_{Z'} & \longrightarrow & \Omega_{Z'}(\log D') & \longrightarrow & \mathcal{O}_{D'_0} & \longrightarrow 0,
\end{array}
\tag{1}
$$

where the middle vertical arrow is defined via Lemma 1. By construction this diagram is commutative and its existence shows that there is a unique morphism $g^*\mathcal{O}_{D_0} \to \mathcal{O}_{D_0'}$ such that the completed diagram

$$
\begin{array}{ccccccc}
g^*\Omega_Z & \longrightarrow & g^*\Omega_Z(\log D) & \longrightarrow & g^*\mathcal{O}_{D_0} & \longrightarrow & 0 \\
\downarrow & & \downarrow & & \downarrow & & \\
0 \longrightarrow \Omega_{Z'} & \longrightarrow & \Omega_{Z'}(\log D') & \longrightarrow & \mathcal{O}_{D_0'} & \longrightarrow & 0
\end{array}
$$

commutes.

Lemma 2. *If D and $g^*(D)$ are normal schemes, then the morphism $g^*\mathcal{O}_{D_0} \to \mathcal{O}_{D_0'}$ just described is an isomorphism.*

Proof of Lemmas 1 and 2. Since Lemmas 1 and 2 are both local statements on Z and Z', we may assume that $Z' = \operatorname{Spec} A$, $Z = \operatorname{Spec} B$ and that the morphism g is induced by a ring morphism $g_0 : A \to B$. We may also suppose that $\Omega_Z(\log D)(Z)$ is free over A and that a basis of $\Omega_Z(\log D)(Z)$ is given by $df_1/f_1, \ldots, df_s/f_s, df_{s+1}, \ldots, df_d$, where the elements $f_1, \ldots, f_s$ are generators of the ideals of the irreducible components $C_1, \ldots, C_s$ of D and the elements $df_1, \ldots, df_d$ form a basis of Ω_A over A. Similarly, we may also suppose that a basis of $\Omega_{Z'}(\log D')(Z')$ over B is given by $df_1'/f_1', \ldots, df_{s'}'/f_{s'}', df_{s'+1}', \ldots, df_{d'}'$, where the elements $f_1', \ldots, f_{s'}'$ are generators of the ideals of the irreducible components $C_1', \ldots, C_{s'}'$ of D' and the elements $df_1, \ldots, df_{d'}$ form a basis of Ω_B over B. We may also suppose that $g_0(f_k) = \prod_{r=1}^{s'} u_{k,r} f_r'^{m_{k,r}}$, where $u_{k,r} \in B^\times$, $m_{k,r} \in \mathbb{Z}^{\geqslant 0}$ and $1 \leqslant k \leqslant s$. (This follows from the fact that the local rings of Z' are regular rings and hence unique factorization domains.)

Notice that we have a canonical isomorphism $j_*\Omega_U(Z) \simeq \Omega_{A, f_1 \cdots f_s}$ (respectively, $j_*'\Omega_{U'}(Z') \simeq \Omega_{B, f_1' \cdots f_{s'}'}$). If we follow the steps of the definition of the morphism $g^*\Omega_Z(\log D) \to j_*'\Omega_{U'}$, we see that it corresponds to the morphism $\Omega_Z(\log D)(Z) \otimes_A B \to \Omega_{B, f_1' \cdots f_{s'}'}$ of B-modules such that

$$
\frac{df_k}{f_k} \mapsto \sum_{r=1}^{s'} \left(\frac{du_{k,r}}{u_{k,r}} + m_{k,r} \frac{df_r'}{f_r'} \right) \tag{2}
$$

for $1 \leqslant k \leqslant s$ and

$$
df_k \mapsto d(g_0(f_k)) \tag{3}
$$

for $s < k \leqslant d$. Since the expressions appearing after the arrows in (2) and (3) are both linear combinations over B of the elements

$$
\frac{df_1}{f_1}, \ldots, \frac{df_{s'}'}{f_{s'}'}, df_{s'+1}', \ldots, df_{d'}',
$$

we have proven Lemma 1.

To prove Lemma 2, notice first that we may assume without loss of generality in the situation of Lemma 2 that D and $g^*(D)$ are integral. We then have $s = 1$,

$s' = 1$ and $m_{k,r} = 1$ for all k, r. The module associated to the coherent sheaf $g^* \mathcal{O}_{D_0}$ (respectively, $\mathcal{O}_{D'_0}$) is then $A/(f_1) \otimes_A B$ (respectively, $B/(g_0(f_1))$). Furthermore, if $a \otimes_A b \in A/(f_1) \otimes_A B$ (respectively, $b' \in B/(g_0(f_1))$), then $a \otimes_A b$ is by definition the image of the element $a \cdot d f_1/f_1 \otimes b \in g^* \Omega_Z(\log D)(Z')$ (respectively, $b' \cdot d f'_1/f'_1 \in \Omega_{Z'}(\log D')(Z')$). Looking at the diagram (1), we see that the morphism $A/(f_1) \otimes_A B \to B/(g_0(f_1))$ sends $a \otimes_A b$ to the image of $g(a)b \cdot (d f'_1/f'_1 + du_{1,1}/u_{1,1})$, i.e., $g(a)b$. Thus the morphism $g^* \mathcal{O}_{D_0} \to \mathcal{O}_{D'_0}$ is given by the natural isomorphism $A/(f_1) \otimes_A B \simeq B/(g_0(f_1))$.   □

2.2 The toroidal compactification of the universal abelian scheme

We shall need the following result, whose proof can be found in [CF, Chapter I, Proposition 2.7].

Proposition 1 (Raynaud). *Let S be a noetherian normal scheme and let $U \subseteq S$ be an open dense subset. Let $\mathcal{B} \to U$ be an abelian scheme. If there is a semi-abelian scheme $\widetilde{\mathcal{B}} \to S$ extending $\mathcal{B}$, then it is unique up to unique isomorphism.*

We now quote a theorem stated in [EV1, Theorem 3.1], which sums up some results that can be found in [CF, Chapter VI, par. 1]. Recall that $n \geqslant 4$ is an even integer.

Theorem 3. *There exists a cartesian diagram of morphisms of schemes*

$$
\begin{array}{ccc}
\mathcal{A} & \hookrightarrow & X \\
f \downarrow & & \downarrow \overline{f} \\
\mathbf{A}_{g,n} & \hookrightarrow & B,
\end{array}
$$

where $f : \mathcal{A} \to \mathbf{A}_{g,n}$ is the universal abelian scheme, such that we have the following:

(1) *The horizontal morphisms are open immersions.*

(2) *The closed set $T := B \backslash \mathbf{A}_{g,n}$, endowed with its reduced induced subscheme structure, is a normal crossings divisor.*

(3) *The closed subscheme $Y := (\overline{f}^* T)_{\mathrm{red}}$ is a normal crossings divisor.*

(4) *X and B are projective smooth varieties over $\mathbb{C}$.*

(5) *There exists a semi-abelian scheme $\widetilde{\mathcal{A}} \to B$ which extends the universal abelian scheme.*

(6) *The n-level structure sections $S_i : \mathbf{A}_{g,n} \to \mathcal{A}$ ($i \in (\mathbb{Z}/n\mathbb{Z})^{2g}$) extend to pairwise disjoint sections of X over B.*

(7) *The action of the inversion on $\mathcal{A}$ extends to an involution α of X over B whose fixed-point scheme factors through $\coprod_{i \in (\mathbb{Z}/n\mathbb{Z})^{2g}} S_i$.*

(8) *Let $\widetilde{e} : B \to \widetilde{\mathcal{A}}$ be the zero-section and let $\widetilde{\mathbb{E}} := \widetilde{e}^* \Omega_{\widetilde{\mathcal{A}}/B}$. There is a natural isomorphism*

$$
\overline{f}^* \widetilde{\mathbb{E}} \simeq \Omega_X(\log Y)/\overline{f}^*(\Omega_B(\log T)) =: \Omega_{X/B}(\log),
$$

where

(9) *there is a natural isomorphism*

$$R^q \overline{f}_*(\mathcal{O}_X) \simeq \wedge^q(\widetilde{\mathbb{E}}^\vee)$$

for all $q \geqslant 0$.

Notice that statement (7) implies that $X_\alpha = \coprod_{i \in (\mathbb{Z}/n\mathbb{Z})^{2g}, 2 \cdot i = 0} S_i$. The conormal sheaf of X_α in X is locally free since both X_α and X are regular and we denote the dual of the conormal sheaf by N.

Proposition 2. *There is a natural isomorphism $N^\vee \simeq \Omega_{X/B}(\log)|_{X_\alpha}$.*

For the proof, we shall need the following lemma.

Lemma 3. *There exists an open neighborhood V of X_α such that $\overline{f}|_V$ is smooth. Furthermore, the natural map $N^\vee \to \Omega_{X/B}|_{X_\alpha}$ is an isomorphism.*

Proof. Recall that there is an exact sequence

$$N^\vee \longrightarrow \Omega_{X/B}|_{X_\alpha} \longrightarrow \Omega_{X_\alpha/X} \longrightarrow 0$$

(see [H, II, Proposition 8.12]). Using the determination of X_α given above we deduce that $\Omega_{X_\alpha/X} = 0$. Furthermore, the restriction of the above sequence to $X_\alpha \cap \mathcal{A}$ is exact, since $\mathcal{A} \to \mathbf{A}_{g,n}$ is smooth (cf. [FL, IV, Part 3, Proposition 3.7 (b)]). Hence $\mathrm{rk}(N) = g$. Now let $r \colon X \to \mathbb{Z}$ be the function $r(x) := \dim_{\kappa(x)} \Omega_{X/B,x} \otimes_{\kappa(x)} \kappa(x)$, where $\kappa(x)$ is the residue field at x. This function is upper semicontinuous (see [H, II, Example 5.8 (a)]) and thus reaches its minimum at g, which is the rank of $\Omega_{X/B}$ on the open dense subset $\mathcal{A}$ of X. The set $V := \{x \in X | r(x) = g\}$ is open and the restriction of $\Omega_{X/B}$ to V is locally free of rank g (see [H, II, Example 5.8 (c)]). The existence of the surjection $N^\vee \to \Omega_{X/B}|_{X_\alpha}$ implies that $r(x) \leqslant g$ when $x \in X_\alpha$ and thus $r(x) = g$ on X_α. Thus $X_\alpha \subseteq V$. The restriction $\overline{f}|_V$ is smooth since V and B are nonsingular varieties over $\mathbb{C}$ and $\dim(V) - \dim(B) = g$ (see [H, III, Proposition 10.4]). The morphism $N^\vee \to \Omega_{X/B}|_{X_\alpha}$ is a surjection of locally free sheaves of the same rank and is thus an isomorphism. This concludes the proof. $\square$

Proof of Proposition 2. Consider the commutative diagram with exact rows on X:

$$
\begin{array}{ccccccc}
\overline{f}^*\Omega_B & \longrightarrow & \overline{f}^*\Omega_B(\log T) & \longrightarrow & \overline{f}^*\mathcal{O}_{T_0} & \longrightarrow & 0 \\
\downarrow & & \downarrow & & \downarrow & & \\
0 \longrightarrow \quad \Omega_X & \longrightarrow & \Omega_X(\log Y) & \longrightarrow & \mathcal{O}_{Y_0} & & \\
\downarrow & & \downarrow & & \downarrow & & \\
\Omega_{X/B} & \longrightarrow & \Omega_{X/B}(\log) & \longrightarrow & \mathcal{O}_{Y_0}/\overline{f}^*\mathcal{O}_{T_0} & & \\
\downarrow & & \downarrow & & \downarrow & & \\
0 & & 0 & & 0, & &
\end{array}
$$

where the morphism $\overline{f}^* \mathcal{O}_{T_0} \to \mathcal{O}_{Y_0}$ is defined by the two vertical morphisms on its left side. Let $V \subseteq T$ be the open subset of T which consists of all the points which do not lie at the intersection of two irreducible components of T. The set V has smooth disjoint irreducible components. Let U_0 be the open neighborhood of X_α provided by Lemma 3. Let now V_B be an open subset of B such that $V_B \cap T = V$ and let $U_B := \mathbf{A}_{g,n} \cup V_B$. Finally, let $U := \overline{f}^{-1}(U_B) \cap U_0$. Using Lemma 2 and the snake lemma, we see that the restriction of the last diagram to U has the following appearance:

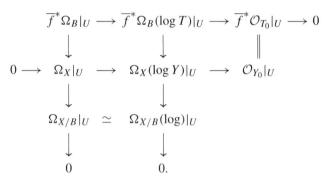

Consider now an n-level section $\sigma: B \to X$ whose image is an irreducible component of X_α. By the last diagram and the second statement in Lemma 3, we have $\sigma|_{U_B}^* \Omega_{X/B}(\log) \simeq \sigma|_{U_B}^* N^\vee$. Since $B \setminus U_B$ has codimension 2 in B by construction and since B is regular, there is a unique extension of the isomorphism $\sigma|_{U_B}^* \Omega_{X/B}(\log) \simeq \sigma|_{U_B}^* N^\vee$ to an isomorphism $\sigma^* \Omega_{X/B}(\log) \simeq \sigma^* N^\vee$. This concludes the proof. □

2.3 A relative Lefschetz fixed-point formula

In this subsection, we shall review a relative fixed-point formula which is a corollary of a formula in Arakelov theory proved in [KR1]. Let S be a noetherian affine scheme. Let Z be a regular scheme which is quasi-projective over S. Let μ_n be the diagonalizable group scheme over S which corresponds to $\mathbb{Z}/n\mathbb{Z}$. Suppose that Z carries a μ_n-action over S; furthermore, suppose that there is an ample line bundle on Z, which carries a μ_n-equivariant structure compatible with the μ_n-equivariant structure of Z (see [T2, par. 1.2] for more details about the latter notion). We shall write $K_0^{\mu_n}(Z)$ for the Grothendieck group of locally free sheaves on Z which carry a compatible μ_n-equivariant structure. Replacing locally free sheaves by coherent sheaves in the latter definition leads to a naturally isomorphic group (see [T2, Lemma 3.3]). If the μ_n-equivariant structure of Z is trivial, then the datum of a (compatible) μ_n-equivariant structure on a locally free sheaf E on Z is equivalent to the datum of a $\mathbb{Z}/n\mathbb{Z}$-grading of E. The group $K_0^{\mu_n}(Z)$ carries a λ-ring structure such that for any μ_n-equivariant locally free sheaf E, the element $\lambda^k(E)$ is represented in $K_0^{\mu_n}(Z)$ by the kth exterior power of E, endowed with its natural μ_n-equivariant structure (see [K, Lemma 3.4]). For any μ_n-equivariant locally free sheaf E on Z, we write

$\lambda_{-1}(E)$ for $\sum_{k=0}^{\mathrm{rk}(E)}(-1)^k\lambda^k(E) \in K_0^{\mu_n}(Z)$. There is a unique isomorphism of rings $K_0^{\mu_n}(S) \simeq K_0(S)[T]/(1-T^n)$ with the following property: it maps the structure sheaf of S endowed with a homogenous $\mathbb{Z}/n\mathbb{Z}$-grading of weight 1 to T and it maps any locally free sheaf carrying a trivial equivariant structure to the corresponding element of $K_0(S)$ $(= K_0^{\mu_1}(S))$.

The functor of fixed points associated to Z is by definition the functor

$$\textbf{Schemes}/S \rightarrow \textbf{Sets}$$

described by the rule

$$T \mapsto Z(T)_{\mu_n(T)}.$$

Here $Z(T)_{\mu_n(T)}$ is the set of elements of $Z(T)$ which are fixed under each element of $\mu_n(T)$. The functor of fixed points is representable by a scheme Z_{μ_n} and the canonical morphism $Z_{\mu_n} \rightarrow Z$ is a closed immersion (see [SGA3, VIII, 6.5 d]). Furthermore, the scheme Z_{μ_n} is regular (see [T, Proposition 3.1]). We shall denote the immersion $Z_{\mu_n} \hookrightarrow Z$ by i. Write $N^\vee$ for the dual of the conormal sheaf of the closed immersion $Z_{\mu_n} \hookrightarrow Z$. It is locally free on Z_{μ_n} and carries a natural μ_n-equivariant structure. This structure corresponds to a μ_n-grading, since Z_{μ_n} carries the trivial μ_n-equivariant structure and it can be shown that the weight-0 term of this grading is 0 (see [T, Proposition 3.1]).

Let W be a regular scheme which is quasi-projective over S and suppose that W carries a μ_n-action over S. Let $h: Z \rightarrow W$ be a projective S-morphism which respects the μ_n-actions and write h_{μ_n} for the induced morphism $Z_{\mu_n} \rightarrow W$. The morphism h induces a direct image map $Rh_*: K_0^{\mu_n}(Z) \rightarrow K_0^{\mu_n}(W)$, which is a homomorphism of groups described by the formula $Rh_*(E) := \sum_{k\geqslant 0}(-1)^k R^k h_*(E)$ for a μ_n-equivariant coherent sheaf E on Z. Here $R^k h_*(E)$ refers to the kth higher direct image sheaf of E under h; the sheaves $R^k h_*(E)$ are coherent and carry a natural μ_n-equivariant structure. The morphism h also induces a pull-back map $Lh^*: K_0^{\mu_n}(W) \rightarrow K_0^{\mu_n}(Z)$; this is a ring morphism which sends a μ_n-equivariant locally free sheaf E on W on the locally free sheaf $h^*(E)$ on Z, endowed with its natural μ_n-equivariant structure. For any elements $z \in K_0^{\mu_n}(Z)$ and $w \in K_0^{\mu_n}(W)$, the projection formula $Rh_*(z \cdot Lh^*(w)) = w \cdot Rh_*(z)$ holds. This implies that the group homomorphism Rh_* is a morphism of $K_0^{\mu_n}(S)$-modules, if the group $K_0^{\mu_n}(Z)$ (respectively, $K_0^{\mu_n}(W)$) is endowed with the $K_0^{\mu_n}(S)$-module structure induced by the pull-back map $K_0^{\mu_n}(S) \rightarrow K_0^{\mu_n}(Z)$ (respectively, $K_0^{\mu_n}(S) \rightarrow K_0^{\mu_n}(W)$).

Let $\mathcal{R}$ be a $K_0^{\mu_n}(S)$-algebra such that $1 - T^k$ is a unit in R for all k such that $1 \leqslant k < n$.

We shall refer to the following hypothesis as (H): S is the spectrum of a Dedekind ring which can be embedded in $\mathbb{C}$, Z and W are flat over S and Z_{μ_n} is flat over S.

Theorem 4. *Let hypothesis* (H) *hold. The element* $\lambda_{-1}(N^\vee)$ *is a unit in the ring* $K_0^{\mu_n}(Z_{\mu_n}) \otimes_{K_0^{\mu_n}(S)} R$. *If the μ_n-equivariant structure on W is trivial, then for any element* $z \in K_0^{\mu_n}(Z)$, *the equality*

$$Rh_*(z) = Rh_{\mu_n,*}((\lambda_{-1}(N^\vee))^{-1} \cdot \mathrm{Li}^*(z))$$

holds in $K_0^{\mu_n}(W) \otimes_{K_0^{\mu_n}(S)} R$.

Proof. The theorem is a consequence of [KR1, Part 6, Theorem 6.1] if the morphism h is an immersion. Furthermore, the theorem is a consequence of [KR1, Part 4, Theorem 4.4] if $W = S$, $Z = \mathbf{P}_S^k$ for some $k \geqslant 0$ and h is the structural morphism $\mathbf{P}_S^k \to S$. These two cases combined with the projection formula and the determination of $K_0^{\mu_n}(\mathbf{P}_S^k)$ given in [T2, Theorem 3.1] imply the full statement.   □

Remarks.

(1) Theorem 4, without hypothesis (H) but with the hypothesis that S is the spectrum of an algebraically closed field of characteristic not dividing n, is proved in [BFM].
(2) Theorem 4, without hypothesis (H) but with the requirement that R is a field is a consequence of [T, Theorem 3.5].
(3) The proof of Theorem 4 given above only apparently refers to Arakelov theory; its underlying structure is purely algebraic and is a variant of the proof of the main result of [BFM]. This variant does not, in fact, use hypothesis (H). In particular, Theorem 4 is true without hypothesis (H).

3 Proof of Theorems 1 and 2

The proofs of Theorems 1 and 2 are similar and proceed in two steps. In the first one, we apply Theorem 4 to a certain geometrical situation and in the second one, we transform the resulting expression using some combinatorics. The first subsection contains the combinatorial statements we shall need and in the second one the computations leading to the proofs are given.

3.1 Combinatorics

Let's consider the two following formal series

$$\exp(x) := \sum_{j=0}^{\infty} \frac{x^j}{j!}$$

and

$$\log(1 + x) := \sum_{j=1}^{\infty} (-1)^{j+1} \frac{x^j}{j}.$$

For $l \geqslant 1$, we shall write $\mathbb{C}[x]^{\leqslant l}$ for the quotient of the ring $\mathbb{C}[[x]]$ by the ideal of formal series divisible by x^r, where $r > l$. We then define $\exp^{\leqslant l}(x) \in \mathbb{Z}[\frac{1}{l!}][x]$ and $\log^{\leqslant l}(1 + x) \in \mathbb{Z}[\frac{1}{l}][x]$ as the only polynomials of degree l representing the above formal series $\exp(x)$ and $\log(1 + x)$ in $\mathbb{C}[x]^{\leqslant l}$.

About these polynomials, we have the following lemma.

Lemma 4. *Let* $u \in \mu_n(\mathbb{C})$, $u \neq 1$. *Then the equality*

$$\log^{\leqslant l}\left(\frac{1 - u \cdot \exp^{\leqslant l}(x)}{1 - u}\right) = -\sum_{j=1}^{l} \zeta_L(u, 1 - j)\frac{x^j}{j!}$$

holds in $\mathbb{C}[x]^{\leqslant l}$. *In particular, the values* $\zeta_L(u, 1 - j)x^j/j!$ *lie in* $\mathcal{O}_{\mathbb{Q}(\mu_n)}[\frac{1}{n \cdot l!}]$ *when* $1 \leqslant j \leqslant l$.

Proof. It will be sufficient to prove the identity

$$\log\left(\frac{1 - u \cdot \exp(x)}{1 - u}\right) = -\sum_{j=1}^{\infty} \zeta_L(u, 1 - j)\frac{x^j}{j!} \tag{4}$$

in $\mathbb{C}[[x]]$. In [MR2, (6), proof of Lemma 3.1], the identity of complex power series

$$\frac{u \cdot \exp(x)}{1 - u \cdot \exp(x)} = \sum_{j=1}^{\infty} \zeta_L(u, -j)\frac{x^j}{j!} \tag{5}$$

is proven. If one takes the formal derivative of both sides of equation (4) (for x), one obtains equation (5). Hence it is sufficient to show that the constant terms of the power series on both sides of (4) coincide. Since both constant terms can be seen to vanish, we are done. $\qquad\square$

The following lemma will be used in the proof of Theorem 1.

Lemma 5. *The equality* $\zeta_L(-1, 1 - l) = -(2^l - 1)B_l/l$ *holds for all* $l \geqslant 2$.

Proof. Let $s \in \mathbb{C}$ be such that $\mathfrak{R}(s) > 1$. By definition, we have

$$\zeta_L(-1, s) = \sum_{k \geqslant 1} \frac{(-1)^k}{k^s}$$

and

$$\zeta_{\mathbb{Q}}(s) = \sum_{k \geqslant 1} \frac{1}{k^s},$$

where $\zeta_{\mathbb{Q}}$ is Riemann's ζ-function. From these equalities, we deduce that $\zeta_L(-1, s) = \zeta_{\mathbb{Q}}(s)(2^{1-s} - 1)$. Now $\zeta_{\mathbb{Q}}(1 - l) = -B_l/l$ (see for example [W, Chapter 4, Theorem 4.2]), whence the lemma. $\qquad\square$

If Z is a scheme which is smooth over a Dedekind ring, we shall write $\mathrm{CH}(Z)^{\leqslant l}$ for the ring $\mathrm{CH}(Z)/\oplus_{j=l+1}^{\infty} \mathrm{CH}^j(Z)$. If E is a locally free sheaf on Z, we shall write $\mathrm{ch}^{\leqslant l}(E)$ ("truncated Chern character") for the element of $\mathrm{CH}(Z)^{\leqslant l} \otimes \mathbb{Z}[\frac{1}{l!}]$ given by the formula $\mathrm{ch}^{\leqslant l}(E) := \sum_{j=0}^{l} \frac{1}{j!} \mathrm{ch}_0^j(E)$. The proof of the following lemma is similar to the proof of the multiplicativity and additivity of the Chern character and we shall omit it.

Lemma 6. *The map* $\mathrm{ch}^{\leqslant l}$ *factors through a ring homomorphism*

$$K_0(Z) \to \mathrm{CH}(Z)^{\leqslant l} \otimes \mathbb{Z}\left[\frac{1}{l!}\right].$$

Let $\mathrm{CH}(Z)^{\leqslant l,*}$ be the multiplicative subgroup of $\mathrm{CH}(Z)^{\leqslant l}$ consisting of elements of the form $1 + z$, where z has the property that its degree-0 part vanishes.

The following lemma is a consequence of the fact that $\log((1 + x)(1 + y)) = \log(1 + x) + \log(1 + y)$ in the ring of power series $\mathbb{C}[[x, y]]$.

Lemma 7. *The polynomial* $\log^{\leqslant l}$ *defines a map* $\mathrm{CH}(Z)^{\leqslant l,*} \to \mathrm{CH}(Z)^{\leqslant l} \otimes \mathbb{Z}[\frac{1}{l}]$ *which is a group homomorphism.*

3.2 Final computations

We shall now prove Theorem 2.

Lemma 8. *Let* ξ *be a primitive nth root of unity in* $\mathcal{O}_{\mathbb{Q}(\mu_n)}$. *Then the elements* $1 - \xi^k$ *are units in* $\mathcal{O}_{\mathbb{Q}(\mu_n)}[\frac{1}{n}]$ *for every integer k such that* $1 \leqslant k < n$.

Proof. Recall the polynomial identity $\prod_{j=1}^{n-1}(X - \xi^j) = X^{n-1} + \cdots + X + 1$. This identity implies that the inverse of $1 - \xi^k$ is given by $n^{-1} \prod_{r=1, r \neq k}^{n-1}(1 - \xi^r)$. □

If A is a $D[\frac{1}{n}]$-algebra such that Spec A is connected and nonempty, then A contains exactly n distinct nth roots of unity, all of which are images of roots of unity contained in $D[\frac{1}{n}]$. This is a consequence of the last lemma and of the Chinese remainder theorem. Fix a primitive root of unity ζ. This choice fixes an isomorphism $\mathbb{Z}/n\mathbb{Z} \simeq \mu_n(D[\frac{1}{n}])$ and hence for each $D[\frac{1}{n}]$-algebra A, there is a canonical isomorphism of groups $\mu_n(A) \simeq \prod_{C \in \mathcal{CC}(A)} \mathbb{Z}/n\mathbb{Z}$, where $\mathcal{CC}(A)$ is the set of connected components of $\mathrm{Spec}(A)$. We have thus described a $D[\frac{1}{n}]$-isomorphism between the constant group scheme over $D[\frac{1}{n}]$ associated to $\mathbb{Z}/n\mathbb{Z}$ and the group scheme μ_n over $D[\frac{1}{n}]$.

Let W be a scheme which is smooth over Spec $D[\frac{1}{n}]$ and which carries the trivial μ_n-equivariant structure. For each μ_n-equivariant locally free sheaf E on C, define

$$\mathrm{ch}_{\mu_n}^{\leqslant l}(E) := \sum_{k \in \mathbb{Z}/n\mathbb{Z}} \zeta^k \cdot \mathrm{ch}^{\leqslant l}(E_k).$$

We can see from the definitions that $\mathrm{ch}_{\mu_n}^{\leqslant l}$ induces a ring morphism

$$\mathrm{ch}_{\mu_n}^{\leqslant l} \colon K_0^{\mu_n}(W) \to \mathrm{CH}^{\leqslant l}(W) \otimes \mathcal{O}_{\mathbb{Q}(\mu_n)}\left[\frac{1}{l!}\right].$$

We shall now apply Theorem 4. Let us denote the morphism $\mathcal{C} \to C$ by c and its relative dimension by d. The automorphism ι defines a $\mathbb{Z}/n\mathbb{Z}$-action on $\mathcal{C}$ over C and using the isomorphism described above, we obtain a μ_n-equivariant structure on C over C. The fixed-point scheme $c_{\mu_n} : C_{\mu_n} \to C$ (which coincides with the fixed-point scheme of ι) is then étale over C. To see this, notice that we have an exact sequence of coherent sheaves

$$N^\vee \longrightarrow \Omega_{C/C}|_{C_{\mu_n}} \longrightarrow \Omega_{C_{\mu_n}/C} \longrightarrow 0,$$

where N is the normal bundle of the immersion $C_{\mu_n} \hookrightarrow C$ and all the maps respect the natural μ_n-actions on the sheaves. The two first sheaves in this sequence are locally free, since C and C_{μ_n} are regular and the map c is smooth. Hence the first morphism in the sequence is injective, because it is injective on the dense open subset where c_{μ_n} is étale. If we now consider the weight-0 part of the sequence, we obtain an isomorphism $(\Omega_{C/C}|_{C_{\mu_n}})_0 \to \Omega_{C_{\mu_n}/C}$ and thus $\Omega_{C_{\mu_n}/C}$ is locally free which in turn implies that $\Omega_{C_{\mu_n}/C} = 0$ since c is finite. Hence c is étale.

We now compute

$$\mathrm{ch}_{\mu_n}^{\leqslant l}\left(Rc_*\left(\sum_{k=0}^{d}(-1)^k \wedge^k (\Omega_{C/C})\right)\right) \overset{(1)}{=} \mathrm{ch}_{\mu_n}^{\leqslant l}\left(\sum_{k=0}^{2d}(-1)^k \wedge^k (\mathcal{H})\right)$$

$$\overset{(2)}{=} \mathrm{ch}_{\mu_n}^{\leqslant l}(Rc_{\mu_n *}((\lambda_{-1}(N^\vee))^{-1}\lambda_{-1}(\Omega_{C/C}|_{C_{\mu_n}})))$$

$$\overset{(3)}{=} \mathrm{ch}_{\mu_n}^{\leqslant l}(Rc_{\mu_n,*}(\lambda_{-1}(\Omega_{C_{\mu_n}/C})))$$

$$\overset{(4)}{=} f_0 \otimes 1 \in \mathrm{CH}^0(C) \otimes \mathcal{O}_{\mathbb{Q}(\mu_n)}\left[\frac{1}{n \cdot l!}\right].$$

Here $f_0 \in \mathrm{CH}^0(C) = \mathbb{Z}$ is the degree of the finite morphism c_{μ_n}. Equality (1) is justified by the fact that the Hodge to de Rham spectral sequence of c degenerates and the fact that there is a natural isomorphism $H_{\mathrm{dR}}^r(C/C) \simeq \wedge^r(H_{\mathrm{dR}}^1(C/C))$ for all $r \in \mathbb{Z}^{\geqslant 0}$ (see [BBM, 2.5.2]). Equality (2) is provided by Theorem 4, applied in the case where $S = \mathrm{Spec}\, D[\frac{1}{n}]$, $Z = C$, $h = c$, the μ_n-equivariant structure on C is the one described above and $z = \lambda_{-1}(\Omega_{C/C})$. Equality (3) is justified by the fact that c_{μ_n} is étale and the multiplicativity of λ_{-1}. Equality (4) derives from the fact that $\Omega_{C_{\mu_n}/C} = 0$. We shall now rewrite the resulting equality

$$\mathrm{ch}_{\mu_n}^{\leqslant l}\left(\sum_{k=0}^{2d}(-1)^k \wedge^k (\mathcal{H})\right) = f_0 \otimes 1 \in \mathrm{CH}^0(C) \otimes \mathcal{O}_{\mathbb{Q}(\mu_n)}\left[\frac{1}{n \cdot l!}\right] \qquad (6)$$

using the combinatorics of the first subsection.

By the splitting principle, we may suppose without restriction of generality that $\mathcal{H} = \sum_{k=1}^{2d} h_k$ in $K_0^{\mu_n}(C)$, where h_k is a line bundle which carries a homogenous $\mathbb{Z}/n\mathbb{Z}$-grading. Write t_k for the first Chern class $c_1(h_k)$ of h_k. Let $w(h_k) \in \mathbb{Z}/n\mathbb{Z}$ be the weight of h_k and let $u_k := \zeta^{w(h_k)}$. Equality (6) implies that

$$\prod_{k=1}^{2d}(1 - u_k) = f_0.$$

In particular, $u_k \neq 1$ for all k. We now have the following reformulation of (6):

$$\frac{1}{f_0} \operatorname{ch}_{\mu_n}^{\leq l} \left(\sum_{k=0}^{2d} (-1)^k \wedge^k (\mathcal{H}) \right) = \prod_{k=1}^{2d} \frac{\operatorname{ch}_{\mu_n}^{\leq l}(1 - h_k)}{(1 - u_k)}$$

$$= \prod_{k=1}^{2d} \frac{1 - u_k \cdot \exp^{\leq l}(t_k)}{1 - u_k} = 1.$$

If we apply the $\log^{\leq l}$ map to the members of the last string of equalities and use Lemma 4, we obtain

$$\sum_{k=1}^{2d} \log^{\leq l} \left(\frac{1 - u_k \cdot \exp^{\leq l}(t_k)}{1 - u_k} \right) = -\sum_{k=1}^{2d} \sum_{j=1}^{l} \zeta_L(u_k, 1 - j) \frac{t_k^j}{j!}$$

$$= -\sum_{j=1}^{l} \sum_{k=1}^{2d} \zeta_L(u_k, 1 - j) \frac{t_k^j}{j!}$$

$$= -\sum_{j=1}^{l} \sum_{u \in \mu_n(D)} \zeta_L(u, 1 - j) \frac{\operatorname{ch}_0^j(\mathcal{H}_u)}{j!}$$

$$= 0, \tag{7}$$

which implies the result.

We now turn to the proof of Theorem 1. We use the notation of Theorem 3. We shall apply Theorem 4 to the situation where $Z = X$, $h = \overline{f}$, $n = 2$, the action of μ_2 is given by the involution α which extends the action of the inversion on $\mathcal{A}$ (notice that over $D[\frac{1}{2}]$ there is a unique isomorphism between μ_2 and the constant group scheme associated to $\mathbb{Z}/2\mathbb{Z}$) and $z = \lambda_{-1}(\Omega_{X/B}(\log))$. Let N be the dual of the conormal sheaf of the immersion $X_\alpha = X_{\mu_2} \hookrightarrow X$.

We compute

$$\operatorname{ch}_{\mu_2}^{\leq l} \left(R\overline{f}_* \left(\sum_{k=0}^{g} (-1)^k \wedge^k (\Omega_{X/B}(\log)) \right) \right)$$

$$\overset{(1)}{=} \operatorname{ch}_{\mu_2}^{\leq l} \left(\sum_{k=0}^{2g} (-1)^k \wedge^k (\widetilde{\mathbb{E}} \oplus \widetilde{\mathbb{E}}^{\vee}) \right)$$

$$\overset{(2)}{=} \operatorname{ch}_{\mu_2}^{\leq l} (R\overline{f}_{\mu_2 *}((\lambda_{-1}(N^{\vee}))^{-1} \lambda_{-1}(\Omega_{X/B}(\log)|_{X_{\mu_2}})))$$

$$\overset{(3)}{=} f_0 \otimes 1 \in \operatorname{CH}(C) \otimes \mathbb{Z} \left[\frac{1}{2 \cdot l!} \right].$$

Here $f_0 \in \operatorname{CH}^0(C) = \mathbb{Z}$ is the degree of the finite morphism $\overline{f}_{\mu_n}$. Equality (1) is justified by Theorem 3(9). Equality (2) is provided by Theorem 4, applied in the situation just described. Equality (3) is justified by Lemma 3. Let us define

$\mathcal{H} := \widetilde{\mathbb{E}} \oplus \widetilde{\mathbb{E}}^\vee$. We can now repeat the computations from (6) to (7) verbatim, setting $n = 2$. We obtain the equation

$$\sum_{j=1}^{l} \sum_{u \in \mu_2(D)} \zeta_L(u, 1 - j) \frac{\mathrm{ch}_0^j(\mathcal{H}_u)}{j!} = 0$$

in $\mathrm{CH}(B) \otimes \mathbb{Z}[\frac{1}{2 \cdot l!}]$. In other words,

$$\sum_{j=1}^{l} \zeta_L(-1, 1 - j) \frac{\mathrm{ch}_0^j(\mathcal{H})}{j!} = 2 \sum_{j=1}^{l} \zeta_L(-1, 1 - j) \frac{\mathrm{ch}_0^j(\widetilde{\mathbb{E}})}{j!} = 0$$

in $\mathrm{CH}(B) \otimes \mathbb{Z}[\frac{1}{2 \cdot l!}])$. Now notice that $\zeta_L(-1, 1 - l) = -(2^l - 1)B_l / l$ by Lemma 5. We have thus proven an analog of Theorem 1, where $b^* \mathbb{E}$ is replaced by $\widetilde{\mathbb{E}}$. To deduce Theorem 1 as stated from it, we shall make the following construction. Let $\Delta : \mathbf{A}_{g,n} \hookrightarrow \widetilde{\mathbf{A}}_{g,n} \times B$ be the diagonal immersion and let $\widetilde{\mathbf{A}}'_{g,n} \to \mathrm{Zar}(\Delta(\mathbf{A}_{g,n}))$ be a desingularization of the Zariski closure of $\Delta(\mathbf{A}_{g,n})$. Let p_1 (respectively, p_2) be the map obtained by composing the natural map $\widetilde{\mathbf{A}}'_{g,n} \hookrightarrow \widetilde{\mathbf{A}}_{g,n} \times B$ and the first (respectively, second) projection map $\widetilde{\mathbf{A}}_{g,n} \times B \to \widetilde{\mathbf{A}}_{g,n}$ (respectively, $\widetilde{\mathbf{A}}_{g,n} \times B \to B$). Let $b' := b \circ p_1$. The map b' is a also a desingularization of $\overline{\mathbf{A}}_{g,n}$. By Proposition 1, we have an isomorphism $p_2^* \mathcal{A} \simeq b'^* G$ on $\widetilde{\mathbf{A}}'_{g,n}$. Hence

$$2 \sum_{j=1}^{l} \zeta_L(-1, 1 - j) \frac{\mathrm{ch}_0^j(p_2^* \widetilde{\mathbb{E}})}{j!} = 2 \sum_{j=1}^{l} \zeta_L(-1, 1 - j) \frac{\mathrm{ch}_0^j(b'^* \mathbb{E})}{j!} = 0$$

in $\mathrm{CH}(\widetilde{\mathbf{A}}'_{g,n}) \otimes \mathbb{Z}[\frac{1}{2 \cdot l!}]$. Now notice that

$$p_{1,*}\left(2 \sum_{j=1}^{l} \zeta_L(-1, 1 - j) \frac{\mathrm{ch}_0^j(b'^* \mathbb{E})}{j!}\right) = p_{1,*} p_1^* \left(2 \sum_{j=1}^{l} \zeta_L(-1, 1 - j) \frac{\mathrm{ch}_0^j(b^* \mathbb{E})}{j!}\right)$$

$$= p_{1,*}(1) \cdot 2 \sum_{j=1}^{l} \zeta_L(-1, 1 - j) \frac{\mathrm{ch}_0^j(b^* \mathbb{E})}{j!}$$

in $\mathrm{CH}(\widetilde{\mathbf{A}}_{g,n}) \otimes \mathbb{Z}[\frac{1}{2 \cdot l!}]$. We have used the projection formula for the last equality. Since p_1 is birational, we have $p_{1,*}(1) = 1$ and we have thus completely proven Theorem 1.

4 Consequences and conjectures

4.1 A corollary of Theorem 2

Let $c : \mathcal{C} \to C$ be a polarized abelian scheme, where C is a regular and quasi-projective variety over $\mathbb{C}$. Let K be a finite abelian extension of $\mathbb{Q}$. Suppose that there is an embedding of rings $\mathcal{O}_K \hookrightarrow \mathrm{End}_C(\mathcal{C})$. Let $\mathcal{H} := H^1_{\mathrm{dR}}(\mathcal{C}/C)$. The coherent sheaf $\mathcal{H}$ carries a ring action of K. Choose an element $k_0 \in K$ such that $K = \mathbb{Q}(k_0)$ (a simple element of K over $\mathbb{Q}$). For each $\sigma \in \mathrm{Hom}(K, \mathbb{C})$, define

$$\mathcal{H}_\sigma := \mathrm{Ker}(k_0 - \sigma(k_0) \cdot \mathrm{Id}).$$

The natural morphism $\oplus_{\sigma \in \mathrm{Hom}(K,\mathbb{C})} \mathcal{H}_\sigma \to \mathcal{H}$ is an isomorphism, as can be seen by considering its restriction to closed points of C. Furthermore, the sheaves $\mathcal{H}_\sigma$ do not depend on the choice of k_0.

Now let $\chi : \mathrm{Gal}(K|\mathbb{Q}) \to S^1$ be a one-dimensional character of K. We shall show that the following proposition is a consequence of Theorem 2.

Proposition 3. *The equality*

$$\sum_{\sigma \in \mathrm{Hom}(K,\mathbb{C})} \chi(\sigma)\,\mathrm{ch}(\mathcal{H}_\sigma) = \sum_{\sigma \in \mathrm{Hom}(K,\mathbb{C})} \chi(\sigma)\mathrm{rk}(\mathcal{H}_\sigma)$$

holds in $\mathrm{CH}(C) \otimes \overline{\mathbb{Q}}$.

Notice that there is a noncanonical isomorphism $\mathrm{Hom}(K, \mathbb{C}) \simeq \mathrm{Gal}(K|\mathbb{Q})$. The equality in the proposition is true for any choice of such an isomorphism.

In the following, the use of $\mathrm{Hom}(\cdot, \mathbb{C})$ instead of $\mathrm{Gal}(\cdot)$ always implies that the corresponding statement is independent of the choice of an identification of $\mathrm{Hom}(\cdot, \mathbb{C})$ and $\mathrm{Gal}(\cdot)$.

Corollary 1. *The equality* $\mathrm{ch}(\mathcal{H}_\sigma) = \mathrm{rk}(\mathcal{H}_\sigma)$ *in* $\mathrm{CH}(C) \otimes \overline{\mathbb{Q}}$ *is true for all* $\sigma \in \mathrm{Hom}(K, \mathbb{C})$.

Proof of Corollary 1. The content of Proposition 3 is that as functions of σ, all the Fourier coefficients of $\mathrm{ch}(\mathcal{H}_\sigma)$ and $\mathrm{rk}(\mathcal{H}_\sigma)$ coincide. Hence the conclusion follows from the uniqueness of the Fourier decomposition.   □

Before coming to the full proof of Proposition 3, we shall prove the following weaker statement.

Proposition 4. *Proposition 3 holds for* $K = \mathbb{Q}(\mu_n)$ *for some* $n \geq 2$.

In the proof of Proposition 4, we shall need the following lemma, which is surprisingly difficult to prove. The hypotheses and the terminology of Proposition 4 are in force.

Lemma 9. *Let* $u_0 := \exp(2i\pi/n)$ *and let* $l \geq 1$ *be an integer.*

(1) *The following equalities of meromorphic functions of* $s \in \mathbb{C}$ *hold. If* χ *is an even character, then*

$$\sum_{\sigma \in \mathrm{Hom}(\mathbb{Q}(\mu_n),\mathbb{C})} \chi(\sigma)\zeta_L(\sigma(u_0), s) = n^{1-s}\pi^{s-1/2}\frac{\Gamma((1-s)/2)}{\Gamma(s/2)}L(\chi, 1-s),$$

while if χ *is an odd character, then*

$$\sum_{\sigma \in \mathrm{Hom}(\mathbb{Q}(\mu_n), \mathbb{C})} \chi(\sigma) \zeta_L(\sigma(u_0), s) = i \cdot n^{1-s} \pi^{s-1/2} \frac{\Gamma(1-s/2)}{\Gamma((s+1)/2)} L(\chi, 1-s).$$

(2) *We have*

$$\sum_{\sigma \in \mathrm{Hom}(K, \mathbb{C})} \chi(\sigma) \zeta_L(\sigma(u_0), 1-l) \neq 0$$

when either (a) χ *is an even character and l is an even integer, or* (b) χ *is an odd character and l is an odd integer.*

Recall that a character χ as above is odd (respectively, even) if the image of complex conjugation under χ is -1 (respectively, 1). The symbol $L(\chi, s)$ refers to the meromorphic function of $s \in \mathbb{C}$ which is defined by the formula

$$L(\chi, s) := \sum_{k=1}^{\infty} \frac{\chi(k)}{k^s}$$

for $\Re(s) > 1$. Notice that the character χ may be nonprimitive. If the character χ is primitive, the equalities in (1) are consequences of the functional equation of Dirichlet L-functions.

Proof of Lemma 9. The second equality in (1) is the content of [KR2, Lemma 5.2]. The proof of the first equality is similar and we shall omit it. Before beginning with the proof of (2) we recall that the function $1/\Gamma(s)$ has zeros at the points $0, -1, -2, \ldots$ and is $\neq 0$ for all the other values of s. Recall also that $L(\chi, s)$ has an Euler product expansion when $\Re(s) > 1$ and thus $L(\chi, s) \neq 0$ when $\Re(s) > 1$. To prove (2)(a), we compute (with χ and l even)

$$\sum_{\sigma \in \mathrm{Hom}(K, \mathbb{C})} \chi(\sigma) \zeta_L(\sigma(u_0), 1-l) = n^l \pi^{1/2-l} \frac{\Gamma(l/2)}{\Gamma((1-l)/2)} L(\chi, l).$$

Using the remarks made before the computation, we can conclude the proof of (2)(a). For the proof of (2)(b), we make a similar computation:

$$\sum_{\sigma \in \mathrm{Hom}(K, \mathbb{C})} \chi(\sigma) \zeta_L(\sigma(u_0), 1-l)$$

$$= n^l \pi^{1/2-l} i \frac{\Gamma(1/2+l/2)}{\Gamma(1-l/2)} L(\chi, l)$$

$$= n^l \pi^{1/2-l} i \frac{\Gamma(1/2+l/2)}{\Gamma(1-l/2)} L(\chi_{\mathrm{prim}}, l) \prod_{p|n} (1 - \chi_{\mathrm{prim}}(p) p^{-l}).$$

Here χ_{prim} is the primitive Dirichlet character associated to χ. It is shown in [W, Chapter 4, Corollary 4.4]) that $L(\chi_{\mathrm{prim}}, 1) \neq 0$ (it is only to treat the case $l = 1$ that we introduced χ_{prim}). Using this fact and again the remarks made before the proof of (2)(a), we can conclude. $\square$

Proof of Proposition 4. Let $u_0 := \exp(2i\pi/n)$. Let $\tau \in \mathrm{Gal}(\mathbb{Q}(\mu_n)|\mathbb{Q})$; the root of unity $\tau(u_0)$ acts on $\mathcal{C}$ as an automorphism of finite order n over C. The fixed-point scheme of $\tau(u_0)$ on $\mathcal{C}$ can be shown to be finite and flat in this situation. We leave this as an exercise to the reader. Applying Theorem 2 to this situation (with similar notations and ι given by $\tau(u_0)$), we obtain the equation

$$\sum_{\sigma \in \mathrm{Hom}(\mathbb{Q}(\mu_n),\mathbb{C})} \zeta_L(\sigma(\tau(u_0)), 1-l)\,\mathrm{ch}^l(\mathcal{H}_\sigma) = 0 \tag{8}$$

in $\mathrm{CH}^l(C) \otimes \overline{\mathbb{Q}}$, for any $l \geqslant 1$. We now identify $\mathrm{Hom}(\mathbb{Q}(\mu_n), \mathbb{C})$ and $\mathrm{Gal}(\mathbb{Q}(\mu_n)|\mathbb{Q})$ via the natural embedding $\mathbb{Q}(\mu_n) \hookrightarrow \mathbb{C}$ and we evaluate at $\overline{\chi} = \chi^{-1}$ the Fourier transform of the left side of (8) for the variable $\tau \in \mathrm{Gal}(\mathbb{Q}(\mu_n)|\mathbb{Q})$. We obtain

$$\sum_{\tau \in \mathrm{Gal}(\mathbb{Q}(\mu_n)|\mathbb{Q})} \overline{\chi}(\tau) \left[\sum_{\sigma \in \mathrm{Gal}(\mathbb{Q}(\mu_n)|\mathbb{Q})} \zeta_L(\sigma(\tau(u_0)), 1-l)\,\mathrm{ch}^l(\mathcal{H}_\sigma) \right]$$

$$= \sum_\tau \chi(\tau) \left[\sum_\sigma \zeta_L(\sigma(u_0), 1-l)\,\mathrm{ch}^l(\mathcal{H}_{\sigma\tau}) \right]$$

$$= \sum_\tau \chi(\tau) \left[\sum_\sigma \zeta_L((\sigma\tau^{-1})(u_0), 1-l)\,\mathrm{ch}^l(\mathcal{H}_\sigma) \right]$$

$$= \sum_\sigma \left(\sum_\tau \chi(\tau)\zeta_L((\sigma\tau^{-1})(u_0), 1-l) \right) \mathrm{ch}^l(\mathcal{H}_\sigma)$$

$$= \left(\sum_\tau \overline{\chi}(\tau)\zeta_L(\tau(u_0), 1-l) \right) \left(\sum_\sigma \chi(\sigma)\,\mathrm{ch}^l(\mathcal{H}_\sigma) \right)$$

$$= 0.$$

Now suppose that l is an even integer (respectively, odd integer) and that χ is an even character (respectively, odd character). Then, using Lemma 9, we deduce that

$$\sum_\sigma \chi(\sigma)\,\mathrm{ch}^l(\mathcal{H}_\sigma) = 0$$

which is the equality to be proven. If l is even (respectively, odd) and χ is odd (respectively, even) then $\mathrm{ch}^l(\mathcal{H}_\sigma)$ is an even function of σ (respectively, odd function of σ), which again implies that

$$\sum_\sigma \chi(\sigma)\,\mathrm{ch}^l(\mathcal{H}_\sigma) = 0.$$

Indeed, the change of variables $\sigma \mapsto \sigma^{-1}$ then changes the sign of the expression $\sum_{\sigma \in \mathrm{Gal}(\mathbb{Q}(\mu_n)|\mathbb{Q})} \chi(\sigma)\,\mathrm{ch}^l(\mathcal{H}_\sigma)$. The fact that $\mathrm{ch}^l(\mathcal{H}_\sigma)$ is an even function of σ (respectively, odd function of σ) when l is even (respectively, odd) follows from the fact that there is a u_0-equivariant isomorphism $\mathcal{H} \simeq \mathcal{H}^\vee$. This in turn follows from relative Lefschetz and Poincaré duality and the fact that there exists an ample invertible sheaf $\mathcal{L}$ on $\mathcal{C}$, which carries a u_0-equivariant structure. To obtain such a sheaf, start with an ample invertible sheaf $\mathcal{L}'$ on $\mathcal{C}$ and let $\mathcal{L} := \otimes_{n=0}^{n-1} (u_0^k)^* \mathcal{L}'$. The sheaf $\mathcal{L}$ carries a natural u_0-equivariant structure.

Combining the two last equations, we can conclude the proof.   □

We shall need the following lemma in the proof of Proposition 3.

Lemma 10. *Let $L'|L$ be a finite extension of number fields such that $\mathcal{O}_{L'}$ is free over $\mathcal{O}_L$ and that L' is abelian over $\mathbb{Q}$. Let χ_L be a one-dimensional character of L. If Proposition 3 holds for $K = L'$ and $\chi = \mathrm{Ind}_L^{L'}(\chi_L)$, then it holds also for $K = L$ and $\chi = \chi_L$.*

Recall that by definition $\mathrm{Ind}_L^{L'}(\chi_L)$ is a one-dimensional character of L' such that $\mathrm{Ind}_L^{L'}(\chi_L)(\sigma_{L'}) = \chi(\sigma_{L'}|_L)$ for all $\sigma_{L'} \in \mathrm{Gal}(L'|\mathbb{Q})$.

Proof. Let $r := [L' : L]$ and let $x_1, \ldots, x_r$ be a basis of $\mathcal{O}_{L'}$ over $\mathcal{O}_L$. The mapping $\varphi \colon \mathcal{O}_{L'} \to M_r(\mathcal{O}_L)$ of $\mathcal{O}_{L'}$ into the $r \times r$-matrices with entries in $\mathcal{O}_L$ which maps an element of $\mathcal{O}_{L'}$ to the matrix representation in this basis of the corresponding $\mathcal{O}_L$ linear map $\mathcal{O}_{L'} \to \mathcal{O}_{L'}$, is an embedding of rings. Via the map φ, we obtain an embedding of rings $\mathcal{O}_{L'} \hookrightarrow \mathrm{End}_C(\mathcal{C}^r)$. There is a natural isomorphism of coherent sheaves

$$\bigoplus_{j=1}^{r} H_{\mathrm{dR}}^1(\mathcal{C}/C) \simeq H_{\mathrm{dR}}^1(\mathcal{C}^r/C),$$

and under this isomorphism, there is a decomposition

$$\bigoplus_{j=1}^{r} H_{\mathrm{dR}}^1(\mathcal{C}/C)_{\sigma_L} \simeq \bigoplus_{\sigma_{L'}|_L = \sigma_L} H_{\mathrm{dR}}^1(\mathcal{C}^r/C)_{\sigma_{L'}}$$

for any $\sigma_L \in \mathrm{Gal}(L|\mathbb{Q})$. Now choose an embedding $L' \hookrightarrow \mathbb{C}$ to identify $\mathrm{Hom}(L', \mathbb{C})$ and $\mathrm{Gal}(L'|\mathbb{Q})$. We compute

$$\sum_{\sigma_{L'}} \mathrm{ch}(H_{\mathrm{dR}}^1(\mathcal{C}^r/C)_{\sigma_{L'}}) \, \mathrm{Ind}_L^{L'}(\chi)(\sigma_{L'})$$

$$= \sum_{\sigma_L} \chi(\sigma_L) \sum_{\sigma_{L'}|_L = \sigma_L} \mathrm{ch}(H_{\mathrm{dR}}^1(\mathcal{C}^r/C)_{\sigma_{L'}})$$

$$= \sum_{\sigma_L} \chi(\sigma_L) \sum_{j=1}^{r} \mathrm{ch}(H_{\mathrm{dR}}^1(\mathcal{C}/C)_{\sigma_L})$$

$$= r \cdot \sum_{\sigma_L} \mathrm{ch}(H_{\mathrm{dR}}^1(\mathcal{C}/C)_{\sigma_L}) \chi(\sigma_L),$$

from which the conclusion follows.   □

Proof of Proposition 3. Class field theory implies that K can be embedded in $\mathbb{Q}(\mu_n)$ for some $n \geqslant 2$. We claim that $\mathcal{O}_{\mathbb{Q}(\mu_n)}$ is free over $\mathcal{O}_K$. To see this, let $u_0 \in \mathcal{O}_{\mathbb{Q}(\mu_n)}$ be a primitive nth root of 1 and let k be the degree of the minimal polynomial of u_0 over K. The elements $1, \ldots, u_0^{k-1}$ are then linearly independent over K, hence over $\mathcal{O}_K$. The minimal polynomial of u_0 over K is of the form $a_0 + a_1 t + \cdots + a_{k-1} t^{k-1} + t^k$, where $\{a_0, \ldots, a_k\} \subseteq \mathcal{O}_K$. Hence $u_0^{k+s} = -a_0 u_0^s - a_1 u_0^{s+1} - \cdots - a_{k-1} u_0^{k-1+s}$ for all integers $s \geqslant 0$. Applying induction over s, we see that all the elements $1, \ldots, u_0^{n-1}$ are contained in the $\mathcal{O}_K$ module generated by $1, \ldots, u_0^{k-1}$. Since the elements $1, \ldots, u_0^{n-1}$ generate $\mathcal{O}_{\mathbb{Q}(\mu_n)}$ as an $\mathcal{O}_K$-module, we see that the elements $1, \ldots, u_0^{k-1}$ generate $\mathcal{O}_{\mathbb{Q}(\mu_n)}$ as an $\mathcal{O}_K$-module and thus form a basis of $\mathcal{O}_{\mathbb{Q}(\mu_n)}$ as an $\mathcal{O}_K$-module.

Now using Lemma 10, we see that we may assume without restriction of generality that $K = \mathbb{Q}(\mu_n)$. In that case, Proposition 4 is equivalent to Proposition 3 and this concludes the proof.   □

4.2 Conjectures and speculations

Let C be a smooth quasi-projective scheme over $\mathbb{C}$. Let furthermore $\mathcal{C} \to C$ be a polarized semi-abelian scheme and let K be a number field which is Galois over $\mathbb{Q}$. Suppose that there is an embedding $K \hookrightarrow \mathrm{End}_C(\mathcal{C}) \otimes \mathbb{Q}$. Let $\mathcal{H} := e^* \Omega_{\mathcal{C}/C} \oplus e^* \Omega_{\mathcal{C}/C}^{\vee}$ where $e \colon C \to \mathcal{C}$ is the zero section. The coherent sheaf $\mathcal{H}$ carries a ring action of K. Choose an element $k_0 \in K$ such that $K = \mathbb{Q}(k_0)$. For each $\sigma \in \mathrm{Hom}(K, \mathbb{C})$, define

$$\mathcal{H}_\sigma := \mathrm{Ker}(k_0 - \sigma(k_0) \cdot \mathrm{Id}).$$

The natural morphism $\oplus_{\sigma \in \mathrm{Hom}(K,\mathbb{C})} \mathcal{H}_\sigma \to \mathcal{H}$ is then an isomorphism, as before and the sheaves $\mathcal{H}_\sigma$ do not depend on the choice of k_0.

Let now $\chi \colon \mathrm{Hom}(K, \mathbb{C}) \to \mathbb{C}$ be a simple Artin character of K. We make the following conjecture.

Conjecture 1. The equality

$$\sum_{\sigma \in \mathrm{Hom}(K,\mathbb{C})} \chi(\sigma) \, \mathrm{ch}(\mathcal{H}_\sigma) = \sum_{\sigma \in \mathrm{Hom}(K,\mathbb{C})} \chi(\sigma) \mathrm{rk}(\mathcal{H}_\sigma)$$

holds in $\mathrm{CH}(C) \otimes \overline{\mathbb{Q}}$.

An even stronger conjecture is the following.

Conjecture 2. The equality $\mathrm{ch}(\mathcal{H}_\sigma) = \mathrm{rk}(\mathcal{H}_\sigma)$ holds for all $\sigma \in \mathrm{Hom}(K, \mathbb{C})$.

Notice that unlike in the case where K is an abelian extension of $\mathbb{Q}$, Conjecture 2 is not a consequence of Conjecture 1.

Conjecture 1 is a consequence of [MR1, Conjecture 2.1], which can be considered as a "lifting" of Conjecture 1 to Arakelov geometry.

We would also like to point out a general conjecture on Gauss–Manin bundles, which overlaps with Conjecture 1 and is a consequence of [MR1, Conjecture 3.1].

Conjecture 3. Let X and Y be smooth quasi-projective varieties over $\mathbb{C}$. Let $f : X \to Y$ be a smooth and projective morphism. Then $\mathrm{ch}(H_{\mathrm{dR}}^l(X/Y)) = \mathrm{rk}(H_{\mathrm{dR}}^l(X/Y))$ in $\mathrm{CH}^l(Y) \otimes \mathbb{Q}$, for all $l \geq 1$.

Conjecture 3 can be related to a conjecture of Bloch and Beilinson.

Suppose that Y is projective over $\mathbb{C}$ and has a model Y_0 over a number field. Let $\mathrm{CH}(Y_0)_0$ be the subgroup of $\mathrm{CH}(Y_0) \otimes \mathbb{Q}$ consisting of homologically trivial cycles. Recall that there is a map from $\mathrm{CH}(Y_0)_0$ to the product of the intermediate Jacobians of $Y(\mathbb{C})$, called the Abel–Jacobi map. The conjecture of Bloch and Beilinson is that the Abel–Jacobi map is injective in this situation (see [BB, after Lemma 5.6]).

Suppose now furthermore that there is a morphism $X_0 \to Y_0$, such that the morphism obtained after a field extension to $\mathbb{C}$ coincides with f. Notice that the classes $\mathrm{ch}(H_{\mathrm{dR}}^l(X_0/Y_0)) - \mathrm{rk}(H_{\mathrm{dR}}^l(X_0/Y_0))$ lie in $\mathrm{CH}(Y_0)_0$, because the bundles $H_{\mathrm{dR}}^l(X_0/Y_0)$ carry an algebraic connection, the Gauss–Manin connection. The Abel–Jacobi map can be described using Cheeger–Simons characteristic classes (see [S, Proposition 2]) and it has been shown by Corlette and Esnault (see [CE]) that the Cheeger–Simons classes of Gauss–Manin bundles vanish. All in all, this implies that the image of the classes $\mathrm{ch}(H_{\mathrm{dR}}^l(X_0/Y_0)) - \mathrm{rk}(H_{\mathrm{dR}}^l(X_0/Y_0))$ under the Abel–Jacobi map vanish and thus Conjecture 3 is implied by the conjecture of Bloch and Beilinson in this situation. The result of Corlette and Esnault could also have been replaced by a general result of Reznikov (see [R]) in this setup.

Finally, we shall indicate how Theorem 2 overlaps with Stickelberger's theorem. Let $K \subseteq \overline{\mathbb{Q}}$ be an abelian extension of $\mathbb{Q}$ and suppose that the conductor of K is n. Let $G := \mathrm{Gal}(\mathbb{Q}(\mu_n)|\mathbb{Q}) \simeq (\mathbb{Z}/n\mathbb{Z})^\times$. By class field theory, we have an inclusion $K \subseteq \mathbb{Q}(\mu_n)$ and the group G thus acts on K by restriction. Via this action, we obtain a $\mathbb{Z}[G]$-module structure on the multiplicative group of the ideals of $\mathcal{O}_K$. If A is an ideal in $\mathcal{O}_K$ and $\upsilon \in \mathbb{Z}[G]$, we write A^υ for the image of A under υ. We write

$$\theta(K) = \theta := \sum_{a \in (\mathbb{Z}/n\mathbb{Z})^\times} \left\{ \frac{a}{n} \right\} \sigma_a^{-1} \in \mathbb{Q}[G],$$

where $\{\cdot\}$ denotes the fractional part of a real number. The element $\theta(K)$ is called the *Stickelberger element*. Let $\beta \in \mathbb{Q}[G]$ and suppose that $\beta \cdot \theta \in \mathbb{Z}[G]$. Stickelberger's theorem asserts that if A is an ideal of $\mathcal{O}_K$, then $A^{\beta\theta}$ is a principal ideal. In particular $A^{n\theta}$ is principal. Let now $\chi : G \to S^1$ be an odd primitive Dirichlet character and let

$$\epsilon_\chi := \sum_{\sigma \in G} \chi(\sigma)\sigma^{-1} \in \overline{\mathbb{Z}}_{\mathrm{ab}}[G].$$

Here $\overline{\mathbb{Z}}_{\mathrm{ab}}$ is the integral closure of $\mathbb{Z}$ in $\overline{\mathbb{Q}}_{\mathrm{ab}}$, the subfield of $\overline{\mathbb{Q}}$ generated by all the roots of unity. Let $L(\chi, s)$ be the L-function of χ, which is a meromorphic function of $s \in \mathbb{C}$ (see Section 4.1 or [W, Chapter 4] for the definition). We have $L(\chi, 1) = -B_{1,\chi} := -\sum_{a \in (\mathbb{Z}/n\mathbb{Z})^\times} \left\{ \frac{a}{n} \right\} \chi(a)$. We compute

$$\epsilon_\chi \theta = \sum_{a \in (\mathbb{Z}/n\mathbb{Z})^\times} \left\{ \frac{a}{n} \right\} \epsilon_\chi \sigma_a^{-1} = \left[\sum_{a \in (\mathbb{Z}/n\mathbb{Z})^\times} \left\{ \frac{a}{n} \right\} \overline{\chi}(\sigma_a) \right] \epsilon_\chi = B_{1,\overline{\chi}} \epsilon_\chi.$$

Now identify $\mathrm{CH}^1(\mathcal{O}_K)$ with the class group of $\mathcal{O}_K$. The last computation shows that Stickelberger's theorem implies that $n B_{1,\overline{\chi}}\epsilon_\chi$ annihilates any element of $CH^1(\mathcal{O}_K)\otimes \overline{\mathbb{Z}}_{\mathrm{ab}}$.

On the other hand, consider the situation of Theorem 2. With $u_0 := \exp(2i\pi/n)$ Theorem 2 says in particular that

$$\sum_{a\in\mathbb{Z}/n\mathbb{Z}} \zeta_L(u_0^a, 0)\, c^1(\mathcal{H}_{u_0^a}) = 0$$

in $\mathrm{CH}^1(C)\otimes\overline{\mathbb{Z}}_{\mathrm{ab}}[\frac{1}{n}]$. More generally, let $b\in(\mathbb{Z}/n\mathbb{Z})^\times$ and apply Theorem 2 again, with ι^{-b} in place of ι. We obtain the identity

$$\sum_{a\in\mathbb{Z}/n\mathbb{Z}} \zeta_L(u_0^{ab}, 0)\, c^1(\mathcal{H}_{u_0^a}) = 0 \tag{10}$$

in $\mathrm{CH}^1(C)\otimes\overline{\mathbb{Z}}_{\mathrm{ab}}[\frac{1}{n}]$.

Define the Gauss sum

$$\tau(\chi) := \sum_{a\in\mathbb{Z}/n\mathbb{Z}} \chi(a)u_0^a.$$

It is shown in [W, Chapter 4, Lemma 4.7] that $\sum_{a\in\mathbb{Z}/n\mathbb{Z}}\chi(a)u_0^{ab} = \tau(\chi)\overline{\chi}(b)$ holds for any $b\in\mathbb{Z}$. This implies that

$$\sum_{b\in\mathbb{Z}/n\mathbb{Z}} \overline{\chi}(b)\zeta_L(u_0^{ab}, 0) = \chi(a)\tau(\overline{\chi})L(\chi, 0).$$

We shall now exploit (10). We compute

$$\sum_{b\in\mathbb{Z}/n\mathbb{Z}} \overline{\chi}(b)\left(\sum_{a\in\mathbb{Z}/n\mathbb{Z}} \zeta_L(u_0^{ab}, 0)\, c^1(\mathcal{H}_{u_0^a})\right) = \sum_{a\in\mathbb{Z}/n\mathbb{Z}}\sum_{b\in\mathbb{Z}/n\mathbb{Z}} \overline{\chi}(b)\zeta_L(u_0^{ab}, 0)\, c^1(\mathcal{H}_{u_0^a})$$

$$= \tau(\overline{\chi})L(\chi, 0)\sum_{a\in\mathbb{Z}/n\mathbb{Z}} \chi(a)\, c^1(\mathcal{H}_{u_0^a})$$

in $\mathrm{CH}^1(C)\otimes\overline{\mathbb{Z}}_{\mathrm{ab}}[\frac{1}{n}]$. Since $\tau(\overline{\chi})\overline{\tau(\overline{\chi})} = |\tau(\overline{\chi})|^2 = n$ (see [W, Chapter 4, Lemma 4.8]), $\tau(\overline{\chi})$ is a unit in $\overline{\mathbb{Z}}_{\mathrm{ab}}[\frac{1}{n}]$. Hence

$$-L(\chi, 0)\sum_{a\in\mathbb{Z}/n\mathbb{Z}} \chi(a)\, c^1(\mathcal{H}_{u_0^a}) = B_{1,\chi}\sum_{a\in\mathbb{Z}/n\mathbb{Z}} \chi(a)\, c^1(\mathcal{H}_{u_0^a}) = 0$$

in $\mathrm{CH}^1(C)\otimes\overline{\mathbb{Z}}_{\mathrm{ab}}[\frac{1}{n}]$. Now suppose furthermore that $D = \mathcal{O}_{\mathbb{Q}(\mu_n)}$ and that the fibration $\mathcal{C}\to C$ and the automorphism ι have models over $\mathbb{Z}[\frac{1}{n}]$. Fix such models. We then obtain

$$B_{1,\chi}\epsilon_{\overline{\chi}}\, c^1(\mathcal{H}_{u_0}) = 0,$$

where $CH^1(C) \otimes \overline{\mathbb{Z}}_{ab}[\frac{1}{n}]$ is considered as $Gal(\mathbb{Q}(\mu_n)|\mathbb{Q})$-module via the given model of C.

Stickelberger's theorem and Theorem 2 thus lead to similar annihilation statements. It is even possible to construct a geometrical situation where Theorem 2 is implied by Stickelberger's theorem. This is left as an exercise to the reader.

One is thus led to speculate whether (the Fourier transform of) Theorem 2 is not a special case of a theorem generalizing Stickelberger's theorem to the Chow groups of the various Shimura varieties classifying abelian varieties with complex multiplications.

Acknowledgments. The authors thank G. van der Geer and B. Moonen for encouraging them to write up the results of this paper and more generally for their interest. Our thanks also go to C. Soulé, for interesting discussions and suggestions. Finally, we are especially grateful to H. Esnault and E. Viehweg for taking the time to discuss their paper [EV1] with us. This paper was the technical basis for the present article.

References

[BFM]   P. Baum, W. Fulton, and G. Quart, Lefschetz-Riemann-Roch for singular varieties, *Acta Math.*. **143** (1979), 193–211.

[BB]   A. A. Beïlinson, Height pairing between algebraic cycles, in *K-Theory, Arithmetic and Geometry* (*Moscow*, 1984–1986), Lecture Notes in Mathematics 1289, Springer-Verlag, Berlin, 1987, 1–25.

[BBM]   P. Berthelot, L. Breen, and W. Messing, *Théorie de Dieudonné cristalline* II, Lecture Notes in Mathematics 930, Springer-Verlag, Berlin, 1982.

[CE]   K. Corlette and H. Esnault, Classes of local systems of hermitian vector spaces, appendix to J.-M. Bismut, "Eta invariants, differential characters and flat vector bundles," preprint, 1995.

[EV1]   H. Esnault and E. Viehweg, Chern classes of Gauss-Manin bundles of weight 1 vanish, *K-Theory*, **26**-3 (2002), 287–305.

[EV2]   H. Esnault and E. Viehweg, *Lectures on Vanishing Theorems*, DMV Seminar 20, Birkhäuser, Basel, 1992.

[CF]   G. Faltings and C.-L. Chai, *Degeneration of Abelian Varieties* (with an appendix by David Mumford), Ergebnisse der Mathematik und ihrer Grenzgebiete 22, Springer-Verlag, Berlin, 1990.

[F]   W. Fulton, *Intersection Theory*, 2nd ed., Ergebnisse der Mathematik und ihrer Grenzgebiete 3: Folge, Series of Modern Surveys in Mathematics 2, Springer-Verlag, Berlin, 1998.

[FL]   W. Fulton and S. Lang, *Riemann-Roch Algebra*, Grundlehren der Mathematischen Wissenschaften 277, Springer-Verlag, New York, 1985.

[SGA3] A. Grothendieck, M. Demazure, M. Artin, J. E. Bertin, P. Gabriel, M. Raynaud, and J. P. Serre, *Schémas en groupes, Tome* III: *Structure des schémas en groupes réductifs*, Séminaire de Géométrie Algébrique du Bois Marie 1962/64 (SGA 3), Lecture Notes in Mathematics 153, Springer-Verlag, Berlin, New York, 1962–1964.

[H]   R. Hartshorne, *Algebraic Geometry*, Graduate Texts in Mathematics 52, Springer-Verlag, New York, Heidelberg, 1977.

[K]   B. Köck, The Grothendieck-Riemann-Roch theorem for group scheme actions, *Ann. Sci. École Norm. Sup.* (4), **31**-3 (1998), 415–458.

[KR1]   K. Köhler and D. Roessler, A fixed point formula of Lefschetz type in Arakelov geometry I: Statement and proof, *Invent. Math.*, **145**-2 (2001), 333–396.

[KR2]   K. Köhler and D. Roessler, A fixed point formula of Lefschetz type in Arakelov geometry IV: The modular height of C.M. abelian varieties, *J. Reine Angew. Math.*, **556** (2003), 127–148.

[L]   S. Lang, *Algebra*, revised 3rd ed., Graduate Texts in Mathematics 211, Springer-Verlag, New York, 2002.

[MR1]   V. Maillot and D. Roessler, Conjectures sur les dérivées logarithmiques des fonctions *L* d'Artin aux entiers négatifs, *Math. Res. Lett.*, **9**-5–6 (2002), 715–724.

[MR2]   V. Maillot and D. Roessler, On the periods of motives with complex multiplication and a conjecture of Gross-Deligne, *Ann. Math.*, **160** (2004), 727–754.

[M]   H. Matsumura, *Commutative Ring Theory* (translated from the Japanese by M. Reid), 2nd ed., Cambridge Studies in Advanced Mathematics 8, Cambridge University Press, Cambridge, UK, 1989.

[R]   A. Reznikov, All regulators of flat bundles are torsion, *Ann. Math.* (2), **141**-2 (1995), 373–386.

[S]   C. Soulé, Classes caractéristiques secondaires des fibrés plats, in *Séminaire Bourbaki*, Vol. 1995/96, Exposés 805–819, Astérisque 241, Société Mathématique de France, Paris, 1997, exposé 819, 411–424.

[T]   R. W. Thomason, Une formule de Lefschetz en *K*-théorie équivariante algébrique, *Duke Math. J.*, **68**-3 (1992), 447–462.

[T2]   R. W. Thomason, Algebraic *K*-theory of group scheme actions, in *Algebraic Topology and Algebraic K-Theory (Princeton, NJ*, 1983), Annals of Mathematics Studies 113, Princeton University Press, Princeton, NJ, 1987, 539–563.

[VDG]   G. van der Geer, Cycles on the moduli space of abelian varieties, in C. Faber and E. Looijenga, eds., *Moduli of Curves and Abelian Varieties: The Dutch Intercity Seminar on Moduli*, Aspects of Mathematics E33, Vieweg, Braunschweig, Germany, 1999, 65–89.

[W]   L. C. Washington, *Introduction to Cyclotomic Fields*, 2nd ed., Graduate Texts in Mathematics 83, Springer-Verlag, New York, 1997.

A Note on the Manin–Mumford Conjecture

Damian Roessler

Institut de Mathématiques de Jussieu
Université Paris 7 Denis Diderot
C.N.R.S.
Case Postale 7012
2 Place Jussieu
F-75251 Paris Cedex 05
France
dcr@math.jussieu.fr

Summary. In [PR1], R. Pink and the author gave a short proof of the Manin–Mumford conjecture, which was inspired by an earlier model-theoretic proof by Hrushovski. The proof given in [PR1] uses a difficult unpublished ramification-theoretic result of Serre. It is the purpose of this note to show how the proof given in [PR1] can be modified so as to circumvent the reference to Serre's result. J. Oesterlé and R. Pink contributed several simplifications and shortcuts to this note.

1 Introduction

Let A be an abelian variety defined over an algebraically closed field L of characteristic 0 and let X be a closed subvariety. If G is an abelian group, write $\mathrm{Tor}(G)$ for the group of elements of G which are of finite order. A closed subvariety of A whose irreducible components are translates of abelian subvarieties of A by torsion points will be called a torsion subvariety. The Manin–Mumford conjecture is the following statement:

The Zariski closure of $\mathrm{Tor}(A(L)) \cap X$ *is a torsion subvariety.*

This was first proved by Raynaud in [R]. In [PR1], R. Pink and the author gave a new proof of this statement, which was inspired by an earlier model-theoretic proof given by Hrushovski in [H]. The interest of this proof is the fact that it relies almost entirely on classical algebraic geometry and is quite short. Its only nonelementary input is a ramification-theoretic result of Serre. The proof of this result is not published and relies (see [Se, pp. 33–34, 56–59]) on deep theorems of Faltings, Nori, and Raynaud. In this note, we show how the reference to Serre's result in [PR1] can be replaced by a reference to a classical result in the theory of formal groups (see Theorem 1(a)).

The structure of the paper is as follows. For the convenience of the reader, the text has been written so as to be logically independent of [PR1]. In particular, no knowledge of [PR1] is necessary to read it. Section 2 recalls various classical results on abelian varieties and also contains two less well known, but elementary propositions (Propositions 1 and 2) whose proofs can be found elsewhere but for which we have included short proofs to make the text more self-contained. The reader is encouraged to proceed directly to Section 3, which contains a complete proof of the Manin–Mumford conjecture and to refer to the results listed in Section 2 as needed.

Notation

— "w.r.o.g." is short for *without restriction of generality*.
— If X is a closed subvariety of an abelian variety A defined over an algebraically closed field L of characteristic 0, then we write $\mathrm{Stab}(X)$ for the stabiliser of X; this is a closed subgroup of A such that $\mathrm{Stab}(X)(L) := \{a \in A(L) | a + X = X\}$; it has the same field of definition as X and A.
— If p is a prime number and G is an abelian group, we write $\mathrm{Tor}^p(G)$ for the set of elements of $\mathrm{Tor}(G)$ whose order is prime to p and $\mathrm{Tor}_p(G)$ for the set of elements of $\mathrm{Tor}(G)$ whose order is a power of p.

2 Preliminaries

Lemma 1. *Let $L \subseteq L'$ be algebraically closed fields of characteristic 0. Let A be an abelian variety defined over L and let X be a closed L-subvariety of A. Then*

(a) *X is a torsion subvariety of A iff $X_{L'}$ is a torsion subvariety of $A_{L'}$;*
(b) *the Manin–Mumford conjecture holds for X in A iff it holds for $X_{L'}$ in $A_{L'}$.*

Proof. We first prove (a). To prove the equivalence of the two conditions, we only need to prove the sufficiency of the second one. The latter is a consequence of the fact that the morphism $\pi : A_{L'} \to A$ is faithfully flat and that any torsion point and any abelian subvariety of $A_{L'}$ has a model in A (see [Mi, Corollary 20.4, p. 146]). To prove (b), let $Z := \mathrm{Zar}(\mathrm{Tor}(A(L)) \cap X)$ (respectively, $Z' := \mathrm{Zar}(\mathrm{Tor}(A(L')) \cap X_{L'})$). Using again the fact that any torsion point in $A_{L'}$ has a model in A and that π is faithfully flat, we see that $\pi^{-1}(\mathrm{Tor}(A(L)) \cap X) = \mathrm{Tor}(A(L')) \cap X_{L'}$. From this and the fact that the morphism π is open [EGA, IV, 2.4.10], we get a set-theoretic equality $\pi^{-1}(Z) = Z'$. Since π is radicial, the underlying set of $\pi^*(Z) := Z_{L'}$ is $\pi^{-1}(Z)$ [EGA, I, 3.5.10]. Since $Z_{L'}$ is reduced [EGA, IV, 4.6.1], we thus have an equality of closed subschemes $Z_{L'} = Z'$. Now by (a), the closed subscheme $Z_{L'}$ is a torsion subvariety of $A_{L'}$ iff Z is a torsion subvariety of A.   □

Proposition 1 (Pink–Roessler). *Let A be an abelian variety over $\mathbb{C}$ and let $F : A \to A$ be an isogeny. Suppose that the absolute value of all the eigenvalues of the pullback map F^* on the first singular cohomology group $H^1(A(\mathbb{C}), \mathbb{C})$ is larger than 1. Then any closed subvariety Z of A such that $F(Z) = Z$ is a torsion subvariety.*

The following proof can be found in [PR1, Remark after Lemma 2.6].

Proof. W.r.o.g., we may replace F by one of its powers and thus suppose that each irreducible component of Z is stable under F. We may thus suppose that Z is irreducible. Notice that $F(\mathrm{Stab}(Z)) \subseteq \mathrm{Stab}(Z)$. Let us first suppose that $\mathrm{Stab}(Z) = 0$.

Write $\mathrm{cl}(Z)$ for the cycle class of Z in $H^*(A(\mathbb{C}), \mathbb{C})$. We list the following facts:

(1) The degree of F is the determinant of the restriction of F^* to $H^1(A(\mathbb{C}), \mathbb{C})$.
(2) Each eigenvalue of F^* on $H^i(A(\mathbb{C}), \mathbb{C})$ is the product of i distinct zeroes (counted with multiplicities) of the characteristic polynomial of F^* on $H^1(A(\mathbb{C}), \mathbb{C})$.

Facts (1) and (2) follow from the fact that for all $i \geq 0$ there is a natural isomorphism $\Lambda^i(H^1(A(\mathbb{C}), \mathbb{C})) \simeq H^i(A(\mathbb{C}), \mathbb{C})$ (see [Mu, p. 3, equation (4)]).

Now notice that since $\mathrm{Stab}(Z) = 0$, the varieties $Z + a$, where $a \in \mathrm{Ker}(F)(\mathbb{C})$, are pairwise distinct. These varieties are thus the irreducible components of $F^{-1}(Z)$. Now we compute

$$\mathrm{cl}(F^*(Z)) = \sum_{a \in \mathrm{Ker}(F)} \mathrm{cl}(Z + a) = \#\,\mathrm{Ker}(F)(\mathbb{C}) \cdot \mathrm{cl}(Z) = \deg(F)\,\mathrm{cl}(Z)$$

and thus $\mathrm{cl}(Z)$ belongs to the eigenspace of $H^*(A(\mathbb{C}), \mathbb{C})$ corresponding to the eigenvalue $\deg(F)$. Facts (1), (2) and the hypothesis on the eigenvalues imply that $\mathrm{cl}(Z) \in H^{2\dim(A)}(A(\mathbb{C}), \mathbb{C})$, which in turn implies that Z is a point. This point is a torsion point since it lies in the kernel of $F - \mathrm{Id}$, which is an isogeny by construction.

If $\mathrm{Stab}(Z) \neq 0$, then replace A by $A/\mathrm{Stab}(Z)$ and Z by $Z/\mathrm{Stab}(Z)$. The isogeny F then induces an isogeny on $A/\mathrm{Stab}(Z)$, which stabilises $Z/\mathrm{Stab}(Z)$. We deduce that $Z/\mathrm{Stab}(Z)$ is a torsion point. This implies that Z is a translate of $\mathrm{Stab}(Z)$ by a torsion point and concludes the proof. $\square$

Corollary 1. *Let A be an abelian variety over an algebraically closed field K of characteristic 0. Let $n \geq 1$ and let M be an $n \times n$-matrix with integer coefficients. Suppose that the absolute value of all the eigenvalues of M is larger than 1. Then any closed subvariety Z of A^n such that $M(Z) = Z$ is a torsion subvariety.*

Proof. Because of Lemma 1(a), we may assume w.r.o.g. that K is the algebraic closure of a field which is finitely generated as a field over $\mathbb{Q}$. We may thus also assume that $K \subseteq \mathbb{C}$. Proposition 1 then implies the result for $Z_{\mathbb{C}}$ in $A^n_{\mathbb{C}}$, and using Lemma 1(a) again we can conclude. $\square$

Proposition 2 (Boxall). *Let A be an abelian variety over a field K of characteristic 0. Let $p > 2$ be a prime number and let $L := K(A[p])$ be the extension of K generated by the p-torsion points of A. Let $P \in \mathrm{Tor}_p(A(\overline{K}))$ and suppose that $P \notin A(L)$. Then there exists $\sigma \in \mathrm{Gal}(\overline{L}|L)$ such that $\sigma(P) - P \in A[p] \setminus \{0\}$.*

A proof of a variant of Proposition 2 can be found in [B]. For the convenience of the reader, we reproduce a proof, which is a simplification by Oesterlé (private communication) of a proof due to Coleman and Voloch (see [Vo]).

Proof. Let $n \geq 1$ be the smallest natural number so that $p^n P \in A(L)$. For all $i \in \{1, \ldots, n\}$, let $P_i = p^{n-i} P$. Let also σ_1 be an element of $\mathrm{Gal}(\overline{L}|L)$ such that $\sigma_1(p^{n-1}P) \neq p^{n-1}P$. Furthermore, let $\sigma_i := \sigma_1^{p^{i-1}}$ and $Q_i := \sigma_i(P_i) - P_i$.

First, notice that we have $pQ_1 = \sigma_1(p^n P) - p^n P = 0$ and $Q_1 = \sigma_1(p^{n-1} P) - p^{n-1} P \neq 0$, hence $Q_1 \in A[p] \backslash \{0\}$. We shall prove by induction on $i \geq 1$ that $Q_i = Q_1$ if $i \leq n$. This will prove the proposition, since $Q_n = \sigma_n(P) - P$.

So assume that $Q_i = Q_1$ for some $i < n$. We have $p^2(\sigma_i - 1)(P_{i+1}) = p(\sigma_i - 1)(P_i) = pQ_i = 0$. Since any p-torsion point of A is fixed by σ, and hence by σ_i, we also have $p(\sigma_i - 1)^2(P_{i+1}) = 0$ and $(\sigma_i - 1)^3(P_{i+1}) = 0$. The binomial formula shows that, in the ring of polynomials $\mathbb{Z}[T]$, T^p is congruent to $1 + p(T - 1)$ modulo the ideal generated by $p(T - 1)^2$ and $(T - 1)^3$ (notice that $p \neq 2!$). We thus have $(\sigma_i^p - 1)(P_{i+1}) = p(\sigma_i - 1)(P_{i+1}) = (\sigma_i - 1)(P_i)$, id est $Q_{i+1} = Q_i$. This completes the induction on i. □

Suppose now that K is a finite extension of $\mathbb{Q}_p$, for some prime number p and let K^{unr} be its maximal unramified extension. Let k be the residue field of K. Suppose that A is an abelian variety over K which has good reduction at the unique nonarchimedean place of K. Denote by A_0 the corresponding special fiber, which is an abelian variety over k.

Theorem 1.

(a) *The kernel of the homomorphism*

$$\mathrm{Tor}(A(K^{\mathrm{unr}})) \to A_0(\bar{k})$$

induced by the reduction map is a finite p-group.
(b) *The equality $\mathrm{Tor}^p(A(K^{\mathrm{unr}})) = \mathrm{Tor}^p(A(\overline{K}))$ holds.*

Proof. For statement (b), see [Mi, Corollary 20.8, p. 147]. Statement (a), which is more difficult to prove, follows from general properties of formal groups over K. See [Oes2, Proposition 2.3 (a)] for the proof. □

Let now $\phi \in \mathrm{Gal}(\bar{k}|k)$ be the arithmetic Frobenius map.

Theorem 2 (Weil). *There is a monic polynomial $Q(T) \in \mathbb{Z}[T]$ with the following properties:*

(a) $Q(\phi)(P) = 0$ *for all $P \in A_0(\bar{k})$.*
(b) *The complex roots of Q have absolute value $\sqrt{\#k}$.*

Proof. See [We]. □

3 Proof of the Manin–Mumford conjecture

Proposition 3. *Let A be an abelian variety over a field K_0 that is finitely generated as a field over $\mathbb{Q}$. Then for almost all prime numbers p, there exists an embedding of K_0 into a finite extension K of $\mathbb{Q}_p$, such that A_K has good reduction at the unique nonarchimedean place of K.*

Proof. Since by assumption K_0 has finite transcendence degree over $\mathbb{Q}$, there is a finite map

$$\mathrm{Spec}\, K_0 \to \mathrm{Spec}\, \mathbb{Q}(X_1, \ldots, X_d)$$

for some $d \geq 0$ (notice that $d = 0$ is allowed). Let $V \to \mathbb{A}^d_{\mathbb{Z}}$ be the normalization of the affine space $\mathbb{A}^d_{\mathbb{Z}}$ in K_0. The scheme V is integral, normal and has K_0 as a field of rational functions. Furthermore, V is finite and surjective onto $\mathbb{A}^d_{\mathbb{Z}}$. There is an open subset $B \subseteq V$ and an abelian scheme $\mathcal{A} \to B$, whose generic fiber is A. Choose B sufficiently small so that its image $f(B)$ is open and so that $f^{-1}(f(B)) = B$ (this can be achieved by replacing B by $f^{-1}(\mathbb{A}^d_{\mathbb{Z}} \setminus f(V \setminus B))$). Let $U := f(B)$. This accounts for the square on the left of the diagram $(*)$ below.

Now notice that $U(\mathbb{Q}) \neq \emptyset$, since $\mathbb{A}^d(\mathbb{Q})$ is dense in $\mathbb{A}^d_{\mathbb{Q}}$ and $U \cap \mathbb{A}^d_{\mathbb{Q}}$ is open and not empty. Thus, for almost all prime numbers p, we have $U(\mathbb{F}_p) \neq \emptyset$. Let p be a prime number with this property. Let $P \in U(\mathbb{F}_p)$ and let $a_1, \ldots, a_d \in \mathbb{F}_p$ be its coordinates. Choose as well elements $x_1, \ldots, x_d \in \mathbb{Q}_p$ which are algebraically independent over $\mathbb{Q}$. The elements $x_1, \ldots, x_d$ remain algebraically independent if we replace some x_i by $1/x_i$ so we may suppose that $\{x_1, \ldots, x_d\} \subseteq \mathbb{Z}_p$. Notice also that any element of the residue field $\mathbb{F}_p$ of $\mathbb{Z}_p$ is the reduction mod p of an element of $\mathbb{Z} \subseteq \mathbb{Z}_p$. Furthermore, the elements $x_1, \ldots, x_d$ remain algebraically independent if some x_i is replaced by $x_i + m$, where m is an integer. Hence, we may also suppose that $x_i \bmod p = a_i$ for all $i \in \{1, \ldots, d\}$. The choice of the x_i induces a morphism $e \colon \mathrm{Spec}\, \mathbb{Z}_p \to \mathbb{A}^d_{\mathbb{Z}}$, which by construction sends the generic point of $\mathrm{Spec}\, \mathbb{Z}_p$ to the generic point of $\mathbb{A}^d_{\mathbb{Z}}$ and hence of U and sends the special point of $\mathrm{Spec}\, \mathbb{Z}_p$ to $P \in U(\mathbb{F}_p)$. Hence $e^{-1}(U) = \mathrm{Spec}\, \mathbb{Z}_p$. This accounts for the lowest square in $(*)$.

The middle square in $(*)$ is obtained by taking the fibre product of $B \to U$ and $\mathrm{Spec}\, \mathbb{Z}_p \to U$. The morphism $B_1 \to \mathrm{Spec}\, \mathbb{Z}_p$ is then also finite and surjective.

To define the arrows in the triangle next to it, consider a reduced irreducible component B_1' of B_1 which dominates $\mathrm{Spec}\, \mathbb{Z}_p$. This exists, because the morphism $B_1 \to \mathrm{Spec}\, \mathbb{Z}_p$ is dominant. The morphism $B_1' \to \mathrm{Spec}\, \mathbb{Z}_p$ will then also be finite and will thus correspond to a finite (and hence integral) extension of integral rings. Let K be the function field of B_1', which is a finite extension of $\mathbb{Q}_p$; the ring associated to B_1' is by construction included in the integral closure $\mathcal{O}_K$ of $\mathbb{Z}_p$ in K and the arrow $\mathrm{Spec}\, \mathcal{O}_K \to B_1$ is defined by composing the morphism induced by this inclusion with the closed immersion $B_1' \to B_1$.

The morphism $\mathrm{Spec}\, K \to \mathrm{Spec}\, \mathbb{Q}_p$ has been implicitly defined in the last paragraph and the morphisms $\mathrm{Spec}\, \mathbb{Q}_p \to \mathrm{Spec}\, \mathbb{Z}_p$ and $\mathrm{Spec}\, K \to \mathrm{Spec}\, \mathcal{O}_K$ are the obvious ones.

We have a commutative diagram $(*)$:

The single-barreled continuous arrows ($\rightarrow$) represent dominant maps; the double-barreled continuous ones ($\Rightarrow$) represent finite and dominant maps; all the schemes in the diagram apart from B_1 are integral; the cartesian squares carry the label "Cart."

Now notice that the map $\operatorname{Spec} K \rightarrow B$ obtained by composing the connecting morphisms sends $\operatorname{Spec} K$ to the generic point of B; to see this notice that the maps $\operatorname{Spec} K \rightarrow \operatorname{Spec} \mathcal{O}_K$, $\operatorname{Spec} \mathcal{O}_K \Rightarrow \operatorname{Spec} \mathbb{Z}_p$ and $\operatorname{Spec} \mathbb{Z}_p \rightarrow U$ are all dominant; hence $\operatorname{Spec} K$ is sent to the generic point of U; since $B \rightarrow U$ is a finite map, this implies that $\operatorname{Spec} K$ is sent to the generic point of B.

Thus the map $\operatorname{Spec} K \rightarrow B$ induces a field extension $K | K_0$. Furthermore, as we have seen, K is a finite extension of $\mathbb{Q}_p$ and by construction, the abelian variety A_K is the generic fiber of the abelian scheme $\mathcal{A} \times_B \operatorname{Spec} \mathcal{O}_K$. In other words A_K is an abelian variety defined over K which has good reduction at the unique nonarchimedean place of K. $\square$

Next, we shall consider the following situation. Let $p > 2$ be a prime number and let K be a finite extension of $\mathbb{Q}_p$. Let k be its residue field. Let A be an abelian variety over K. Suppose that A has good reduction at the unique nonarchimedean place of K. Let A_0 be the corresponding special fiber, which is an abelian variety over k.

Recall that K^{unr} refers to the maximal unramified extension of K. Let $\phi \in \operatorname{Gal}(\bar{k}|k)$ be the arithmetic Frobenius map and let $\tau \in \operatorname{Gal}(K^{\mathrm{unr}}|K)$ be its canonical lift.

Proposition 4. *Let X be a closed K-subvariety of A. Then the Zariski closure of $X_{\overline{K}} \cap \operatorname{Tor}(A(K^{\mathrm{unr}}))$ is a torsion subvariety.*

Proof. W.r.o.g., we may suppose that $\operatorname{Tor}(A(K^{\mathrm{unr}}))$ is dense in $X_{\overline{K}}$ (otherwise, replace X by the natural model of $\operatorname{Zar}(X_{\overline{K}} \cap \operatorname{Tor}(A(K^{\mathrm{unr}})))$ over K). By Theorem 1(a), the kernel of the reduction homomorphism $\operatorname{Tor}(A(K^{\mathrm{unr}})) \rightarrow A_0(\bar{k})$ is a finite p-group. Let p^r be its cardinality and let $Y := p^r \cdot X$. Let $Q(T) := T^n - (a_n T^{n-1} + \cdots + a_0) \in \mathbb{Z}[T]$ be the polynomial provided by Theorem 2 (i.e., the characteristic polynomial of ϕ on $A_0(\bar{k})$). Let F be the matrix

$$
\begin{pmatrix}
0 & 1 & \dots & 0 & 0 \\
\vdots & \vdots & & \vdots & \vdots \\
0 & 0 & \dots & 0 & 1 \\
a_0 & a_1 & \dots & a_{n-2} & a_{n-1}
\end{pmatrix}.
$$

For any $a \in A(K^{\mathrm{unr}})$, write $u(x) := (x, \tau(x), \tau^2(x), \dots, \tau^{n-1}(x)) \in A^n(K^{\mathrm{unr}})$. Let $\widetilde{Y} := \operatorname{Zar}(\{u(a)|a \in (p^r \cdot \operatorname{Tor}(A(K^{\mathrm{unr}}))) \cap Y_{\overline{K}}\})$. Theorems 2(a) and 1(a) imply that

$$
F(u(a)) = u(\tau(a))
$$

for all $a \in p^r \cdot \operatorname{Tor}(A(K^{\mathrm{unr}}))$. Furthermore, by construction,

$$
\tau(p^r \cdot \operatorname{Tor}(A(K^{\mathrm{unr}}))) \subseteq p^r \cdot \operatorname{Tor}(A(K^{\mathrm{unr}})).
$$

Hence $F(\tilde{Y}) = \tilde{Y}$. Now Theorem 2(b) implies that the absolute values of the eigenvalues of the matrix F are larger than 1 and Corollary 1 then implies that $\tilde{Y}$ is a torsion subvariety of $A_{\overline{K}}$. The variety $Y_{\overline{K}}$ is the projection of $\tilde{Y}$ on the first factor and is thus also a torsion subvariety. Finally, this implies that $X_{\overline{K}}$ is a torsion subvariety.   □

Proposition 5. *Let X be a closed K-subvariety of A. Then the Zariski closure of $X_{\overline{K}} \cap \mathrm{Tor}(A(\overline{K}))$ is a torsion subvariety.*

Proof. We may suppose w.r.o.g. that $K = K(A[p])$, that X is geometrically irreducible and that $X_{\overline{K}} \cap \mathrm{Tor}(A(\overline{K}))$ is dense in $X_{\overline{K}}$. We shall first suppose that $\mathrm{Stab}(X) = 0$. Let $x \in X_{\overline{K}} \cap \mathrm{Tor}(A(\overline{K}))$ and suppose that $x \notin A(K^{\mathrm{unr}})$. Write $x = x^p + x_p$, where $x^p \in \mathrm{Tor}^p(A(\overline{K}))$ and $x_p \in \mathrm{Tor}_p(A(\overline{K}))$. By Theorem 1(b) $x^p \in A(K^{\mathrm{unr}})$ and thus $x_p \notin A(K^{\mathrm{unr}})$. By Proposition 2, there exists $\sigma \in \mathrm{Gal}(\overline{K}|K^{\mathrm{unr}})$ such that

$$\sigma(x_p) - x_p = \sigma(x) - x \in A[p] \setminus \{0\}.$$

Now notice that for all $y \in X(\overline{K})$ and all $\tau \in \mathrm{Gal}(\overline{K}|K^{\mathrm{unr}})$, we have $\tau(y) \in X(\overline{K})$. Hence if the set $\{x \in X_{\overline{K}} \cap \mathrm{Tor}(A(\overline{K}))|x \notin A(K^{\mathrm{unr}})\}$ is dense in $X_{\overline{K}}$ then $\mathrm{Stab}(X)(\overline{K})$ contains an element of $A[p] \setminus \{0\}$. Since $\mathrm{Stab}(X) = 0$, we deduce that the set $\{x \in X_{\overline{K}} \cap \mathrm{Tor}(A(\overline{K}))|x \notin A(K^{\mathrm{unr}})\}$ is not dense in $X_{\overline{K}}$ and thus the set $X_{\overline{K}} \cap \mathrm{Tor}(A(K^{\mathrm{unr}}))$ is dense in $X_{\overline{K}}$. Proposition 4 then implies that $X_{\overline{K}}$ is a torsion point. If $\mathrm{Stab}(X) \neq 0$, then we may apply the same reasoning to $X/\mathrm{Stab}(X)$ and $A/\mathrm{Stab}(A)$ to conclude that $X_{\overline{K}}$ is a translate of $\mathrm{Stab}(X)_{\overline{K}}$ by a torsion point.   □

We shall now prove the Manin–Mumford conjecture. Let the terminology of the introduction hold. By Lemma 1(b), we may assume w.r.o.g. that L is the algebraic closure of a field K_0 that is finitely generated as a field over $\mathbb{Q}$ and that A (respectively, X) has a model $\mathbf{A}$ (respectively, $\mathbf{X}$) over K_0. By Proposition 3, there is an embedding of K_0 into a field K, with the following properties: K is a finite extension of $\mathbb{Q}_p$, where p is a prime number larger than 2 and $\mathbf{A}_K$ has good reduction at the unique nonarchimedean place of K. Proposition 5 now implies that the Manin–Mumford conjecture holds for $\mathbf{X}_{\overline{K}}$ in $\mathbf{A}_{\overline{K}}$, and using Lemma 1(b) we deduce that it holds for X in A.

Remark. Let the notation of the introduction hold. Proposition 2 *alone* implies the statement of the Manin–Mumford conjecture, with $\mathrm{Tor}(A(L))$ replaced by $\mathrm{Tor}_p(A(L))$, for any prime number $p > 2$. To see this, we may w.r.o.g. assume that X is irreducible and that $\mathrm{Tor}_p(A(L)) \cap X$ is dense in X. By an easy variant of Lemma 1(b), we may w.r.o.g. assume that L is the algebraic closure of a field K that is finitely generated as a field over $\mathbb{Q}$ and that A (respectively, X) has a model $\mathbf{A}$ (respectively, $\mathbf{X}$) over K. Finally, we may assume w.r.o.g. that $K = K(\mathbf{A}[p])$. Suppose first that $\mathrm{Stab}(X) = 0$. By the same argument as above, the set $\{a \in \mathrm{Tor}_p(A(L))|a \notin \mathbf{A}(K), a \in X\}$ is not dense in X. Hence the set $\{a \in \mathrm{Tor}_p(A(L))|a \in \mathbf{A}(K), a \in X\}$ must be dense in X; the theorem of Mordell–Weil (for instance) implies that this set is finite and thus X consists of a single torsion point. If $\text{Stab}(X) \neq 0$, then we deduce by the same reasoning that $X / \text{Stab}(X)$ is a torsion point in $A / \text{Stab}(X)$ and hence X is a translate of $\text{Stab}(X)$ by a torsion point. This proof of a special case of the Manin–Mumford conjecture is outlined in [B, Remarque 3, p. 75].

Acknowledgments. We want to thank J. Oesterlé for his interest and for suggesting some simplifications in the proofs of [PR1] (see [Oes]) which have inspired some of the proofs given here. Also, the proof of Proposition 2 in its present form is due to him (see the explanations before the proof). I am also very grateful to R. Pink, who carefully read several versions of the text and suggested many improvements and simplifications. In particular, Proposition 3 was suggested by him. Many thanks as well to J. Boxall, who read the final version of the paper carefully and suggested generalizations. I am also grateful to T. Ito for his remarks and corrections. See his recent preprint *On the Manin-Mumford conjecture for abelian varieties with a prime of supersingular reduction* (ArXiv math.NT/0411291), which is partially inspired by this paper. Finally, my thanks go to the referee for a careful reading of the article.

References

[B]   J. Boxall, Sous-variétés algébriques de variétés semi-abéliennes sur un corps fini, in S. David, ed., *Number Theory, Paris* 1992–3, London Mathematical Society Lecture Notes Series 215, Cambridge University Press, Cambridge, UK, 1995, 69–89.

[EGA] A. Grothendieck, *Élements de Géométrie Algébrique*, Publications Mathématiques IHES 4, 8, 11, 17, 20, 24, 28, 32, Institut des Hautes Études Scientifiques, Bures-sur-Yvette, France, 1960–1967; see also Grundlehren 166, Springer-Verlag, Berlin, Heidelberg, 1971.

[H]   E. Hrushovski, The Manin-Mumford conjecture and the model theory of difference fields, *Ann. Pure Appl. Logic*, **112**-1 (2001), 43–115.

[Mi]   J. S. Milne, Abelian varieties, in *Arithmetic Geometry* (*Storrs, Connecticut*, 1984), Springer-Verlag, New York, 1986, 103–150.

[Mu]   D. Mumford, *Abelian Varieties*, Tata Institute of Fundamental Research Studies in Mathematics 5, Oxford University Press, London, 1970.

[Oes] J. Oesterlé, Lettre à l'auteur dated December 20, 2002.

[Oes2] J. Oesterlé, Courbes sur une variété abélienne (d'après M. Raynaud), in *Séminaire Bourbaki*, Vol. 1983/84, Astérisque 121–122, Société Mathématique de France, Paris, 1985, 213–224.

[PR1] R. Pink and D. Roessler, On Hrushovski's proof of the Manin-Mumford conjecture, in *Proceedings of the International Congress of Mathematicians: Vol.* I (*Beijing*, 2002), Higher Education Press, Beijing, 2002, 539–546.

[R]   M. Raynaud, Sous-variétés d'une variété abélienne et points de torsion, in *Arithmetic and Geometry*, Vol. I, Progress in Mathematics 35, Birkhäuser Boston, Cambridge, MA, 1983, 327–352.

[Se]   J.-P. Serre, *Oeuvres, Vol.* IV (1985–1998), Springer-Verlag, New York, 2000.

[Vo]   J.-F. Voloch, Integrality of torsion points on abelian varieties over p-adic fields, *Math. Res. Lett.*, **3**-6 (1996), 787–791.

[We]   A. Weil, *Variétés abéliennes et courbes algébriques*, Hermann, Paris, 1948.

Progress in Mathematics

Edited by:

Hyman Bass
Dept. of Mathematics
University of Michigan
Ann Arbor, MI 48109
USA
hybass@umich.edu

Joseph Oesterlé
Equipe de théorie des nombres
Université Paris 6
175 rue du Chevaleret
75013 Paris
FRANCE
oesterle@math.jussieu.fr

Alan Weinstein
Dept. of Mathematics
University of California
Berkeley, CA 94720
USA
alanw@math.berkeley.edu

Progress in Mathematics is a series of books intended for professional mathematicians and scientists, encompassing all areas of pure mathematics. This distinguished series, which began in 1979, includes authored monographs and edited collections of papers on important research developments as well as expositions of particular subject areas.

We encourage preparation of manuscripts in some form of TEX for delivery in camera-ready copy which leads to rapid publication, or in electronic form for interfacing with laser printers or typesetters.

Proposals should be sent directly to the editors or to: Birkhäuser Boston, 675 Massachusetts Avenue, Cambridge, MA 02139, USA or to Birkhauser Verlag, 40-44 Viadukstrasse, CH-4051 Basel, Switzerland.